Informatics and Machine Learning

Informatics and Machine Learning

From Martingales to Metaheuristics

Stephen Winters-Hilt
Meta Logos Systems
Albuquerque, NM, USA

Registered Office
John Wiley & Sons, Inc., 111 River Street, Hoboken, NJ 07030, USA

Editorial Office
111 River Street, Hoboken, NJ 07030, USA

For details of our global editorial offices, customer services, and more information about Wiley products visit us at www.wiley.com.

Wiley also publishes its books in a variety of electronic formats and by print-on-demand. Some content that appears in standard print versions of this book may not be available in other formats.

Library of Congress Cataloging-in-Publication Data applied for:

ISBN: 9781119716747

Cover Design: Wiley
Cover Image: © agsandrew/Shutterstock

Set in 9.5/12.5pt STIXTwoText by Straive, Pondicherry, India

10 9 8 7 6 5 4 3 2 1

This book is dedicated to my family: Cindy, Nathaniel, Zachary, Sybil, Teresa, Eric, and Josh.

Contents

Preface

The material in this book draws from undergraduate and graduate coursework taken while I was a student at Caltech, and from further graduate coursework and studies at Oxford, University of Wisconsin, and University of California, Santa Cruz. The material also draws upon my teaching experience and research efforts while a tenured professor of computer science at the University of New Orleans and jointly appointed as principal investigator and director of a protein channel biosensor lab at the Research Institute for Children at Children's Hospital in New Orleans.

1

Introduction

Informatics provides new avenues of understanding and inquiry in any medium that can be captured in digital form. Areas as diverse as text analysis, signal analysis, and genome analysis, to name a few, can be studied with informatics tools. Computationally powered informatics tools are having a phenomenal impact in many fields, including engineering, nanotechnology, and the biological sciences (Figure 1.1).

In this text I provide a background on various methods from Informatics and Machine Learning (ML) that together comprise a "complete toolset" for doing data analytics work at all levels – from a first year undergraduate introductory level to advanced topics in subsections suitable for graduate students seeking a deeper understanding (or a more detailed example). Numerous prior book, journal, and patent publications by the author are drawn upon extensively throughout the text [1–68]. Part of the objective of this book is to bring these examples together and demonstrate their combined use in typical signal processing situations. Numerous other journal and patent publications by the author [69–100] provide related material, but are not directly drawn upon this text. The application domain is practically everything in the digital domain, as mentioned above, but in this text the focus will be on core methodologies with specific application in informatics, bioinformatics, and cheminformatics (nanopore detection, in particular). Other disciplines can also be analyzed with informatics tools. Basic questions about human origins (anthrogenomics) and behavior (econometrics) can also be explored with informatics-based pattern recognition methods, with a huge impact on new research directions in anthropology, sociology, political science, economics, and psychology. The complete toolset of statistical learning tools can be used in any of these domains.

In the chapter that follows an overview is given of the various information processing stages to be discussed in the text, with some highlights to help explain the

Informatics and Machine Learning: From Martingales to Metaheuristics, First Edition.
Stephen Winters-Hilt.
© 2022 John Wiley & Sons, Inc. Published 2022 by John Wiley & Sons, Inc.

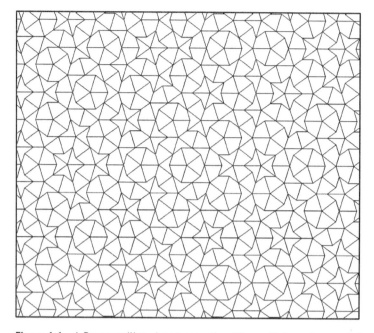

Figure 1.1 A Penrose tiling. A non-repeating tiling with two shapes of tiles, with 5-point local symmetry and both local and global (emergent) golden ratio.

order and connectivity of topics, as well as motivate their presentation in further detail in what is to come.

1.1 Data Science: Statistics, Probability, Calculus ... Python (or Perl) and Linux

Knowledge construction using statistical and computational methods is at the heart of data science and informatics. Counts on data features (or events) are typically gathered as a starting point in many analyses [101, 102]. Computer hardware is very well suited to such counting tasks. Basic operating system commands and a popular scripting language (Python) will be taught to enable doing these tasks easily. Computer software methods will also be shown that allow easy implementation and understanding of basic statistical methods, whereby the counts, for example, can be used to determine event frequencies, from which statistical anomalies can be subsequently identified. The computational implementation of basic statistics methods then provides the framework to perform more sophisticated knowledge construction and discovery by use of information theory and basic ML methods. ML can be thought of as a specialized branch of statistics where there is minimal assumption of a statistical "model" based on prior human learning. This book shows how to use computational, statistical, and informatics/ algorithmic methods to analyze any data that is captured in digital form, whether it be text, sequential data in general (such as experimental observations over time, or stock market/econometric histories), symbolic data (genomes), or image data. Along the way there will be a brief introduction to probability and statistics concepts (Chapter 2) and basic Python/Linux system programming methods (Chapter 2 and Appendix A).

1.2 Informatics and Data Analytics

It is common to need to acquire a signal where the signal properties are not known, or the signal is only suspected and not discovered yet, or the signal properties are known but they may be too much trouble to fully enumerate. There is no common solution, however, to the acquisition task. For this reason the initial phases of acquisition methods unavoidably tend to be *ad hoc*. As with data dependency in non-evolutionary search metaheuristics (where there is no optimal search method that is guaranteed to always work well), here there is no optimal signal acquisition method known in advance. In what follows methods are described for bootstrap optimization in signal acquisition to enable the most general-use, almost "common," solution possible. The bootstrap algorithmic method involves repeated passes over the data sequence, with improved priors, and trained filters, among other things, to have improved signal acquisition on subsequent passes. The signal acquisition is guided by statistical measures to recognize anomalies. Informatics methods and information theory measures are central to the design of a good finite state automata (FSAs) acquisition method, and will be reviewed in signal acquisition context in Chapters 2–4. Code examples are given in Python and C (with introductory Python described in Chapter 2 and Appendix A). Bootstrap acquisition methods may not automatically provide a common solution, but appear to offer a process whereby a solution can be improved to some desirable level of general-data applicability.

The signal analysis and pattern recognition methods described in this book are mainly applied to problems involving stochastic sequential data: power signals and genomic sequences in particular. The information modeling, feature selection/ extraction, and feature-vector discrimination, however, were each developed separately in a general-use context. Details on the theoretical underpinnings are given in Chapter 3, including a collection of *ab initio* information theory tools to help "find your way around in the dark." One of the main *ab initio* approaches is to search for statistical anomalies using information measures, so various information measures will be described in detail [103–115].

The background on information theory and variational/statistical modeling has significant roots in variational calculus. Chapter 3 describes information theory ideas and the information "calculus" description (and related anomaly detection methods). The involvement of variational calculus methods and the possible parallels with the nascent development of a new (modern) "calculus of information" motivates the detailed overview of the highly successful physics development/applications of the calculus of variations (Appendix B). Using variational calculus, for example, it is possible to establish a link between a choice of information measure and statistical formalism (maximum entropy, Section 3.1). Taking the maximum entropy on a distribution with moment constraints leads to the classic distributions seen in mathematics and nature (the Gaussian for fixed mean and variance, etc.). Not surprisingly, variational methods also help to establish and refine some of the main ML methods, including Neural Nets (NNs) (Chapters 9, 13) and Support Vector Machines (SVM) (Chapter 10). SVMs are the main tool presented for both classification (supervised learning) and clustering (unsupervised learning), and everything in between (such as bag learning).

1.3 FSA-Based Signal Acquisition and Bioinformatics

Many signal features of interest are time limited and not band limited in the observational context of interest, such as noise "clicks," "spikes," or impulses. To acquire these signal features a time-domain finite state automaton (tFSA) is often most appropriate [116–124]. Human hearing, for example, is a nonlinear system that thereby circumvents the restrictions of the Gabor limit (to allow for musical geniuses, for example, who have "perfect pitch"), where time-frequency acuity surpasses what would be possible by linear signal processing alone [116], such as with Nyquist sampled linear response recording devices that are bound by the limits imposed by the Fourier uncertainty principle (or Benedick's theorem) [117]. Thus, even when the powerful Fourier Transform or Hidden Markov Model (HMM) feature extraction methods are utilized to full advantage, there is often a sector of the signal analysis that is only conveniently accessible to analysis by way of FSAs (without significant oversampling), such that a parallel processing with both HMM and FSA methods is often needed (results demonstrating this in the context of channel current analysis [1–3] will be described in Chapter 14). Not all of the methods employed at the FSA processing stage derive from standard signal processing approaches, either, some are purely statistical such as with oversampling [118] (used in radar range oversampling [119, 120]) and dithering [121] (used in device stabilization and to reduce quantization error [122, 123]).

All of the tFSA signal acquisition methods described in Chapters 2–4 are O(L), i.e. they scan the data with a computational complexity no greater than that of simply seeing the data (via a "read" or "touch" command, O(L) is known as "order of," or "big-oh," notation). Because the signal acquisition is only O(L) it is not significantly costly, computationally, to simply repeat the acquisition analysis multiple times with a more informed process with each iteration, to have arrived at a "bootstrap" signal acquisition process. In such a setting, signal acquisition is often done with bias to very high specificity initially (and sensitivity very poor), to get a "gold standard" set of highly likely true signals that can be data mined for their attributes. With a filter stage thereby trained, later scan passes can pass suspected signals with very weak specificity (very high sensitivity now) with high specificity then recovered by use of the filter. This then allows a bootstrap process to a very high specificity (SP) and sensitivity (SN) at the tFSA acquisition stage on the signals of interest.

An example of a bootstrap FSA from genomic analysis is to first scan through a genome base-by-base and obtain counts on nucleotide pairs with different gap sizes between the nucleotides observed [1, 3]. This then allows a mutual information analysis on the nucleotide pairs taken at the different gap sizes (shown in Chatpers 3 and 4). What is found for prokaryotic genomes, with their highly dense gene placement, that is mostly protein coding (i.e. where there is little "junk" deoxyribonucleic acid (DNA) and no introns), is a clear signal indicating anomalous statistical linkages on bases three apart [1, 3, 60]. What is discovered thereby is codon structure, where the coding information comes in groups of three bases. Knowing this, a repeated pass (bootstrap) with frequency analysis of the 64 possible 3-base groupings can then be done, at which point the anomalously low counts on "stop" codons is then observed. Upon identification of the stop codons their placement (topology) in the genome can then be examined and it is found that their counts are anomalously low because there are large stretches of regions with no stop codon (e.g. there are stop codon "voids," known as open reading frames, or "ORF"s). The codon void topologies are examined in a comparative genomic analysis in [60] (and shown in Chapter 3). The stop codons, which should occur every 21 codons on average if DNA sequence data was random, are sometimes not seen for stretches of several hundred codons. For the genomic data we are finding the longer genes, whose anomalous non-random DNA sequence is more distinctive the longer the gene-coding region. This basic analysis can provide a genefinder on prokaryotic genomes that comprises a one-page Python script that can perform with 90–99% accuracy depending on the prokaryotic genome (shown in Chapter 3). A second page of Python coding to introduce a "filter," along the lines of the bootstrap learning process mentioned above, leads to an *ab initio* prokaryotic gene-predictor with 98.0–99.9% accuracy. Python code to accomplish this is shown in Chapter 4. In this bootstrap acquisition process all that is used is the raw genomic data (with its highly structured intrinsic statistics) and methods for identifying statistical anomalies and informatics structural anomalies: (i) anomalously high mutual information is identified (revealing codon structure); (ii) anomalously high (or low) statistics on an attribute or event is then identified (low stop codon counts, lengthy stop codon voids); then anomalously high sub-sequences (binding site motifs) are found in the neighborhood of the identified ORFs (used in the filter).

Ad hoc signal acquisition refers to finding the solution for "this" situation (whatever "this" is) without consideration of wider application. The solution is strongly *data dependent* in other words. Data dependent methodologies are, by definition, not defined at the outset, but must be invented as the data begins to be understood. As with data dependency in non-evolutionary search metaheuristics, where there is no optimal search method that is guaranteed to always work well, here there is no optimal signal acquisition method known in advance. This is simply restating a fundamental limit from non-evolutionary search metaheuristics in another form [1, 3]. What can be done, however, is assemble the core tools and techniques from which a solution can be constructed and to perform a bootstrap algorithmic learning process with those tools (examples in what follows) to arrive at a functional signal acquisition on the data being analyzed. A universal, automated, bootstrap learning process may eventually be possible using evolutionary learning algorithms. This is related to the co-evolutionary Free Lunch Theorem [1, 3], and this is discussed in Chapter 12.

"Bootstrap" refers to a method of problem solving when the problem is solved by seemingly paradoxical measures (the name references Baron von Munchausen who freed the horse he was riding from a bog by pulling himself, and the horse with him, up by his bootstraps). Such algorithmic methods often involve repeated passes over the data sequence, with improved priors, or a trained filter, among other things, to have improved performance. The bootstrap amplifier from electrical engineering is an amplifier circuit where part of the output is used as input, particularly at start-up (known as bootstrapping), allowing proper self-initialization to a functional state (by amplifying ambient circuit noise in some cases). The bootstrap FSA proposed here is a meta-algorithmic method in that performance "feedback" with learning is used in algorithmic refinements with iterated meta-algorithmic learning to arrive at a functional signal acquisition status.

Acquisition is often all that is needed in a signal analysis problem, where a basic means to acquire the signals is sought, to be followed by a basic statistical analysis on those signals and their occurrences. Various methods for signal acquisition using FSA constructs are described in what follows, with focus on statistical anomalies to identify the presence of signal and "lock on" [1, 3]. The signal acquisition is initially only guided by use of statistical measures to recognize anomalies. Informatics methods and information theory measures are central to the design of a good FSA acquisition method, however, and will be reviewed in the signal acquisition context [1, 3], along with HMMs.

Thus, FSA processes allow signal regions to be identified, or "acquired," in O(L) time. Furthermore, in that same order of time complexity, an entire panoply of statistical moments can also be computed on the signals (and used in a bootstrap learning process). The O(L) feature extraction of statistical moments on the signal region acquired may suffice for *localized* events and structures. For sequential information or events, however, there is often a *non-local*, or extended structural, aspect to the signal sought. In these situations we need a general, powerful, way to analyze sequential signal data that is stochastic (random, but with statistics, such as average, that may be unchanging over time if "stationary," for example). The general method for performing stochastic sequential analysis (SSA) is via HMMs, as will be extensively described in Chapters 6 and 7, and briefly summarized in Section 1.5 that follows. HMM approaches require an identification of "states" in the signal analysis. If an identification of states is difficult, such as in situations where there can be changes in meaning according to context, e.g. language, then HMMs may not be useful. Text and language analytics are described in Chapters 5 and 13, and briefly outlined in the next section.

1.4 Feature Extraction and Language Analytics

The FSA sequential-data signal processing, and extraction of statistical moments on windowed data, will be shown in Chapter 2 to be O(L) with L the size of the data (double the data and you double the processing time). If HMMs can be used, with their introduction of states (the sequential data is described as a sequennce of "hidden" states), then the computational cost goes as $O(LN^2)$. If $N = 10$, then this could be 100 times more computational time to process than that of a FSA-based O(L) computation, so the HMMs can generally be a lot more expensive in terms of computational time. Even so, if you can benefit from a HMM it is generally possible to do so, even if hardware specialization (CPU farm utilization, etc.) is required. The problem is if you do not have a strong basis for a HMM application, e.g. when there is no strong basis for delineating the states of the system of communication under study. This is the problem encounterd in the study of natural languages (where there is significant context dependency). In Chapter 5 we look into FSA analysis for language by doing some basic text analytics.

Chapter 5 shows some (very) basic extensions to an FSA analysis in applications to text. This begins with a simple frequency analysis on words, which for some classics (in their original languages) reveal important word-frequency results with implied meanings meant by the author (polysemy word usage by Machiavelli, for example). The frequency on word groupings in a given text can be studied as well, with some useful results from texts of sufficient size with clear stylistic conventions by the author. Authors that structure their lines of text according to iambic pentameter (Shakespeare, for example) can also be identified according to the profile (histogram) of syllables used on each line (i.e. 10 for iambic pentameter will dominate).

Text analytics can also take what is still O(L) processing into mapping the mood or sentiment of text samples by use of word-scored sentiment tables. The generation and use of such sentiment tables is its own craft, usually proprietary, so only minimal examples are given. Thus Chapter 5 shows an elaboration of FSA-based analysis that might be done when there is no clear definition of state, such as in language. NLP processing in general encompasses a much more complete grammatical knowledge of the language, but in the end the NLP and the FSA-based "add-on" still suffer from not being able to manage word context easily (the states cannot simply be words since the words can have different meaning according to context). The inability to use HMMs has been a blockade to a "universal translator" that has since been overcome with use of Deep Learning using NNs (Chapter 13) – where immense amounts of translation data, such as the massive corpus of dual language Canadian Government proceedings, is sufficient to train a translator (English–French). Most of the remaining Chapters focus on situations where a clear delinaeation of signal state can be given, and thus benefit from the use of HMMs.

1.5 Feature Extraction and Gene Structure Identification

HMMs offer a more sophisticated signal recognition process than FSAs, but with greater computational space and time complexity [125, 126]. Like electrical engineering signal processing, HMMs usually involve preprocessing that assumes linear system properties or assumes observation is frequency band limited and not time limited, and thereby inherit the time-frequency uncertainty relations, Gabor limit, and Nyquist sampling relations. FSA methods can be used to recover (or extract) signal features missed by HMM or classical electrical engineering signal processing. Even if the signal sought is well understood, and a purely HMM-based approach is possible, this is often needlessly computationally intensive (slow), especially in areas where there is no signal. To address this there are numerous hybrid FSA/HMM approaches (such as BLAST [127]) that benefit from the

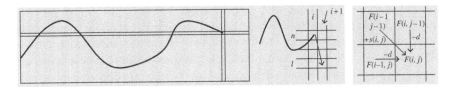

Figure 1.2 The Viterbi path. (Left) The Viterbi path is recursively defined, thus tabulatable, with one column only, recursively, dependent on the prior column. (Right) A related recursive algorithm used to perform sequence alignment extensions with gaps (the Smith–Waterman algorithm) is provided by the neighbor-cell recursively-defined relation shown.

O(L) complexity on length L signal with FSA processing, with more targeted processing at $O(LN^2)$ complexity with HMM processing (where there are N states in the HMM model).

HMMs, unlike tFSAs, have a straightforward mathematical and computational foundation at the nexus where Bayesian probability and Markov models meet dynamic programming. To properly define or choose the HMM model in a ML context, however, further generalization is usually required. This is because the "bare-bones" HMM description has critical weaknesses in most applications, which are described in Chapter 7, ***along with their "fixes."*** Fortunately, each of the standard HMM weaknesses can be addressed in computationally efficient ways. The generalized HMMs described in Chapter 7 allow for a generalized Viterbi Algorithm (see Figure 1.2) and a generalized Baum–Welch Algorithm. The generalized algorithms retain path probabilities in terms of a sequence of likelihood ratios, which satisfy Martingale statistics under appropriate circumstances [102], thereby having Martingale convergence properties (where here convergence is associated with "learning" in this context). Thus, HMM learning proceeds via convergence to a limit state that provably exists in a similar sense to that shown with the Hoeffding inequality [59], via its proven extension to Martingales [108]. The Hoeffding inequality is a key part of the VC Theorem in ML, whereby convergence for the Perceptron learning process to a solution in an infinite solution space is proven to exist in a finite number of learning steps [109]. Further details on the Fundamental Theorems [102, 103, 108, 109] are summarized in Appendix C.

HMM tools have recently been developed with a number of computationally efficient improvements (described in detail in Chapter 7), where application of the HMM methods will be described for gene-finding, alt-splice gene-finding, and nanopore-detector signal analysis.

HMM methods are powerful, especially with the enhancements described in Chapter 7, but this would all be for naught in a real-time, O(L), processing on L size data if the core $O(LN^2)$ algorithm (N states in the HMM) could not be distributed, onto $O(N^2)$ nodes, say, to get back to an overall distributed process involving HMM feature extraction with $O(L)$ processing (to be part of our real-time signal processing pipeline). So, a way is needed to distribute the core algorithms for HMM learning: Viterbi and Baum–Welch. It turns out distributed processing, or "chunking," is possible for the single sweep Viterbi algorithm (ignoring the trivial traceback optimal path recovery that does not cause table alteration). The key to having this chunking capability on the other core learning algorithm, Baum–Welch, is to have a similar single-pass table production. The standard Baum–Welch requires a forward and a backward sweep across the table during production of the result (with algorithms named accordingly for this purpose in Chapter 3). As this would disallow the chunking solution, what is needed is a single sweep Baum Welch algorithm, which has been discovered and is described in Chapter 3, where it is known as the Linear Memory HMM (at the time of its discovery it was most notable to reviewers due to another oddity, that it required only linear space memory during computation – but memory is cheap, while being able to perform distributed processing with massive speed-up operationally is much more important). With distributability (asynchronous), computational time is directly reduced by ~N on a cluster with N nodes (see Figure 1.2). The HMM with single-sweep Linear Memory (distributable) for expectation/maximization (EM) also allows distribution (massive parallel asynchronous processing) for the generalized Viterbi and Baum–Welch algorithms on the Meta-HMM and Hidden Markov model-with-duration (HMMD) variants described in Chapter 3 with distribution shown in (Figure 1.3).

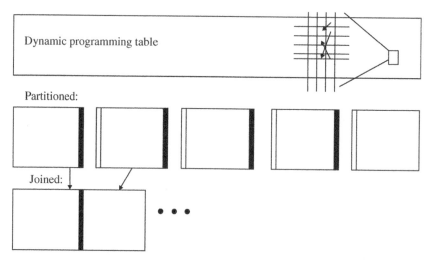

Figure 1.3 Chunking on a dynamic table. Works for a HMM using a simple join recovery.

1.5.1 HMMs for Analysis of Information Encoding Molecules

The main application areas for HMMs covered in this book are power signal analysis generally, and bioinformatics and cheminformatics specifically (the main reviews and applications discussed are from [128–134]). For bioinformatics, we have information encoding molecules that are polymers, giving rise to sequential data format, thus HMMs are well suited for analysis. To begin to understand bioinformatics, however, we need to know not only the biological encoding rules, largely rediscovered on the basis of their statistical anomalies in Chapters 1–4, but also the idiosyncratic structures seen (genomes and transcriptomes) that are full of evolutionary artifacts and similarities to evolutionary cousins. To know the nature of the statistical imprinting on the polymeric encodings also requires an understanding of the biochemical constraints that give rise to the statistical biases seen. Once taken altogether, bioinformatics offers a lot of clarity on why Nature has settled on the particular genomic "mess," albeit with optimizations, that it has selectively arrived at. See [1, 3] for further discussion of bioinformatics.

1.5.2 HMMs for Cheminformatics and Generic Signal Analysis

The prospect of having a HMM feature extraction in the streaming signal processing pipeline (O(L), for size L data process) offers powerful real-time feature extraction capabilities and specialized filtering (all of which is implemented in the Nanoscope, Chapter 14). One such processing method, described in Chapter 6, is HMM/Expectation Maximization (EM) EVA (Emission Variance Amplification) Projection which has application in providing simplified automated tFSA Kinetic Feature Extraction from channel current signal. What is needed is the equivalent of low-pass filtering on blockade levels while retaining sharpness on the timing of the level changes. This is not possible with the standard low-pass filter because the edges get blurred out in the local filtering process, but notice how this does not happen with the HMM-based filter, for the data shown in Figure 1.4.

HMM is a common *intrinsic* statistical sequence modeling method (implementations and applications are mainly drawn from [135–158] in what follows), so the question naturally arises – how to optimally incorporate *extrinsic* "side-information" into a HMM? This can be done by treating duration distribution information *itself* as side-information and a process is shown for incorporating side-information into a HMM. It is thereby demonstrated how to bootstrap from a HMM to a HMMD (more generally, a hidden semi-Markov model or HSMM, as it will be described in Chapter 7).

In many applications, the ability to incorporate the state duration into the HMM is very important because conventional HMM-based, Viterbi and Baum-Welch

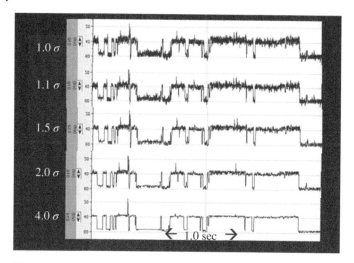

Figure 1.4 Edge feature enhancement via HMM/EM EVA filter. The filter "projects" via a Gaussian parameterization on emissions with variance boosted by the factor indicated. From prior publications by the author [1–3]. *Source:* Based on Winters-Hilt [1–3].

algorithms are otherwise critically constrained in their modeling ability to distributions on state intervals that are geometric (this is shown in Chapter 7). This can lead to a significant decoding failure in noisy environments when the state-interval distributions are not geometric (or approximately geometric). The starkest contrast occurs for multimodal distributions and *heavy-tailed* distributions, the latter occurring for exon and intron length distributions (thus critical in gene finders). The hidden Markov model with binned duration (HMMBD) algorithm eliminates the HMM geometric distribution modeling constraint, as well as the HMMD maximum duration constraint, and offers a significant reduction in computational time for all HMMBD-based methods *to be approximately equal to the computational time of the HMM-process alone.*

In adopting any model with "more parameters," such as a HMMBD over a HMM, there is potentially a problem with having sufficient data to support the additional modeling. This is generally not a problem in any HMM model that requires thousands of samples of non-self transitions for sensor modeling, such as for the gene-finding that is described in what follows, since knowing the boundary positions allows the regions of self-transitions (the durations) to be extracted with similar sample number as well, which is typically sufficient for effective modeling of the duration distributions in a HMMD.

Improvement to overall HMM application rests not only with the aforementioned improvements to the HMM/HMMBD, but also with improvements to the hidden state model and emission model. This is because standard HMMs are at low Markov order in transitions (first) and in emissions (zeroth), and transitions are decoupled from emissions (which can miss critical structure in the model, such as state transition probabilities that are sequence dependent). This weakness is eliminated if we generalize to the largest state-emission clique possible, fully interpolated on the data set, as is done with the generalized-clique HMM, where gene finding is performed on the *Caenorhabditis elegans* genome. The clique generalization improves the modeling of the critical signal information at the transitions between exon regions and noncoding regions, e.g. intron and junk regions. In doing this we arrive at a HMM structure identification platform that is novel, and robustly performing, in a number of ways.

Prior HMM-based systems for SSA had undesirable limitations and disadvantages. For example, the speed of operation made such systems difficult, if not impossible, to use for real-time analysis of information. In the SSA Protocol described here, distributed generalized HMM processing together with the use of the SVM-based Classification and Clustering Methods (described next) permit the general use of the SSA Protocol free of the usual limitations. After the HMM and SSA methods are described, their synergistic union is used to convey a new approach to signal analysis with HMM methods, including a new form of stochastic-carrier wave (SCW) communication.

1.6 Theoretical Foundations for Learning

Before moving on to classification and clustering (Chapter 10), a brief description is given of some of the theoretical foundations for learning, starting with the foundation for the choice of information measures used in Chapters 2–4, and this is shown in Chapter 8. In Chapter 9 we then describe the theory of NNs. The Chapter 9 background is not meant to be a complete exposition on NN learning (the opposite), but merely goes through a few specific analyses in the area of Loss Bounds analysis to give a sense of what makes a good classification method.

1.7 Classification and Clustering

SVMs can be used for classification and clustering (to be described in detail in Chapter 10), as well as aiding with signal analysis and pattern recognition on stochastic sequential data. The signal processing material described next, and in detail later, mainly draw from prior journal publications [159–189]. Analysis tools for stochastic sequential data have broad-ranging application by making any device producing a sequence of measurements more sensitive, or "smarter," by efficient learning of measured signal/pattern characteristics. The SVM and HMM/ SVM application areas described in this book include cheminformatics, biophysics, and bioinformatics. The cheminformatics application examples pertain to channel current analysis on the alpha-hemolysin nanopore detector (Chapter 14).

The biophysics and "information flows" associated with the nanopore transduction detector (NTD) in Chapter 14 are analyzed using a generalized set of HMM and SVM-based tools, as well as *ad hoc* FSAs-based methods, and a collection of distributed genetic algorithm methods for tuning and selection. Used with a nanopore detector, the channel current cheminformatics (CCC) for the stationary signal channel blockades (with "stationary statistics") enables a method for a highly sensitive nanopore detector for single molecule biophysical analysis.

The SVM implementations described involve SVM algorithmic variants, kernel variants, and chunking variants; as well as SVM classification tuning metaheuristics; and SVM clustering metaheuristics. The SVM tuning metaheuristics typically enable use of the SVM's confidence parameter to bootstrap from a strong classification engine to a strong clustering engine via use of label changes, and repeated SVM training processes with the new label information obtained.

SVM Methods and Systems are given in Chapter 10 for classification, clustering, and SSA in general, with a broad range of applications:

- sequential-structure identification
- pattern recognition
- knowledge discovery
- bioinformatics
- nanopore detector cheminformatics
- computational engineering with information flows
- "SSA" Architectures favoring Deep Learning (see next section)

SVM binary discrimination outperforms other classification methods with or without dropping weak data (while many other methods cannot even identify weak data).

1.8   Search

All of the core methods described thus far (FSA, HMM, SVM) require some amount of parameter "tuning" for good performance. In essence, tuning is a search through parameter space (of the method) for best performance (according to a variety of metrics). The tuning on acquisition parameters in an FSA, or choice of states in a HMM, or SVM Kernels and Kernel parameters, is often not terribly complicated allowing for a "brute-force" search over a set of parameters, choosing the best from that set. On occasion, however, a more elaborate, and fully automated, search-optimization is sought (or just search problem in general), For more complex search tasks it is good to know the modern search methodologies and what they are capable of, so these are described in Chapter 11.

1.9 Stochastic Sequential Analysis (SSA) Protocol (Deep Learning Without NNs)

The SSA protocol is shown in Figure 1.5 (from prior publications and patent work, see [1–3]) and is a general signal-processing flow topology and database schema (Left Panel), with specialized variants for CCC (Center) and kinetic feature

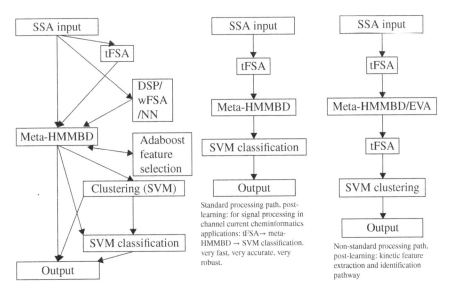

Figure 1.5 (Left) The general stochastic sequential analysis flow topology. (Center) The general signal processing flow in performing channel current analysis is typically Input → tFSA → Meta-HMMBD → SVM → Output. (Right) Notable differences occur in channel current cheminformatics during state discovery when EVA-projection (emission variance amplification projection), or a similar method, is used to achieve a quantization on states, then have Input → tFSA → HMMBD/EVA (state discovery) → meta-HMMBD-side → SVM → Output. While, in gene-finding just have: Input → meta-HMMBD-side → Output. In gene-finding, however, the HMM internal "sensors" are sometimes replaced, locally, with profile-HMMs [1, 3] (equivalent to position-dependent Markov Models, or pMM's, see Chapter 7), or SVM-based profiling [1, 3], so the topology can differ not only in the connections between the boxes shown, but in their ability to embed in other boxes as part of an internal refinement. *Source:* Based on Winters-Hilt [1, 3].

extraction based on blockade-level duration observations (Right). The SSA Protocol allows for the discovery, characterization, and classification of localizable, approximately-stationary, statistical signal structures in channel current data, or genomic data, or sequential data in general. The core signal processing stage in Figure 1.5 is usually the feature extraction stage, where central to the signal processing protocol is a generalized HMM. The SSA Protocol also has a built-in recovery protocol for weak signal handling, outlined next, where the HMM methods are complemented by the strengths of other ML methods.

The sequence of algorithmic methods used in the SSA Protocol, for the information-processing flow topology shown in Figure 1.5, comprise a weak signal handling protocol as follows: (i) the weakness in the (fast) Finite State Automaton (FSA) methods will be shown to be their difficulty in nonlocal structure identification, for which HMM methods (and tuning metaheuristics) are the solution; (ii) for the HMM, in turn, the main weakness is in local sensing "classification" due to conditional independence assumptions. Once in the setting of a classification problem, however, the problem can be solved via incorporation of generalized SVM methods [1, 3]. If facing only classification task (data already preprocessed), the SVM will also be the method of choice in what follows. (iii) The weakness of the SVM, whether used for classification or clustering, but especially for the latter, is the need to optimize over algorithmic, model (kernel), chunking, and other process parameters during learning. This is solved via use of metaheuristics for optimization such as simulated annealing, and genetic algorithm optimization in (iv). The main weaknesses in the metaheuristic tuning effort is partly resolved via use of the "front-end" methods, like the FSA, and partly resolved by a knowledge discovery process using the SVM clustering methods. The SSA Protocol weak signal acquisition and analysis method thereby establishes a robust signal processing platform.

The HMM methods are the central methodology or stage in the SSA Protocol, particularly in the gene finders, and sometimes with the CCC protocol or implementation, in that the other stages can be dropped or merged with the HMM stage in many incarnations. For example, in some CCC analysis situations the tFSA methods could be totally eliminated in favor of the more accurate (but time consuming) HMM-based approaches to the problem, with signal states defined or explored in more or less the same setting, but with the optimized Viterbi path solution taken as the basis for the signal acquisition.

The HMM features, and other features (from NN, wavelet, or spike profiling, etc.) can be fused and selected via use of various data fusion methods, such as a modified Adaboost selection (from [1, 3], and Chapter 11). The HMM-based feature extraction provides a well-focused set of "eyes" on the data, no matter what its nature, according to the underpinnings of its Bayesian statistical representation. The key is that the HMM not be too limiting in its state definition, while there is the typical engineering trade-off on the choice of number of states, N, which impacts the order of computation via a quadratic factor of N in the various dynamic programming calculations (comprising the Viterbi and Baum–Welch algorithms among others).

The HMM "sensor" capabilities can be significantly improved via switching from profile-Markov Model (pMM) sensors to pMM/SVM-based sensors, as indicated in [1, 3] and Chapter 7, where the improved performance and generalization capability of this approach is demonstrated.

In standard band-limited (and not time-limited) signal analysis with periodic waveforms, sampling is done at the Nyquist rate to have a fully reproducible signal. If the sample information is needed elsewhere, it is then compressed (possibly lossy) and transmitted (a "smart encoder"). The received data is then decompressed and reconstructed (by simply summing wave components, e.g. a "simple" decoder). If the signal is sparse or compressible, then compressive sensing [190] can be used, where sampling and compression are combined into one efficient step to obtain compressive measurements (the simple encoding in [190] since a set of random projections are employed), which are then transmitted (general details on noise in this context are described in [191, 192]). On the receiving end, the decompression and reconstruction steps are, likewise, combined using an asymmetric "smart" decoding step. This progression toward asymmetric compressive signal processing can be taken a step further if we consider signal sequences to be equivalent if they have the same stationary statistics. What is obtained is a method similar to compressive sensing, but involving stationary-statistics generative-projection sensing, where the signal processing is non-lossy at the level of stationary statistics equivalence. In the SCW signal analysis the signal source is generative in that it is describable via use of a HMM, and the HMM's Viterbi-derived generative projections are used to describe the sparse components contributing to the signal source. In SCW encoding the modulation of stationary statistics can be man-made or natural, with the latter in many experimental situations involving a flow phenomenology that has stationary statistics. If the signal is man-made, usually the underlying stochastic process is still a natural source, where it is the changes in the stationary statistics that is under the control of the man-made encoding scheme. Transmission and reception are then followed by generative projection via Viterbi-HMM template matching or via Viterbi-HMM feature extraction followed by separate classification (using SVM). So in the SCW approach the encoding is even simpler (possibly non-existent, other than directly passing quantized signal) and is applicable to any noise source with stationary statistics (e.g. a stationary signal with reproducible statistics, the case for many experimental observations). The decoding must be even "smarter," on the other hand, in that generalized Viterbi algorithms are used, and possibly other ML methods as well, SVMs in particular. An example of the stationary statistics sensing with a ML-based decoder is described in application to CCC studies in Chapter 14.

1.9.1   Stochastic Carrier Wave (SCW) Analysis – Nanoscope Signal Analysis

The Nanoscope described in Chapter 14 builds from nanopore detection with introduction of reporter molecules to arrive at a nanopore transduction detection paradigm. By engineering reporter molecules that produce stationary statistics (a SCW) together with ML signal analysis methods designed for rapid analysis of such signals, we arrive at a functioning "nanoscope."

Nanopore detection is made possible by the following well-established capabilities: (i) classic electrochemistry; (ii) pore-forming protein toxin in a bilayer; and (iii) patch clamp amplifier. Nanopore *transduction* detection leverages the above detection platform with (iv) an event-transducer pore-blockader that has stationary statistics and (v) ML tools for real-time SCW signal analysis. The meaning of "real-time" is dependent on the application. In the Nanoscope implementation discussed in Chapter 14, each signal is usually identified in less than 100 ms, where calling accuracy is 99.9% if rejection is employed, and improved even further if signal sample duration, when a call is forced, is used with duration greater than 100 ms.

Nanopore transduction detection offers prospects for highly sensitive and discriminative biosensing. The NTD "Nanoscope" functionalizes a single nanopore with a channel current modulator that is designed to transduce events, such as binding to a specific target. Nanopore event transduction involves single-molecule biophysics, engineered information flows, and nanopore cheminformatics. In the NTD functionalization the transducer molecule is drawn into the channel by an applied potential but is too big to translocate, instead becoming stuck in a bistable capture such that it modulates the channel's ion-flow with stationary statistics in a distinctive way. If the channel modulator is bifunctional in that one end is meant to be captured and modulated while the other end is linked to an aptamer or antibody for specific binding, then we have the basis for a remarkably sensitive and specific biosensing capability.

In the NTD Nanoscope experiments [2], the molecular dynamics of a (single) captured non-translocating transducer molecule provide a unique stochastic reference signal with stable statistics on the observed, single-molecule blockaded channel current, somewhat analogous to a carrier signal in standard electrical engineering signal analysis. Discernible changes in blockade statistics, coupled with SSA signal processing protocols, enable the means for a highly detailed characterization of the interactions of the transducer molecule with binding targets (cognates) in the surrounding (extra-channel) environment.

The transducer molecule is engineered to generate distinct channel blockade signals depending on its interaction with target molecules [2]. Statistical models are trained for each binding mode, bound and unbound, for example, by exposing the transducer molecule to zero or high (excess) concentrations of the target molecule. The transducer molecule is engineered so that these different binding states generate distinct signals with high resolution. Once the signals are characterized, the information can be used in a real-time setting to determine if trace amounts of the target are present in a sample through a serial, high-frequency sampling, and pattern recognition, process.

Thus, in Nanoscope applications of the SSA Protocol, due to the molecular dynamics of the captured transducer molecule, a unique reference signal with strongly stationary (or weakly, or approximately stationary) signal statistics is engineered to be generated during transducer blockade, analogous to a carrier signal in standard electrical engineering signal analysis. In these applications a signal is deemed "strongly" stationary if the EM/EVA projection (HMM method from Chapter 6) on the entire dataset of interest produces a discrete set of separable (non-fuzzy domain) states. A signal is deemed "weakly" stationary if the EM/EVA projection can only produce a discrete set of states on subsegments (windowed sections) of the data sequence, but where state-tracking is possible across windows (i.e. the non-stationarity is sufficiently slow to track states – similar to the adiabatic criterion in statistical mechanics). A signal is approximately stationary, in a general sense, if it is sufficiently stationary to still benefit, to some extent, from the HMM-based signal processing tools (that assume stationarity).

The adaptive SSA ML algorithms, for real-time analysis of the stochastic signal generated by the transducer molecule can easily offer a "lock and key" level of signal discrimination. The heart of the signal processing algorithm is a generalized Hidden Markov Model (gHMM)-based feature extraction method, implemented on a distributed processing platform for real-time operation. For real-time processing, the gHMM is used for feature extraction on stochastic sequential data, while classification and clustering analysis are implemented using a SVM. In addition, the design of the ML-based algorithms allow for scaling to large datasets, via real-time distributed processing, and are adaptable to analysis on any stochastic sequential dataset. The ML software has also been integrated into the NTD Nanoscope [2] for "real-time" pattern-recognition informed (PRI) feedback [1–3] (see Chapter 14 for results). The methods used to implement the PRI feedback include *distributed* HMM and SVM implementations, which enable the processing speedup that is needed.

1.9.2 Nanoscope Cheminformatics – A Case Study for Device "Smartening"

The Nanoscope example can also be considered as a case study for device "smartening," whereby device state is tracked in terms of easily measured device characteristics, such as the ambient device "noise." A familiar example of this would be the sound of your car engine. In essence, you could eventually have an artificial intelligence (AI) listening to the sound of your engine to similarly track state and issue warnings like an expert mechanic with that car, without the need for sensors, or to supplement sensors (reducing expense, providing secondary fail-safe). Such an AI might even offer predictive fault detection.

1.10 Deep Learning using Neural Nets

ML provides a solution to the "Big Data" problem, whereby a vast amount of data is distilled down to its information essence. The ML solution sought is usually required to perform some task on the raw data, such as classification (of images) or translation of text from one language to another. In doing so, ML solutions are strongly favored where a clear elucidation of the features used in the classification are also revealed. This then allows a more standard engineering design cycle to be accessed, where the stronger features thereby identified may play a stronger role, or guide the refinement of related strong features, to arrive at an improved classifier. This is what is accomplished with the previously mentioned SSA Protocol.

So, given the flexibility of the SSA Protocol to "latch on" to signal that has a reasonable set of features, you might ask what is left? (Note that, all communication protocols, both natural (genomic) and man-made, have a "reasonable" set of features.) The answer is simply when the number of features is "unreasonable" (with enumeration not even known, typically). So instead of 100 features, or maybe 1000, we now have a situation with 100 000 to 100s of millions of features (such as with sentence translation or complex image classification). Obviously Big Data is necessary to learn with such a huge number of features present, so we are truly in the realm of Big Data to even begin with such problems, but now have the Big Features issue (e.g. Big Data with Big Features, or BDwBF). What must occur in such problems is a means to wrangle the almost intractable large feature set of information to a much smaller feature set of information, e.g. an intial layer of processing is needed just to compress the feature data. In essence, we need a form of compressive feature extraction at the outset in order to not overwhelm the acquisition process. An example from the biology of the human eye, is the layer of local neural processing at the retina before the nerve impulses even travel on to the brain for further layers of neural processing.

For translation we have a BDwBF problem. The feature set is so complex the best approach is NN Deep Learning where we assume no knowledge of the features but rediscover/capture those features in compressed feature groups that are identified in NN learning process at the first layer of the NN architecture. This begins a process of tuning over NN architectures to arrive at a compressive feature acquisitiuon with strong classification performance (or translation accuracy, in this example). This learning approach began seeing widespread application in 2006 and is now the core method for handling the Big Feature Set (BFS) problem. The BFS problem may or may not exist at the initial acquisition ("front-end") of your signal processing chain. NN Deep Learning to solve the BFS problem will be described in detail in Chapter 13, where examples using a Python/TensorFlow application to translation will be given. In the NN Deep Learning approach, the features are not implicitly resolvable, so improvements are initially brute force (even bigger data) since an engineering cycle refinement would involve the enormous parallel task of explicitly resolving the feature data to know what to refine.

1.11 Mathematical Specifics and Computational Implementations

Throughout the text an effort is made to provide mathematical specifics to clearly understand the theoretical underpinnings of the methods. This provides a strong exposition of the theory but the motivation for this is not to do more theory, but to then proceed to a clearly defined computational implementation. This is where mathematical elegance meets implementation/computational practicality (and the latter wins). In this text, the focus is almost entirely on elegent methods that also have highly efficient computational implementations.

2

Probabilistic Reasoning and Bioinformatics

In this chapter, a review is given of statistics and probability concepts, with implementation of many of the concepts in Python. Python scripts are then used to do a preliminary examination of the randomness of genomic (virus) sequence data. A short review of Linux OS setup (with Python automatically installed) and Python syntax is given in Appendix A.

Numerous prior book, journal, and patent publications by the author [1–68] are drawn upon extensively throughout the text. Almost all of the journal publications are open access. These publications can typically be found online at either the author's personal website (www.meta-logos.com) or with one of the following online publishers: www.m-hikari.com or bmcbioinformatics.biomedcentral.com.

2.1 Python Shell Scripting

A "fair" die has equal probability of rolling a 1, 2, 3, 4, 5, or 6, i.e. a probability of 1/6 for each of the outcomes. Notice how the sum of all of the discrete probabilities for the different outcomes all add up to 1, this is always the case for probabilities describing a complete set of outcomes.

A "loaded" die has a non-uniform distribution, for prob = 0.5 to roll a "6" and uniform on the other die rolls you have loaded die_roll_probability = (1/10,1/10,1/10,1/10,1/10,1/2).

The first program to be discussed is named prog1.py and will introduce the notion of discrete probability distributions in the context of rolling the familiar six-sided die. Comments in Python are the portion of a line to the right of any "#" symbol (except for the first line of code with "#!.....", that is explained later).

The Shannon entropy of a discrete probability distribution is the measure of its amount of randomness, with the uniform probability distribution having the greatest randomness (e.g. it is most lacking in any statistical "structure" or

Informatics and Machine Learning: From Martingales to Metaheuristics, First Edition.
Stephen Winters-Hilt.
© 2022 John Wiley & Sons, Inc. Published 2022 by John Wiley & Sons, Inc.

"information"). Shannon entropy is the sum of each outcome probability times its log probability, with an overall negative placed in front to arrive at a definition involving a positive value. Further details on the mathematical formalism will be given in Chapter 3, but for now we can implement this in our first Python program:

———————————————— prog1.py ————————————

```
#!/usr/bin/python

import numpy as np
import math
import re
arr = np.array([1.0/10,1.0/10,1.0/10,1.0/10,1.0/10,1.0/2])
# print(arr[0])

shannon_entropy = 0
numterms = len(arr)
print(numterms)

index = 0
for index in range(0, numterms):
    shannon_entropy += arr[index]*math.log(arr[index])

shannon_entropy = -shannon_entropy
print(shannon_entropy)
```

———————————————— end prog1.py ————————————

The maximum Shannon entropy on a system with six outcomes, uniformly distributed (a fair die), is log(6). In the prog1.py program above we evaluate the Shannon entropy for a loaded die: (1/10.1/10,1/10,1/10,1/10,1/2). Notice in the code, however, that I use "1.0" not "1". This is because if the expression only involves integers the mathematics will be done as integer operations (returning an integer, thus truncation of some sort). An expression that is mixed, some integer terms, some floating point (with a decimal), will be evaluated as a floating point number. So, to force recognition of the numbers as floating point the "1" value in the terms is entered as "1.0". Further tests are left to the Exercises (Section 2.7).

A basic review of getting a linux system running, with it is standard Python installed, is described in the Appendix, along with a discussion of how to install added Python modules (added code blocks with very useful, pre-built, data structures, and subroutines), particularly "numpy," which is indicated as a module to be imported (accessed) by the program by the first Python command: "import numpy as np." (We will see in the Appendix that the first line is not a Python command but a shell directive as to what program to use to process the commands that follow, and this is the mechanism whereby a system level call on the Python script can be done.)

Let us now move on to some basic statistical concepts. How do we know the probabilities for the outcomes of the die roll? In practice, you would observe numerous die rolls and get counts of how many times the various outcomes were observed. Once you have counts, you can divide by the total counts to have the frequency of occurrence of the different outcomes. If you have enough observational data, the frequencies then become better and better estimates of the true underlying probabilities for those outcomes for the system observed (a result due to the law of large numbers (LLN), which is rederived in Section 2.6.1). Let us proceed with adding more code in prog1.py that begins with counts on the different die rolls:

––––––––––––––––––– prog1.py addendum 1 –––––––––––––––––––

```
rolls = np.array([3435.0,3566,3245,3600,3544,3427])
numterms = len(rolls)
total_count = 0
for index in range(0,numterms):
    total_count += rolls[index]

print(total_count)

probs = np.array([0.0,0,0,0,0,0])
for index in range(0,numterms):
    probs[index] = rolls[index]/total_count;

print(probs)
```
––––––––––––––––––– end prog1.py addendum 1 –––––––––––––––––––

Some notes on syntax: "len" is a Python function that returns the length of (number of items in) an array (from the numpy module). Notice how the probs array initialization has one entry as 1.0 and the others just 1. Again, this is an instance where the data structure must have components of the same type and if presented with mixed type will promote to a default type that represents the least loss of information (typically), in this instance, the "1.0" forces the array to be an array of floating point (decimal) numbers, with floating point arithmetic (for the division in the frequency evaluation used as the estimate of the probability in the "for loop").

At this point we can estimate a new probability distribution based on the rolls observed, for which we are interested in evaluating the Shannon entropy. To avoid repeatedly copying and pasting the above code for evaluating the Shannon entropy, let us create a subroutine, called "shannon" that will do this standard computation. This is a core software engineering process, whereby tasks that are done repeatedly become recognized as such, and become rewritten as subroutines, and then need no longer be rewritten. Subroutines also avoid clashes in variable usage, compartmentalizing their variables (whose scope is only in their subroutine), and more clearly delineate what information is "fed in" and what information is returned (e.g. the application programming interface, or API).

———————————————— prog1.py addendum 2 ————————————

```
def shannon( probs ):
    shannon_entropy = 0
    numterms = len(probs)
    print(numterms)

    for index in range(0, numterms):
        print(probs[index])
        shannon_entropy += probs[index]*math.log(probs[index])

    shannon_entropy = -shannon_entropy
    print(shannon_entropy)
    return shannon_entropy

shannon(probs)
value = shannon(probs)
print(value)
```

———————————————— end prog1.py addendum 2 ————————————

If we do another set of observations, getting counts on the different rolls, we then need to repeat the process of converting those counts to frequencies... so it is time to elevate the count-to-frequency computation to subroutine status as well, as is done next. The standard syntactical structure for defining a subroutine in Python is hopefully starting to become apparent (more detailed Python notes are in Appendix A).

---------------------------- prog1.py addendum 3 ----------------------------

```
def count_to_freq( counts ):
    numterms = len(counts)
    total_count=0
    for index in range(0,numterms):
        total_count+=counts[index]

    probs = counts # to get memory allocation
    for index in range(0,numterms):
        probs[index] = counts[index]/total_count

    return probs

probs = count_to_freq(rolls)
print(probs)
```

---------------------------- end prog1.py addendum 3 ----------------------------

Is genomic DNA random? Let us read thru a dna file, consisting of a sequence of a,c,g, and t's, and get their counts... then compute the shannon entropy vs. random (uniform distribution, e.g. $p = 1/4$ for each of the four possibilities). In order to do this we must learn file input/output (i/o) to "read" the data file:

---------------------------- prog1.py addendum 4 ----------------------------

```
fo = open("Norwalk_Virus.txt", "r+")
str = fo.read()
# print(str)
fo.close()
```

---------------------------- end prog1.py addendum 4 ----------------------------

Notes on syntax: the example above shows the standard template for reading a data file, where the datafile's name is Norwalk_Virus.txt. The subroutine

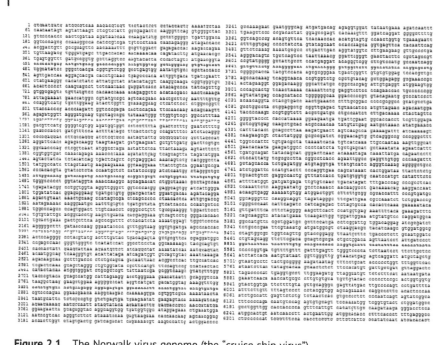

Figure 2.1 The Norwalk virus genome (the "cruise ship virus").

"open" is a Python command that handles file i/o. As its name suggests, it "opens" a datafile.

The Norwalk virus file has nonstandard format and is shown in its entirety in Figure 2.1 (split into two columns). The *Escherichia coli* genome (Figure 2.2), on the other hand, has standard FASTA format. (FASTA is the name of a program (~1985), where a file format convention was adopted, allowing "flat-file" record access that has been used in similar form ever since.)

The *E. coli* genome file is shown only for the first part (it is 4.6 Mb) in Figure 2.2, where the key feature of the FASTA file is apparent on line 1, where a ">" symbol should be present, indicating a label (or comment – information that will almost always be present):

Python has a powerful regular expression module (named "re" and imported in the first code sample of prog1.py). Regular expression processing of strings of characters is a mini programming language in its own right, thus a complete list of the functionalities of the re module will not be given here. Focusing on the "findall" function, it does as its name suggests, it finds all entries matching the search string characters in an array in the order they are encountered in the string. We begin with the string comprising the data file read in the previous example. We now traverse the string searching for characters matching those specified in the pattern field. The

>gi|556503834|ref|NC_000913.3| Escherichia coli str. K-12 substr. MG1655,
complete gene

AGCTTTTCATTCTGACTGCAACGGGCAATATGTCTCTGTGTGGATTAAAAAAAGAGTGTCTGATAGCAGC
TTCTGAACTGGTTACCTGCCGTGAGTAAATTAAAATTTATTGACTTAGGTCACTAAATACTTTAACCAA
TATAGGCATAGCGCACAGACAGATAAAAATTACAGAGTACAACATCCATGAAACGCATTAGCACCACC
ATTACCACCACCATCACCATTACCACAGGTAACGGTGCGGGCTGACGCGTACAGGAAACACAGAAAAAG
CCCGCACCTGACAGTGCGGGCTTTTTTTTTTTTCGACCAAAGTAACGAGGTAACAACCATGCGAGTGTTGAA
GTTCGGCGGTACATCAGTGGCAAATGCAGAACGTTTTCTGCGTGTTGCCGATATTCTGGAAAGCAATGCC
AGGCAGGGGCAGGTGGCCACCGTCCTCTCTGCCCCCGCCAAAATCACCAACCACCTGGTGGCGATGATTG
AAAAAACCATTAGCGGCCAGGATGCTTTACCCAATATCAGCGATGCCGAACGTATTTTTGCCGAACTTTT
GACGGGACTCGCCGCCGCCCAGCCGGGGTTCCCGCTGGCCAATTGAAAACTTTCGTCGATCAGGAATTT
GCCCAAATAAAACATGTCCTGACGGCATTAGTTTGTTGGGGCAGTGCCCGATAGCATCAACGCTGCGC
TGATTTGCCGTGGCGAGAAAATGTCGATCGCCATTATGGCCGGCGTATTAGAAGCGCGGCGGTCACAACGT
TACTGTTATCGATCCGGTCGAAAAAACTGCTGGCAGTGGGCATTACCTCGAATCTCGATATTGCT
GAGTCCACCCGCGTATTGCGGCAAGCCGCATTCCGGCTGATCACATGGTGCTGATGGCAGGTTTCACCG
CCGGTAATGAAAAGGCGAACTGGTGGTGCTTGGACGCAACGGTTCCGACTACTCTCGCTGCGTGCTGGC
TGCCTGTTTACGCGCCGATTGTTGCGAGATTTGACGGACGCAGTTGACGGGGTCTATACCTGCGACCCGCGT
CAGGTGCCCGATGCGAGGTTGTTGAAGTCGATGTCCTACCAGGAAGCGATGAGCTTTCCTACTTCGGCG

Figure 2.2 The start of the *E. coli* genome file, FASTA format.

array of a,c,g,t's that results has conveniently stripped any numbers, spaces, or line returns in this process, and a straightforward count can be done:

─────────────────────── prog1.py addendum 5 ───────────────────

```
pattern = '[acgt]'
result = re.findall(pattern, str)
seqlen = len(result)
# sequence = ""
# sequence = sequence.join(result)
# print(sequence)

print("The sequence length of the Norwalk genome is: ")
print(seqlen)

a_count=0
c_count=0
g_count=0
t_count=0

for index in range(0,seqlen):
    if result[index] == 'a':
        a_count+=1.0
    elif result[index] == 'c':
        c_count+=1.0
    elif result[index] == 'g':
        g_count+=1.0
    elif result[index] == 't':
        t_count+=1.0
    else:
        print("bad char\n")

norwalk_counts = np.array([a_count, c_count, g_count,
t_count])
print(norwalk_counts)

norwalk_probs = np.array([0.0,0,0,0])
norwalk_probs = count_to_freq(norwalk_counts)
value = shannon(norwalk_probs)
print(value)
```

──────────────────── end prog1.py addendum 5 ───────────────

We now traverse the array of single acgt's extracted from the raw genome data file, and increment counters associated with the acgt's as appropriate. At the end we have gotten the needed counts, and can then use our subroutines to see what Shannon entropy occurs.

Note on the informatics result: notice how the Shannon entropy for the frequencies of {a,c,g,t} in the genomic data differs only slightly from the Shannon entropy that would be found for a perfectly random, uniform, probability distribution (e.g. log(4)). This shows that although the genomic data is not random, i.e. the genomic data holds "information" of some sort, but we are only weakly seeing it at the level of single base usage.

In order to see clearer signs of nonrandomness, let us try evaluating frequencies at the base-pair, or dinucleotide, level. There are 16 (4×4) dinucleotides that we must now get counts on:

─────────────── progl.py addendum 6 ───────────────

```
di_uniform = [1.0/16]*16
stats = {}
for i in result:
    if i in stats:
        stats[i]+=1
    else:
        stats[i]=1

#for i in sorted(stats, key=stats.get):
for i in sorted(stats):
    print("%dx'%s'" % (stats[i],i))

stats = {}
for index in range(1,seqlen):
    dinucleotide = result[index-1] + result[index]
    if dinucleotide in stats:
        stats[dinucleotide]+=1
    else:
        stats[dinucleotide]=1

for i in sorted(stats):
    print("%dx'%s'" % (stats[i],i))
```

─────────────── end progl.py addendum 6 ───────────────

In the above example we see our first use of hash variables ("stats") in keeping tabs on counts of occurrences of various outcomes. This is a fundamental way to perform such counts without enumerating all of the outcomes beforehand (which results in what is known as the "enumeration problem," which is not really a problem, just a poor algorithmic approach). Further discussion of the enumeration "problem" and how it can be circumvented with use of hash variables will be described in Section 2.2.

The sequence information is traversed in a manner such that each of the dinucleotides is counted in the order seen, where the dinucleotide is extracted as a "window" of width two bases is *slid* across the genomic sequence. Each dinucleotide is entered into the count hash variable as a "key" entry, with the associated "value" being an increment on the count already seen and held as the old "value." These counts are then transferred to an array to make use of our prior subroutines count_to_freq and Shannon.

In the results for Shannon entropy on dinucleotides, we still do not see clear signs of nonrandomness. Similarly, let us try trinucleotide level. There are 64 ($4 \times 4 \times 4$) trinucleotides that we must now get counts on:

```
———————————— prog1.py addendum 7 ————————————
stats = {}
order = 3
for index in range(order-1,seqlen):
    xmer = ""
    for xmeri in range(0,order):
        xmer+=result[index-(order-1)+xmeri]

    if xmer in stats:
        stats[xmer]+=1
    else:
        stats[xmer]=1

for i in sorted(stats):
    print("%dx'%s'" % (stats[i],i))

———————————— end prog1.py addendum 7 ————————————
```

Still do not see real clear signs of non-random at tribase-level! So let us try 6-nucleotide level. There are 4096 6-nucleotides that we must now get counts on:

──────────── prog1.py addendum 8 ────────────

```
def shannon_order( seq, order ):
    stats = {}
    seqlen = len(seq)
    for index in range(order-1,seqlen):
        xmer = ""
        for xmeri in range(0,order):
            xmer+=result[index-(order-1)+xmeri]

        if xmer in stats:
            stats[xmer]+=1
        else:
            stats[xmer]=1

    nonzerocounts = len(stats)
    print("nonzerocounts=")
    print(nonzerocounts)

    counts = np.empty((0))
    for i in sorted(stats):
        counts = np.append(counts,stats[i]+0.0)

    probs = count_to_freq(counts)
    value = shannon(probs)
    print "The shannon entropy at order", order, "is:",
value, "."

shannon_order(result,6)
```
──────────── end prog1.py addendum 8 ────────────

2.1.1 Sample Size Complications

The 6-nucleotide statistics analyzed in prog1.py in the preceding is typically called a hexamer statistical analysis. Where the window-size for extracting the substrings has "-mer" appended, thus six-mer or hexamer. The term "-mer" comes from oligomer, a polymer containing a small number of monomers in its specification. In the case of the hexamers we saw that there were 4096 possible hexamers, or length six substrings, when the "alphabet" of monomer types consists of four elements: a,c,g, and t. In other words, there are $4^6 = 4096$ such substrings. In the Norwalk virus analysis this large number of different things to count, versus sample size overall, raises sampling questions. The Norwalk virus has a genome that is only 7654 nucleotides long. As we sweep the six-base window over that string to extract all of the hexamer counts we then obtain only $7654 - 5 = 7649$ hexamer samples! Even with uniform distribution we will be getting barely two counts for most of the different hexamer types! Limitations due to sample size play a critical role in these types of analysis.

The Norwalk virus genome is actually smaller than the typical viral genome, which ranges between 10 000 and 100 000 bases in length. Prokaryotic genomes typically range between 1 and 10 million bases in length. While the human genome is approximately three billion bases in length (3.23 Gb per haploid genome, 6.46 Gb total diploid). To go forward with a "strong" statistical analysis in the current discussion, the key as with any statistical analysis, is sample size, which is obviously dictated in this analysis by genome size. So to have "good statistics" meaning to have sufficient samples that frequencies of outcomes provide a good estimation of the underlying probabilities for those outcomes, we will apply the methods developed thus far to a bacterial genome in Chapter 3 (the classic model organism, *E. coli.*). In this instance the genome size will be approximately four and a half million bases in length, so much better counts should result than with the 7654 base Norwalk virus genome.

2.2 Counting, the Enumeration Problem, and Statistics

In the example in the previous section we left off with counts on all 4096 hexamers seen in a given genome. If we go from counts on substrings of length 6 to substrings of length 30 we run into a problem – there are now a million million million (10^{18}) substrings to get counts on. No genome is even remotely this large, so when getting counts on substrings in this situation most substring counts will necessarily be zero. Due to the large number of substrings, this is often referred to as "the enumeration problem," but since counts need only be maintained that are nonzero, we are bounded by genome size, for which there is no enumeration problem. The main mechanism for capturing count information on substrings without dedicated (array) memory, is by use of associative memory constructs, such as the hash variable, and this technique is employed in the code examples.

2.3 From Counts to Frequencies to Probabilities

The conventional relations on probabilities say nothing as to their interpretation. According to the Frequentist (frequency-based) interpretation, probabilities are defined in terms of fractions of a set of observations, as the number of observations tends to infinity (where the LLN works to advantage). In practice, infinite observations are not done, and often only one observation is done (predicting the winner of a marathon, for example). In the case of one race, however, it seems intuitive that prior information would still be beneficial to predicting winners. With the formal introduction of prior probabilities we then arrive at the Bayesian interpretation. From the Bayesian perspective, prior probabilities can be encoded as "pseudocounts" in the frequentist framework (i.e. observation counts do not necessarily initialize from zero). In the computer implementations used here there are typically tuned/selected psuedocounts and minimum/maximum probability cutoffs, thus the implementations can be formally described on a Bayesian footing [1, 3].

Whenever you can list all the outcomes for some situation (like rolls on a six-sided die), it is natural to think of the "probabilities" of those outcomes, where it is also natural for the outcome probabilities sum to one. So, with probability we assume there are "rules" (the probability assignments), and using those rules we make predictions on future outcomes. The rules are a mathematical framework, thus probability is a mathematical encapsulation of outcomes.

How did we get the "rules," the probability assignments on outcomes? This is the realm of statistics, where we have a bunch of data and we want to distill any rules that we can, such as a complete set of outcomes (observed) and their assigned (estimated) probabilities. If the analysis to go from raw data to a probability model was somehow done in one step, then it could be said that statistics is whatever takes you from raw data to a probability model, and hopefully do so without dependency on a probability model. In practice, however, the statistical determination of a probability model suitable for a collection of data is like the identification of a physical law in mathematical form given raw data – it is math and a lot more, including an iterative creative/inventive process where models are attempted and discarded, and built from existing models.

2.4 Identifying Emergent/Convergent Statistics and Anomalous Statistics

Expectation, $E(X)$, of random variable (r.v.) X:

$$E(X) \equiv \sum_{i=1}^{L} x_i \, p(x_i) \text{ if } x_i \in \Re$$

X is the total of rolling two six-sided dice: $X = 2$ can occur in one way, rolling "snake eyes," while rolling $X = 7$ can be done in six ways, etc. $E(X) = 7$. Now consider the expectation for rolling a single die, now $E(X) = 3.5$. Notice that the value of the expectation need not be one of your possible outcomes (it is really hard to roll a 3.5).

The expectation, $E(g(X))$, of a function g of r.v. X:

$$E(g(X)) \equiv \sum_{i=1}^{L} g(x_i)\, p(x_i) \text{ if } x_i \in \Re$$

Consider special case $g(X)$ where $g(x_i) = -\log(p(x_i))$:

$$H(X) \equiv E[g(X)] = -\sum_{i=1}^{L} p(x_i) \log(p(x_i)) \text{ if } \quad p(x_i) \in \Re+,$$

which is Shannon Entropy for the discrete distribution $p(xi)$. For Mutual Information, similarly, use $g(X,Y) = \log(p(x_i, y_i)/p(x_i)p(y_i))$:

$$I(X;Y) \equiv E[g(X,Y)] \equiv \sum_{i=1}^{L} p(x_i, y_i) \log(p(x_i, y_i)/p(x_i)p(y_i))$$

if $p(x_i)$, $p(y_i)$, $p(x_i, y_i)$ are all $\in \Re^+$, which is the Relative Entropy between a joint distribution and the same distribution if r.v.'s independent: $D(\, p(x_i, y_i) \,\|\, p(x_i)p(y_i)\,)$.

Jensen's Inequality:

Let $\varphi(\cdot)$ be a convex function on a convex subset of the real line: $\varphi: \chi \to \Re$. Convexity by definition: $\varphi(\lambda_1 x_1 + ... y_n x_n) \leq \lambda_1 \varphi(x_1) + ... + \lambda_n \varphi(x_n)$, where $\lambda_i \geq 0$ and $\Sigma \lambda_i = 1$. Thus, if $\lambda_1 = p(x_1)$, we satisfy the relations for line interpolation as well as discrete probability distributions, so can rewrite in terms of the Expectation definition:

$$\varphi(E(X)) \leq E(\varphi(X))$$

Since $\varphi(x) = -\log(x)$ is a convex function:

$$\log(E(X)) \geq E(\log(X)) = -H(X)$$

Variance:

$$\text{Var}(X) \equiv E\big([X - E(X)]^2\big) = \sum_{i=1}^{L} (x_i - E(X))^2 p(x_i) = E(X^2) - (E(X))^2$$

Chebyshev's Inequality:

For $k > 0$, $P(|X - E(X)| > k) \leq \text{Var}(X)/k^2$

$$\text{Var}(X) = \sum_{i=1}^{L} (x_i - E(X))^2 p(x_i)$$

Proof:
$$= \sum_{\{x_i|\ |x_i - E(X)| > k\}} (x_i - E(X))^2 p(x_i)$$
$$+ \sum_{\{x_i|\ |x_i - E(X)| \leq k\}} (x_i - E(X))^2 p(x_i)$$
$$\geq k^2 P(|X - E(X)| > k)$$

2.5 Statistics, Conditional Probability, and Bayes' Rule

So far we have counts and probabilities, but what of the probability of X when you know Y has occurred (where X is dependent on Y)? How to account for a greater state of knowledge? It turns out the answer to this was not put on a formal mathematical footing until half way thru the twentieth century, with the Cox derivation [101].

2.5.1 The Calculus of Conditional Probabilities: The Cox Derivation

The rules of probability, including those describing conditional probabilities, can be obtained using an elegant derivation by Cox [101]. The Cox derivation uses the rules of logic (Boolean algebra) and two simple assumptions. The first assumption is in terms of "$b|a$," where $b|a \equiv$ "likelihood" of proposition b when proposition a is known to be true. (The interpretation of "likelihood" as "probability" will fall out of the derivation.) The first assumption is that likelihood c-and-$b|a$ is determined by a function of the likelihoods $b|a$ *and* $c|b$-and-a:

(**Assumption 1**) c-and-$b|a = F(c|b$-and-$a, b|a)$,

for some function F. Consistency with the Boolean algebra then restricts F such that (**Assumption 1**) reduces to:

$$Cf(c\text{-and-}b \mid a) = f(c \mid b\text{-and-}a)f(b \mid a)$$

where f is a function of one variable and C is a constant. For the trivial choice of function and constant there is:

$$p(c, b \mid a) = p(c \mid b, a)p(b \mid a)$$

which is the conventional rule for conditional probabilities (and c-and-$b|a$ is rewritten as $p(c,b|a)$, etc.). The second assumption relates the likelihoods of propositions b and $\sim b$ when the proposition a is known to be true:

(**Assumption 2**) $\sim b|a = S(b|a)$,

for some function S. Consistency with the Boolean algebra of propositions then forces two relations on S:

$$S[S(x)] = x \text{ and } xS[S(y)/x] = yS[S(x)/y]$$

which together can be solved to give:

$$S(p) = (1 - p^m)^{1/m}$$

where m is an arbitrary constant. For $m = 1$ we obtain the relation $p(b|a) + p(\sim b|a) = 1$, the ordinary rule for probabilities. In general, the conventions for **Assumption 1** can be matched to those on **Assumption 2**, such that the likelihood relations reduce to the conventional relations on probabilities. Note: conditional probability relationships can be grouped:

$$p(b \mid a) = p(a \mid b)p(b)/p(a)$$

to obtain the classic Bayes Theorem.

2.5.2 Bayes' Rule

The derivation of Bayes' rule is obtained from the property of conditional probability:

$$p\left(x_i, y_j\right) = p\left(x_i \mid y_j\right) p\left(y_j\right) = p\left(y_j \mid x_i\right) p(x_i)$$

$$p\left(x_i \mid y_j\right) = p\left(y_j \mid x_i\right) p(x_i)/p\left(y_j\right) = \frac{p\left(y_j \mid x_i\right) p(x_i)}{\sum_{i=1}^{L} p\left(y_j \mid x_i\right) p(x_i)}$$

$$p\left(x_i \mid y_j\right) = \frac{p\left(y_j \mid x_i\right) p(x_i)}{\sum_{i=1}^{L} p\left(y_j \mid x_i\right) p(x_i)}$$

Bayes' Rule provides an update rule for probability distributions in response to observed information. Terminology:

$p(x_i)$ is referred to as the "prior distribution on X" in this context.
$p(x_i \mid y_j)$ is referred to as the "posterior distribution on X given Y."

2.5.3 Estimation Based on Maximal Conditional Probabilities

There are two ways to do an estimation given a conditional problem. The first is to seek a maximal probability based on the optimal choice of outcome (maximum a posteriori [MAP]), versus a maximal probability (referred to as a "likelihood" in this context) given choice of conditioning (maximum likelihood [ML]).

MAP Estimate:

Provides an estimate of r.v. X given that $Y = y_j$ in terms of the posterior probability:

$$\hat{X}_{\text{MAP}} = \operatorname*{argmax}_{x \in X} \ p\left(x \mid y_j\right)$$

ML Estimate:

Provides an estimate of r.v. X given that $Y = y_j$ in terms of the maximum likelihood:

$$\hat{X}_{\text{ML}} = \operatorname*{argmax}_{x \in X} \ p\left(y_j \mid x\right)$$

2.6 Emergent Distributions and Series

In this section we consider a r.v., X, with specific examples where those outcomes are fully enumerated (such as 0 or 1 outcomes corresponding to a coin flip). We review a series of observations of the r.v., X, to arrive at the LLN. The emergent structure to describe a r.v. from a series of observations is often described in terms of probability distributions, the most famous being the Gaussian Distribution (a.k. a. the Normal, or Bell curve).

2.6.1 The Law of Large Numbers (LLN)

The LLN will now be derived in the classic "weak" form. The "strong" form is derived in the modern mathematical context of Martingales in Appendix C.1.

Let X_k be independent identically distributed (iid) copies of X, and let X be the real number "alphabet." Let $\mu = E(X)$, $\sigma^2 = \text{Var}(X)$, and denote

$$\bar{x}_N = \frac{1}{N} \sum_{k=1}^{N} X_K$$

$$E(\bar{x}_N) = \mu$$

$$\text{Var}(\bar{x}_N) = \frac{1}{N^2} \sum_{k=1}^{N} \text{Var}(X_k) = \frac{1}{N} \sigma^2$$

From Chebyshev: $P(|\bar{x}_N - \mu| > k) \leq \text{Var}(\bar{x}_N)/k^2 = \frac{1}{Nk^2}\sigma^2$

As $N \to \infty$ get the LLN (weak):

If X_k are iid copies of X, for $k = 1,2,...$, and X is a real and finite alphabet, and $\mu = E(X)$, $\sigma^2 = \text{Var}(X)$, then: $P(|\bar{x}_N - \mu| > k) \to 0$, for any $k > 0$. Thus, the arithmetic mean of a sequence of iid r.v.s converges to their common expectation. The weak form has convergence "in probability," while the strong form has convergence "with probability one."

2.6.2 Distributions

2.6.2.1 The Geometric Distribution(Emergent Via Maxent)

Here, we talk of the probability of seeing something after k tries when the probability of seeing that event at each try is "p." Suppose we see an event for the first time after k tries, that means the first $(k-1)$ tries were nonevents (with probability $(1-p)$ for each try), and the final observation then occurs with probability p, giving rise to the classic formula for the geometric distribution:

$$P(X = k) = (1-p)^{(k-1)} p$$

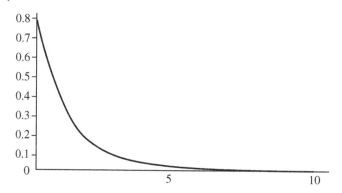

Figure 2.3 The Geometric distribution, $P(X = k) = (1 - p)^{(k-1)}p$, with $p = 0.8$.

As far as normalization, i.e. do all outcomes sum to one, we have:

Total Probability $= \Sigma_{k=1}(1-p)^{(k-1)}p = p[1 + (1-p) + (1-p)^2 + (1-p)^3 + ...] = p[1/(1-(1-p))] = 1$

So total probability already sums to one with no further normalization needed. In Figure 2.3 is a geometric distribution for the case where $p = 0.8$:

2.6.2.2 The Gaussian (aka Normal) Distribution (Emergent Via LLN Relation and Maxent)

$$N_x(\mu, \sigma^2) = \exp\left(-(x-\mu)^2/(2\sigma^2)\right)/(2\pi\sigma^2)^{(1/2)}$$

For the Normal distribution the normalization is easiest to get via complex integration (so we'll skip that). With mean zero and variance equal one (Figure 2.4) we get:

2.6.2.3 Significant Distributions That Are Not Gaussian or Geometric

Nongeometric duration distributions occur in many familiar areas, such as the length of spoken words in phone conversation, as well as other areas in voice recognition. Although the Gaussian distribution occurs in many scientific fields (an observed embodiment of the LLN, among other things), there are a huge number of significant (observed) skewed distributions, such as heavy-tailed (or long-tailed) distributions, multimodal distributions, etc.

Heavy-tailed distributions are widespread in describing phenomena across the sciences. The log-normal and Pareto distributions are heavy-tailed distributions that are almost as common as the normal and geometric distributions in descriptions of physical phenomena or man-made phenomena. Pareto distribution was originally used to describe the allocation of wealth of the society, known as the famous 80–20 rule, namely, about 80% of the wealth was owned by a small amount of people, while "the tail," the large part of people only have

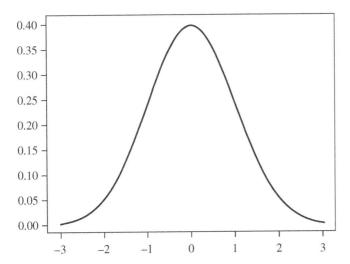

Figure 2.4 The Gaussian distribution, aka Normal, shown with mean zero and variance equal to one: $N_x(\mu,\sigma^2) = N_x(0,1)$.

the rest 20% wealth. Pareto distribution has been extended to many other areas. For example, internet file-size traffic is a long-tailed distribution, that is, there are a few large sized files and many small sized files to be transferred. This distribution assumption is an important factor that must be considered to design a robust and reliable network and Pareto distribution could be a suitable choice to model such traffic. (Internet applications have many other heavy-tailed distribution phenomena.) Pareto distributions can also be found in a lot of other fields, such as economics.

Log-normal distributions are used in geology and mining, medicine, environment, atmospheric science, and so on, where skewed distribution occurrences are very common. In Geology, the concentration of elements and their radioactivity in the Earth's crust are often shown to be log-normal distributed. The infection latent period, the time from being infected to disease symptoms occurs, is often modeled as a log-normal distribution. In the environment, the distribution of particles, chemicals, and organisms is often log-normal distributed. Many atmospheric physical and chemical properties obey the log-normal distribution. The density of bacteria population often follows the log-normal distribution law. In linguistics, the number of letters per words and the number of words per sentence fit the log-normal distribution. The length distribution for introns, in particular, has very strong support in an extended heavy-tail region, likewise for the length distribution on exons or open reading frames (ORFs) in genomic deoxyribonucleic acid (DNA). The anomalously long-tailed aspect of the ORF-length distribution is the key distinguishing feature of this distribution, and has been the key attribute used by biologists using ORF finders to identify likely protein-coding regions in genomic DNA since the early days of (manual) gene structure identification.

2.6.3 Series

A series is a mathematical object consisting of a series of numbers, variables, or observation values. When observations describe equilibrium or "steady state," emergent phenomenon familiar from physical reality, we often see series phenomena that are martingale. The martingale sequence property can be seen in systems reaching equilibrium in both the physical setting and algorithmic learning setting.

A discrete-time martingale is a stochastic process where a sequence of r.v. $\{X_1, \ldots, X_n\}$ has conditional expected value of the next observation equal to the last observation: $E(X_{n+1} \mid X_1, \ldots X_n) = X_n$, where $E(|X_n|) < \infty$. Similarly, one sequence, say $\{Y_1, \ldots, Y_n\}$, is said to be martingale with respect to another, say $\{X_1, \ldots, X_n\}$, if for all n: $E(Y_{n+1} \mid X_1, \ldots X_n) = Y_n$, where $E(|Y_n|) < \infty$. Examples of martingales are rife in gambling. For our purposes, the most critical example is the likelihood-ratio testing in statistics, with test-statistic, the "likelihood ratio" given as: $Y_n = \Pi^n_{i=1} g(X_i)/f(X_i)$, where the population densities considered for the data are f and g. If the better (actual) distribution is f, then Y_n is martingale with respect to X_n. This scenario arises throughout the hidden Markov models (HMM) Viterbi derivation if local "sensors" are used, such as with profile-HMM's or position-dependent Markov models in the vicinity of transition between states. This scenario also arises in the HMM Viterbi recognition of regions (versus transition out of those regions), where length-martingale side information will be explicitly shown in Chapter 7, providing a pathway for incorporation of any martingale-series side information (this fits naturally with the clique-HMM generalizations described in Chapter 7). Given that the core ratio of cumulant probabilities that is employed is itself a martingale, this then provides a means for incorporation of side-information in general (further details in Appendix C).

2.7 Exercises

2.1 Evaluate the Shannon Entropy, by hand, for the fair die probability distribution: (1/6,1/6,1/6,1/6,1/6,1/6), for the probability of rolling a 1 thru a 6 (all are the same, 1/6, for uniform prob. Dist). Also evaluate for loaded die: (1/10,1/10,1/10,1/10,1/10,1/2).

2.2 Evaluate the Shannon Entropy for the fair and loaded probability distribution in Exercise 2.1 computationally, by running the program described in Section 2.1.

2.3 Now consider you have two dice, where each separately rolls "fair," but together they do not roll "fair," i.e. each specific pair of die rolls does not have probability 1/36, but instead has probability:

Die 1 roll	Die 2 roll	Probability
1	1	(1/6)∗(0.001)
1	2	(1/6)∗(0.125)
1	3	(1/6)∗(0.125)
1	4	(1/6)∗(0.125)
1	5	(1/6)∗(0.124)
1	6	(1/6)∗(0.5)
2	Any	(1/6)∗(1/6)
3	Any	(1/6)∗(1/6)
4	Any	(1/6)∗(1/6)
5	Any	(1/6)∗(1/6)
6	1	(1/6)∗(0.5)
6	2	(1/6)∗(0.125)
6	3	(1/6)∗(0.125)
6	4	(1/6)∗(0.125)
6	5	(1/6)∗(0.124)
6	6	(1/6)∗(0.001)

What is Shannon Entropy for the Die 1 outcomes? (call $H(1)$) What is the Shannon entropy of the Die 2 outcomes (refer to as $H(2)$)? What is the Shannon entropy on the two-dice outcomes with probabilities shown in the table above (denote $(H(1,2))$?

Compute the function MI(Die 1,Die 2) $= H(1) + H(2) - H(1,2)$. Is it positive?

2.4 Go to genbank (https://www.ncbi.nlm.nih.gov/genbank/) and select the genome of a small virus (~10 kb). Using the Python code shown in Section 2.1, determine the base frequencies for {a,c,g,t}. What is the shannon entropy (if those frequencies are taken to be the probabilities on the associated outcomes)?

2.5 Go to genbank (https://www.ncbi.nlm.nih.gov/genbank/) and select the genome of three medium-sized viruses (~100 kb). Using the Python code shown in Section 2.1, determine the trinucleotide frequencies. What is the Shannon entropy of the trinucleotide frequencies for each of the three virus genomes? Using this as a distance measure phylogenetically speaking, which two viruses are most closely related?

2.6   Repeat (Exercise 2.5) but now use symmetrized relative entropy between the trinucleotide probability distributions as a distance measure instead (reevaluate pairwise between the three viruses). Using this as a distance measure phylogenetically speaking, which two viruses are most closely related?

2.7   Prove that relative entropy is always positive (hint: use Jensen's Inequality from Section 2.4).

2.8   What is the Expectation for the two-dice roll with pair outcome probabilities listed in (Exercise 2.3)?

2.9   What is the Expectation for the two-dice roll with fair dice? Is this expectation an actual outcome possibility? What does it mean if it is not?

2.10   Survey the literature and write a report on common occurrences of distributions of the type: uniform, geometric, exponential, Gaussian, log-normal, heavy-tail.

2.11   Survey the literature and write a report on common occurrences of series of the type: Martingale.

2.12   Consider the loaded die example, where the probability of rolling a 1,2,3,4, or 5, is 0.1, and the probability of rolling a 6 is 0.5.
   i) What is the expectation for the loaded die?
   ii) What is its variance?
   iii) What is its mode?
   iv) What is its median?
   v) The LLN for the loaded die above indicates that a sequence of rolls could be done and if its average tends toward 4.5, you know it is loaded, and if it goes toward 3.5, you know it is fair. So it comes down to how soon you can resolve that its converging on these two possible expectations differing by 1.0. Suppose someone is rolling a die that is either fair or loaded as described above, how many rolls do you think you will need to see before it will be obvious how the average is trending? Is the better way to spot the anomaly? Like frequency of seeing three sixes in a row is notably skewed?

2.13   You have a genomic sequence of length L. (For DNA genomes you have approximately 10**4 for viruses, 10**6 for bacteria, and 10**9 for mammals.) A typical analysis is to get counts on subsequences of length N within the full sequence L, where there are $L - N + 1$ subsequences of length N (by sliding a window of width N across the length L sequence and taking the window samples accordingly). The number of possible subsequences of length N grows exponentially with increase in that length. For DNA subsequences of length six bases, the 6mers, with four base possibilities, {a,c,g,t}, there are thus $4**6 = 4096$ possible 6mers. If the 6mers are equally probable, then in the approximate 10 000 length of a virus each 6mer might be seen a couple times (10 000/4096 to be precise), while a particular 6mer can be seen millions of times in mammalian genomes. Sounds fine so far, but now consider an analysis of 25mers.... The possible 25mers number $4**25 = 2**50 = (2**10)** 5 = 1024**5 = 10**15$. So, a million billion possibilities.... It turns out that DNA information does not have subsequences with approximately equal statistical counts (equal probabilities), but, instead, is highly structured with a variety of overlapping encoding schemes, so has subsequences with very unequal statistics. The vast majority of the 25mer subsequences, in fact, will have zero counts such that enumeration of the possibilities ahead of time in an array data-structure is not useful or even possible in some cases, which then leads to associative arrays in this context as shown in the sample code. Do a 25mer analysis on bacterial genome (get from genbank, like *E. coli*). What is the highest count 25mer subsequence?

2.14 A rare genetic disease has probability $P(\text{disease}) = 0.0000001$. Suppose you have a test for this condition with sensitivity that is perfect ($SN = 1.0$) and with specificity that is 99.99% correct ($SP = 0.9999$) (i.e. false positives occur with probability 0.0001). Is this test useful in practice?

2.15 Prove $P(X,Y|Z) = P(X|Z)\, P(Y|X,Z)$.

2.16 Prove the Bonferroni inequality: $P(X,Y) \geq P(X) + P(Y) - 1$.

2.17 Suppose you are on a game show (with Monty Hall), and you are given the choice of three doors: Behind one door is a car; behind the others, goats. You pick a door, say No. 1, and the host, who knows what is behind the doors, opens another door, say No. 3, which has a goat. He then says to you, "Do you want to change your pick door to No. 2?" Is it to your advantage to switch your choice? Prove by tabulation of possibilities, and then prove using Bayes' Rule with appropriate choice of variables.

2.18 Prove $E(Y/X) \geq E(Y)/E(X)$.

3

Information Entropy and Statistical Measures

In this chapter, we start with a description of information entropy and statistical measures (Section 3.1). Using these measures we then examine "raw" genomic data. No biology or biochemistry knowledge is needed in doing this analysis and yet we almost trivially rediscover a three-element encoding scheme that is famous in biology, known as the codon. Analysis of information encoding in the four element {a, c, g, t} genomic sequence alphabet is about as simple as you can get (without working with binary data), so it provides some of the introductory examples that are implemented. A few (simple) statistical queries to get the details of the codon encoding scheme are then straightforward (Section 3.2). Once the encoding scheme is known to exist, further structure is revealed via the anomalous placement of "stop" codons, e.g. anomalously large open reading frames (ORFs) are discovered. A few more (simple) statistical queries from there, and the relation of ORFs to gene structure is revealed (Section 3.3). Once you have a clear structure in the sequential data that can be referenced positionally, it is then possible to gather statistical information for a Markov model. One example of this is to look at the positional base statistics at various positions "upstream" from the start codon. We thereby identify binding sites for critical molecular interaction in both transcription and translation. Since the Markov model is needed in analysis of sequential processes in general for what is discussed in later chapters (Chapters 6 and 7 in particular), a review of Markov models, and some of their specializations, are given in Section 3.4 (Chapters 6 and 7 covers *Hidden* Markov models, or HMMs).

Numerous prior book, journal, and patent publications by the author are drawn upon throughout the text [1–68]. Almost all of the journal publications are open access. These publications can typically be found online at either the author's personal website (www.meta-logos.com) or with one of the following online publishers: www.m-hikari.com or bmcbioinformatics.biomedcentral.com.

Informatics and Machine Learning: From Martingales to Metaheuristics, First Edition.
Stephen Winters-Hilt.
© 2022 John Wiley & Sons, Inc. Published 2022 by John Wiley & Sons, Inc.

3.1 Shannon Entropy, Relative Entropy, Maxent, Mutual Information

If you have a discrete probability distribution P, with individual components p_k, then the rules for probabilities requires that the sum of the probabilities of the individual outcomes must be 1 (as mentioned in Chapter 2). This is written in math shorthand as:

$$\Sigma_k p_k = 1$$

Furthermore, the individual outcome probabilities must always be positive, and by some conventions, nonzero. In the case of hexamers, there are 4096 types, thus the index variable "k" ranges from 1 to $4^6 = 4096$. If we introduce a second discrete probability distribution Q, with individual components q_k, those components sum to 1 as well:

$$\Sigma_k q_k = 1$$

The definition of Shannon Entropy in this math notation, for the P distribution, is:

$$\text{Shannon entropy of } P = -\Sigma_k p_k \log(p_k)$$

The degree of randomness in a discrete probability distribution P can be measured in terms of Shannon entropy [106].

Shannon entropy appears in fundamental contexts in communications theory and in statistical physics [100]. Efforts to derive Shannon entropy from some deeper theory drove early efforts to at least obtain axiomatic derivations, with the one used by Khinchine given in the next section being the most popular. The axiomatic approach is limited by the assumptions of its axioms, however, so it was not until the fundamental role of relative entropy was established in an "information geometry" context [113–115], that a path to show that Shannon entropy is uniquely qualified as a measure was established (c. 1999). The fundamental (extremal optimum) aspect of relative entropy (and Shannon entropy as a simple case) is found by differential geometry arguments akin to those of Einstein on Riemannian spaces (here involving spaces defined by the family of exponential distributions). Whereas the "natural" notion of metric and distance locally is given by the Minkowski metric and Euclidean distance, a similar analysis on comparing distributions (evaluating their "distance" from eachother) indicates the natural measure is relative entropy (which reduces to Shannon entropy in variational contexts when the relative entropy is relative to the uniform probability distribution). Further details on this derivation are given in Chapter 8.

3.1.1 The Khinchin Derivation

In his now famous 1948 paper [106], Claude Shannon provided a qualitative measure for entropy in connection with communication theory. The Shannon entropy measure was later put on a more formal footing by A. I. Khinchin in an article where he proves that with certain assumptions the Shannon entropy is unique [107]. (Dozens of similar axiomatic proofs have since been made.) A statement of the theorem is as follows:

Khinchine Uniqueness Theorem: Let $H(p_1, p_2, ..., p_n)$ be a function defined for any integer n and for all values $p_1, p_2, ..., p_n$ such that $p_k \geq 0$ ($k = 1, 2, ..., n$), and $\Sigma_k p_k = 1$. If for any function n this function is continuous with respect to its arguments, and if the function obeys the three properties listed below, then $H(p_1, p_2, ..., p_n) = -\lambda \Sigma_k p_k \log(p_k)$, where λ is a positive constant (with Shannon entropy recovered for convention $\lambda = 1$). The three properties are:

1) For given n and for $\Sigma_k p_k = 1$, the function takes its largest value for $p_k = 1/n$ ($k = 1, 2, ..., n$). This is equivalent to Laplace's principle of insufficient reason, which says if you do not know anything assume the uniform distribution (also agrees with Occam's Razor assumption of minimum structure).
2) $H(ab) = H(a) + H_a(b)$, where $H_a(b) = -\Sigma_a p(a) \log(p(b|a))$, is the conditional entropy. This is consistent with $H(ab) = H(a) + H(b)$, for probabilities of a and b independent, with modifications involving conditional probability being used when not independent.
3) $H(p_1, p_2, ..., p_n, 0) = H(p_1, p_2, ..., p_n)$. This reductive relationship, or something like it, is implicitly assumed when describing any system in "isolation."

Note that the above axiomatic derivation is still "weak" in that it assumes the existence of the conditional entropy in property (2).

3.1.2 Maximum Entropy Principle

The law of large numbers (Section 2.6.1), and related central limit theorem, explain the ubiquitous appearance of the Gaussian (a.k.a., Normal) distribution in Nature and statistical analysis. Even when speaking of a probability distribution purely in the abstract, the Gaussian distribution (amongst a collection) still stands out in a singular way. This is revealed when seeking the discrete probability distribution that maximizes the Shannon entropy subject to constraints. The Lagrangian optimization method is a mathematical formalism to solve problems of this type, where you want to optimize something, but must do so subject to constraints. Lagrangians are described in detail in Chapters 6, 7, and 10. For our purposes here, once you know how to group the terms to create the Lagrangian expression appropriate to your problem, the problem is then reduced to simple differential calculus and algebra (you take a derivative of the Lagrangian and solve for it being zero – the classic way to find an extremum from calculus). I will skip most of the math here, and just state the Lagrangians and their solutions in the small examples that follow.

If no constraint on probabilities, other than that they sum to 1, the Lagrangian form for the optimization is as follows:

$$L(\{p_k\}) = -\Sigma(p_k \log(p_k)) - \lambda(1 - \Sigma(p_k))$$

where, $\partial L/\partial p_k = 0 \rightarrow p_k = e^{-(1+\lambda)}$ for all k, thus $p_k = 1/n$ for system with n outcomes. Thus, the maximum entropy hypothesis in this circumstance results in Laplace's Principle of Insufficient Reasoning, a.k.a., principle of indifference, where if you do not know any better, use the uniform distribution.

If you have as prior information the existence of the mean, μ, of some quantity x, then you have the Lagrangian:

$$L(\{p_k\}) = -\Sigma(p_k \log(p_k)) - \lambda(1 - \Sigma(p_k)) - \delta(\mu - \Sigma(p_k x_k))$$

where, $\partial L/\partial p_k = 0 \rightarrow p_k = A \exp(-\delta x_k)$, leading to the exponential distribution. If for the latter we had the mean of the function, $f(x_k)$, of some random variable X, then a similar derivation would again yield the exponentional distribution $p_k = A \exp(-\delta f(x_k))$, where now A is not simply a normalization factor, but is known as the partition function and it has a variety of generative properties vis-à-vis statistical mechanics and thermal physics.

If you have as prior information the existence of the mean and variance of some quantity (the first and second statistical moments), then you have the Lagrangian:

$$L(\{p_k\}) = -\Sigma(p_k \log(p_k)) - \lambda(1 - \Sigma(p_k)) - \delta(\mu - \Sigma(p_k x_k)) - \gamma(\nu - \Sigma(p_k(x_k)^2))$$

where, $\partial L/\partial p_k = 0 \rightarrow$ the Gaussian distribution (see Exercise 3.3).

With the introduction of Shannon entropy above, c. 1948, a reformulation of statistical mechanics was indicated (Jaynes [112]) whereby entropy could be made the starting point for the entire theory by way of maximum entropy with whatever system constraints – immediately giving rise to the classic distributions seen in nature for various systems (itself an alternate derivation starting point for statistical mechanics already noted by Maxwell over 100 years ago). So instead of introducing other statistical mechanics concepts (ergodicity, equal *a priori* probabilities, etc.) and matching the resulting derivations to phenomenological thermodynamics equations to get entropy, with the Jaynes derivation we start with entropy and maximize it directly to obtain the rest of the theory.

3.1.3 Relative Entropy and Its Uniqueness

Relative entropy ($\rho = \Sigma_x \, p(x) \, \log(p(x)/q(x)) = D(P||Q)$) uniquely results from a geometric (differentiable manifold) formalism on families of distributions – the Information Geometry formalism was first described by Amari [113–115]. Together with Laplace's principle of insufficient reason on the choice of "reference" distribution in the relative entropy expression, this will reduce to Shannon entropy, and thus uniqueness on Shannon entropy from a geometric context. The parallel with geometry is the Euclidean distance for "flat" geometry (simplest assumption of structure), vs. the "distance" between distributions as described by the Kullback–Leibler divergence.

When comparing discrete probability distributions P and Q, both referring to the same N outcomes, the measure of their difference is sometimes measured in terms of their *symmetrized* relative entropy [105] (a.k.a. Kullback–Leibler divergence), $D(P,Q)$:

$$D(P, Q) = \left[D\left(P \middle\| Q \right) + D\left(Q \middle\| P \right) \right] / 2$$

where,

$$D\left(P \middle\| Q \right) = \Sigma_k p_k \log \left(p_k / q_k \right)$$

where, P and Q have outcome probabilities $\{p_k\}$ and $\{q_k\}$.

Relative entropy has some oddities that should be explained right away. First, it does not have the negative sign in front to make it a positive number (recall this was done for the Shannon entropy definition since all of the log factors are always negative). Relative entropy does not need the negative sign, however, since it is provably always positive as is! (The proof uses Jensen's Inequality from Section 2.6.1, see Exercise 3.12.) For relative entropy there is also the constraint to the convention mentioned above where all the outcome probabilities are *nonzero* (otherwise have a divide by zero or a log(0) evaluation, either of which is undefined). Relative entropy is also asymmetric in that $D(P||Q)$ is not equal to $D(Q||P)$.

3.1.4 Mutual Information

One of the most powerful uses of Relative entropy is in the context of evaluating the statistical linkage between two sets of outcomes, e.g. in determining if two random variables are independent are not. In probability we can talk about the probability of two events happening, such as the probabilities for the outcomes of rolling two dice $P(X, Y)$, where X is the first die, with outcomes $x_1 = 1, ...,$ $x_6 = 6$, and similarly for the second die Y. If they are both fair dice, they act independently of each other, then their joint probability reduces to: $P(X, Y) = P(X)P(Y)$, if $\{X, Y\}$ are independent of each other.

If using loaded dice, but with dice that have no interaction, then they are still independent of eachother, and their probabilities are thus still independent, reducing to the product of two simpler probability distributions (with one argument each) as shown above. In games of dice where two dice are rolled (craps) it is possible to have dice that individually roll as fair, having uniform distribution on outcomes, but that when rolled together interact such that their combined rolls are biased. This can be accomplished with use of small bar magnets oriented from the "1" to "6" faces, such that the dice tend to come up with their magnets anti-aligned one showing its "1" face, the other showing its "6" face, for a total roll count of "7" (where the roll of a "7" has special significance in the game of craps). In the instance of the dice with magnets, the outcomes of the individual die rolls are not independent, and the simplification of $P(X, Y)$ to $P(X)P(Y)$ cannot be made.

In evaluating if there is a statistical linkage between two events we are essentially asking if the probability of those events are independent, e.g. does $P(X, Y) = P(X)P(Y)$? In this situation we are again in the position of comparing two probability distributions, $P(X, Y)$ and $P(X)P(Y)$, so if relative entropy is best for such comparisons, then why not evaluate $D(P(X, Y) \| P(X)P(Y))$? This is precisely what should be done and in doing so we have arrived at the definition of what is known as "mutual information" (finally a name for an information measure that is perfectly self-explanatory!).

$$\mathrm{MI}(X, Y) = \text{mutual information between } X \text{ and } Y = D\left(P(X,Y) \middle\| P(X)P(Y)\right)$$

The use of mutual information is very powerful in bioinformatics, and informatics in general, as it allows statistical linkages to be discovered that are not otherwise apparent. In Section 3.2 we will start with evaluating the mutual information between genomic nucleotides at various degrees of separation. If we see nonzero mutual information in the genome for bases separated by certain, specified, gap distances, we will have uncovered that there is "structure" of some sort.

3.1.5 Information Measures Recap

The fundamental information measures are, thus, Shannon entropy, mutual information, and relative entropy (also known as the Kullback–Leibler divergence, especially when in symmetrized form). Shannon entropy, $\sigma = -\Sigma_x p(x) \log(p(x))$, is a measure of the information in distribution $p(x)$. Relative entropy (Kullback–Leibler divergence): $\rho = \Sigma_x p(x) \log(p(x)/q(x))$, is a measure of distance between two probability distributions. MI(X, Y), $\mu = \Sigma_x \Sigma_y p(xy) \log(p(xy)/p(x)p(y))$, is a measure of information one random variable has about another random variable. As shown above, Mutual Information is a special case of relative entropy. Let us now write code to implement these measures, and then apply them to analysis of genomic data.

The next program, cleverly named prog2.py, will build off the code devised previously, with the file i/o operation now "lifted" into a subroutine for safer encapsulation (to avoid scope errors, etc.) and to avoid the confusing clutter of copying and pasting such a large block of code repeatedly that would be required otherwise. By now, this has hopefully made a convincing case for why subroutines are a big deal in the evolution of software engineering constructs (and the computer languages that implement them). Further discussion is given in the comments in the code.

──────────────────── prog2.py ────────────────────

```
#!/usr/bin/python

import numpy as np
import math
import re

# from prior code we carry over the subroutines:
# shannon, count_to_freq, Shannon_order;
# with prototypes:
# def shannon( probs ) with usage:
# value = shannon(probs)
# print(value)
#
# def count_to_freq( counts ) with usage
# probs = count_to_freq(rolls)
# print(probs)
#
# def shannon_order( seq, order ) with usage:
# order = 8
# maxcounts = 4**order
# print "max counts at order", order, "is =", maxcounts
# val = math.log(maxcounts)
# shannon_order(sequence,order) # shannon_order prints
entropy
# print "The max entropy would be log(4**order) = ", val


# New code is now created to have subroutines for text
handling.
# There are two types of text-read, one for genome data in
"fasta"
# format (gen_fasta_read) and one for generic format
(gen_txt_read):
```

```python
def gen_txt_read( text ):
    if (text == ""):
        text = "Norwalk_Virus.txt"

    fo = open(text, "r+")
    str = fo.read()
    fo.close()

    pattern = '[acgt]'
    result = re.findall(pattern, str)
    # seqlen = len(result)
    return result

#usage
null=""
gen_array = gen_txt_read(null) # defaults, uses
Norwalk_Virus.txt genome
sequence = ""
sequence = sequence.join(gen_array)
Norwalk_sequence = sequence;
seqlen = len(gen_array)
print "The sequence length of the Norwalk genome is:",
seqlen

def gen_fasta_read( text ):
    if (text == ""):
        text = "EC_Chr1.fasta.txt"

    slurp = open(text, 'r')
    lines = slurp.readlines()

    sequence = ""
    for line in lines:
        pattern = '>'
        test = re.findall(pattern,line)
        testlen = len(test)
        if testlen>0:
            print "fasta comment:", line.strip()
        else:
            sequence = sequence + line.strip()
```

```
slurp.close()

pattern = '[acgtACGT]'
result = re.findall(pattern, sequence)
return result
```

```
#usage
null=""
fasta_array = gen_fasta_read(null) # defalus, uses e.coli
genome
sequence = ""
sequence = sequence.join(fasta_array)
EC_sequence = sequence
seqlen = len(fasta_array)
print "The sequence length of the e. coli Chr1 genome is:",
seqlen
print "Doing order 8 shannon analysis on e-coli:"
order = 6
shannon_order(EC_sequence,order)
```
———————————————— prog2.py end ——————————————

As mentioned previously, when comparing two probability distributions on the same set of outcomes, it is natural to ask if they can be compared in terms of the difference in their scalar-valued Shannon entropies. Similarly, there is the standard manner of comparing multicomponent features by treating them as points in a manifold and performing the usual Euclidean distance calculation generalized to whatever dimensionality of the feature data. Both of these approaches are wrong, especially the latter, when comparing discrete probability distributions (of the same dimensionality). The reason being, when comparing two discrete probability distributions there are the additional constraint on the probabilities (sum to 1, etc.), and the provably optimal difference measure under these circumstances, as described previously, is relative entropy. This will be explored in Exercise 3.5, so some related subroutines are included in the first addendum to prog2.py:

———————————————— prog2.py addendum 1————————————

```
# We can use the Shannon_order subroutine to return a
probability array
# for a given sequence. Here are the probability arrays on
3-mers
# (ordered alphabetically):
```

```
Prob_Norwalk_3mer = shannon_order(Norwalk_sequence,3)
Prob_EC_3mer = shannon_order(EC_sequence,3)

# the standard Euclidean distance and relative entropy are
given next

def eucl_dist_sq ( P , Q ):
    Pnum = len(P)
    Qnum = len(Q)
    if Pnum != Qnum:
        print "error: Pnum != Qnum"
        return -1

    euclidean_distance_squared = 0
    for index in range(0, Pnum):
       euclidean_distance_squared += (P[index]-Q[index])**2

    return euclidean_distance_squared

# usage
value = eucl_dist_sq(Prob_Norwalk_3mer,Prob_EC_3mer)
print "The euclidean distance squared between EC and Nor
3mer probs is", value

# P and Q are probability arrays, meaning components are
positive definite
# and sum to one. If P and Q are proability arrays, can
compare them in
# terms of relative entropy (not Euclidean distance):

def relative_entropy ( P , Q ):
    Pnum = len(P)
    Qnum = len(Q)
    if Pnum != Qnum:
        print "error: Pnum != Qnum"
        return -1

    rel_entropy = 0
    for index in range(0, Pnum):
       rel_entropy += P[index]*math.log(P[index]/Q[index])

    return rel_entropy
```

```
# usage
value1 = relative_entropy(Prob_Norwalk_3mer,Prob_EC_3mer)
print "The relative entropy between Nor and EC 3mer probs
is", value1
value2 = relative_entropy(Prob_EC_3mer,Prob_Norwalk_3mer)
print "The relative entropy between EC and Nor 3mer probs
is", value2
sym = (value1+value2)/2
print "The symmetrized relative entropy between EC and Nor
3mer probs is", sym
```

——————————————— prog2.py addendum 1end ———————————————

Recall that the definition of mutual information between two random variables,
{X, Y} is simply the relative entropy between P(X, Y) "P" and P(X)P(Y) "p", and this
is implemented in addendum #2 to prog2.py:

——————————————— prog2.py addendum 2 ———————————————

```
# the mutual information between P(X,Y) and p(X)p(Y)
requires passing two
# prob arrays: P and p, where |P| = |p|^2 is relation
between # of terms:

def mutual_info ( P , p ):
    Pnum = len(P)
    pnum = len(p)
    if Pnum != pnum*pnum:
        print "error: Pnum != pnum*pnum"
        return -1

    mi = 0
    for index in range(0, Pnum):
        row = index/pnum
        column = index%pnum
        mi += P[index]*math.log(P[index]/(p[row]*p
[column]))

    return mi
```

```
#usage
Prob_EC_2mer = shannon_order(EC_sequence,2)
Prob_EC_1mer = shannon_order(EC_sequence,1)
mutual_info(Prob_EC_2mer,Prob_EC_1mer)
```

———————————————— prog2.py addendum 2 end ————————————————

3.2 Codon Discovery from Mutual Information Anomaly

As mentioned previously, mutual information allows statistical linkages to be discovered that are not otherwise apparent. Consider the mutual information between nucleotides in genomic data when different gap sizes are considered between the nucleotides as shown in Figure 3.1a. When the MI for different gap sizes is evaluated (see Figure 3.1b), a highly anomalous long-range statistical linkage is seen, consistent with a three-element encoding scheme (the codon structure is thereby revealed) [1, 3].

To do the computation for Figure 3.1, the next subroutine (prog2.py addendum 3) is a recycled version of the code for the dinucleotide counter described previously (prog1.py addendum 6), except that now the two bases in the sample window have a fixed gap size between them of the indicated size. Before, with no gap, the gapsize was zero.

———————————————— prog2.py addendum 3 ————————————————

```
def gap_dinucleotide( seq, gap ):
    stats = {}
    pattern = '[acgtACGT]'
    result = re.findall(pattern, seq)
    seqlen = len(seq)
    probs = np.empty((0))

    if (1+gap>seqlen):
        print "error, gap > seqlen"
        return probs

    for index in range(1+gap,seqlen):
        xmer = result[index-1-gap]+result[index]
        if xmer in stats:
            stats[xmer]+=1
        else:
            stats[xmer]=1
```

(a)

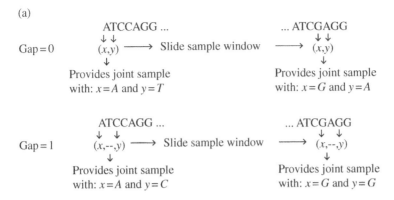

(b)

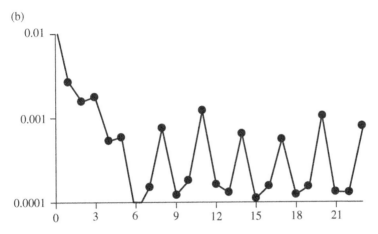

Figure 3.1 Codon structure is revealed in the *V. cholera* genome by mutual information between nucleotides in the genomic sequence when evaluated for different gap sizes.

```
counts = np.empty((0))
for i in sorted(stats):
    counts = np.append(counts,stats[i]+0.0)

probs = count_to_freq(counts)
return probs

# usage:
for i in range(0,25):
```

```
gap_probs = gap_dinucleotide(EC_sequence,i)
val=mutual_info(gap_probs,Prob_EC_1mer)
print "Gap", i, "has mi=", val
```

——————————— prog2.py addendum 3 end ———————————

In Figure 3.1, at first the mutual information falls off as we look at statistical linkages at greater and greater distance, which makes sense for any information construct that is "local," e.g. it should have less linkage with structure further away. After a certain point, however, the mutual information no longer falls off, instead cycling back to a certain level of mutual information with a cycle period of three bases. This suggests that a long-range three-element encoding scheme might exist (among other things), which can easily be tested. In doing so we ask Nature "the right question" and the answer is the rediscovery of the codon encoding scheme, as will be shown in what follows.

So, to clarify before proceeding, suppose we want to get information on a three-element encoding scheme for the *Escherichia coli* genome (Chromosome 1), say, in file EC_Chr1.fasta.txt. We, therefore, want an order = 3 oligo counting, but on 3-element windows seen "stepping" across the genome, e.g. a "stepping" window for sampling, not a sliding window, resulting in three choices on stepping, or framing, according to how you take your first step:

case 0: agttagcgcgt --> (agt)(tag)(cgc)gt
case 1: agttagcgcgt --> a(gtt)(agc)(gcg)t
case 2: agttagcgcgt --> ag(tta)(gcg)(cgt)

In the code that follows we get codon counts for a particular frame-pass (prog2. py addendum 4):

——————————— prog2.py addendum 4 ———————————
```
# so suspect existence of three-element coding scheme, the
codon,
# so need stats (anomolous) on codons....

# 'frame' specifies the frame pass as case 0, 1, or 2 in
the text.

# getting codon counts for a specified framing and
specified sequence
# will now be shown in two ways, one built from re-use of
code blocks,
# one from re-use of an entire subroutine:
```

```python
def codon_counter ( seq, frame ):
    codon_counts = {}
    pattern = '[acgtACGT]'
    result = re.findall(pattern, seq)
    seqlen = len(seq)
    # probs = np.empty((0))

    for index in range(frame,seqlen-2):
        if (index+3-frame)%3!=0:
            continue

        codon = result[index]+result[index+1]+result[index+2]
        if codon in codon_counts:
            codon_counts[codon]+=1
        else:
            codon_counts[codon]=1

    counts = np.empty((0))
    for i in sorted(codon_counts):
        counts = np.append(counts,codon_counts[i]+0.0)
        print "codon", i, "count =", codon_counts[i]

    probs = count_to_freq(counts)
    return probs

codon_counter(EC_sequence,0)

# could also get codon counts by shannon_order with
modification to step
# (and have order=3 for codon:

def shannon_codon( seq, frame ):
    order=3
    stats = {}
    pattern = '[acgtACGT]'
    result = re.findall(pattern, seq)
    seqlen = len(seq)
    for index in range(order-1+frame,seqlen):
        if index%3!=2:
            continue
```

```
xmer = ""
for xmeri in range(0,order):
    xmer+=result[index-(order-1)+frame+xmeri]

if xmer in stats:
    stats[xmer]+=1
else:
    stats[xmer]=1

for i in sorted(stats):
    print("%d %s" % (stats[i],i))

counts = np.empty((0))
for i in sorted(stats):
    counts = np.append(counts,stats[i]+0.0)

probs = count_to_freq(counts)
return probs


shannon_codon(EC_sequence,0)
```

———————————— prog2.py addendum 4 end ————————————

In running prog2.py addendum 4 we find that the codon "tag" has much lower counts, and similarly for the codon "cta":

frame 0 have tag with 8970 and cta with 8916
frame 1 have tag with 9407 and cta with 8821
frame 2 have tag with 8877 and cta with 9033

The tag and cta trinucleotides happen to be related – they are reverse compliments of each other (the first hint of information encoding via *duplex* deoxyribonucleic acid (DNA) with Watson–Crick base-pairing). There are two other notably rare codons: taa and tga (and their reverse compliment in this all-frame genomewide study as well).

Now that we have identified an interesting feature, such as "tag," it is reasonable to ask about this feature's placement across the genome. Having done that, the follow-up is to identify any anomalously recurring feature proximate to the feature of interest. Such an analysis would need a generic subroutine for getting counts on sub-strings of indicated order on an indicated reference, to genome sequence data, and that is provided next as an addendum #5 to prog2.py.

———————————————— prog2.py addendum 5 ————————————————

```
# see that 'tag' is anomolous, want to get sense of the
distribution on
# gaps between 'tag' (still satepping 'in-frame').

def codon_gap_counter ( seq, frame, delimiter ):
    counts = {}
    pattern = '[acgtACGT]'
    result = re.findall(pattern, seq)
    seqlen = len(seq)
    # probs = np.empty((0))

    oldindex=0
    for index in range(frame,seqlen-2):
        if (index+3-frame)%3!=0:
            continue

        codon = result[index]+result[index+1]+result
        [index+2]
        if codon!=delimiter:
            continue
        else:
            gap = index - oldindex
            quant = 100
            bin = gap/quant
            if oldindex!=0:
                if bin in counts:
                    counts[bin]+=1
                else:
                    counts[bin]=1

            oldindex=index

    npcounts = np.empty((0))
    for i in sorted(counts):
        npcounts = np.append(npcounts,counts[i]+0.0)
        print "gapbin", i, "count =", counts[i]

    probs = count_to_freq(npcounts)
    return probs
```

```
# usage:
delimiters =
("AAA","AAC","AAG","AAT","ACA","ACC","ACG","ACT",

"AGA","AGC","AGG","AGT","ATA","ATC","ATG","ATT",

"CAA","CAC","CAG","CAT","CCA","CCC","CCG","CCT",

"CGA","CGC","CGG","CGT","CTA","CTC","CTG","CTT",

"GAA","GAC","GAG","GAT","GCA","GCC","GCG","GCT",

"GGA","GGC","GGG","GGT","GTA","GTC","GTG","GTT",

"TAA","TAC","TAG","TAT","TCA","TCC","TCG","TCT",

"TGA","TGC","TGG","TGT","TTA","TTC","TTG","TTT")
for delimiter in delimiters:
    print "\n\ndelimiter is", delimiter
    codon_gap_counter(EC_sequence,0,delimiter)
─────────────── prog2.py addendum 5 end ───────────────
```

Upon running the above code with codon delimiter set to "tag," we arrive at Table 3.1, which shows the distribution on (tag) gap sizes. Bin size is 100. So

Table 3.1 (tag) Gap sizes, with bin size 100.

Gap bin	Count
0	2115
1	1428
2	1066
3	829
4	696
5	484
6	399
7	293
8	241
9	222

Table 3.2 (aaa) Gap sizes, with bin size 100.

Gap bin	Count
0	21 256
1	7843
2	3375
3	1665
4	827
5	480
6	287
7	163
8	86
9	70

gap bin 0 has the count on all gaps seen sized anywhere from 1 to 99. Bin 1 has counts on occurrences of gaps in the domain 100–199, etc.

In order to see how strongly the (tag) distribution is skewed, we consider some other codon to evaluate, such as for the "aaa" gap, where aaa is most common. The aaa gaps, shown in Table 3.2, tend to be much smaller, with a standard exponential distribution fall-off indicative of no long-range encoding linkages:

Thus the codon tag is clearly very different from aaa, it is as if tag roughly marks the boundaries of regions, and aaa is just scattered throughout. Are any other codons similar to tag? The frequency analysis blurs counts so more subtle differences not as obvious have to run gap counter for each to directly see, and how to easily "see"? Notice how the tag distribution has a long tail. The gap bins only go to 9 in the figures, but for the full dataset the last nonzero gap bin for "tag" is at a remarkable bin 70. For "aaa" the last nonzero bin is much earlier, at bin number 23 (even though there are 10 times as many (aaa) codons as (tag) codons). For "taa" the last nonzero bin is at 60, while for tga it is at 53. The codons taa, tga, and tag are known as the stop codons and the gaps between them are known as ORFs, or ORFs. A subtlety in the statistical analysis is that the stop codons do not have to match to define such anomalously large regions (according to observation). Thus, a biochemical encoding scheme must exist that works with any of the three stop codons seen as equivalent, thus the naming for this group as "stop" codons (and their grouping as such in Figure 3.2). For more nuances of the naming convention "stop" codon when delving into the encoding biochemistry see [1, 3].

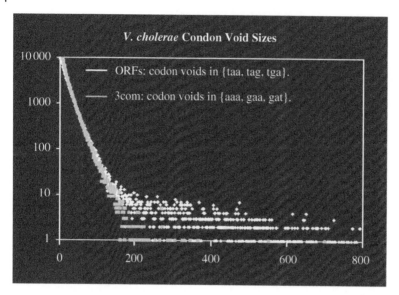

Figure 3.2 ORF encoding structure is revealed in the V. cholera genome by gaps between stop codons in the genomic sequence. *X*-axis shows the size of the gap in codon count between reference codons (stops for conventional ORFs, or 3com set for comparisons in table), *Y*-axis shows the counts.

3.3 ORF Discovery from Long-Tail Distribution Anomaly

Once codon grouping is revealed, where a frequency analysis on codons on the stop codons (TAA, TAG, TGA) shows they are rare. Focusing on the stop codons it is easily found that the gaps between stop codons can be quite anomalous compared to the gaps between other codons (see prog2.py addendum 6):

```
———————————————— prog2.py addendum 6 ————————————————
# need gap stats between codons, the stop codon group
(orf_finder), and
# the 'common' reference group (corf_finder):

def orf_finder ( seq, frame ):
    gapcounts = {}
    edgecounts = {}
    pattern = '[acgtACGT]'
    result = re.findall(pattern, seq)
    seqlen = len(seq)
```

```
output_fh = open("orf_output", 'w')
output_fh.close()
oldindex=0
oldcodon=""
for index in range(frame,seqlen-2):
    rem = (index+3-frame)%3
    if rem!=0:
        continue

    codon = result[index]+result[index+1]+result
    [index+2]
    if (codon!="TAA" and codon!="TAG" and codon!="TGA"):
        continue
    else:
        gap = index - oldindex
        if gap%3!=0:
            print "gap=", gap, "index=", index
            break

        quant = 100
        bin = gap/quant
        if oldindex!=0:
            if bin in gapcounts:
                gapcounts[bin]+=1
            else:
                gapcounts[bin]=1

        if oldcodon!="":
            edge=oldcodon + codon
            if edge in edgecounts:
                edgecounts[edge]+=1
            else:
                edgecounts[edge]=1

        slice = result[oldindex: index+2+1]
        output_fh = open("orf_output", 'a')
        slicejoin = ""
        slicejoin = slicejoin.join(slice)
        orfline = slicejoin + '\n'
        output_fh.write(orfline)
```

```
                oldindex=index
                oldcodon=codon

        npcounts = np.empty((0))
        for i in sorted(gapcounts):
            npcounts = np.append(npcounts,gapcounts[i]+0.0)
            print "gapbin", i, "count =", gapcounts[i]

        ecounts = np.empty((0))
        for i in sorted(edgecounts):
            ecounts = np.append(ecounts,edgecounts[i]+0.0)
            print "edgecodon", i, "count =", edgecounts[i]

        probs = count_to_freq(npcounts)

#usage:
orf_finder(EC_sequence,0)




# def corf_finder ( seq, frame ):
# same except not 'stop' boundariy condition but 'common':
# if (codon!="AAA" and codon!="GAA" and codon!="GAT")

#usage:
#corf_finder(EC_sequence,0)
```

———————————— prog2.py addendum 6 end ————————————

ORFs are "open reading frames," where the reference to what is open is lack of encounter with a stop codon when traversing the genome with a particular codon framing , e.g. ORFs are regions devoid of stop codons when traversed with the codon framing choice of the ORF. When referring to ORFs in most of the analysis we refer to ORFs of length 300 bases or greater. The restriction to larger ORFs is due to their highly anomalous occurrences and likely biological encoding origin (see Figure 3.2), e.g. the long ORFs give a strong indication of containing the coding region of a gene. By restricting to transcripts with ORFs >= 300 in length we have a resulting pool of transcripts that are mostly *true* coding transcripts.

The above example shows a bootstrap finite state automaton (FSA) process on genomic data: first scan through the genomic data base-by-base and obtain counts on nucleotide pairs with different gap sizes between the nucleotides observed [1, 3]. This then allows a mutual information analysis on the nucleotide pairs taken at the different gap sizes. What is found for prokaryotic genomes (with their highly dense gene placement), is a clear signal indicating anomalous statistical linkages on bases three apart [1, 3]. What is discovered thereby is codon structure, where the coding information comes in groups of three bases. Knowing this, a bootstrap analysis of the 64 possible 3-base groupings can then be done, at which point the anomalously low counts on "stop" codons is then observed. Upon identification of the stop codons their placement (topology) in the genome can then be examined and it is found that their counts are anomalously low because there are large stretches of regions with no stop codon (e.g. there are stop codon "voids," known as "ORFs"). The codon void topologies are examined in a comparative genomic analysis in [1, 3]. As noted previously, the stop codons, which should occur every 21 codons on average if DNA sequence data was random, are sometimes not seen for stretches of several hundred codons (see Figure 3.2).

Not surprisingly, longer genes stand out clearly in this process, since their anomalous, clearly nonrandom DNA sequence, is being maintained as such, and not randomized by mutation, (as this would be selected against in the survival of the organism that is dependent on the gene revealed).

The preceding basic analysis can provide a gene-finder on prokaryotic genomes that comprises a one-page Python script that can perform with 90–99% accuracy depending on the prokaryotic genome. A second page of Python coding to introduce a "filter," along the lines of the bootstrap learning process mentioned above, leads to an *ab initio* prokaryotic gene-predictor with 98.0–99.9% accuracy. Python code to accomplish this is shown in what follows (Chapter 4). In this process, all that is used is the raw genomic data (with its highly structured intrinsic statistics) and methods for identifying statistical anomalies and informatics structural anomalies: (i) anomalously high mutual information is identified (revealing codon structure); (ii) anomalously high (or low) statistics on an attribute or event is then identified (low stop codon counts, lengthy stop codon voids); then anomalously high sub-sequences (binding site motifs) are found in the neighborhood of the identified ORFs (used in the filter).

3.3.1 *Ab initio* Learning with smORF's, Holistic Modeling, and Bootstrap Learning

In work on prokaryotic gene prediction (*V. cholera* in what follows), a program (smORF) was developed for an extended ORF characterization (to characterize "some more ORFs" with different trinucleotide delimiters than stops). Using that software with a simple start-of-coding heuristic it was possible to establish good gene prediction for ORFs of length greater than 500 nucleotides. The smORF gene identification was used in a bootstrap gene-annotation process (where no initial training data was provided). Part of the functionality for smORF is encompassed in prog2.py program described thus far. The strength of the gene identification was then improved by use of a gap-interpolating-Markov-model (gIMM's to be described in Section 3.4). When applied to the identified coding regions (most of the >500 length ORFs), six gIMMs were used (one for each frame of the codons, with forward and backward read senses). If poorly gIMM-scoring coding regions were rejected, performance improved, with results slightly better than those of the early Glimmer gene-prediction software [125], where an interpolating Markov model was used (but not generalized to permit gaps). More recent versions of Glimmer incorporate start-codon modeling in order to strengthen predictions. One of the benefits of the gap-interpolating generalization is that it permits regulatory motifs to be identified, particularly those sharing a common positional alignment with the start-of-coding region. Using the bootstrap-identified genes from the smORF-based gene-prediction (including mis-calls) as a training set permitted an unsupervised search for upstream regulatory structure. The classic Shine-Dalgarno sequence (the ribosome binding site) was found to be the strongest signal in the 30-base window upstream from the start codon. Similar results will be found with the full gene-finder example in Chapter 4.

Before moving on to more sophisticated gene structure identification (Chapter 4), let us first consider the multi-frame and two-strand aspect of the genomic information and what this might mean for the "topology" or overlap placement of coding regions. To recap, *smORF* offers information about ORFs, and tallies information about other such codon void regions (an ORF is a void in three codons: TAA, TAG, TGA). This allows for a more informed selection process when sampling from a genome, such that non-overlapping gene starts can be cleanly and unambiguously sampled. Furthermore, overlapping ORF coding regions can be identified and enumerated (see Figures 3.3 and 3.4).

The goal with smORF was, initially, to identify key gene structures (e.g. stop codons, etc.) and use only the highest confidence examples to train profilers. Once this was done, Markov models (MMs) were (bootstrap) constructed on the suspected start/stop regions and coding/noncoding regions. The algorithm then iterated again, informed with the MM information, and partly relaxes the high fidelity sampling restrictions (essentially, the minimum allowed ORF length is made smaller). A crude gene-finder was then constructed on the high fidelity ORFs by use of a very simple heuristic: scan from the start of an ORF and stop at the first in-frame "atg" (to be implemented in Chapter 4). This analysis was applied to the *Vibrio cholerae* genome (Chr. I). 1253 high fidelity ORFs were identified out of 2775 known genes. This first-"atg" heuristic provided a gene prediction accuracy of 1154/1253 (92.1% of predictions of gene regions were exactly correct). If small shifts are allowed in the predicted position of the start-codon relative to the

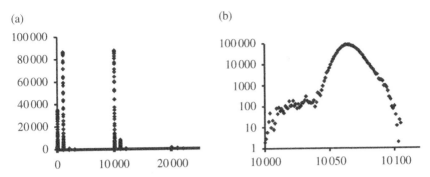

Figure 3.3 **(a) Topology index histograms shown for the *V. cholerae* CHR. I genome**, where the *x*-axis is the topology index, and the *y*-axis shows the event counts (i.e. occurrence of that particular topology index in the genome). The topology index is computed by the following scheme: (i) initialize index for all bases in sequence to zero. (ii) Each base in a forward sense ORF, with length greater than a specified cutoff, is incremented by +10 000 for each such ORF overlap. Similarly, bases in reverse sense ORFs are incremented by +1000 for each such overlap. Voids larger than the cutoff length in the nonstandard smORFs each give rise to an increment of +1. The top panel above shows that *V. cholerae* only has a small portion of its genome involved in multiple gene encodings. The (b) panel shows a "blow-up" of the 10 000 peak.

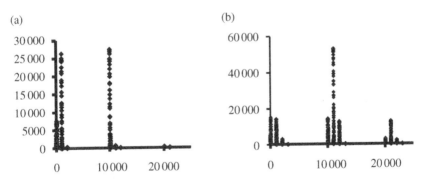

Figure 3.4 **Topology-index histograms are shown for the *Chlamydia trachomatis* genome**, (a), and *Deinococcus radiodurans* genome, (b) *C. trachomatis*, like *V. cholerae*, shows very little overlapping gene structure. *D. radiodurans*, on the other hand, is dominated by genes that overlap other genes (note the strong 11 000 peak).

first-"atg" (within 25 bases on either side), then prediction accuracy improves to 1250/1253 (99.8%). This actually elucidates a key piece of information needed to improve such a prokaryotic gene-finder: information is needed to help identify the correct start codon in a 50-base window from the first ATG. Such information exists in the form of DNA motifs corresponding to the binding footprint of regulatory biomolecules (that play a role in transcriptional or translational control). Further bootstrap refinements along these lines are done in Chapter 4 to produce an *ab initio* prokaryotic gene finder with 99.9% or better accuracy.

Ab initio gene-finding can identify the stop codons and, thus, (standard) ORFs. A generalization to codon void regions, with all six frame passes, also leads to recognition of different, *overlapping*, potential gene regions (and then doubled given the two orientations). A genome-topology scoring as shown in Figure 3.3 can clearly show differences between bacteria (Figure 3.4) – and is thus a possible "fingerprinting" tool.

The prokaryotic genome analysis is similar to both the prokaryotic *and eukaryotic transciptome* analysis (where eukaryotic *transcriptome* analysis is similar since the introns have been removed). The analysis tools for prokaryotic genomes, described thus far, are primarily what are needed for either prokaryotic or eukaryotic transcriptome analysis. Surprisingly, the same overlapping void topologies, with reverse overlap orientation ("duals"), are seen at transcriptome level *in eukaryotes* as in prokaryotes. For eukaryotic transcripts with overlaps that are "dual", however, this has special significance. Recall that a transcript that encodes overlapping read direction "duality" (with regulatory regions intact and lengthy ORF size, so highly likely functional), is only from a *single* genome-level pre-messenger ribonucleic acid (mRNA) due to intron splicing in eukaryotes. This is a very odd arrangement (artifact) for eukaryotes unless they evolved from an ancient prokaryote as hypothesized in a number of theories where such an overlap topology would already be in place to "imprint thru." The specific nature of this transcriptome artifact, however, is best explained via the viral eukaryogenesis hypothesis (see [1, 3]).

3.4 Sequential Processes and Markov Models

Just as Chapter 2 finished with a Math review, we do the same again here in the context of sequential processes. The core mathematical tool for describing a sequential process (where limited memory suffices) is the Markov chain, so that will be defined first. In the context of genome analysis, however, the standard Markov chain based feature extraction is no longer optimal (especially given the nature of the computational resources). Thus, novel mathematical generalization of the Markov chain description, interpolated Markov models, will be given as well. The gap/hash interpolated Markov model, in particular, can be used to "vacuum-up" all motif information in specified regions. This could be used and directly integrated into an HMM-based gene finder (Chapters 7 and 8), or, alternatively, provide identification of a typical motif set for some circumstance (as will be done in Chapter 4).

3.4.1 Markov Chains

A Markov chain is a sequence of random variables S_1; S_2; S_3; ... with the Markov property of limited memory, where a first-order Markov assumption on the probability for observing a sequence "$s_1 s_2 s_3 s_4 ... s_n$" is:

$$P(S_1 = s_1, ..., S_n = s_n) = P(S_1 = s_1) P(S_2 = s_2 \mid S_1 = s_1) ... P(S_n = s_n \mid S_{n-1} = s_{n-1})$$

In the Markov chain model, the states are also the observables. For a HMM we generalize to where the states are no longer directly observable (but still first-order Markov), and for each state, say S_1, we have a statistical linkage to a random variable, O_1, that has an observable base emission, with the standard (0th-order) Markov assumption on prior emissions.

The key "short-term memory" property of a Markov chain is: $P(x_i \mid x_{i-1}, ..., x_1) = P(x_i \mid x_{i-1}) = a_{i-1, i}$, where $a_{i-1, i}$ are sometimes referred to as "transition probabilities", and we have: $P(x) = P(x_L, x_{L-1}, ..., x_1) = P(x_1) \prod_{i=2...L} a_{i-1, i}$. If we denote C_y for the count of events y, C_{xy} for the count of simultaneous events x and y, T_y for the count of strings of length one, and T_{xy} for the count of strings of length two: $a_{i-1, i} = P(x \mid y) = P(x, y)/P(y) = [C_{xy}/T_{xy}]/[C_y/T_y]$. Note: since $T_{xy} + 1 = T_y \rightarrow T_{xy} \cong T_y$ (sequential data sample property if one long training block), $a_{i-1, i} \cong C_{xy}/C_y = C_{xy}/\sum_x C_{xy}$, so counts C_{xy} is complete information for determining transition probabilities.

For prokaryotic gene prediction much of the problem with obtaining high-confidence training data can be circumvented by using a bootstrap gene-prediction approach. This is possible in prokaryotes because of their simpler and more compact genomic structure: simpler in that long ORFs are usually long genes, and compact in that motif searches upstream usually range over hundreds of bases rather than thousands (as in human).

Interpolated Markov Model (IMM): the order of the MM can be interpolated according to a *globally* imposed cutoff criterion (see Figure 3.5), such as a minimum sub-sequence count: fourth-order passes if Counts (x_0; x_{-1}; x_{-2}; x_{-3}; x_{-4}) >cutoff for all $x_{-4}...x_0$ sub-sequences (100, for example), the utility of this becomes apparent with the following reexpression:

$$P(x_0 \mid x_{-1}; x_{-2}; x_{-3}; x_{-4}) = P(x_0; x_{-1}; x_{-2}; x_{-3}; x_{-4})/P(x_{-1}; x_{-2}; x_{-3}; x_{-4})$$
$$= \text{Counts}(x_0; x_{-1}; x_{-2}; x_{-3}; x_{-4})/\text{Counts}(x_{-1}; x_{-2}; x_{-3}; x_{-4})$$

TotalCounts(length5)/TotalCounts(length4)

$$= \text{Counts}(x_0; x_{-1}; x_{-2}; x_{-3}; x_{-4})/\text{Counts}(x_{-1}; x_{-2}; x_{-3}; x_{-4}) [(L-4)/(L-3)]$$
$$\cong \text{Counts}(x_0; x_{-1}; x_{-2}; x_{-3}; x_{-4})/\text{Counts}(x_{-1}; x_{-2}; x_{-3}; x_{-4})$$

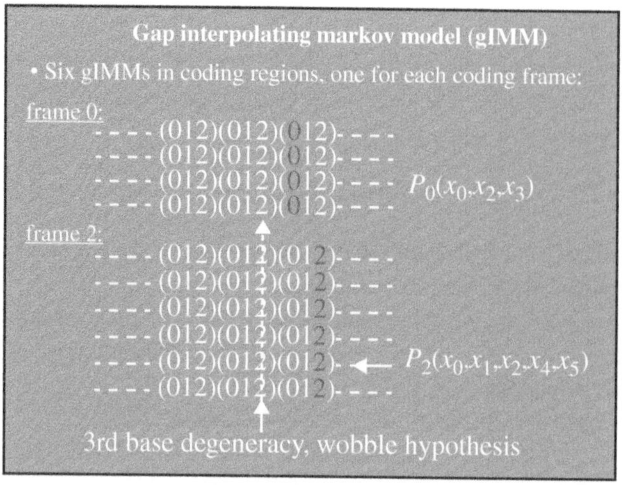

Gap interpolating markov model (gIMM)

• Six gIMMs in coding regions, one for each coding frame:

frame 0:

```
- - - - (012)(012)(012)- - - -
- - - - (012)(012)(012)- - - -
- - - - (012)(012)(012)- - - -        P_0(x_0,x_2,x_3)
- - - - (012)(012)(012)- - - -
```

frame 2:

```
- - - - (012)(012)(012)- - - -
- - - - (012)(012)(012)- - - -
- - - - (012)(012)(012)- - - -
- - - - (012)(012)(012)- - - -
- - - - (012)(012)(012)-  ←  P_2(x_0,x_1,x_2,x_4,x_5)
- - - - (012)(012)(012)- - - -
```

3rd base degeneracy, wobble hypothesis

Figure 3.5 Hash interpolated Markov model (hIMM) and gap/hash interpolated Markov model (ghIMM): now no longer employ a global cutoff criterion – count cutoff criterion applied at the sub-sequence level. Given the current state and the emitted sequence as $x_1, ..., x_L$; compute: $P(x_L \mid x_1, ..., x_{L-1}) \approx \text{Count}(x_1, ..., x_L)/\text{Count}(x_1, ..., x_{L-1})$. If $\text{Count}(x_1, ..., x_{L-1}) \geq$ 400 i.e. only if the parental sequence shows statistical significance (consider $400 = 4 \times 100$, or requiring >100 counts on observations assuming uniform distribution for count cutoff determination), store $P(x_L \mid x_1, ..., x_{L-1})$ in the hash. (If gene finding, have at least five states, e.g. need to maintain a separate hash for each of the following states – Junk, Intron, and Exon0, 1, 2.)

Suppose Counts $(x_0; x_{-1}; x_{-2}; x_{-3}; x_{-4}; x_{-5})$ <cutoff for some $x_{-5}...x_0$ subsequence, then the interpolation would halt (globally), and the order of MM used would be fourth-order.

Gap Interpolated Markov Model (gIMM): like IMM with its count cutoff, but when going to higher order in the interpolation there is no constraint to contiguous sequence elements – i.e. "gaps" are allowed. The resolution of what gap-size to choose when going to the next higher order is resolved by evaluating the Mutual Information. i.e. when going to 3rd order in the Markov context, $P(x_0 \mid x_{-5}; x_{-2}; x_{-1})$ is chosen over $P(x_0 \mid x_{-3}; x_{-2}; x_{-1})$ if

$$\text{MI}(\{x_0; x_{-1}; x_{-2}\}, \{x_{-5}\}) > \text{MI}(\{x_0; x_{-1}; x_{-2}\}, \{x_{-3}\}).$$

Or, in terms of Kullback–Leibler divergences, if

$$D\Big[P(x_0; x_{-1}; x_{-2}; x_{-5}) \big\| P(x_0; x_{-1}; x_{-2})P(x_{-5})\Big]$$
$$> D\Big[P(x_0; x_{-1}; x_{-2}; x_{-3}) \big\| P(x_0; x_{-1}; x_{-2})P(x_{-3})\Big].$$

3.5   Exercises

3.1   In Section 3.1, the Maximum Entropy Principle is introduced. Using the Lagrangian formalism, find a solution that maximizes on Shannon entropy subject to the constraint of the "probabilities" sum to one.

3.2   Repeat the Lagrangian optimization of (Exercise 3.1) subject to the added constraint that there is a mean value, $E(X) = \mu$.

3.3   Repeat the Lagrangian optimization of (Exercise 3.2) subject to the added constraint that there is a variance value, $\text{Var}(X) = E(X^2)(E(X))^2 = \sigma^2$.

3.4   Using the two-die roll probabilities from (Exercise 2.3) compute the mutual information between the two die using the relative entropy form of the definition. Compare to the pure Shannon definition: $\text{MI}(X, Y) = H(X) + H(Y) - H(X, Y)$.

3.5   Go to genbank (https://www.ncbi.nlm.nih.gov/genbank) and select the genomes of three medius-sized bacteria (~1 Mb), where two bacteria are closely related. Using the Python code shown in Section 2.1, determine their hexamer frequencies (as in Exercise 2.5 with virus genomes). What is the Shannon entropy of the hexamer frequencies for each of the three bacterial genomes? Consider the following three ways to evaluate distances between the genome hexamer-frequency profiles (denoted Freq(genome1), etc.), try each, and evaluate their performance at revealing the "known" (that two of the bacteria are closely related):
   i) distance = Shannon difference = | H(Freq(genome1))−H(Freq (genome2))|.
   ii) distance = Euclidean distance = d(Freq(genome1),Freq(genome2)).
   iii) distance = Symmetrized Relative Entropy = [D(Freq(genome1)‖Freq (genome2))+D(Freq(genome2)‖Freq(genome1))]/2
   Which distance measure provides the clearest identification of phylogenetic relationship? Typically it should be (iii).

3.6   Exercise 3.5, if done repeatedly, will eventually reveal that the best distance measure (between distributions) is the symmetrized relative entropy (case (iii)). Notice that this means that when comparing two distributions we quantify their difference not by a difference on Shannon entropies, case (i). In other words, we choose:
   Difference$(X, Y) = \text{MI}(X, Y) = H(X) + H(Y) - H(X, Y)$,
   Not Difference $= | H(X) - H(Y)|$.

The latter case satisfies the metric properties, including triangle inequality, in order to be a "distance" measure, is this true for the mutual information difference as well?

3.7 Go to genbank (https://www.ncbi.nlm.nih.gov/genbank) and select the genome of the K-12 strain of *E. coli*. (The K-12 strain was obtained from the stool sample of a diphtheria patient in Palo Alto, CA, in 1922, so that seems like a good one.) Reproduce the MI codon discovery described in Figure 3.1.

3.8 Using the *E. coli* genome (the one described above) and using the codon counter code, get the frequency of occurrence of the 64 different codons genome-wide (without even restricting to coding regions or to a particular "framing," these are still unknowns, initially, in an *ab initio* analysis). This should reveal oddly low counts for what will turn out to be the "stop" codons.

3.9 Using the code examples with stops used to mark boundaries, identify long ORF regions in the *E. coli* genome (the one described in (Exercise 3.7)). Produce an ORF length histogram like that shown in Figure 3.2.

3.10 Create an overlap-encoding topology scoring method (like that described for Figure 3.3), and use it to obtain topology histograms like those shown in Figure 3.3a and Figure 3.4. Do this for the following genomes: *E. coli* (K-12 strain); *V. cholera*; and *Deinococcus Radiodurans*.

3.11 In a highly trusted coding region, such as the top 10% of the longest ORFs in the *E. coli* genome analysis, perform a gap IMM analysis, which should approximately reproduce the result shown in Figure 3.5.

3.12 Prove that relative entropy is always positive.

4

Ad Hoc, Ab Initio, and Bootstrap Signal Acquisition Methods

In this chapter, concepts introduced in Chapters 2 and 3 are combined with efficient implementation details and objective measures of performance, as well as signal analysis tools from electrical engineering and statistics. This permits *ad hoc, ab initio*, signal acquisition. Signal analysis definitions and terminology are explained in [189–196] and are discussed further in Section 4.6.

In Section 4.1, we discuss the critical speed advantages of Finite State Automoton (FSA) methods that are $O(L)$, and in Section 4.2 we briefly return to the gene finder problem from Chapter 3, where the bootstrap methodology is shown in practice (named after the paradoxical Baron von Munchausen story [194]). In Section 4.3 objective measures of performance evaluation are described. Section 4.4 delves into signal analytics, where Section 4.4.2 time-domain Finite State Automoton (tFSA) gives methods for channel current cheminformatics (CCC) when there is stable channel baseline reference signal and Section 4.4.3 describers tFSA methods for CCC with unstable baseline. In Section 4.5 are efficient implementations of the FSAs and core statistical tools.

Numerous prior book, journal, and patent publications by the author are drawn upon extensively throughout [1–68]. Almost all of the journal publications are open access. These publications can typically be found online at either the author's personal website (www.meta-logos.com) or with one of the following online publishers: www.m-hikari.com or bmcbioinformatics.biomedcentral.com.

4.1 Signal Acquisition, or Scanning, at Linear Order Time-Complexity

All of the tFSA signal acquisition methods described in this chapter are $O(L)$, i.e. they scan the data with a computational complexity no greater than that of simply seeing the data (via a "read" or "touch" command). Because the signal acquisition is only $O(L)$ it is not significantly costly, computationally, to simply repeat the acquisition analysis multiple times with a more informed process with each iteration, to arrive at a "bootstrap" signal acquisition process. In such a setting, signal acquisition is often done with bias to very high specificity (SP) initially (and sensitivity very poor), to get a "gold standard" set of highly likely true signals that can be data mined for their attributes. With a filter stage thereby trained, later scan passes can pass suspected signals with very weak SP (very high sensitivity now) with high SP then recovered by use of the filter. This then allows a bootstrap process to a very high SP and sensitivity (SN) at the tFSA acquisition stage on the signals of interest.

The same procedure outlined for the above signal discovery and acquisition on genomic data can be applied to the analysis of any set of stochastic sequential data. If the data is numerical, e.g. observational data on an electrical current, or stock price, or time series in general, then even more tools from statistics can be brought to bear (e.g. the expectation of a real number observational sequence can be calculated, while the same cannot be done, directly, with the symbol-based $\{a, c, g, t\}$ genomic data). In what follows a description of a time-domain FSA on real-number observational data (channel current readings) will be given. Details include practical implementation considerations such as working with raw integer encoded data, not floating point representation until absolutely needed. Working with integer representation, in some settings, allows a significant speedup for every processing loop working with integer multiplications instead of floating point multiplications.

In [1, 3], a time-domain finite state automaton with eight states is used for signal identification and acquisition (based on the first 100 ms of channel current blockade signal in Figure 4.1). Two states, sequentially connected, were used for reset and initialization on the FSA. Transition between the two states, from reset-start to reset-ready, was accomplished upon measuring a short section of acceptable baseline current (200 μs). An abrupt drop in current to 70% residual current (determined by the holistic tuning that is described in what follows), or less, then triggered transition from the reset-ready state to the signal-active state. From the signal-active state, processing advanced to one of two states (good- and bad-end-level states) according to an end-of-signal profile. The profile rule simply required that the last end-level-range observations had to have current above minimum-end-level-value. Satisfying the rule led to the good-end-level state, otherwise the bad-end-level state was reached. If there was a normal return to baseline (good-end-level state), or a signal-blockade scan exited due to truncation (bad-end-level state), the signal complete state was reached, otherwise further scanning was performed. Further scanning involved transition through the internal active state, where local signal properties, observation less than maximum-cutoff and observation greater than minimum-cutoff, were used to decide whether to exit (to the reset-end state) or continue the blockade scan (return to the signal-active state). Similar to the local

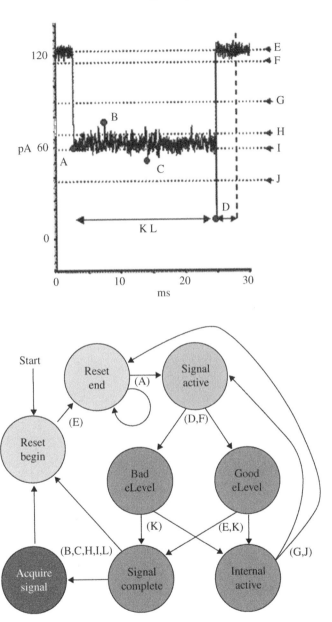

Figure 4.1   Schematic for the finite state automaton used for acquisition of six base-pair DNA hairpin blockade signals observed in [1, 3], where a sample signal is shown to the left. The letters label various types of feature extraction parameters and their placement in the FSA diagram indicate where the decision-making or thresholding is dependent on those parameters. *Source:* Based on Winters-Hilt [1]; Winters-Hilt [3].

blockade signal properties that determined how to transition from the internal-active state, transition to the acquire-signal state from the signal-complete state was based on several global properties of the signal trace: maximum blockade sample less than maximum-cutoff and greater than min–max-internal, minimum blockade sample greater than minimum-cutoff and less than max–min-internal, and signal duration greater than or equal to minimum-duration.

4.2 Genome Analytics: The Gene-Finder

Let us now return, briefly, to the gene-finder developments begun in Chapter 3. We now continue the "bootstrap" evolution of the FSA-based prokaryotic gene-finder code given in Chapter 3. Variations on the FSA-based open reading frame (ORF)-finder code for prokaryotic genomes can be applied to both prokaryotic and eukaryotic transcriptomes, but these are described and used elsewhere [1, 3].

Let us begin prog3.py with an attempt to data-mine the orf_output file to identify the "start" codon (further comments in code). We will find that the mysterious "start" codon is "atg":

```
——————————————— prog3.py ———————————————
#!/usr/bin/python

import numpy as np
import math
import re


# from prior code we carry over the subroutines:
# shannon, count_to_freq, shannon_order, gen_txt_read,
gen_fasta_read,
# relative_entropy, mutual_info, gap_dinucleotide,
codon_counter,
# codon_gap_counter, and orf_finder; with prototypes:
#
# def shannon( probs ) with usage:
# value = shannon(probs)
# print(value)
#
# def count_to_freq( counts ) with usage:
# probs = count_to_freq(rolls)
# print(probs)
#
```

```
# def shannon_order( seq, order ) with usage:
# order = 8
# maxcounts = 4**order
# print "max counts at order", order, "is =", maxcounts
# val = math.log(maxcounts)
# shannon_order(sequence,order) # shannon_order prints
entropy
# print "The max entropy would be log(4**order) = ", val
#
# def gen_txt_read( text ) with usage:
# null=""
# gen_array = gen_txt_read(null) # defaults, uses
Norwalk_Virus.txt genome
# sequence = ""
# sequence = sequence.join(gen_array)
# Norwalk_sequence = sequence;
# seqlen = len(gen_array)
# print "The sequence length of the Norwalk genome is:",
seqlen
#
# def gen_fasta_read( text ) with usage:
# null=""
# fasta_array = gen_fasta_read(null) # defalus, uses e.coli
genome
# sequence = ""
# sequence = sequence.join(fasta_array)
# EC_sequence = sequence
# seqlen = len(fasta_array)
# print "The sequence length of the e. coli Chr1 genome
is:", seqlen
# print "Doing order 8 shannon analysis on e-coli:"
# order = 6
# shannon_order(EC_sequence,order)
#
# def relative_entropy ( P , Q ) with usage: (P and Q are
prob. Arrays)
# value1 = relative_entropy
(Prob_Norwalk_3mer,Prob_EC_3mer)
# print "The relative entropy between Nor and EC 3mer probs
is", value1
# value2 = relative_entropy
```

```
(Prob_EC_3mer,Prob_Norwalk_3mer)
# print "The relative entropy between EC and Nor 3mer probs
is", value2
# sym = (value1+value2)/2
# print "The symmetrized relative entropy is", sym
#
# def mutual_info ( P , p ) with usage: (pro. arrays with |
P|=|p|^2)
# Prob_EC_2mer = shannon_order(EC_sequence,2)
# Prob_EC_1mer = shannon_order(EC_sequence,1)
# mutual_info(Prob_EC_2mer,Prob_EC_1mer)
#
# def gap_dinucleotide( seq, gap ) with usage:
# for i in range(0,25):
#     gap_probs = gap_dinucleotide(EC_sequence,i)
#     val=mutual_info(gap_probs,Prob_EC_1mer)
#     print "Gap", i, "has mi=", val
#
# def codon_counter ( seq, frame )with usage:
# codon_counter(EC_sequence,0)
#
# def codon_gap_counter ( seq, frame, delimiter ) with usage:
# delimiters = ("TAA","TAG","TGA")
# for delimiter in delimiters:
#    print "\n\ndelimiter is", delimiter
#    codon_gap_counter(EC_sequence,0,delimiter)
#
# def orf_finder ( seq, frame ) with usage:
# orf_finder(EC_sequence,0)  # produces orfs with frame
framing
# output file: orf_output

# New code is now created to have subroutines for text
datafile handling, # where the data is one orf per line.
The objective is to have the
# subroutine, line_codon_counter, scan each line in file
orf_output and
# tabulate the codon frequencies in various windows of
codons (in search
# of an anomolous codon usage at a particular position near
the beginning
```

```
# of the orf, the discovery of which will identify the
"start" codon to be
# "atg" (typically, sometimes "gtg").

def line_codon_counter ( filename, zone ):
    if filename=="":
        filename="orf_output"

    fh = open(filename, 'r')
    lines = fh.readlines()

    codon_counts = {}
    for line in lines:
        sequence = ""
        pattern = '>'
        test = re.findall(pattern,line)
        testlen = len(test)
        if testlen>0:
            print "fasta comment:", line.strip()
            continue
        sequence = sequence + line.strip()

        pattern = '[acgtACGT]'
        result = re.findall(pattern, sequence)

        seqlen = len(result)
        if seqlen<300:
            continue

        start=3
        end=seqlen-2-3
        if zone=="full": #clipping stop codon boundaries
            start=3
            end=seqlen-2-3
        elif zone=="front":
            start=3
            end=18
        elif zone=="back":
            end=seqlen-2-3
            start=end-15
        elif zone=="first3codons":
            start=3
            end=12
```

```
        elif zone=="second3codons":
            start=12
            end=21
        elif zone=="third3codons":
            start=21
            end=30
        elif zone=="codons10to21":
            start=30
            end=66
        elif zone=="codons345":
            start=9
            end=18
        elif zone=="codons456":
            start=12
            end=21
        elif zone=="codons567":
            start=15
            end=24
        elif zone=="codons678":
            start=18
            end=27

        for index in range(start,end):
            if index%3!=0:
                continue

            codon = result[index]+result[index+1]+result
[index+2]
            if codon in codon_counts:
                codon_counts[codon]+=1
            else:
                codon_counts[codon]=1

    print "zone=", zone
    counts = np.empty((0))
    for i in sorted(codon_counts):
        counts = np.append(counts,codon_counts[i]+0.0)
        print "codon", i, "count =", codon_counts[i]

    probs = count_to_freq(counts)
    return probs
```

```
line_codon_counter("orf_output","full")
line_codon_counter("orf_output","front")
line_codon_counter("orf_output","back")
line_codon_counter("orf_output","first3codons")
line_codon_counter("orf_output","second3codons")
line_codon_counter("orf_output","third3codons")
line_codon_counter("orf_output","codons10to21")
line_codon_counter("orf_output","codons345")
line_codon_counter("orf_output","codons456")
line_codon_counter("orf_output","codons567")
line_codon_counter("orf_output","codons678")
```

———————————— prog3.py end ————————————

The code above shows how a rapid prototyping might be done to learn what "the rule" might be for the start of a gene. What is found is that the appearance of "ATG" in-frame in the first several codons of an ORF is typical, and failing that it is the first "ATG" or "GTG" typically found after that. This will give rise to the crude "first ATG" rule for start of coding, which will eventually be relaxed under conditions allowing for nonstandard start codons that are identified by means of their surroundings (motif placement context).

The code that follows is a continuation of a bootstrap learning process, beginning with a subroutine that finds ORFs longer than a specified cutoff, and moves in-frame from the left to the first "ATG" codon, for which a tentative coding region is identified, as well as the upstream "*cis*" region and downstream "*trans*" regions. Extracts of the *coding, cis,* and *trans* regions are taken. The crude longORFfirstATG_geneFinder indicated will catch many true gene regions, but the "first atg" heuristic for the start of the coding region will fail 1–10% of the time depending on the genome. This is where a filter is obtained by looking for anomalous high count motif structures in the *cis* region. Even if 10% of the genes so indicated are incorrect, the strongly recurring motifs tallied from the 90% correctly delineated genes will provide a clear indication of the *cis* regulatory motifs found in valid *cis* regions. This then provides a filter test for a second-pass (bootstrap) genefinder that has a validation test on the *cis* region. This boosts the genefinder accuracy to 99% for many prokaryotes genomes (if not sufficiently accurate for a genome of interest, similar refinements could be made for validation on *trans* motifs and on codon usage).

Let us proceed under the assumption that the gene starts with the first "atg" in the ORF and try to find associated structures that will offer a means to validate the first "atg" as the start, so begin effort by capturing the "first atg genes" in the long-ORFs, along with their pre (*cis*) and post (*trans*) regions as in the Python subroutine shown in the first addendeum to prog3.py:

```
———————————————— prog3.py addendum 1————————————————
def longORFfirstATG_geneFinder( seq, frame, start, minlen,
window ):
    pattern = '[acgtACGT]'
    result = re.findall(pattern, seq)
    seqlen = len(result)

    gene_fh = open("genefile", 'w')
    cis_fh = open("cisfile", 'w')
    trans_fh = open("transfile", 'w')
    gene_fh.close()
    cis_fh.close()
    trans_fh.close()
    oldindex=0
    for index in range(frame,seqlen-2):
        rem = (index+3-frame)%3
        if rem!=0:
            continue

        codon = result[index]+result[index+1]+result[index+2]
        if (codon!="TAA" and codon!="TAG" and codon!="TGA"):
            continue
        else:
            gap = index - oldindex
            if gap<minlen:
                oldindex=index
                continue

            orf = result[oldindex: index+2+1]
            orflen = len(orf)
            atg_index=0
            no_atg=0
            for sindex in range(3,orflen-2):
                rem = sindex%3
                if rem!=0:
                    continue

                if sindex>100:
                    no_atg = 1
                    break
```

```
            cdn = orf[sindex]+orf[sindex+1]+orf[sindex+2]
            if cdn!="ATG":
                continue
            else:
                atg_index=sindex
                break


        if no_atg==1:
            oldindex=index
            continue

        gene = result[oldindex+atg_index: index+2+1]
        gene_fh = open("genefile", 'a')
        slicejoin = ""
        slicejoin = slicejoin.join(gene)
        geneline = slicejoin + '\n'
        gene_fh.write(geneline)

        cis = result[oldindex+atg_index-window:
oldindex+atg_index]
        cis_fh = open("cisfile", 'a')
        slicejoin = ""
        slicejoin = slicejoin.join(cis)
        cisline = slicejoin + '\n'
        cis_fh.write(cisline)

        if (index+2+1+window)>seqlen:
            continue

        trans = result[index+2+1: index+2+1+window]
        trans_fh = open("transfile", 'a')
        slicejoin = ""
        slicejoin = slicejoin.join(trans)
        transline = slicejoin + '\n'
        trans_fh.write(transline)

        oldindex=index


#usage:
longORFfirstATG_geneFinder( EC_sequence, 0, "ATG", 300, 15)
———————————— prog3.py addendum 1end ————————————
```

By running longORFfirstATG_geneFinder we generate output files for the hypothesized gene region and the preceding (*cis*) region and the following (*trans*) region. We now want to look through the collection of *cis* regions to see if there are any anomalously occurring DNA substrings (motifs) that may associate with the hypothesized start of gene region, as shown in the line_oligo_counter subroutine in prog3.py addendum 2:

──────────────── prog3.py addendum 2────────────────

```
def line_oligo_counter ( filename, order, anom_mult ):
    if filename=="":
        filename="cisfile"

    fh = open(filename, 'r')
    lines = fh.readlines()

    oligo_counts = {}
    oligo_sample_count=0
    for line in lines:
        sequence = ""
        sequence = sequence + line.strip()
        pattern = '[acgtACGT]'
        result = re.findall(pattern, sequence)
        seqlen = len(result)
        start=0
        end=seqlen-order
        for index in range(start,end):
            xmer = ""
            for xmeri in range(0,order):
                xmer+=result[index+xmeri]

            oligo_sample_count+=1

            if xmer in oligo_counts:
                oligo_counts[xmer]+=1
            else:
                oligo_counts[xmer]=1

    oligo_type_count = len(oligo_counts)
    oligo_count_expectation = oligo_sample_count/
oligo_type_count
```

```
anom_count_cutoff = oligo_count_expectation*anom_mult
acc_oligo = {}
acc_count=0
for i in sorted(oligo_counts):
    if (oligo_counts[i]+0.0)>anom_count_cutoff:
        acc_oligo[i]=oligo_counts[i]
        acc_count+=1

print "acc_count=", acc_count
for i in sorted(acc_oligo):
        print "oligo", i, "count =", acc_oligo[i]


return acc_oligo


# ver_30 has window=30 and anom_mult=4
# ver_15 has window=15 and anom_mult=3
acc_oligo_counts=line_oligo_counter("cisfile",6,3)


# have acc_count=140 out of 4096, returned in motif hass,
use these as
# validators at next stage of gene identification
```

———————————— prog3.py addendum 2 end ————————————

In the above while loop, we are getting counts on the xmers (oligomers, or subsequences) of the specified order (length). In what follows we then estimate the number of occurences of a particular type of xmer when the data is random (uniform probability distribution) – this is referred to as the expected_average_-count. A cutoff is then introduced for when the number of occurrences of a particular motif is anomalous by use of a multiplier, anom_mult. All motifs with counts above the indicated cutoff are deemed anomalous and thus potentially are of interest as a signaling motif that is paired with the "atg" start of coding.

There is found to be a lengthy list of anomalous 6mers occurring in the window 15 bases prior to the start codon – going forward, we can simply look for an occurrence of one of these 6mers to validate any hypothesized start codon. Failure to validate on the first "atg" would then lead to looking for the next, in-frame, "atg" as a possible start. It turns out that "atg" is used for start codon only 90–99% of the time, the other main start codon being "gtg". The percentage of genes starting with "atg" is genome (thus organism) specific. For some strains of *Escherichia coli*, about 99% of the genes start with "atg," for the most common strain of *Vibrio cholerae*, only about 93% of the genes start with "atg," 6.9% starting with "gtg." 0.1% starting with "ttg," and very rarely, some genes starting with "ctg." To handle this, if we do not have validation via an occurrence of the Shine–Dalgarno or other *cis* motifs in the window 15-based prior to the hypothesized start, then we want to continue to step codon-by-codon (in-frame) into the ORF (from left to right) until the next "atg" or "gtg" is encountered, where the validation is attempted again. This is then repeated until either validation is achieved, or the remaining length of the ORF drops below our cutoff. In the code thus far we have focused on ORFs > 500 in length, to ensure the very likely collection of true gene regions. Now that we have a validation test, we can relax the length-anomaly filter to not be quite so stringent, to 300 initially, and repeat the datarun (see prog3.py with addendum 3 next).

———————————— prog3.py addendum 3 ————————————

```
def geneFinder ( seq, frame, start, minlen, window, motifs ):
    pattern = '[acgtACGT]'
    result = re.findall(pattern, seq)
    seqlen = len(result)

    gene_fh = open("genefile2", 'w')
    cis_fh = open("cisfile2", 'w')
    trans_fh = open("transfile2", 'w')
    gene_fh.close()
    cis_fh.close()
    trans_fh.close()
    oldindex=0
    gene300_count=0
    gene300wm_count=0
    for index in range(frame,seqlen-2):
        rem = (index+3-frame)%3
        if rem!=0:
            continue

        codon = result[index]+result[index+1]+result[index+2]
        if (codon!="TAA" and codon!="TAG" and codon!="TGA"):
            continue
        else:
            gap = index - oldindex
            if gap<minlen:
```

```
        oldindex=index
        continue

    orf = result[oldindex: index+2+1]
    orflen = len(orf)
    atg_index=0
    no_atg=0
    for sindex in range(3,orflen-2):
        rem = sindex%3
        if rem!=0:
            continue

        if sindex>100:
            no_atg = 1
            break
      cdn = orf[sindex]+orf[sindex+1]+orf[sindex+2]
        if cdn!="ATG":
            continue
        else:
            atg_index=sindex
            break


    if no_atg==1:
        oldindex=index
        continue

    gene300_count+=1
    motif_hit=0
    cis = result[oldindex+atg_index-window:
oldindex+atg_index]
    for cindex in range(0,window-order):
        xmer = ""
        for xmeri in range(0,order):
            xmer+=cis[cindex+xmeri]

        if xmer in motifs:
#           print "hit on motif=", xmer
            motif_hit=1
            break
```

```
              if motif_hit!=1:
#                     print "no motif hit"
                      oldindex=index
                      continue

              gene300wm_count+=1
              gene = result[oldindex+atg_index: index+2+1]
              gene_fh = open("genefile", 'a')
              slicejoin = ""
              slicejoin = slicejoin.join(gene)
              geneline = slicejoin + '\n'
              gene_fh.write(geneline)
              cis = result[oldindex+atg_index-window:
     oldindex+atg_index]
              cis_fh = open("cisfile", 'a')
              slicejoin = ""
              slicejoin = slicejoin.join(cis)
              cisline = slicejoin + '\n'
              cis_fh.write(cisline)
              if (index+2+1+window)>seqlen:
                  continue

              trans = result[index+2+1: index+2+1+window]
              trans_fh = open("transfile", 'a')
              slicejoin = ""
              slicejoin = slicejoin.join(trans)
              transline = slicejoin + '\n'
              trans_fh.write(transline)
              oldindex=index

    print "gene300_count=", gene300_count
    print "gene300wm_count=", gene300wm_count

geneFinder( EC_sequence, 0, "ATG", 300, 15,
acc_oligo_counts)
```

———————————— prog3.py addendum 3 end ————————————

At this point we have a complete genefinder, with *cis* motif discovery code and code for use of those motifs for start-of-gene validation. What is missing however are the repeated passes over the genomic data for the different frames (referred to as frame 0, 1, and 2 in what follows). As noted in the Chapter 3 discussions on this matter, there are actually three possible framings that can occur on the genomic data when you have a three-element encoding scheme. As indicated there, the other framings can be obtained by repeating the analysis just done with the first base removed (for the frame "1" case), or the first two bases removed (for the frame "2" case). Having obtained the gene predictions for the three possible framings there is a further subtlety that must be dealt with having to do with the fact that the genomic sequence information provided is for only "half" of the genome. This is because the sequence information is obtained from double stranded DNA genomes, such that we need to repeat the entire analysis for the other strand. The other strand is fully specified by the first, however, since the second strand pairs with the first strand according to the Watson–Crick base-pairing scheme, where you have A pairing with T (and vice versa) and C pairing with G. The read direction of the other strand is also in the opposite direction to the read direction of the first (an explanation of the biochemistry is in [1]). This means that the gene count is actually approximately six times the amount indicated from the frame 0 "positive" strand analysis done thus far. If we use the motif results from just the frame 0 positive strand above, to keep thing simple we can repeat the analysis directly for the other strands with some simple alterations on the processing of the DNA data for the repeated passes with different framing.

4.3   Objective Performance Evaluation: Sensitivity and Specificity

Now that we have a gene-finder, a predictive algorithm as to whether a gene has streamed into view, we want to know how well it works. A similar need for performance measure for classifiers in general will be needed (in Chapter 10 especially), so will be covered with sufficient detail to cover the three conventions (shown in Figures 4.2 and 4.3) and why the one chosen is optimal for prediction (or classification), when not wanting to use, or be concerned about, counts on true negatives (TNs).

4.4   Signal Analytics: The Time-Domain Finite State Automaton (tFSA)

The FSA shown in Figure 4.1 [1, 3] was tuned to operate such that it would rarely miss signal acquisitions (low false negatives) by allowing for large numbers of mistaken signal acquisitions (to be initiated), followed by filtering to achieve high SP (see Figure 4.4). The acquisition bias was accomplished by imposing constraints on

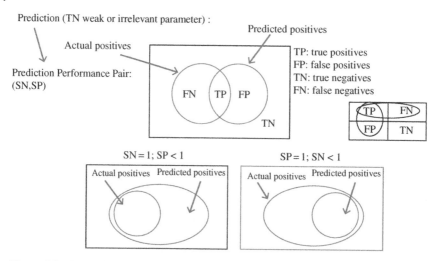

Figure 4.2 Sensitivity (SN) and Specificity (SP). For the predictor evaluator convention (with Venn diagram set correspondence) we have SN = TP/(TP + FN) and SP = TP/(TP + FP). The Venn diagram correspondence is shown with the above {SN,SP} conventions. To report a score, both {SN,SP} values must be given (according to any of the three conventions), as indicated by the extreme cases whereby either SN = 1 or SP = 1 is possible, as indicated in the lower Venn diagrams, but where the other score parameter was very poor. Often, score = (SN + SP)/2 is used.

valid starts that were weak while maintaining constraints on valid interior and ends that were strong. The bias towards high SN for *initiating* acquisition permitted tuning on FSA parameters with a simplified objective (part of the benefit of a multi-pass bootstrap tuning process). For the blockade signatures studied, the FSA parameters for maximal signal acquisition shared a broad, common range, allowing one set of FSA parameters (a single generic FSA) to acquire all signals.

The FSA described in Figure 4.1 enables acquisition of localizable channel current signals using "holistic" tuning and "emergent grammar" tuning. (Emergent grammar tuning, and use of wavelets, is described in [1, 3] and would not be discussed further here.) When attempting to tune the FSA it can be viewed as a "holistic engine" of a multiply connected set of variables and states. For acquisition we seek minimal feature identification comprising identification of signal beginnings and ends (and thus durations as well). *Holistic tuning is mainly done by testing global features for anomalous changes, or "phase transitions."* One of the main global features of the acquisition process is the number of acquisitions itself, made under a particular set of tuning parameters. In Figure 4.5 is shown the result of a holistic tuning process on the signal acquisition count for different start_drop_value parameter. A critical requirement for holistic tuning is having a viable initial

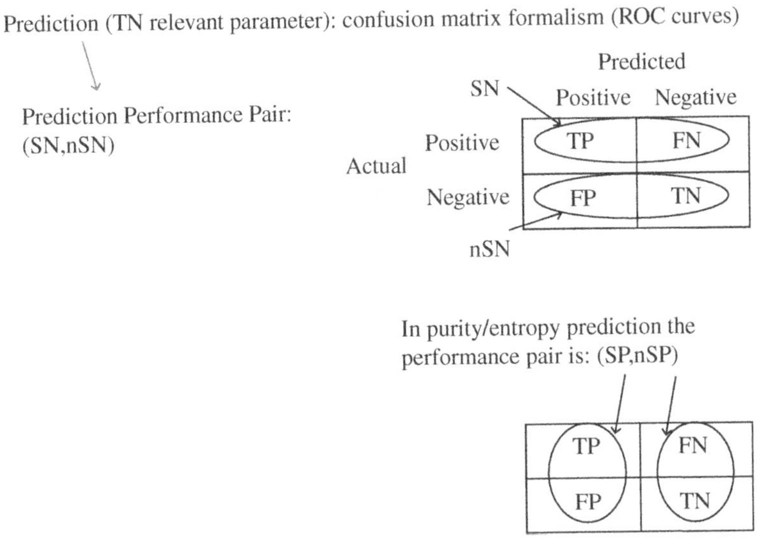

Prediction (TN relevant parameter): confusion matrix formalism (ROC curves)

Predicted

Prediction Performance Pair:
(SN,nSN)

SN — Positive Negative

Actual

	Positive	Negative
Positive	TP	FN
Negative	FP	TN

nSN

In purity/entropy prediction the
performance pair is: (SP,nSP)

TP	FN
FP	TN

Figure 4.3 **Sensitivity (SN) and Specificity (SP) for other two conventions** (used in EE with ROC curves, and purity/entropy in clustering analysis), we have nSN = TN/(TN + FP) and nSP = TN/(TN + FN). The alternative predictive performance pairs, {SN,nSN} and {SP,nSP} are shown diagrammatically in terms of the EE "confusion matrix."

tuning state to initialize the process, e.g. multiple parameters must be within their live "lock range" on tuning parameters analogous to the phase-locked loop (PLL) lock-range constraint [1, 3]. The code description with the core tuning parameters highlighted is given in Section 4.4.2.

The $O(L)$ time-complexity feature identification "scan" process can be employed for simultaneous feature extraction on various statistical moments, as mentioned previously. Identification of sharply localizable "spike" behavior can also be done in the scan process (still with only $O(L)$ time complexity) based on a nonparametric method that is described next.

4.4.1 tFSA Spike Detector

A channel current spike detector algorithm can be used to characterize the brief, very strong, blockade "spike" behavior observed for duplex DNA molecular termini that occasionally fray in the region exposed to the limiting aperture's strong electrophoretic force region. (See [2] for details, where nine base-pair hairpins were studied, the spike events were attributed to a fray/extension event on the terminal base-pair.) A complication with the spike feature extraction is the blockade level from which the spike event occurs is not known, or too variable to use to identify the spike blockade event. To have a robust feature extraction a test-level-crossing heuristic was used,

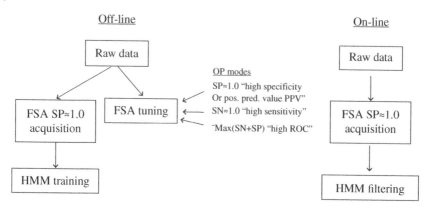

Figure 4.4   FSA with alternating SP:SN optimized tuning. Step 1: Acquire signals with high specificity (SP = 1, SN = whatever), obtain a "gold standard" reference set (if you have an expert, have them provide as much of this as they can manually). Step 2: Extract feature information from the gold standard set, to know what "it" looks like. (HMMs will often be used for this in what follows.) Step 3: Do acquisition with high sensitivity, followed by a specificity filter learned at Step 2. In other words, have (SN = 1, SP = whatever)→(filter boosts to SP = 1 with minimal drop in SN).

where for a fixed blockade level the number of signal crossings at that level are counted (such as from spikes). The test level used in the crossing analysis is then shifted to higher levels, with increasing crossing counts as the level passes thru the signal region. What results is linear increase in crossing count for actual spike features as the test level used in the crossing analysis is increased, until the main signal region is reached. In the case of the channel current analysis the various levels of blockade seen for a particular molecular blockade typically have Gaussian noise about the average of each level. Thus, as the line-crossing sweeps thru the signal blockade level and probes the tail of the Gaussian noise distribution about that signal blockade level, an exponential increase in level crossings is seen (see Figure 4.6). Focusing on the linearly increasing count region, and extrapolating to the counts up to the average of the signal blockade level from which the spike deflections are seen, a count on spike events (or a frequency on spike events) can then be robustly ascertained.

The spike detector software is designed to count "anomalous" spikes, i.e. spike noise not attributable to the Gaussian fluctuations about the mean of the dominant blockade-level. The extrapolations provide an estimate of "true" anomalous spike counts. Together, the formulation of hidden Markov model (HMM)-expectation/maximization (EM), FSAs and spike detector provide a robust method for analysis of channel current data [1–3]. In Figure 4.6 the plot is automatically generated for spike characteristics for blockade data for DNA hairpins examined: one with cross-linking radiation damage and one without damage. The plots are also automatically fit with extrapolations of their linear

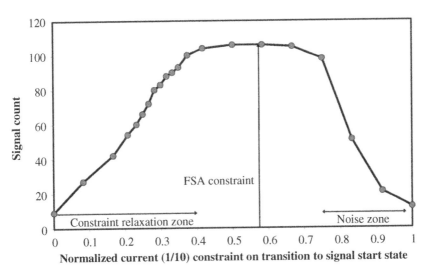

Figure 4.5 Tuning on "start_drop_value for a collection of blockade signals resulting from channel captures of DNA hairpins with 6 base-pair stem length. For baseline-normalized current constrained to drop to 0 channel current to trigger possible acquisition we see that very few acquisitions succeed (approximately 10 signal acquisitions shown). As we relax this start of acquisition constraint on possible signal acquisitions, we steadily see more signal counts until it plateaus starting at a baseline-normalized current of 0.4 to a baseline-normalized current of 0.7. The paradoxical seeming drop in signal acquisitions for the more hair-trigger acquisitions for baseline-normalized current drop, to only 0.8 or greater, is due to the FSA often triggering on noise, and eventually rejecting the indicated signal as invalid, but in doing so sometimes missing a valid signal start, resulting in fewer overall signal acquisitions. The holistic tuning process seeks the plateau region (that is not directly responsive to change in cutoff over a broad range) as an indication of a robust acquisition setting, with the 0.57 value chosen in the example shown.

phases. By this method, the non-radiated DNA exhibited a full-blockade "spike" from its lower-level blockade with a frequency of 3.58 spikes per second (indicating a fraying of the blunt ended terminus of the molecule at that rate). For the radiated molecule the frequency of spikes was 17.6 spikes per second, indicating a much greater fraying rate (and associated dissociation of the terminal base-pair), consistent with that molecule being weakened by radiation such that its terminal base-pair frays more frequently.

The additional "spike" frequency feature is found to improve classification accuracy between two species of DNA hairpins by approximately 5% in the hairpin discrimination support vector machine (SVM) tuning that is scored for various kernel parameters in Figure 4.7. This is an example of how non band limited signal features can be extracted without the limitations of a HMM state quantization preprocessing (or Fourier transform method feature extraction from electrical

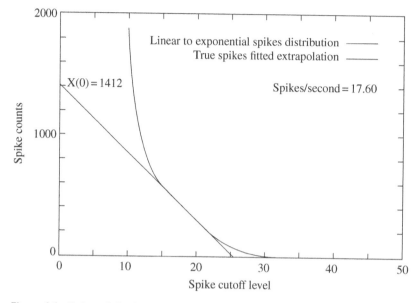

Figure 4.6 Robust Spike feature extraction: radiated DNA. A time-domain FSA is used to extract fast time-domain features, such as "spike" blockade events. Automatically generated "spike" profiles are created in this process. One such plot is shown here for a radiated nine base-pair hairpin, with a fraying rate indicated by the spike events per second (from the lower level sub-blockade). Results: the radiated molecule has more "spikes" which are associated with more frequent "fraying" of the hairpin terminus – the radiated molecules were observed with 17.6 spike events per second resident in the lower sub-level blockade.

engineering signal processing) to arrive at a more informed process than that seems possible given the usual constraint of the Gabor limit, as mentioned previously.

Once the lifetimes of the various levels are obtained, information about a variety of other kinetic properties is accessible. If the experiment is repeated over a range of temperatures, a full set of kinetic data is obtained (including the aforementioned "spike" feature frequency analysis). This data may be used to calculate k_{on} and k_{off} rates for binding events, as well as indirectly calculate forces by means of the van't Hoff Arrhenius equation (see [1–3] for details).

4.4.2 tFSA-Based Channel Signal Acquisition Methods with Stable Baseline

The tFSA program begins with State = "Reset Begin" (see Figure 4.1 and comments in Code snippet 1 below) with a loop to self, a minimum of 10 times on the sample data being scanned, where the data in [1–3] was sampled at 20 μs, thus minimum time to advance from the "Reset Begin" state is 0.2 ms in the code shown (via baseline_to_reset=10). In order to only do a reset loop 10 times, the observed

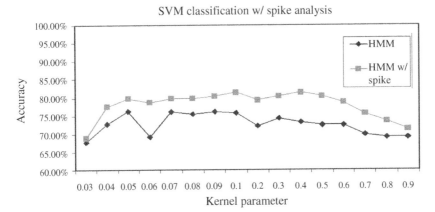

Figure 4.7 SVM classification results with and without spike analysis. Adding a spike feature significantly improves classification accuracy, by approximately 5%, over a wide range of kernel parameters.

blockade value must exceed the open_channel_avg value on each sample of the 10 observations. Until 10 such observations exceeding the open_channel_avg value are tallied, whether consecutively or not, the loop will not advance from "Reset Begin" to "Reset End". The baseline_to_reset parameter is reset to 10 after each possible signal acquisition is resolved (with acquisition or rejection). The value of 10 is itself chosen by a "holistic" tuning process.

————— Code Snippet 1: FSA signal scan loop for Fig. 1 —————

```
while (index<length):
    # data is read from (binary) datafile, or taken from
streaming (buffered)
    # data from live experiment, and placed in the variable
"rescale"

    if (baseline_to_reset>0):
        if (rescale>open_channel_avg):
            baseline_to_reset-=1
        index+=1
        continue
    # now at state = "Reset End"
    if (start_active<1 and rescale<start_drop_value and
rescale>start_drop_lmit):
        # initialize for possible signal acquisition
        signal_start[j] = index
```

```
            signal_max[j] = rescale
            signal_min[j] = rescale
            sigindex = 0
            sigdata[sigindex] =rescale
            start_active = 1
            get_base_lead = 0
            index+=1
            continue
    if (start_active<1):
        index+=1
        continue
    else:
        # now at state = "Signal Active"
        sigindex+=1
        sigdata[sigindex] = rescale
        bad_end_level=0
        for i in range(0,end_level_range):
            if (data[i]<end_level_value):
                bad_end_level=1
                i=end_level_range
        signal_end[j] = index-1-end_level_range
        signal_length = signal_end[j]-signal_start[j]+1
        if (signal_length>max_length):
            signal_end[j] += 1+end_level_range
            signal_length += 1+end_level_range
        if ((bad_end_level<1 and rescale>open_channel_avg) or
            (signal_length>max_length)):
            # exit condition obtained
            if (signal_length>min_length and signal_min[j]
<max_min_internal):
                #now at state = "Acquire Signal"
                t = sigindex - end_level_range
                do_simple_profile(sigfile,signal_start,
signal_end,
                    signal_max,signal_min,t, sigdata,j)
                sigindex=0 #resets signal info
                j++
            baseline_to_reset=10
            start_active=0 # resets
            # now reset to state = "Reset Begin"
            get_base_lead = 1
            index+=1
```

```
            continue
        #now at state = "Bad eLevel"
        elif (((index-signal_start[j]>end_level_range) and
               (data[end_level_range-1]>max_internal)) or
rescale<min_internal):
            start_active = 0 # resets
            # now at state = "Rest End". Note, not a full
reset to
            # state = "Reset Start" since baseline_to_reset
not reset to 10.
            get_base_lead = 1
            index+=1
            continue
        # Now have fall-thru to state = "Signal Active", for
another blockade
        # sample iteration, after some min and max
evaluations and sweep
        # boundary avoidance (according to datatype and
only if nonzero
        # mod_index given)
        elif (mod_index>0 and ((index%mod_index<mod_index_
range) or
                               (index%mod_index>mod_index-
mod_index_range))):
            start_active=0 # resets
            get_base_lead = 1
            index+=1
            continue
        else:
            if ((index-signal_start[j]>end_level_range) and
                (data[end_level_range-1]>signal_max[j])):
                signal_max[j]=data[end_level_range-1]
            if (rescale<signal_min[j]):
                signal_min[j] = rescale
            index+=1
            continue
        index+=1
    #closes else
#closes while
```

———— Code Snippet 1: FSA signal scan loop for Fig. 1 ————

Once at the State "Reset End", the blockade sample values are checked (shown as self-loop in Figure 4.1) to see if they have dropped significantly from the reset condition (e.g. dropped below baseline). This will be the first of a series of instances where weak conditions are used on initiating signal acquisitions, while much stricter conditions must be satisfied later (when better informed about the signal) in order to complete, and fully acquire, the signal. A blockade sample observation is deemed to have "dropped significantly" from its reset condition if it drops below a cutoff named the "start_drop_value" in Figure 4.1, to arrive at the next state, "signal active", for initiating signal acquisition. Sometimes a blockade sample drops right through the floor, however, to large negative values, etc., due to noise or a shock, etc. These falsely triggered signals are excluded by excluding start drops that go below the "start_drop_limit" value. (Again, all parameters are tuned.) Once at the State "Signal Active," each subsequent blockade sample is read into an array, for possible signal acquisition and recording, and for use by $O(1)$ data analysis algorithms (keeping the overall FSA operation $O(L)$ on L observation samples). Such algorithms are used to calculate simple statistical properties of the blockade region (in an $O(1)$ process), such as the maximum, minimum, duration, and (running) average of the blockade signal, and the (running) standard deviation of the blockade signal. The notion of a "running" statistical evaluation is that the initialization of the statistical parameter may be $O(N)$, for N length observation in the signal scan window, but that as the windowing on data used in the scan operation is slid along the data observation sequence, further updates on that sliding-window statistical parameter is only $O(1)$. This sliding-window, or "running," evaluation, then allows higher order statistical moments to be computed at $O(L)$ on the full observation sequence under study (code implementations for this will be shown in detail for the first few statistical moments in what follows).

As a preliminary step for each new sample acquisition, to minimize bad-acquisition blocking on good signal that might immediately follow, exit conditions are tested for channel blockade completion (i.e. a return to the baseline, open channel, current readings). In Figure 4.1, the exit condition is obtained either by a return to baseline (bad_end_level=0, or State="Good eLevel"), or due to acquisition truncation (case with State="Bad eLevel", due to truncation, even though good for acquisition). Once an exit condition is reached without rejection, a proper signal acquisition has occurred (with data already loaded into the acquisition array), and we now arrive at the State "Signal Complete." A collection of signal conditions are then tested on the total signal data for the final acceptance or rejection decision.

Recall from Figure 4.1 that we had onset of possible signal acquisition triggered by a significant drop in the blockade level average (evaluated $O(\text{windowsize})$ on data, so "real-time" with minimal memory buffering needs). We could also trigger

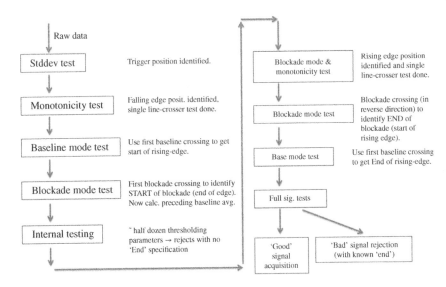

Figure 4.8 FSA acquisition flowchart.

on change of blockade level standard deviation in that same window evaluation (still just O(windowsize) evaluation). A diagram showing the latter, and related methods, for acquisition of unstable signals, is given in Figure 4.8.

4.4.3 tFSA-Based Channel Signal Acquisition Methods Without Stable Baseline

In channel current blockade analysis, and electrical signal analysis in general, the tFSA signal acquisition is much more difficult if the reference baseline is not stable. For electrical signals one type of unstable baseline that typically results is from capacitive discharge/charge giving rise to an exponential rise (or fall) in the baseline current when event sampling is necessary before system relaxation can occur (to steady baseline). If the exponential rise/fall in the channel current signal was the same this could simply be factored out, but typically the charge/discharge reset is incomplete, and the capacitive properties themselves variable under load, giving rise to (effectively) different exponential rise/fall baseline references with every device reset. Even this extreme case can be handled with a properly designed tFSA (one making use of filters analogous to boxcar filters from electrical engineering, for example). A tFSA signal acquisition for stable baseline scenarios, but still with very challenging noisy data, is described first, followed by the enhancements needed for handling unstable baseline.

Using the start_drop_value parameter introduced already one seeks to identify onset of blockade events by their deviation, typically expressed by some multiple of standard deviations, from the baseline mean (see Figure 4.8). A "three-sigma" rule is often used, i.e. event acquisition onset is triggered when a channel current reduction by more than three standard deviations of baseline noise, from the baseline mean, is observed. The falling edge of the blockade onset can then be precisely fixed (down to a specific sample observation in many cases) by performing a monotonicity test on the falling edge, which can be done with an $O(L)$ line-crosser analysis like that used in the spike analysis [1–3]. Fixing the start of blockade then depends on the data processing conventions adopted (often one chooses whatever convention yields the best classification/clustering at later SVM processing stages, if in use). Whatever the convention, the core, "stable," information that guides the blockade level identification is a modal analysis on the different blockade levels seen. What is revealed by this is a mode identifying the baseline level (at least in the region just prior to the triggered signal acquisition) and the blockade level. If there is stable baseline, or even stable capacitive discharge baseline (i.e. a single exponential profile occurring with each sampling reset), then the modal analysis will directly reveal that baseline, or identification of the asymptotic, "relaxed," exponential baseline level. Once a suspected blockade signal onset has occurred, passing the standard deviation, monotonicity, and mode tests, signal acquisition commences on further streaming signal, with ongoing "running" $O(L)$ measurement of statistical moments and max, min sample values, with rejection if the "internal" signal statistics does not fall within the desired range of possibilities. Identification of end-of-blockade, i.e. identification of the rising edge back to baseline current with no blockade, is then much the same as the falling edge identification, with use of standard deviation, monotonicity and modal tests. Typically identification of return to baseline requires additional measurement of a minimum number of baseline samples after return to baseline to avoid truncation on signal acquisitions that are sufficiently noisy that their internal blockade fluctuations occasionally "spike" back to baseline level.

Sample code for a modal scan is described in [1], where use is also made of integer-based variables for speedup on multiplications with integer variables instead of floating point variables. Working with integer-valued data is not as lossy as it might seem at first since the data encoding schemes used by third-party digital-to-analog converters (DAQs) and amplifier developers, and by their efficient binary datafile encodings (if saved to file), are themselves integer-based with shift and float multiplication operations to bring the data back to the floating value and precision originally observed, with the resolution (number of significant figures) used in the recording or quantization process. Note: the hardest part of this process is often "cracking" the binary file to learn the data encoding scheme and the shift and float-multiplication conversion values needed to recover the original data values direct from the binary data (using a "known ciphertext" attack is often the quickest way to proceed).

If the channel current blockade signals have an unstable baseline, then the window-based FSA shown in Code Snippet 2, and elaborations on it, can be critical to locking onto the signal, but may not be optimal at calling the edges, thus multiple passes may be needed (bootstrap acquisition again), with early passes involving the window-based tFSA to get a preliminary lock so that the unstable base-line moving average can be subtracted, shifting to a simpler acquisition where a sample-based tFSA can then take over for the final signal acquisition with the most precise edge recognition possible.

— Code Snippet 2: FSA signal scan for unstable baseline —

```
while (sample_index<total_sample_count):
    sample_index+=1
    sample_value = data[sample_index]
    if (not reset_start):
        if (sample_value>reset_value):
            continue
        else:
            reset_start=1 #old notation where 1 used for
true, 0 for false
            continue
    elif (not reset_end):
        if (sample_value<reset_value):
            continue
        else:
            reset_end=1
            active[start][sweep_count]=sample_index
            print "Active signal region starting at",
sample_index
            continue
    #do look-ahead for end of active signal region
    lookahead_sample_index = sample_index+10
    lookahead_sample_value = data[lookahead_sample_index]
    if (lookahead_sample_value>total_sample_count-1):
        lookahead_sample_value = signal_end_cutoff+1
    if (lookahead_sample_value>signal_end_cutoff ):
        active[end][sweep_count]=sample_index
        print "and ending at", sample_index
```

```
            reset_start=0
            reset_end=0
            sweep_count+=1
            lowpass_initialized=0
            continue
    # only here if ready to process active signal region
    # identify rising edge from baseline, this references
the current
    # lowpassdiff_value
    # call to get current lowpassdiff_value not shown,
rising_diff_cutoff and
    # falling_diff_cutoff are parameters defined earlier
and determined by
    # tuning offline

    if (lowpassdiff_value>rising_diff_cutoff):
        if (sample_index-prior_cutoff_index>20): #hard
20 to prevent multi-trigger
            prior_cutoff_index = sample_index
            # current data window access via data ref "dref"
            # the "factor" variable is for integer math,
with 20-fold
            # speedup, if the encoding scheme is known
            std_dev = Get_Std_Dev(dref,samnple_index,-30,
window_size,factor)
            if (std_dev>std_dev_cutoff):
                fail_count+=1
            else:
                edge_start=sample_index
                signal[start][j]=sample_index-5
                signal[start_baseline][j]=lowpass_baseref
                start_flag=1
                running_signal_length=0
                running_signal_max=100   #part of tuning
                running_signal_min=400   #part of tuning
                running_signal_sum=0
        else:
            prior_cutoff_index=sample_index
    # identify falling edge to baseline
    if (lowpassdiff_value<-falling_diff_cutoff):
        lowpass_baseref=Get_Lowpass(dref,sample_index,
```

```
pref,window_size,factor)
        if (lowpass_baseref<(factor*highest_count_asympt_
mode+100) ): #from tuning
            std_dev2 = Get_Std_Dev(dref,sample_index,50,
window_size,factor)
    # now perform running signal evaluation
    if (start_flag==1):
        running_signal_length = sample_index-signal[start]
[j]-5
        if (running_signal_length>0):
            rescaled_data = sample_value*factor-
signal[start_baseline][j]
                if (rescaled_data<running_signal_min):
                    running_signal_min=rescaled_data
                if (rescaled_data>running_signal_max):
                    running_signal_max=rescaled_data
                running_signal_sum+=rescaled_data
                running_signal_avg=running_signal_sum/
running_signal_length
```

- Code Snippet 2: FSA signal scan for unstable baseline end -

4.5 Signal Statistics (Fast): Mean, Variance, and Boxcar Filter

Sometimes the more sophisticated window-based tFSA methods can be avoided entirely by use of a boxcar filter (a form of lowpass notch filter) as a preprocessing stage, which is shown in Code Snippet 3. In the worst case scenario, all of these methods need to be used, with the boxcar filter used in a post-processing validation method (as well as throwing the kitchen sink at the problem) in order to get the signal acquisition to work. The process that might be undertaken on a challenging signal acquisition, thus, might go as follows:

1) scan for asymptotic baseline statistics
2) do a preliminary window-based FSA scan to get a handle on the baseline
3) estimate the baseline signal
4) perform a sample-based tFSA scan on the baseline-subtracted signal

5) perform repeated tFSA scans (since fast) with different biases to lock onto all signal regions
6) perform boxcar filter on raw signal in indicated signal regions with identified baseline attributes used to determine the optimal boxcar filter
7) perform merge on signal acquisitions indicated at steps (5) and (6)

```
───────────── Code Snippet 3: The Boxcar Filter ─────────────
def boxcar_filter(data,total_sample_count,notch_window,
buffer_window):
    # var inits not shown
    for sample_index in range(0,total_sample_count):
        sample_value=data[sample_index]
        if (sample_index < notch_window):
            pop = notch_window_array[0]
            for i in range(0,notch_window-1):
              notch_window_array[i] = notch_window_array[i+1]
            notch_window_array[notch_window-1]=sample_value
            notch_window_sum += (sample_value - pop)
            notch_window_avg = notch_window_sum/
(sample_index+1)
            lowpass_filter_data[sample_index]
=notch_window_avg
        else:
            pop = data[sample_index-notch_window]
            push = data[sample_index]
            lowpass_filter_data[sample_index]=
            lowpass_filter_data[sample_index-1]+(push-pop)/
notch_window
    for sample_index in range(0,total_sample_count):
        sample_value=data[sample_index]
        if (sample_index < notch_window+buffer_window):
            notch_filter_data[sample_index]=data
[sample_index]-
            lowpass_filter_data[sample_index+2*notch_window
+buffer_window]
        else:
            notch_filter_data[sample_index]=data
[sample_index]-
```

```
          0.5*(lowpass_filter_data[sample_index-
notch_window-buffer_window]
+lowpass_filter_data[sample_index
+2*notch_window+buffer_window])
     return notch_filter_data
```

—————— Code Snippet 3: The Boxcar Filter end ——————

4.5.1 Efficient Implementations for Statistical Tools (*O(L)*)

Working with the native integer encoded binary representation of the data is faster on multiple levels. This would not be of much benefit, however, if the subroutines for the statistical moments (mean, standard deviation, etc.) could not operate at the integer variable level for most of their evaluation. An implementation of statistical methods for evaluating the mean and standard deviation at integer-variable level is shown in Code Snippet 4. The window-based implementation is more robust with variable, unstable, baseline, but is less precise at identifying falling edges and other sharp transitions. Since the window based method often involves sums over the window, it is sometimes called an integration (or calculus-based) tFSA.

—————— Code Snippet 4: Mean and Standard Deviation ——————

```
def Get_Lowpass(data,sample_index,offset,window_size,
factor):
    window_sum=0
    for i in range(0,window_size):
        window_sum+=data[i]
    mean = factor*window_sum/window_size
    return mean

def Get_Std_Dev(data,sample_index,offset,window_size,factor):

mean=Get_Lowpass(data,sample_index,offset,window_size,
factor)
    sum_squared_central_moment=0
    factorlessmean = int(mean/factor+0.5)
    for i in range(0,window_size):
```

```
diff=data[i]-factorlessmean
sum_squared_central_moment+=diff*diff;
variance=sum_squared_central_moment/window_size
std_dev=factor*math.sqrt(variance)
return std_dev
```

———————— Code Snippet 4: Mean and Std Dev end ————————

4.6 Signal Spectrum: Nyquist Criterion, Gabor Limit, Power Spectrum

Thus far, from a signal processing perspective, the impact of compressive sensing [189], noise [190], and signal communication [191] have been briefly discussed. Search theorems [192, 193] will be discussed in Chapter 11. Terminology like "bootstrapping" has been explained as to what it means in this text and where the terminology originates [194, 195]. Let us now focus on the noise properties where spectral analysis plays a large role [190]. The fundamental tool in spectral analysis is the Fourier transform (FT), which gives the frequency decomposition of the transformed signal [196]. For noise fluctuations in an electrical signal, attention is usually focused on the FT of the signal squared. This is because the square of a voltage or current signal is proportional to the power. Depending on the incorporation of that proportionality constant (i.e. impedance value) the spectral densities obtained are known as voltage, current, or power spectral density [190]. Due to properties of FTs, convolution, such as in the definition of the autocorrelation function, transforms to multiplication. This provides a FT relationship between a signal's power spectral density and its autocorrelation function (the Weiner–Khinchine theorem).

4.6.1 Nyquist Sampling Theorem

Let $x(t)$ be a band limited signal with $X(\omega) = 0$ for $|\omega| > \omega_M$. Then $x(t)$ is uniquely determined by its samples $x(nT)$, $n = 0, \pm 1, \pm 2, \ldots$ if $\omega_S > 2\omega_M$, where $\omega_S = 2\pi/T$. The frequency $2\omega_M$ is known as the Nyquist rate and must be exceeded by the sampling frequency to satisfy the sampling theorem [196].

4.6.2 Fourier Transforms, and Other Classic Transforms

The response of a linear (i.e. superposition property) time-invariant system (time-shift in input leads to same output but with that time-shift) to a complex exponential input (a phasor) is the same phasor with a change in

amplitude: $e^{i\omega t} \rightarrow H(\omega)e^{i\omega t}$. This motivates phasor reconstruction of a periodic signal
$x(t)$, with fundamental period $T = 2\pi/\omega$, using $x(t) = \Sigma_k a_k e^{ik\omega t}$, where k summation
is over both positive and negative integers. Evaluation of the Fourier series components a_k is via: $a_k = 1/T \int x(t)e^{-ik\omega t}dt$ [196]. (Similar form for continuous time transform.) Other classic transforms include the Laplace, Mellin, Hankel, and Z-transform. There are also a variety of (non-lossy) data-compression methods that can be used as transforms insofar as feature extraction purposes.

4.6.3 Power Spectral Density

The power spectral density, $S(f)$, of a signal, $x(t)$, is a real, even, nonnegative function of frequency. Integration over $S(f)$ gives total average power per ohm: $P = \int S(f)$ $df = <x^2(t)>$ (frequency integration $-\infty$ to ∞ unless specified otherwise). The autocorrelation function, $R(\tau) = <x(t)x(t + \tau)>$, of a power signal, $x(t)$, is defined as the time average $<x(t)x(t + \tau)> = \lim_{T\to\infty} (1/2T) \int x(t)x(t + \tau)dt$. For an ergodic process, $S(f)$ and $R(\tau)$ are a FT pair (Weiner–Khinchine theorem): $R(\tau) = \int S(f) e^{i2\pi ft}$ df and $S(f) = \int R(\tau) e^{-i2\pi ft} df$ [191].

4.6.4 Power-Spectrum-Based Feature Extraction

Typical power spectra for captured nine-base-pair DNA hairpins are shown in Chapter 14, along with a spectrum for the open channel. Below 10 kHz, the current fluctuation caused by the captured DNA molecule (i.e. the blockade noise) is greater than all other noise sources. Such blockade noise typically arises from changes in transient bonds with the protein channel, DNA conformational changes in molecular structure or overall orientation vis-a-vis the channel, changes in DNA conplexation/solvation (involving waters of hydration and salt ions), and changes in internal chemical bonds (terminus fraying, for example). The power spectra for all the signals examined in [67] had approximately Lorentzian profiles, indicative of a predominately two-state switching process (seen as random telegraph noise). Discriminating between the DNA hairpins on the basis of their power spectral (or other FT properties, or wavelet properties) is possible for small sets of hairpins. For larger sets of hairpins, or for very similar hairpins like here, the HMM-based feature extraction proved critical, due to their strengths at extracting features from aperiodic (stochastic) sequential data. HMMs can be used for classification as well as feature extraction. In Chapter 14, HMMs are mainly used for feature extraction in conjunction with a SVM. The resulting signal processing and pattern recognition architecture enabled real-time single molecule classification on blockade samplings of only 100 ms [67].

4.6.5 Cross-Power Spectral Density

For ergodic processes, time and ensemble averages are interchangeable, in particular, $R(\tau) = <x(t)x(t + \tau)> = E\{x(t)x(t + \tau)\}$. If the ergodic processes for two power signals are present the net power signal (noise voltage, for example) is $z(t) = x(t) + y(t)$ and $P_z = E\{z^2(t)\} = P_x + P_y + 2P_{xy}$, where $P_{xy} = E\{x(t)y(t)\}$. The latter quantity is the $\tau = 0$ case of the cross-correlation function: $R_{xy}(\tau) = E\{x(t)y(t+\tau)\}$. Similarly, cross-power spectral density, $S_{xy}(f)$, is the FT of $R_{xy}(\tau)$ [191].

4.6.6 AM/FM/PM Communications Protocol

Amplitude modulation involves addition of a DC bias to a message signal and using this as an amplitude modulation factor on some carrier frequency [191]: $x(t) = [A + m(t)] \cos (\omega t)$, this is often rewritten as:

$$x(t) = A[1 + am_N(t)] \cos (\omega t), \text{where } m_N(t) = m(t)/ \mid \min m(t), \mid$$
$$\text{and } a = \mid \min m(t) \mid /A.$$

The parameter "a" is the modulation index and envelope detection can only be used if $a < 1$. The total power in the AM modulated signal is proportional to: $< x^2(t) > = <[A+m(t)]^2 \cos^2(\omega t)>$. If $m(t)$ is more slowly varying than $\cos (\omega t)$, then the latter time integration can be performed to yield a factor of ½: $<x^2(t) > = [A^2 + 2A<m(t)> + <m^2 (t)>]/2$, which typically reduces with $<m(t)> = 0$ to: $<x^2(t) > = [A^2 + <m^2 (t)>]/2$. Note: the AM signal power with maximum information content is for square-wave signal max $m(t) = 1$ and min $m(t) = -1$, efficiency is 50%. For sinusoidal, efficiency is 33%.

AM does not require a coherent reference for demodulation, this leads to AM radios that are simple and inexpensive. Similarly, this is one point for a branching in the communications theory to other instances where there is not necessarily a coherent reference, such as with the stochastic carrier wave methods.

4.7 Exercises

4.1 Objective is to identify gene regions via the long open reading frame (long ORF) anomaly. It is found that there is an anomalous spike in ATG codon frequency just inside the left ORF boundary – we use the first (in-frame) ATG seen from the left as our purported gene "start." The gene "end" is the stop codon at the right boundary of the ORF. The first-ATG heuristic, thereby defined, turns out to be true 90% or more of the time! (for prokaryotes). We now want to boost the performance by looking for motifs to the left ("upstream") of the purported "start," at the first ATG of the ORF (from the left). The idea from biology is that there are regulatory molecules that would bind upstream, thereby having a binding-site "footprint" in the statistics that we could possibly pick up. So we look for anomalous hexamer motifs upstream from the ATG for the region extending 30 bases upstream (to the left). We then use the presence of these anomalous motifs to validate a purported start. So consider what happens if initially we are 90% accurate, that means that we are 10% wrong, thus 10% of the count data on hexamers (there are 4096) would be mostly random nonsense. By the nature of the counting on individual hexamer frequencies, however, the anomolous counts coming from the other 90% of the data clearly wins out for still identifying anything anomalous. In fact, the "noise" could be the significantly greater and this method still work. As is, those anomolous motifs (counts significantly higher than would occur if random) can be used to validate. So, here is the assignment:

Run `longORFfirstATG_geneFinder`, to identify "cis" region 30 bases upstream of ATG start

Run `line_oligo_counter`, to identify anomalous hexamer motifs in cis regions

List the top 20 anomolous motifs. Does "aaggaa" show among them?

4.2 Building on (Exercise 4.1), now that you have identified motifs to use to validate a start, then proceed with modification to boost to genefinder performance that is approximately 99% correct. The modified heuristic: start = first ATG unless does not have motif validation, if validation fails, take try next ATG or GTG (moving in-frame), repeat until validation. This is what is done in genefinder, so here's the next assignment:

Run `geneFinder`, to get gene predictions (for frame 0). Repeat for other two forward frames and three reverse frames to get full gene prediction. How do the counts of genes observed (above length cutoff of 300 bases) compare on the different frame passes?

(Ex. 4.3–4.6)

The non-HMM bootstrap approach is now carried over to analysis of eukaryotic genomes (*Caenorhabditis elegans*), where we will be able to "discover" introns (and their rules), but not much more will be easy (this will motivate moving to an HMM representation in Chapters 6 and 7 to fully solve this):

In Figure 4.9 is the prokaryotic gene-structure we have "discovered" thus far, where the coding region maps to a complete protein product and usually starts with the codon (ATG):

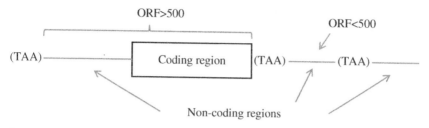

Figure 4.9 Prokaryotic gene structure discovered thus far.

Figure 4.10 Two types of "stop" codon.

Notice in Figure 4.10 how we distinguish between two types of "stop codon" delimiters in the ORF-finder approach: (i) those that delimit regions of coding (upstream) from noncoding (Center "TAA" and downstream); and (ii) those that delimit regions of noncoding from other noncoding:

In what follows we will want to leverage this ability to identify "true" stop codons in a eukaryotic setting, and will thereby attempt to rediscover eukaryotic gene structure by starting from recognized stop-codon structures.

4.3 **Scan the eukaryotic genome of C. elegans to identify the "true" stop codons on ORF's>500.** This requires a means to differentiate non-codons from coding (to know a "true stop" has coding on left and noncoding on right.

4.4 Suppose the splice signal just upstream of the true stop codon was unknown and we wanted to discover it. A reasonable guess would be that it would border the coding region at its upstream end and further upstream from there would be the ORF's leftmost (TAA) boundary. Note that such a picture (shown below) would indicate the mystery 5′ splice signal motif (shown as "AG") would necessarily be separated by a *noncoding* region from its upstream stop (Figure 4.11).

Given 3 stops and 64 codons, and random sequence in the noncoding (non-true, but certainly random when compared to coding), we'd expect to "see" a stop every 21 codons or so. Thus, if we take a window of 60 bases

Figure 4.11 Hypothesized splice signal upstream.

Figure 4.12 Hypothesized splice signal downstream of stop upstream from true stop.

downstream from the stop codon that is itself just upstream from a "true" stop (see Figure 4.12), then we'd probably capture much of the anomalous statistics from the signal we are seeking in the window (shown overlaid on picture):

Collect all of the 60-base windows, relative to the "true" stops as indicated. Then obtain a histogram of the 2-base elements ("AG" should be a significant peak) and 3-base elements ("ATG" will be a significant peak if there are a lot of "singleton" genes that have only one exon).

4.5 Suppose "AG" is the most common 2-base element in your 60-base window. There might be too many "AG"s in a 60-base window, however, in which case the question becomes which one is associated with a splice site ... in other words, you will need to validate that you have a splice-signal AG by performing a test to confirm that the region downstream is coding and upstream is noncoding.

4.6 Once the "AG" consensus is identified for the intron boundary, we can examine the bulk intron statistics (and find that it is similar to junk away from the splice-boundaries). Also, upstream from the "AG" signaling area we expect to see a series of stop codons (every 21 codons or so). Eventually, with possibly a different absolute framing, we encounter a ORF>500 bases (we crossed the intron moving upstream and now have encountered the >500 ORF because it is enclosing a coding region). We now face the reverse of the argument that motivated our AG search – now the >500 ORF is bordered on its rightmost side by a stop codon that is, on average, 21 codons from whatever coding boundary resides further upstream. So, again, we try a 60 base window (you may need to adjust, a window of 30, or a window of 120, etc.) on the region upstream from the >500 ORF's rightmost boundary. **Again, need histograms on 2-base and 3-base elements, and here we aim to rediscover the "GT" consensus signaling.**

In doing Exercises 4.3–4.6 you get to re-discover the core gene-structure elements at the eukaryotic level. If you then wanted to cobble it all together to make a gene-finder you would have an optimization task on how to bring the various components together for best performance. If this optimization was pushed to the highest level of refinement (base-level coding/noncoding identification) you would effectively arrive at an HMM, to be described in Chapter 7.

(Ex. 4.7–4.10)

The exercises that follow involve generating signal data ("Throw" section) using the `tfsa_generate.pl` Perl script below (Exercise 4.7), rewriting it as a Python script and acquiring signal data (Exercise 4.8, "Catch" script – a toy version of the tFSA shown in Figure 4.1 called `tfsa_acquire.pl` below), tuning for improved acquisition (Exercise 4.9). Also, a brief analysis of stock market data (CSCO) is done using the FSA to capture events and outcome information (good or bad, used to label as positive or negative), which can then be used to train a classifier (whose classification in the moment can be used to determine buy orders).

4.7 Throw: Rewrite `tfsa_generate.pl` as Python and generate the datafile (in "gendata2") and annot via redirection of printed output (so run as "`./tfsa.py > annot`" to get annotation file).

4.8 Catch: Rewrite `tfsa_catch.pl` as Python to catch the signals in the gendata2 file and place predicted signal regions in "predict" file (start with argument start_drop_percentage = 0.8). Rewrite `scoring.pl` as Python to score predict against annot. Indicate the accuracy score (SN + SP)/2 for the given start_drop_percentage chosen. Indicate the signal counts for the given start_drop_percentage chosen.

4.9 Repeat (Exercise 4.8) to get plots of accuracy and sig counts for different choice of start_drop_percentage – produce these tuning plots using whatever plotting method you want.

4.10 Repeat 4.7–4.9 (e.g. generate data again) for the more challenging case of overlapping noise bands – perhaps modify `tfsa_generate.pl` to where you hard code changes to the gaussians to have the same mean but wider noise bands (greater sigmas), so the new {mean,sigma} is {70,15} for the baseline level and {40,15} for the blockade level. (If this does not work well, try less noise, 10 instead of 15).

---------- Code Samples for Homework Problems -----------

Throw:

tfsa_generate.pl

```perl
use strict;
use FileHandle;
my $data_range=100;
my $output_fh2 = new FileHandle ">gendata2";
my $pi = 3.1415;

my $event_number = 100;
my $event_index=0;
for $event_index (0..$event_number) {

# no indent>>>>>>>>>>>>
my $mean = 70;
my $sigma = 3;
my $data_length = 500*rand();
for $index (0..$data_length-1) {
    # use Box-Muller transform
    my $gen_value1 = rand()+0.00000001;
    my $gen_value2 = rand()+0.00000001;
    my $Z1 = sqrt(-2*log($gen_value1)) * sin(2*$pi*
$gen_value2);
    my $Z2 = sqrt(-2*log($gen_value1)) * cos(2*$pi*
$gen_value2);
    my $X1 = $mean + $Z1 * $sigma;
    my $X2 = $mean + $Z2 * $sigma;

    my $int_gen_val = int($X1);
    print "$int_gen_val\n";
    $output_fh2->print("$int_gen_val\n");

    my $int_gen_val = int($X2);
    print "$int_gen_val\n";
    $output_fh2->print("$int_gen_val\n");
}
```

```perl
    my $mean = 40;
    my $sigma = 6;
    my $data_length = 200*rand();
    for $index (0..$data_length-1) {
        # use Box-Muller transform
        my $gen_value1 = rand()+0.00000001;
        my $gen_value2 = rand()+0.00000001;
        my $Z1 = sqrt(-2*log($gen_value1)) * sin(2*$pi*
$gen_value2);
        my $Z2 = sqrt(-2*log($gen_value1)) * cos(2*$pi*
$gen_value2);
        my $X1 = $mean + $Z1 * $sigma;
        my $X2 = $mean + $Z2 * $sigma;

        my $int_gen_val = int($X1);
        print "$int_gen_val\n";
        $output_fh2->print("$int_gen_val\n");

        my $int_gen_val = int($X2);
        print "$int_gen_val\n";
        $output_fh2->print("$int_gen_val\n");
    }
    # end no indent

}
```

Catch:

tfsa_acquire.pl

```perl
#!/usr/bin/perl

use strict;
use FileHandle;
my $input_fh = new FileHandle "gendata2";
my $output_fh = new FileHandle ">predict";

my $start_drop_percentage=$ARGV[0];


my @data;
my $data_index=-1;
while (<$input_fh>) {
```

```perl
    $data_index++;
    chop;
    $data[$data_index]=$_;
}

my $data_length = scalar(@data);
my $sig_count;
my $sig_avg;
my $start_drop_seen=0;
my $baseline;
my $index;
my $last_index=0;
for $index (0..$data_length-1) {
    my $value = $data[$index];
    my $past_avg=0;
    if ($index>4) {
        $past_avg += $data[$index-5];
        $past_avg += $data[$index-4];
        $past_avg += $data[$index-3];
        $past_avg += $data[$index-2];
        $past_avg += $data[$index-1];
    }
    $past_avg=$past_avg/5;
    if ($value<$start_drop_percentage*$past_avg && !
$start_drop_seen) {
        if ($index>$last_index+20) {
            print "possible start region at index=$index\n";

            my $start = $index+1;
            $output_fh->print("$start\t");


            $start_drop_seen=1;
            $baseline = $past_avg;
        }
        $last_index=$index;
    }

    if ($start_drop_seen) {
        $sig_avg+=$value;
    }
```

```perl
    my $fut_avg=0;
    if ($index<$data_index-6) {
        $fut_avg += $data[$index+5];
        $fut_avg += $data[$index+4];
        $fut_avg += $data[$index+3];
        $fut_avg += $data[$index+2];
        $fut_avg += $data[$index+1];
    }
    $fut_avg=$fut_avg/5;

    my $end_drop_percentage=$start_drop_percentage;
    if ($value>$end_drop_percentage*$baseline &&
$start_drop_seen
        && $fut_avg<1.1*$baseline && $fut_avg>0.9*
$baseline) {
        print "possible end region at index=$index\n";
        my $end = $index;
        $output_fh->print("$end\t");

        my $signal_duration = $index-$last_index;

print "baseline=$baseline\n";
print "index=$index\n";
print "last_index=$last_index\n";

        print "signal_duration = $signal_duration\n";
        if ($signal_duration> 10 ) {
            $sig_avg = $sig_avg/$signal_duration;
          if ($sig_avg>30 && $sig_avg<50 && $baseline>65)
{
                $sig_count++;
                $output_fh->print("good\n");
            }
        }
        else {
            $output_fh->print("bad\n");
        }

print "sig_avg=$sig_avg\n";
        $sig_avg=0;
```

```
            $start_drop_seen=0;
        }
    }
}
print "signal count = $sig_count\n";
```

Stock Scanner:
```perl
#!/usr/bin/perl

use strict;
use FileHandle;
my $input_fh = new FileHandle "CSCO.txt";
my $output_fh = new FileHandle ">CSCO_events";
my @closes;
my @opens;
my $index=-2;

while (<$input_fh>) {
    my ($date,$open,$high,$low,$close,$adjclose,$volume) =
split;
    $index++;
    if ($index<0) { next; }
    $closes[$index]=$close;
    $opens[$index]=$open;
}
my $length = scalar(@closes);
my $i = 0;
my $window=7;
my @past_opens;
my $start_drop_percentage = 0.90;
my $past_open_sum=0;
for ($i=0; $i<$length; $i++) {
    unshift @past_opens, $opens[$i];
    if (scalar(@past_opens)>$window) {
        my $pop_val = pop @past_opens;
        $past_open_sum += $opens[$i];
        $past_open_sum -= $pop_val;
    }
    else {
        $past_open_sum += $opens[$i];
    }
```

```perl
        my $past_open_avg = $past_open_sum/$window;

#     print @past_opens;
#     print "\t $past_open_avg";
#     print "\n";

    my $j = $i +1;
    if ($j<$length) {
        my $new_open = $opens[$j];
        if ($new_open < $start_drop_percentage*
$past_open_avg) {
            print "triggered at index $j\n";
            my $label;
            if ($closes[$j]>$opens[$j]) { $label = 1; }
            else {$label = -1; }
            my @data_instance = ($label,$new_open,
@past_opens);
            my $inst_length = scalar(@data_instance);
            my $k;
            for ($k=0; $k<$inst_length; $k++) {
                $output_fh->print("$data_instance[$k]");
                if ($k<$inst_length-1) {
                    $output_fh->print("\t");
                }
                else {
                    $output_fh->print("\n");
                }
            }
        }
    }
}
```

Scoring:
scoring.pl
```perl
#!/usr/bin/perl
use strict;
use FileHandle;
my $annot_fh = new FileHandle "annot";
my $predict_fh = new FileHandle "predict";
my %annot;
```

```perl
my %pred;
my $TP;
my $FP;
my $FN;
while (<$annot_fh>) {
    my ($start,$end) = split;
    $annot{$start}{$end} = 1;
}
close($annot_fh);

while (<$predict_fh>) {
    my ($start,$end) = split;
    $pred{$start}{$end} = 1;
    if ($annot{$start}{$end} == 1) { $TP++; }
    else { $FP++; }
}
my $annot_fh = new FileHandle "annot";
while (<$annot_fh>) {
    my ($start,$end) = split;
    if ($pred{$start}{$end} != 1) { $FN++; }
}
print "TP=$TP\t FP=$FP\t FN=$FN\n";
my $SN = $TP/($TP+$FN);
my $SP = $TP/($TP+$FP);
print "SN=$SN\t SP=$SP\n";
```

5

Text Analytics

A brief description of text analysis is given in this chapter, where we make use of the basic informatics tools introduced thus far. The description begins with how to get the text data of interest, including a variety of internet sources and internet website "scraping" methods. Python-based code is then used to perform text analysis in the context of words (Section 5.1), short phrases (Section 5.2), and long phrases (Section 5.3). These are very basic tools that when coupled with Google translate and other Deep Learning (Chapter 13), as well as Natural Language Processing (NLP) tools, provide an analytical basis for automated language processing.

5.1 Words

The basic unit of language, words, are reflected by clear separation into word-units in text. In this section, we will explore methods to do text analysis at the (single) word level, starting at word-frequency analysis (Section 5.1.2), followed by a "meta-analysis," using a sentiment table that scores individual words according to positive or negative "sentiment" to identify sentiment in a passage and associate it with keywords identified and present in that passage (Section 5.1.3). Before getting into the implementation, however, we need some data to analyze. In Section 5.1.1, examples are given of basic text acquisition from repositories of many classic written works (the Gutenberg project), or a local copy of rapid-searchable Wikipedia (using kiwix), or of text-scraping methods for data off of any website of interest (such as weather or stock data).

5.1.1 Text Acquisition: Text Scraping and Associative Memory

5.1.1.1 Kiwix, Gutenberg Project, Wikipedia
Kiwix (www.kiwix.com) is an offline reader that allows the storage and retrieval of anything from the web, including the entire repositories of Wikipedia and the

Informatics and Machine Learning: From Martingales to Metaheuristics, First Edition.
Stephen Winters-Hilt.
© 2022 John Wiley & Sons, Inc. Published 2022 by John Wiley & Sons, Inc.

entire Project Gutenberg Library (more than 50 000 books). The offline version can be searched faster, securely, and with much less bandwidth overall. All of Wikipedia, for example, can be downloaded at less than 100 GB. When downloading massive repositories, the fastest Kiwix download option entails using BitTorrent, with significant (10-fold to a 100-fold) increase in download speed.

5.1.1.2 Library of Babel

The Gutenberg Library at ~50 000 books is minuscule compared to the Babel Library (10^{4677} books). The latter is more a literary representation of an information compression process than an actual library. As seen in Chapter 4, various forms of signal acquisition involve compressive sensing, i.e., have an active role for compression in the signal processing (especially if transmission latency impacts performance otherwise). The description of the Library of Babel that follows is given with a little detail in hopes of conveying the oddities that can be accomplished with compression.

The Library of Babel (www.libraryofbabel.info) is an online/offline reader with everything written and everything that will ever be written (with the indicated character set and book page size indicated, e.g., 3200 characters/page). To find each page of your book, you just have to know where to look.... The idea originated in 1941 when Jorge Luis Borges wrote "The Library of Babel." Borges describes this universal library containing books with every possible combination of 410 pages of letters with the theory that every book written and that could be written would be buried in the enormous collection of letters. In his dream of such a library, he explains, "Everything will be in its blind volumes... the unwritten chapters of Edwin Drood, those same chapters translated into the language of the Garamantes..." (Borges).

The current implementation of The Library of Babel includes all possible pages of 3200 characters. This accounts for about 10^{4677} books (Library of Babel). By including every possible permutation of lower-case letters, spaces, commas, and periods any text can be found within the pool of letters and punctuation. For the next version, the inclusion of numbers would be even better. Consider this paragraph thus far, for example, I was hoping the Library of Babel page matching this initial paragraph would give me ideas how to finish with an example, but none was forthcoming. First, I had to search on the section of the paragraph after the numerical characters and parentheses (matching for all lowercase and starting at "By including..."), and having done that, there were "a *lot*" of matches, and all mostly gibberish.

The Library is composed of hexagonal "galleries" (allows infinite tiling and good connectivity when four of six sides are used as shelves). Each hexagonal gallery consists of "twenty bookshelves, five to each side, line four of the hexagon's six sides... each bookshelf holds thirty-two books identical in format; each book contains four hundred ten pages; each page, forty lines; each line, approximately eighty black letters" (Library of Babel). In other words:

$$\text{Page address} = (\text{hexagonal chamber})(\text{choice of four walls})$$
$$(\text{choice of five shelves per wall})$$
$$(\text{choice of 32 volumes on shelf})(\text{page in text})$$
$$\text{Line address} = (\text{Page address})(\text{choice of forty lines})$$
$$\text{Character address} = (\text{Line address})(\text{choice of eighty letters})$$

This amounts to a remarkable compression (call it "Babel compression"). The Library of Babel idea offers a means for data compression via a local-copy "Library-of-Books" Cipher, like the Book (Ottendorf) Cipher but with a publicized Cipher Library (the construction of a local-copy Library of Babel).

The Babel library idea need not be restricted to character sets on pages comprising "books," consider a Babel image archive comprising every image that has been created and could ever be created within the indicated color palette and dimensions. Consider an image archive created by permuting the 4096 colors with images of pixel grids with 416 rows and 640 columns. The universal slideshow that cycles thru every possible image would take 10^{961748} years.

5.1.1.3 Weather Scraper

Suppose you do not have the data downloaded (as with Wikipedia, a book of interest, the Canadian Parliaments Dual Language Record, etc.). The data may be changing or streaming and available on a website, such as weather data. In such an instance it is most convenient to have a simple script to access the website, a "web scraper," and select out the datafields of interest and provide them as output. The example that follows is in Perl, in the Exercises a Python version is assigned:

───────── weather_scraper.pl ─────────

```perl
#!/usr/bin/perl
use LWP::Simple;
use strict;
my @city;
$city[0]="boston%2Cma";
foreach (@city) {
    my $webpage=LWP::Simple::get(
    "http://www.wund.com/cgi-bin/findweather/getForecast?
    query=$_");
    $webpage=~/\<h1\>(\w+, )(\w+)/;
    print "\n$1 $2";
```

```perl
$webpage=~/class="hi">H <span>(\d+)/;
print "\n$1 degrees Fahrenheit";
$webpage=~/class="b">(\d+)/;
print "\nhumidity $1%";
$webpage=~/gif" alt="(\w+ )(\w+)/;
print "\ncurrent conditions: $1 $2";
$webpage=~/Visibility:<\/label><div><span>(\d+)/;
print "\nvisibility is $1 miles\n";
}
```

──────────────── weather_scraper.pl end ────────────────

5.1.1.4 Stock Scraper – New-Style with Cookies

Sometimes web scraping requires "cookie management," as is the case getting stock market data from Yahoo. The example that follows is in Perl, in the Exercises a Python version is assigned:

──────────────── stock_scraper.pl ────────────────

```perl
#!/usr/bin/perl
package SingleTrack;
use LWP::Simple;
use WWW::Curl::Easy;
use strict;

sub new {
    my ($class) = @_;
    my $self = bless {}, $class;
    return $self;
}

my $crumb = undef;
my $cookiejar = "/tmp/cookiejar";

# Get the crumb value
sub get_crumb {
    my ($self,$sym) = @_;

    my $curl = WWW::Curl::Easy->new;
    my $url = "https://finance.yahoo.com/quote/$sym/?p=$sym";

    $curl->setopt(CURLOPT_URL, $url);
    unlink($cookiejar);
```

```perl
    $curl->setopt(CURLOPT_COOKIEJAR, $cookiejar);
    $curl->setopt(CURLOPT_COOKIEFILE, $cookiejar);

    # Perform curl
    my $data;
    $curl->setopt(CURLOPT_WRITEDATA, \$data);
    $curl->perform;

    # Extract CrumbStore
    (my $crumb) = $data =~ m/{"crumb":"([a-zA-Z0-9\.]*)"}/g;
    print "crumb=$crumb\n";
    return $crumb;
}


sub get_yahoo_quotes {
    my ($self, $sym, $window) = @_;
    # Try twice before giving up
    $crumb = $self->get_crumb($sym) unless defined $crumb;
    $crumb = $self->get_crumb($sym) unless defined $crumb;

    if(!defined $crumb){
        print "Unable to get crumb for $sym\n";
        return -1;
    }

    # Build URL
    # earlier = current time in sec minus 3*$window days in
    sec
    my $earliertime = time()-$window*3*86400;
    my $url =
        "https://query1.finance.yahoo.com/v7/finance/
        download/$sym?".
            "period1=$earliertime" . "&period2=".time().
            "&interval=1d&events=history&". "crumb=$crumb";

    my $curl = WWW::Curl::Easy->new;
    $curl->setopt(CURLOPT_URL, $url);
    $curl->setopt(CURLOPT_COOKIEFILE, $cookiejar);
```

```
    # Perform curl
    my $data;
    $curl->setopt(CURLOPT_WRITEDATA, \$data);
    $curl->perform;

    return $data;
}


#usage:
# my $webpage = $self->get_yahoo_quotes($symbol,$window);
————————————— stock_scraper.pl end ——————————————
```

5.1.2 Word Frequency Analysis: Machiavelli's Polysemy on *Fortuna* and *Virtu*

Text analysis is done for Il Principe (The Prince) by Machiavelli. The analysis of Il Principe is done on the original Italian text and reveals clear use of polysemy by Machiavelli for the concepts *fortuna* and *virtu*. Furthermore, the polysemy usage for *virtu* is clearly linked to *fortuna* (e.g., it is *virtu* to recognize *fortuna* and be prepared to seize it). The importance of polysemy and the errors in translation of texts that use it, such as translations of Machiavelli's Il Principe (from ancient medieval Italian) have been noted in variety of translation contexts, including recently in Spanish [197]: "During the Middle Ages the term virtu experienced an important change in meaning, from the initial idea of 'maturity' or 'excellence', to that of 'perfection' in a moral sense. Machiavelli made the idea of virtu a key concept, but bestowing its initial polysemy on it and, therefore, disregarding its moral dimension. Translations into Spanish that did not take into account that polysemy would betray a fundamental idea in works as relevant as The Prince."

The script that follows (word_freq.pl) performs a word-frequency analysis on the text specified. The example is in Perl, and in the Exercises a Python version is assigned:

```
————————————————— word_freq.pl ———————————————
#!/usr/bin/perl
use strict;
use FileHandle;
my $data_input_fh = new FileHandle "ilprincipe.txt";
my $sequence;
my %count;
while (<$data_input_fh>) {
    s/[[:punct:]]//g;
    my @words = split;
```

```
    my $word;
    FOR1: foreach $word (@words) {
        my $size = length($word);
        if ($size<5) { next FOR1; }
        $count{$word}++;
    }
}
my @keyvals = keys %count;
my $key;
my $index=0;
my @unordered_nonzero_count_strings;
foreach $key (@keyvals) {
    if ($count{$key}>10) {
        my $count_string = "$count{$key} count on $key";
        print "$count_string\n";
        $unordered_nonzero_count_strings[$index] =
        $count_string;
        $index++;
    }
}

my @ordered_nonzero_count_strings = sort
@unordered_nonzero_count_strings;
my $i;
for $i (0..$index-1) {
    print "$ordered_nonzero_count_strings[$i]\n";
}
```
———————————— word_freq.pl end ———————————

Sorting the output from the word_freq.pl program into a table, we get the following (where words with fewer than 30 counts are dropped):

The counts in Table 5.1 reveal four large classes of words other than generic language constructs (e.g., conjunction, pronouns, prepositions, adjectives, and adverbs):

Type I: power, and large entities of power, such as The Prince, The People, The State, and power: Principe, popolo/uomini, Stato/stato, potere.

Type II: concerns opportunity, preparedness, and timing: fortuna, virtu, tempo, tempi, volta.

Table 5.1 High frequency word counts from Il principle.

COUNT	Italian	English
215	Principe	Prince
209	perché	because
144	della	of
136	questo	this
127	essere	be
104	potere	power
96	quando	when
91	sempre	always
89	hanno	have
83	altri	others
79	delle	of the
72	popolo	people
69	tutti	all
68	Stato	state
68	senza	without
65	uomini	men
61	stato	state
59	contro	against
59	quali	Which
57	questi	these
54	fortuna	luck
53	parte	part
53	tempo	time
52	difficoltà	difficulties
52	prima	before
52	principi	principles
51	tutto	all
50	dunque	therefore
49	quelle	those
49	questa	this
48	dalla	from
48	quello	one
47	anche	also

Table 5.1 (Continued)

COUNT	Italian	English
47	così	so
47	degli	from
46	aveva	he had
46	nelle	in
46	quale	which
45	quelli	those
44	coloro	they
44	necessario	necessary
43	quella	that
43	sudditi	subjects
42	essendo	being
42	soldati	soldiers
41	fosse	was
41	tanto	much
40	possono	can
40	sarebbe	would be
40	tutte	all
39	nuovo	new
39	queste	these
38	altro	other
37	fatto	done
37	tempi	times
36	Alessandro	Alexander
36	nella	in
36	potenti	powerful
36	virtù	virtue
35	erano	they were
35	esercito	army
34	Francia	France
34	azioni	actions
34	città	city
34	qualche	some

(*Continued*)

Table 5.1 (Continued)

COUNT	Italian	English
34	quanto	how much
33	altre	other
33	detto	say
33	regno	kingdom
32	mantenere	keep
31	avrebbe	would
31	principato	principality

Type III: players/agents/groups such as subjects, soldiers, powerful, city, (small) principality, army, enemies, friends: sudditi, soldati, potenti, citta, principato, esercito, nemici, amici.

Type IV: actions such as "actions" (azioni), war (guerra), force (forza), cruelty (crudelta).

Discussion surrounding type I and type III words is to delineate and understand the groups and entities present. This is a structural part of the political science discourse and makes frequent reference to the role of the (new) Prince, and in advising the new Prince (the stated objective of the text). Discussion surrounding type II words focuses on how to be successful through preparedness, awareness, initiative, and use of timing. Discussion surrounding type IV and some type III words were concerned with how to manage risk (regardless of appearances in some circumstances).

Omitting generic language and showing type (with greater than 10 count) we have the results shown in Table 5.2:

In association with word frequency we can examine word contexts to identify further meaning. In doing this we also see the proximity linkages between the anomalous words. The script modifications to accomplish this are shown next (word_freq.pl addendum 1):

Table 5.2 Keyword types (I power; II opportunity; III parties; IV actions).

Count	Italian	English	Type
215	Principe	Prince	I
104	potere	power	I
72	popolo	people	I
68	Stato	state	I
65	uomini	men	I
61	stato	state	I
54	fortuna	"luck"	II
53	tempo	time	II
43	sudditi	subjects	III
42	soldati	soldiers	III
37	tempi	times	II
36	potenti	powerful	III
36	virtù	"virtue"	II
35	esercito	army	III
31	principato	principality	III
30	guerra	war	IV
29	principati	principalities	III
26	amici	friends	III
26	nemici	enemies	III
26	volta	time	II
23	crudeltà	cruelty	IV
23	forza	power	IV
23	popoli	peoples	I
22	stati	States	III
21	Chiesa	Church	I
20	mercenarie	mercenary	III
19	imprese	companies	III
18	nemico	enemy	III
18	pericoli	dangers	IV
17	truppe	troops	III
16	potente	powerful	III
15	milizie	militias	III

(*Continued*)

Table 5.2 (Continued)

Count	Italian	English	Type
14	potenza	power	I
13	sicurezza	safety	IV
12	fazioni	factions	III
12	l'occasione	the occasion	II
12	militare	military	III
12	milizia	militia	III
11	autorità	authority	IV
11	crudele	cruel	IV
11	governo	government	I
11	istituzioni	institutions	III
11	lealtà	loyalty	IV
11	l'esercito	the army	III
11	ministro	minister	III
11	nobili	nobles	III

```
——————————————— word_freq.pl addendum 1 ———————————
my $scan_word = "virtu";
my $scan_window = 5; # number of words taken before and
after scan word
my %oldcount = %count;
my %count;
my $total_count;
my $i;
for $i (0..$word_count-1) {
    my $word = $word_array[$i];
    if ($word ne $scan_word) { next; }
    my $left_index=$i-$scan_window;
    if ($left_index<0) {$left_index=0;}
    my $right_index=$i+$scan_window;
    if ($right_index>$word_count-1) {$right_index=
$word_count-1;}
    my $j;
```

```
    FOR2: for $j ($left_index..$right_index) {
        my $wword = $word_array[$j];
        my $size = length($wword);
        if ($size<5) { next FOR2; }
        $count{$wword}++;
        $total_count++;
    }
}

my @keyvals = keys %count;
my $key;
my $index=0;
my @unordered_nonzero_count_strings;
foreach $key (@keyvals) {
    if ($count{$key}>0) {
        my $count_string = "$count{$key} count on $key";
        $unordered_nonzero_count_strings[$index] =
        $count_string;
        $index++;
    }
}

my @ordered_nonzero_count_strings = reverse sort
@unordered_nonzero_count_strings;
my $i;
for $i (0..$index-1) {
    if ($ordered_nonzero_count_strings[$i]>1) {
        print "$ordered_nonzero_count_strings[$i]\n";
    }
}
```

Consider the 13 highest frequency words (Table 5.3):

Let us examine each of these words and what other high frequency words are proximate (within five words, before or after) (Table 5.4):

To explore further, let us get a visual on the high frequency "important words," where high-count anomalous word parings will be visually apparent. We will simply map each word of the text to be simply "-" unless it is one of the high frequency words listed above. The first part of Il Principe book then compresses to:

Table 5.3 The three highest frequency words.

Count	Italian	English	Type
215	Principe	Prince	I
104	potere	power	I
72	popolo	people	I
68	Stato	state	I
65	uomini	men	I
61	stato	state	I
54	fortuna	"luck"	II
53	tempo	time	II
43	sudditi	subjects	III
42	soldati	soldiers	III
37	tempi	times	II
36	potenti	powerful	III
36	virtù	"virtue"	II

Table 5.4 The proximate high frequency words

Scan word	Nearby high frequency words (count)
Principe	Principe(223); nuovo(27)
potere	potere(106); mantenere(12)
popolo	popolo(80); Principe(6)
Stato	Stato(68); Principe(7)
uomini	uomini(67); sempre(12); eccellenti(5)
stato	stato(63); sarebbe(11); Principe(4)
fortuna	fortuna(54); virtu(11); potere(4)
tempo	tempo(53); Principe(2)
tempi	tempi(37); Principe(2)
virtu	virtu(36); fortuna(11); Principe(3)

```
_____stato__tempo_____
stato_____
fortuna_____uomini_____
_____
_____tempo_____
_____
_____
_____
_____uomini_____
_____
_____uomini_____
_____Principe_____Principe_____
_____stato_____tempo_____
_____stato_____stato_____
tempo_____Stato_____
_____stato_____
_____Principe_____Principe_____
_____uomini_____
_____tempo_____
_____
_____fortuna_____
_____fortuna__ _____ _____potere_
uomini_____Principe_____tempo_potere__Stato__Principe
___Principe_____fortuna__virtù_____
```

For the above output we end with the first of the fortuna-virtu word pairings, there are 5 with that order, and 6 with the reverse order. Translations of these passages using Google Translate, with the untranslated fortuna and virtu terms (to retain polysemy), results in an interesting distillation of one of the main messages of Machiavelli (further details left to the exercises).

5.1.3 Word Frequency Analysis: Coleridge's Hidden Polysemy on *Logos*

Word frequency analysis can catch polysemy, as with Machiavelli's Il Principe, but can also identify high-frequency word subsets indicative of the topic discussed. In this section, an attempt to define the concept of the Logos using such an analysis by analyzing the entire works of Coleridge will be done for this purpose as will be explained in what follows.

Early Stoicism describes the Logos as the active and vivifying power. Philo, A Hellenistic Jew (c. 30 BC) thought of the Logos as the aspect of the divine that creates the world and is "the bond of everything." The most famous description/ definition of the Logos is described in two lines in the Book of John 1 : 1 and 1 : 14 (where Jesus is the incarnation of the Logos). Elsewhere in the Book of John, however, all other translations of logos are for unrelated common words, such as "saying." Thus, there are roughly four sentences of information to describe the core concept of "Logos." Centuries later, Coleridge (1772–1834) was not satisfied with this and sought a more clear exposition of the Logos.

To Coleridge the Logos presented a unifying philosophy/theology [198]. In Coleridge's earliest known mention of efforts to examine the Logos [199] he describes the Logos as "the communicative intelligence in nature and man." With the latter he seems to connect with earlier Kabbalistic notions of flow of information in nature and man. Coleridge also spoke of Logos "Human and Divine," where he argued for Faith as a form of human Logos: "Faith must be a Light originating in the Logos" [200]. For the last twenty years of his life Coleridge put time on his "Opus Maximum" on the Logos, but it was never completed, only being published as a collection of writings in 2002 [200]. Aside from his philosophical refinements to the Logos just mentioned, Coleridge was not able to advance his plan for an "Opus Maximum," on the Logos as a rational basis for Christianity, much beyond stating his objectives and restating previous ideas.

Coleridge describes his discovery of the concept of the Logos and his efforts to study it as occurring only after being a prolific writer and poet for decades [200]. Coleridge then makes an interesting point about his prior career and body of work in relation to his current efforts to explore the Logos – he says that all of his prior work was about the Logos in some manner without ever knowing of the explicit unifying concept. In essence, he is claiming that there is a "hidden polysemy" in his work where the Logos concept is scattered over many words, themselves repeated in various ways..... *So let's do a word frequency analysis on the entire body of work by Coleridge and see if we can distill the concept of "Logos" down to a collection of high frequency words with strong context associations that are "the Logos".*

In Table 5.5, the word counts are listed in order with the highest count at the top. The first result that becomes apparent is the strong romantic literary style with its heavy use of the subjunctive and vagueness (without ambiguity). This is not surprising as Coleridge was the founder of the Romantic movement, along with Wordsworth.

In Table 5.6, the high-frequency words are listed, but now excluding words in the category "subjunctive+" or "romantic."

Table 5.5 High frequency up to first word that is not subjunctive+ or romantic, where subjunctive+ = subjunctive and subjunctive-related phrase linkage words; romantic = romantic style word usage, especially vagueness with subjunctive; and heart+ = heart category: heart, sense, feeling.

Count	Word	Category
11855	which	subjunctive+
4816	their	romantic
4367	would	subjunctive+
3441	other	romantic
3003	first	romantic
2937	there	romantic
2819	should	subjunctive+
2576	these	romantic
2191	without	romantic
2081	could	subjunctive+
2047	shall	subjunctive+
1984	those	romantic
1945	great	romantic
1843	heart	heart+

Categories:

- heart+ = {heart, sense, feelings, feeling} (Sense)
- power+ = {power, whole, nature, world, natural} (Nature)
- word+ = {words, thought, think, truth, reason, written, language, thoughts, poetry, knowledge, genius, understanding, number} (Word)
- light+ = {light, spirit} (emanation)
- three+ = {three, faith, character, believe, Church, Christ, moral, Christian, religion, purpose, father, Spirit} (trinity)
- human+ = {human, death} (emanation incarnate)

The indications about Coleridge's earlier work being in-line with a study of the aspects of the Logos is correct. The categories "word+" and "power+" contain all the standard Greek translations of "logos" relevant to the concept. The categories "light+" and "human+" describe emanation phenomena in line with ideas of logos as a divine emanation. The category "three+" encompasses Coleridge's efforts to affirm Trinitarian Christianity on a rational basis (later an explicit effort in his Opus Maximus [200], where he uses the Logos concept to provide a foundation for that rational basis). The category that does not fit is the first encountered (highest

Table 5.6 High frequency words given in terms of six categories: heart, power, word, light, three, human.

Count	Word	Category
1843	heart	heart+
1600	sense	heart+
1528	power	power+
1527	whole	power+
1508	nature	power+
1416	words	word+
1327	thought	word+
1298	think	word+
1283	truth	word+
1228	reason	word+
1104	light	light+
1082	spirit	light+
1008	three	three+
974	faith	three+
958	character	three+
942	written	word+
938	world	power+
936	believe	three+
896	human	human+
Count	Word	Category
877	Church	three+
873	Christ	three+
799	language	word+
797	moral	three+
719	feelings	heart+
683	death	human+
676	Christian	three+
669	feeling	heart+
660	thoughts	word+
635	religion	three+
572	poetry	word+
571	knowledge	word+

Table 5.6 (Continued)

Count	Word	Category
569	genius	word+
567	purpose	three+
567	natural	power+
560	father	three+
527	Spirit	three+
501	understanding	word+
500	number	word+

count), where Coleridge speaks of the subjective senses via the terms {heart, sense, feelings, feeling}. The "heart+" category may be merged with "human+", thereby describing both the subjective and objective perspectives of the emanation incarnate. This is consistent with his later writings on this in [200].

It would appear Coleridge's comments on his early work being entirely focused on the Logos, without his knowing it, is consistent with the word frequency analysis just shown. Furthermore, Coleridge's early works indicated the main outline of Coleridge's contributions to the Logos concept that would follow decades later (adding a few more sentences to the core description of the Logos to the four indicated at the start of this section).

5.1.4 Sentiment Analysis

Rather than simply incrementing a count by one for every word seen, we could add to a running total of "sentiment," where each word has an associated sentiment value. The sentiment value is typically designed to range between −1 and +1 for words with sentiment negative such as "bad," or positive, such as "good," as shown in Table 5.7.

Table 5.7 Sample sentiment table values.

Word	Sentiment value
bad	−0.625
badness	−0.875
fortuna	0.625
fortunate	0.75
good	0.875
goodness	0.75

Sentiment table lists nonzero scoring sentiment values in [−1,1].

Let us evaluate the overall sentiment for Machiavelli's The Prince (an English translation of Il Principe) for words >5 letters in length. The same code is then reused without the lines coding for the >5 to find the overall sentiment including all words. This is then repeated to evaluate the overall sentiment of the entire works of Shakespeare. The Perl script to do this is shown as sentiment.pl, while Python versions are left to the Exercises.

———————————————— sentiment.pl ————————————

```perl
#!/usr/bin/perl
use strict;
use FileHandle;

my $sent_file="sentiments.txt";
my $sent_fh=new FileHandle "$sent_file";
my %sentiment_score;
while (<$sent_fh>) {
    my ($word,$score) = split;
    $sentiment_score{$word}= $score;
}
print "sentiment_score{good}=$sentiment_score{good}\n";

my $text_file="Prince.txt";
my $data_input_fh=new FileHandle "$text_file";
my $sequence;
my %count;
my $total_count;
my $word_index=0;
my @word_array;
my $overall_sentiments;while (<$data_input_fh>) {
    s/{{:punct]] //g;
    my @words=split;
    my $word;
    FOR1: foreach $word (@words){
        $word_array [$word_index] = $word;
        $word_index++;
        my $size = length ($word);
        if ($size<5) {next FOR1;}
        $count{word}++;
        $total_count++;
        $overall_sentiments+=$sentiment_score{$word};
    }
}
```

```
print "total sentiments=$overall_sentiments\n";
```
———————————— sentiment.pl end ————————————

Sentiment score of The Prince

Overall sentiment score: 204.194
Sentiment score words >5: 0.875

Sentiment score of the entire works of Shakespeare

Overall sentiment score: 3480.428
Sentiment score words >5: 263.053

5.2   Phrases – Short (Three Words)

Let us consider the frequency on (consecutive) three-word groupings across the entire works of Shakespeare (Taken from the Project Gutenberg Library). The example that follows is in Perl, in the Exercises a Python version is assigned:

———————————— threeword.pl ————————————

```perl
#!/usr/bin/perl
use strict;
use FileHandle;

my $sent_file="sentiments2.txt";
my $sent_fh=new FileHandle "$sent_file";
my %sentiment_score;
while (<$sent_fh>) {
    my ($word,$score) = split;
    $sentiment_score{$word}= $score;
}
print "sentiment_score{good}=$sentiment_score{good}\n";

my $text_file="shakespeare.txt";
my $data_input_fh=new FileHandle "shakespeare.txt";
my $sequence;
my %count;
my $total_count;
my $word_index=0;
my @word_array;
my $overall_sentiments;
while (<$data_input_fh>) {
    s/{{:punct]] //g;
```

```perl
    my @words=split;
    my $word;
    FOR1: foreach $word (@words){
        $word_array [$word_index] = $word;
        $word_index++;
    }
}


my $word1;
my $word2;
my $word3;
my $word_count = scalar(@word_array);
my $index;
my %tripwrd_counts;
my $maxcount=0;
my $maxcountword;
for $index (0..$word_count-1) {
    my $word1 = $word_array [$index];
    my $word2 = $word_array [$index+1];
    my $word3 = $word_array [$index+2];
    my $trpwrd = "$word1$word2$word3";
    $tripwrd_counts{$trpwrd}++;
    if ($tripwrd_counts{$trpwrd}>$maxcount) {
        $maxcount = $tripwrd_counts {$trpwrd};
        $maxcountword = $trpwrd;
    }
}
print "maxcountword = $maxcountword\n";
print "count = $maxcount\n";

my@keyvals=keys %count;
my $key;
my $index=0;
my @unordered_nonzero_count_strings;
foreach $key(@keyvals) {
    if ($count{$key}>10) {
        my $count_string="$count{$key} count on $key";
        print "$count_string\n";
        $unordered_nonzero_count_strings[$index] =
$count_string;
        $index++;
    }
}
```

```
my @ordered_count_strings = sort @unordered_nonzero_count_
strings;
```
———————————— threeword.pl end ——————————

The top five most used three-word phrases:

1) 192 count on Iwillnot
2) 188 count on Iprayyou
3) 138 count on Idonot
4) 137 count on Iama
5) 132 count on Iamnot

Notice how the three word phrases are also all three-syllable phrases. In analysis on specific plays, not based on all of Shakespeare's works (including the Sonnets, etc.), the most common three-word phrase is "I pray you.". One of the most common two-syllable phrases (analysis not shown) is "Prithee." Shakespeare has many lines in plays that begin with the three-syllable, three-word phrase "I pray Thee," or with the similar meaning two-syllable "Prithee." This stylistic convention is one of Shakespeare's methods for "keeping the meter," i.e., for maintaining iambic pentameter.

Iambic pentameter is where you have five two-syllable steps per line of dialogue or verse, i.e., 10 syllables per line. In practice 11-syllable lines were common. Furthermore, just as important as the meter was the accent on the syllables in that meter. Here, distinctive styles emerge even more clearly between authors. Accenting rules on top of the iambic pentameter were strictly obeyed by some authors, like Pope and Shakespeare, and not at all by others like Donne (Ben Johnson is quoted as having said that Donne deserved hanging for this [201], so the artistic sensibilities regarding this matter could be quite heated). For the analysis to follow we will content ourselves with the task of identifying the iambic pentameter structure in the works of Shakespeare without further discussion of accent. Even so, this is not a simple matter as there is not a straightforward syllable look-up table for words, further efforts along these lines are discussed in Section 5.3.1.

5.2.1 Shakespearean Insult Generation – Phrase Generation

Instead of extracting three-word phrases by a frequency analysis, we could generate three-word expressions instead. Perhaps the most entertaining way to do this is to construct a Shakespeare Insult Generator. Consider the three columns of words in Table 5.8:

Consider code shown below to take the values from the three columns to construct and insult according to the template: Thou [Column 1 word] [Culumn2 word] [Column 3 word].

Table 5.8 Shakespeare insult kit (internet author anonymous).

Column 1	Column 2	Column 3
artless	base-court	apple-john
bawdy	bat-fowling	baggage
beslubbering	beef-witted	barnacle
bootless	beetle-headed	bladder
churlish	boil-brained	boar-pig
cockered	clapper-clawed	bugbear
clouted	clay-brained	bum-bailey
craven	common-kissing	canker-blossom
currish	crook-pated	clack-dish
dankish	dismal-dreaming	clotpole
dissembling	dizzy-eyed	coxcomb
droning	doghearted	codpiece
errant	dread-bolted	death-token
fawning	earth-vexing	dewberry
fobbing	elf-skinned	flap-dragon
froward	fat-kidneyed	flax-wench
frothy	fen-sucked	flirt-gill
gleeking	flap-mouthed	foot-licker
goatish	fly-bitten	fustilarian
gorbellied	folly-fallen	giglet
impertinent	fool-born	gudgeon
infectious	full-gorged	haggard
jarring	guts-griping	harpy
loggerheaded	half-faced	hedge-pig
lumpish	hasty-witted	horn-beast
mammering	hedge-born	hugger-mugger
mangled	hell-hated	joithead
mewling	idle-headed	lewdster

Table 5.8 (Continued)

Column 1	Column 2	Column 3
paunchy	ill-breeding	lout
pribbling	ill-nurtured	maggot-pie
puking	knotty-pated	malt-worm
puny	milk-livered	mammet
qualling	motley-minded	measle
rank	onion-eyed	minnow
reeky	plume-plucked	miscreant
roguish	pottle-deep	moldwarp
ruttish	pox-marked	mumble-news
saucy	reeling-ripe	nut-hook
spleeny	rough-hewn	pigeon-egg
spongy	rude-growing	pignut
surly	rump-fed	puttock
tottering	shard-borne	pumpion
unmuzzled	sheep-biting	ratsbane
vain	spur-galled	scut
venomed	swag-bellied	skainsmate
villainous	tardy-gaited	strumpet
warped	tickle-brained	varlot
wayward	toad-spotted	vassal
weedy	unchin-snouted	whey-face
yeasty	weather-bitten	wagtail

Insult is constructed by arranging "Thou [column 1 word] [column 1 word] [column 1 word]."

———————————— insult_generator.pl ————————————

```perl
#!/usr/bin/perl
use strict;
use FileHandle;

my $data_input_fh=new FileHandle "three_columns.txt";
```

```
my @column_one;
my @column_two;
my @column_three;
while (<$data_input_fh>) {
    s/{{:punct]] //g;
    my @words=split;
    my $word;
    push @column_one, @words[0];
    push @column_two, @words[1];
    push @column_three, @words[2];
}

for my $index(1..10){
    my $int_1=rand(scalar @column_one);
    my $int_2=rand(scalar @column_two);
    my $int_3=rand(scalar @column_three);
    print "Thou @column_one[$int_1] @column_two[$int_2]
@column_three[$int_3]\n";
}
```
———————————— insult_generator.pl end ————————

Some Generated phrases

1) Thou impertinent fen-sucked varlot
2) Thou roguish rump-fed vassal
3) Thou mangled pox-marked whey-face

5.3 Phrases – Long (A Line or Sentence)

5.3.1 Iambic Phrase Analysis: Shakespeare

In this section, text analysis is done for Henry VI by William Shakespeare (and Christopher Marlowe). The analysis of Henry VI involves iambic profiling for Parts 1 and 2, revealing a shift in iambic profile, possibly associated with a shift in authorship. The application of the script and production of histograms showing syllable counts on lines is left to the exercises. Other than a main peak at 10 syllables, there are strong side-peaks at 9 and 11 syllables, and the difference in the histograms based on Henry VI Part 1 and Part 2 indicate a shift in authorship as expected (since this is known to be true in this case as Marlowe officially covered for Shakespeare at this time due to the latter's illness).

──────────────── line_syllable_counter.pl ────────────────

```perl
#!/usr/bin/perl
use strict;
use FileHandle;

my $syll_input_fh = new FileHandle "syllable.txt";
my %syllables;
while (<$syll_input_fh>) {
    chomp;
    my ($word,$parts,$syllable_count) = split /\t/;
    $syllables{"$word"} = $syllable_count;
}

my $data_input_fh = new FileHandle "shakespeare.txt";
my $sequence;
my %count;
while (<$data_input_fh>) {
    chomp;
    chop;
    tr/[A-Z]/[a-z]/;
    my @words = split;
    my $syllable_count=0;
    my $word;
    FOR1: foreach $word (@words) {
        $count{$word}++;
        # have weak syllable.txt table, so now proceed to
        determine
        # number of syllables by an approximate algorithm
        if (!$syllables{"$word"}) {
            my @syllables = split /[a,e,i,o,u,y]+/, $word;
            print "@syllables\n";
            my $numsyll = scalar(@syllables);
            my $syll_count=0;
            my $syll_index;
            for $syll_index (0..$numsyll-2) {
                if ($syllables[$syll_index] ne "") {
                    $syll_count++;
                }
            }
            if ($syll_count == 0) { $syll_count = 1; }
            $syllables{"$word"} = $syll_count;
```

```
        }
        $syllable_count += $syllables{"$word"};
    }
    print "$_\t$syllable_count\n";
}
```

———————————— line_syllable_counter.pl end ——————————

5.3.2 Natural Language Processing

NLP in the broader sense includes language processing with initial signal processing (Chapters 1–4 techniques) and speech recognition (Chapters 6 and 7 HMM methods) to arrive at a "written" text, if not already text-based. The language analytics begins with the basics: syntactic and semantic analysis. Practical applications, however, typically require further analysis of context reference and discourse response. Some of this processing may be captured using Deep Learning methods for translation (to be described in Chapter 13).

5.3.3 Sentence and Story Generation: Tarot

The recurrence of two main types of classification, generative methods and discriminative methods, is a recurring theme in Machine Learning. If we can process paragraphs according to methods outlined in Section 5.3.2, there then arises the possibility to generate such paragraphs. Furthermore, a simple Google-translate (Chapter 13) on the semi-coherent paragraph from English → French → German → English, for example, would "smooth out" the linguistic kinks to generate an even more semi-coherent dialogue, albeit without long-range dialog planning (until that's addressed as well). See the Exercises in Chapter 13 for more along these lines.

Websites offering short story generation are easily found on the internet, where story elements are entered via a web-interface. Some story generators are based on Dungeons & Dragons (D&D) style setting and character generation, which underlies many game engines. Some recent story generation relates to one of the most ancient forms of story generation via the Tarot Deck (e.g., from Fortune Telling) [202]. The Tarot Deck has 22 Major Arcana Cards and 56 Minor Arcana cards, roughly equivalent to 22 archtypes and 56 forms of action in the storytelling. The Tarot story generation archtypes are consistent with archtype theories by Jung [203], and the $22 + 56 = 78$ numerology is consistent with the "language" of physics, where an objects motion is its "story" (in the areas of string theory and emanator theory [100]).

5.4 Exercises

5.1 Redo the Weather Scraper program in Python.

5.2 Redo the Stock Scraper program in Python.

5.3 Write Python code to do word frequency analysis as shown in the word_freq. pl example. Using statistical arguments and results from the code, what is "virtu" to Machiavelli in his text *Il Principe*?

5.4 Redo the hidden polysemy analysis on the Works of Coleridge. Using statistical arguments and results from the code, what is "logos" to Coleridge? (base the Coleridge analysis on his life works listed on Gutenberg Project).

5.5 Redo the sentiment analysis program in Python, apply to *Il Principe* and the Works of Shakespeare.

5.6 Do text analysis on a text of your choice. Generally this will include word frequency and sentiment analysis (get sentiment table off of internet). Sometimes punctuation usage, such as the exclamation point can be very distinctive (Tolkien). So include a punctuation frequency analysis as well. The default text, if you cannot decide on one, is to use Machiavelli's *Il Principe* in the original Italian.

5.7 Find the most common three-word grouping in the works of Shakespeare. Submit your code with your answer. Why did Shakespeare do this, was it to "stay true to the meter"?

5.8 Write Python to map words to syllables, then get histogram of syllable counts per line of Shakespeare. There should be a peak at 10 syllables per line (for iambic pentameter).

5.9 Get an "iambic pentameter profile" on Part I and Part II of Shakespeare's Henry VI. Is there a notable shift in the iambic signature of the two parts? (the second is known to have actually been written by Marlowe). By profile, get a histogram of counts on syllables in the lines of Shakespeare – it should peak at 10, with a side-peak at 11.

6

Analysis of Sequential Data Using HMMs

Generalized Hidden Markov model (HMM) methods are described for both signal feature extraction and structure identification. The generalized HMMs described also enable a new form of carrier-based communication, where the carrier is stationary but not periodic (further details in Chapter 12). HMM-with-binned-duration, and meta-HMM generalizations, shown in Chapter 7, enable practical stochastic carrier wave encoding/decoding, where the generalized HMM methods have generalized Viterbi algorithms with all of the inherent benefits of an efficient dynamic programming implementation, as well as Martingale convergence properties when used for filtering and robust feature extraction.

Numerous prior book, journal, and patent publications by the author are drawn upon extensively throughout the text [1–68]. Almost all of the journal publications are open access. These publications can typically be found online at either the author's personal website (www.meta-logos.com) or with one of the following online publishers: www.m-hikari.com or bmcbioinformatics.biomedcentral.com.

6.1 Hidden Markov Models (HMMs)

6.1.1 Background and Role in Stochastic Sequential Analysis (SSA)

HMMs have been used in speech recognition since the 1970s [128], and in bioinformatics since the 1990s [134], and have an extensive, and growing, breadth of applications in other areas (especially as more computational resources become available). Other areas of HMM application include gesture recognition [146–148], handwriting and text recognition [149–151], image processing [151–153], computer vision [154], communication [155], climatology [156], and acoustics [157, 158]. An HMM is the central method in all of these approaches because it is the simplest, most efficient, modeling approach that is obtained when you

Informatics and Machine Learning: From Martingales to Metaheuristics, First Edition.
Stephen Winters-Hilt.
© 2022 John Wiley & Sons, Inc. Published 2022 by John Wiley & Sons, Inc.

combine a Bayesian statistical foundation for Markovian stochastic sequential analysis (SSA) [103] with the efficient dynamic programming table constructions possible on a computer.

In automated gene finding there are two types of approaches, based on data intrinsic to the genome under study [129], or extrinsic to the genome (e.g. homology, and EST data). Since c. 2000, the best gene finders have been based on combined intrinsic/extrinsic statistical modeling [130–133]. The most common intrinsic statistical model is an HMM, so the question naturally arises – how to optimally incorporate extrinsic side-information into an HMM? We resolve that question in Section 7.2 by treating duration distribution information *itself* as side-information and demonstrate a process for incorporating that side-information into an HMM. We thereby bootstrap from an HMM formalism to a HMM-with-duration (HMMD) formulation (more generally, a hidden semi-Markov model or HSMM).

In many applications, the ability to incorporate the state duration into the HMM is very important because the standard, HMM-based, Viterbi and Baum–Welch algorithms are otherwise critically constrained in their modeling ability to distributions on state intervals that are geometric (see Section 7.2 for derivation of this restrictive property). This can lead to a significant decoding failure in noisy environments when the state-interval distributions are not geometric (or approximately geometric). The starkest contrast occurs for multimodal distributions and heavy-tailed distributions. The hidden Markov model with binned duration (HMMBD) algorithm, presented in Section 7.2 and [36], eliminates the HMM geometric distribution constraint, as well as the HMMD maximum duration constraint, and offers a significant reduction in computational time for all HMMD-based methods to approximate the computational time of the HMM-process alone. In adopting any model with "more parameters", such as an HMMD over an HMM, there is potentially a problem with having sufficient data to support the additional modeling. This is generally not a problem in any HMM model that requires thousands of samples of nonself transitions for sensor modeling, such as for the gene-finding that is described in what follows, since knowing the boundary positions allows the regions of self-transitions (the durations) to be extracted with similar sample number as well, which is typically sufficient for effective modeling of the duration distributions in a HMMD.

Critical improvement to overall HMM application rests not only with the aforementioned generalizations to the HMM/HMMD, but also with generalizations to the hidden state model and emission model. This is because standard HMMs are at low Markov order in transitions (first) and in emissions (zeroth), and transitions are decoupled from emissions (which can miss critical structure in the model, such as state transition probabilities that are sequence dependent). This weakness is eliminated if we generalize to the largest state-emission clique possible, fully interpolated on the data set, as is done with the generalized-clique HMM described in

Section 7.1 and in [33], where gene finding is performed on the *Caenorhabditis elegans* genome. The objective with the clique generalization is to improve the modeling of the critical signal information at the transitions between exon regions and noncoding regions, e.g. intron and junk regions. In doing this we arrive at a HMM structure identification platform that is novel, and robustly performing, in a number of ways.

The generalized clique HMM ("meta-HMM") application to gene-finding begins by enlarging the primitive hidden states associated with the individual base labels (as exon, intron, or junk) to substrings of primitive hidden states or *footprint* states. The emissions are likewise expanded to higher order in the fundamental joint probability that is the basis of the generalized-clique, or "meta-State," HMM. In [33] we show how a meta-state HMM significantly improves the strength of coding/noncoding-transition contributions to gene-structure identification when compared to similar, intrinsic-statistics-only, geometric models. We describe situations where the coding/noncoding-transition modeling can effectively "recapture" the exon and intron heavy tail distribution modeling capability as well as manage the exon-start "needle-in-the-haystack" problem. In analysis of the *C. elegans* genome, the sensitivity and specificity (SN,SP) results for both the individual-state and full-exon predictions are greatly enhanced over the standard HMM when using the generalized-clique HMM [33]. These meta-HMMBD developments provide a foundation from which to explore a core new paradigm, the holographic HMM, where generalization to multiple labels are possible at each emission. Holographic HMMs are explored theoretically and in implementations, such as alternative-splice gene finding (as described in what follows).

The improved signal resolution possible via a meta-HMMBD signal processing method will allow for reduced signal processing overhead, thereby reducing power usage. This directly impacts satellite communications where a minimal power footprint is critical, and cell phone construction, where a low-power footprint allows for smaller cell phones, or cell phones with smaller battery requirements, or cell phones with less expensive power system methodologies. For real-time signal processing, meta-HMMBD signal processing permits much more accurate signal resolution and signal de-noising than current, HMM-based, methods. This impacts real-time operational systems such as voice recognition hardware implementations, over-the-horizon radar detection systems, sonar detection systems, and receiver systems for streaming low-power digital signal broadcasts (such an enhancement could improve receiver capabilities on various high-definition radio and TV broadcasts). For batch (off-line) signal resolution, the meta-HMMBD signal processing allows for significantly improved gene-structure resolution in genomic data, and extraction of binding/conformational kinetic feature data from nanopore detector channel current data. For scientific and engineering endeavors in general, where there is any data analysis that can be related to a sequence of measurements or observations, the meta-HMMBD signal processing systems that can be implemented permit improved signal resolution and speed of signal processing.

In a "holographic" HMM we extend to a multitrack label-sequence (e.g. a multistate labeling framework at each observation instance). The simplest example of this is to have two label sequences for one observation sequence, and this is the implementation used in the alternative-splice gene-finding in the *C. elegans* genome effort described in [1, 3, 18], where it is shown that there is sufficient statistical support for a two-track label model.

In instances of 2-D and higher order dimensional data, such as 2-D images, the data can be reduced to a single-track sequence of measurements via a rastering process, as has been done with HMM methods in the past, *or* the reduction from the rastering could be to a multi-track hidden-label state to better track the local 2-D information, e.g. a 3×3 window with hidden states corresponding to the different 3×3 windows that can be seen, etc., in a self-consistent tiling (the center of the 3×3 grid could be the former, single sample, datum used in the 1-D reduced rasterization, for example). This can be extended for larger 2-D "windows,", or n-D neighborhoods (the latter reducible to a 2-D representation in an extension of the holographic hypothesis [40, 204]). (Another area impacted by the multitrack HMM method is protein folding and conformational analysis. This is because now contemporanious information can be absorbed into the multitrack HMM. This is an extensive application area in its own right.)

All of the HMM generalizations and feature extraction methods discussed in what follows can be optimized for speed with binned durations and through dynamic ("null") binning, distributed table-chunking, and GPU-usage. This allows the limiting speed constraint on the core HMMBD component in the SSA protocol (Figure 6.1) to be controlled as much as possible. The SSA protocol outlined in what follows is for the discovery, characterization, and classification of localizable, approximately-stationary, statistical signal structures in channel current data, or genomic data, or stochastic sequential data in general, and changes between such structures.

The core signal processing stage in Figure 6.1 is usually the feature extraction and feature selection stage, where central to the signal processing protocol is the HMM. The HMM methods are the central methodology/stage in the channel current cheminformatics (CCC) protocol in that the other stages can be dropped or merged with the HMM stage in many incarnations. For example, in some data analysis situations the time-domain finite state automaton (tFSA) methods could be totally eliminated in favor of the more accurate HMM-based approach to the problem, with signal states defined/explored in much the same setting, but with the optimized Viterbi path solution taken as the basis for the signal acquisition structure identification. The reason this is not typically done is that the finite state

Figure 6.1 The most common stochastic sequential analysis flow topology. The main signal processing flow is typically Input → tFSA → Meta-HMMBD → SVM → Output. Notable differences occur in channel current cheminformatics (CCC) where there is use of EVA-projection, or similar method, to achieve a quantization on states, then have Input → tFSA → HMM/EVA → meta-HMMBD-side → SVM → Output. While, in gene-finding just have: Input → meta-HMMBD-side → Output. In gene-finding, however, the HMM internal "sensors" are sometimes replaced, locally, with profile-HMMs or SVM-based profiling, so topology can differ not only in the connections between the boxes shown, but in their ability to embed in other boxes as part of an internal refinement. *Source:* Based on Winters-Hilt [1]; Winters-Hilt [3]; Roux and Winters-Hilt [44].

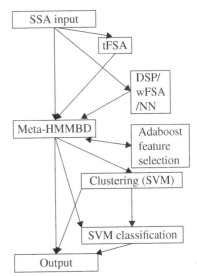

automaton (FSA) methods are usually only $O(T)$ computational expense, where "T" is the length of the stochastic sequential data that is to be examined, and "$O(T)$" denotes an order of computation that scales as "T" (linearly in the length of the sequence). The typical HMM Viterbi algorithm, on the other hand, is $O(TN^2)$, where "N" is the number of states in the HMM. So, use of the tFSA provides a faster, and often more flexible, means to acquire signal, but it is more hands-on. If the core HMM/Viterbi method can be approximated such that it can run at $O(TN)$ or even $O(T)$ in certain data regimes, for example, then the non-HMM methods can be phased out.

The HMM emission probabilities, transition probabilities, and Viterbi path sampled features, among other things, provide a rich set of data to draw from for feature extraction (to create "feature vectors"). The choice of features in the SSA Protocol is optimized along with the classification or clustering method that will make use of that feature information. In typical operation of the protocol in Chapter 12, the feature vector (f.v.) information is classified using a support vector machine (SVM) [32, 82, 83]. Once again, however, the separate classification step could be totally eliminated in favor of the HMM's log likelihood ratio classification capability, for example, when a number of template HMMs are employed (one for each signal class). This classification approach is weaker and slower than the (off-line trained) SVM methodology in many respects, but, depending on the data, there are circumstances where it may provide the best performing implementation of the protocol.

The HMM features, and other features (from neural net, wavelet, or spike profiling, etc.) can be fused and selected via use of various data fusion methods, such as a modified Adaboost selection (from [1, 3, 87, 176], and described in what follows). The HMM-based feature extraction provides a well-focused set of "eyes" on the data, no matter what its nature, according to the underpinnings of its Bayesian statistical representation. The key is that the HMM not be too limiting in its state definition, while there is the typical engineering trade-off on the choice of number of states, N, which impacts the order of computation via a quadratic factor of N in the various dynamic programming calculations used (comprising the Viterbi and Baum–Welch algorithms among others).

The HMM "sensor" capabilities can be significantly improved via switching from profile-HMM sensors to profile Markov model (pMM)/SVM-based sensors, as indicated in [44], where the superior performance and generalization capability of this approach was demonstrated. A martingale f.v. is described in this context in [32, 82, 83].

Preliminary work with HMMD binning [36], clique-generalized HMMs [33], and HMMs with side information [34], lays the foundation for an intrinsic-statistics optimized HMM and shows how to optimally incorporate extrinsic side-information (if available). This is briefly described in what follows (details in Chapter 7) and may provide a transformative platform for gene-structure identification. The methods described will also show how to collect data for the statistics to validate the strength of a multi-track state labeling statistical model. An analysis of the *C. elegans* genome's alternative-splice state-space complexity shows that a two-track alt.-splice modeling is possible. If successful with high accuracy, alternatively spliced regions could be analyzed with much greater automation, possibly leading to further breakthroughs as new tracts of genomic data are then understood. The HMM speed optimizations could have a profound effect on HMM real-time applications, as well, although such optimizations may have data-dependent complexity, and this is one of the matters that will be explored. In what follows we describe how to "upgrade" from a pMM sensor to a pMM/SVM sensor. This could significantly boost standard HMM performance, especially when used to lift the general log-likelihood ratio (LLR) terms that arise in the meta-HMM Viterbi-type algorithm into an SVM classifier.

6.1.2 When to Use a Hidden Markov Model (HMM)?

Suppose you have a sequence of observations (or measurements, or samplings, etc.) and take "$b_1 b_2 \ldots b_L$" to denote an observation sequence of length L. Introduce as Bayesian parameter a hidden label associated with each observation, denote the label sequence as "$\lambda_1 \lambda_2 \ldots \lambda_L$". The joint probability of the observations and a particular label-sequence is given via:

$$P(B; \Lambda) = P(b_1 b_2 \ldots b_L; \lambda_1 \lambda_2 \ldots \lambda_L) = P(b_1 b_2 \ldots b_L \mid \lambda_1 \lambda_2 \ldots \lambda_L) P(\lambda_1 \lambda_2 \ldots \lambda_L).$$

Markov assumptions for a standard first-order HMM then allow reduction to:

$$P(B; \Lambda; 1\text{st} - \text{order HMM}) = [P(b_1 \mid \lambda_1)...P(b_L \mid \lambda_L)] [P(\lambda_1) P(\lambda_2 \mid \lambda_1)...P(\lambda_L \mid \lambda_{L-1})]$$

If there are 50 labels and 50 observations (as in the quantized power signal analysis considered in that follows), then the full joint probability has $(50)^{2L}$ possibilities for labeled observation sequence of length L, which is unmanageable for typical sequences of interest, while, after reduction to the first-order HMM Markov assumptions on the conditional probabilities we have a set of $2 \times (50)^2$ possibilities (independent of sequence length). If nth-order Markov models (MMs) are employed, the set of possibilities grows as $(50)^{n+1}$, which can be easily enumerated for n up to 3 or 4, and still be accessed via hash-indexing for $n > 4$ (as in the hIMM approach described in [60]).

In the Viterbi algorithm we seek the λ-sequence that maximizes the joint probability above for given observation sequence. In order to consider all of the possible λ-sequences in some direct enumeration for the above N-label case, we would have $(N)^L$ possible L length λ-sequences. Here we employ the classic dynamic programming solution employed by Viterbi instead, to perform a HMM Viterbi table calculation that retains information such that $O(LN^2)$ computations are needed to consider all paths.

HMMs are, thus, an amazing tool at the nexus where Bayesian probability and MMs meet dynamic programming. To properly define/choose the HMM model in a machine learning context, however, further generalization is usually required. This is because the "bare-bones" HMM description has critical weaknesses in most applications, which are summarized below. Fortunately, these weaknesses can be addressed, and in computationally efficient ways, as will be described in what follows.

6.1.3 Hidden Markov Models (HMMs) – Standard Formulation and Terms

We define the 1st order HMM as consisting of the following:

- A hidden state alphabet, Λ, with "Prior" Probabilities $P(\lambda)$ for all $\lambda \in \Lambda$, and "Transition" Probabilities $P(\lambda_2 | \lambda_1)$ for all $\lambda_1 \lambda_2 \in \Lambda$ – where the standard transition probability is denoted $a_{kl} = P(\lambda_n = 1 | \lambda_{n-1} = k)$ for a 1st order MM on states with homogenous stationary statistics (i.e. no dependence on position "n").
- An observable alphabet, B, with "Emission" Probabilities $P(b|\lambda)$ for all $\lambda \in \Lambda$ $b \in B$ – where the standard emission probability is $e_k(b) = P(b_n = b | \lambda_n = k)$, i.e. a 0th order MM on bases with homogenous stationary statistics.

There are three classes of problems that the HMM can be used to solve [126, 128]:

1) *Evaluation* – Determine the probability of occurrence of the observed sequence.

2) *Learning (Baum–Welch)* – Determine the most likely emission and transition probabilities for a given set of observational data.
3) *Decoding (Viterbi)* – Determine the most probable sequence of states emitting the observed sequence.

Most of the examples focus on the third problem, the Viterbi decoding problem, but full gHMM solutions for both Viterbi and Baum–Welch have been derived and implemented, and are shown in the sections that follow.

The probability of a sequence of observables $B = b_0 \, b_1 ... \, b_{n-1}$ being emitted by the sequence of hidden states $\Lambda = \lambda_0 \, \lambda_1 ... \, \lambda_{n-1}$ is solved by using $P(B, \Lambda) = P(B|\Lambda) \, P(\Lambda)$ in the standard factorization, where the two terms in the factorization are described as the *observation model* and the *state model*, respectively. In the first-order HMM, the state model has the first-order Markov property and the observation model is such that the current observation, b_n, depends only on the current state, λ_n:

$$P(B \mid \Lambda) \, P(\Lambda) = P(b_0 \mid \lambda_0) \, P(b_1 \mid \lambda_1)...P(b_{n-1} \mid \lambda_{n-1}) \times P(\lambda_0) \, P(\lambda_1 \mid \lambda_0)$$
$$P(\lambda_2 \mid \lambda_0, \lambda_1)...P(\lambda_{n-1} \mid \lambda_0...\lambda_{n-2})$$

With first order Markov assumption in the state-model this becomes:

$$P(B \mid \Lambda) \, P(\Lambda) = P(b_0 \mid \lambda_0) \, P(b_1 \mid \lambda_1)...P(b_{n-1} \mid \lambda_{n-1}) \times P(\lambda_0)$$
$$P(\lambda_1 \mid \lambda_0) P(\lambda_2 \mid \lambda_1)...P(\lambda_{n-1} \mid \lambda_{n-2})$$

6.2 Graphical Models for Markov Models and Hidden Markov Models

In this section, we reexpress MMs and HMMs in terms of graphical models (Figures 6.2 and 6.3, respectively). This will then be extended to a "full-clique" graphical model in Figure 6.6 and Figure 7.1.

6.2.1 Hidden Markov Models

For (hidden) Markov state sequence $\Pi = $ "$\pi_1 \, \pi_2 \, \pi_3 ... \, \pi_L$," where the states take on values (e.g. $\pi_n = k$) in a finite alphabet, with transition probability $a_{kl} = P(\pi_i = l | \pi_{i-1} = k)$, and with an associated observation sequence $X = $ "$x_1 \, x_2 \, x_3 ... \, x_L$", and "emission"

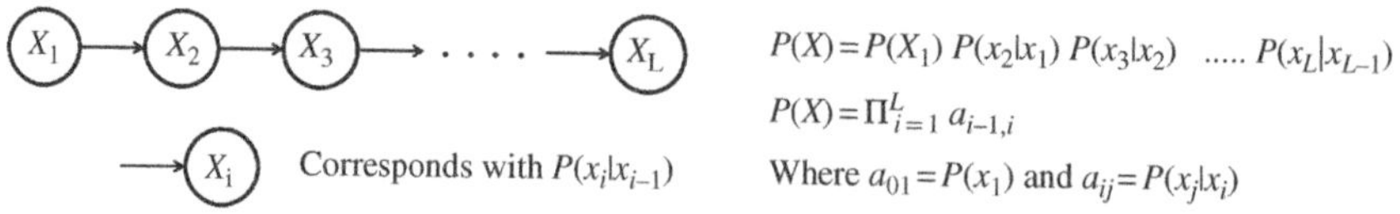

$$P(X) = P(X_1) \, P(x_2|x_1) \, P(x_3|x_2) \; ..... \; P(x_L|x_{L-1})$$
$$P(X) = \Pi_{i=1}^{L} \, a_{i-1,i}$$
Where $a_{01} = P(x_1)$ and $a_{ij} = P(x_j|x_i)$

Corresponds with $P(x_i|x_{i-1})$

Figure 6.2 Graphical model for a first-order Markov model.

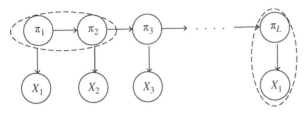

$$P(X; \Pi) = P(X|\Pi) \cdot P(\Pi) = P(X|\Pi) \cdot P(\pi_1) \, P(\pi_2|\pi_1) \, P(\pi_3|\pi_2) \, \ldots\ldots \, P(\pi_L|\pi_{L-1})$$

$$P(X; \Pi) = \prod_{i=1}^{L} P(x_i = x_i | \pi_i = \pi_i) \cdot \prod_{i=1}^{L} P(\pi_i = \pi_i | \pi_{i-1} = \pi_{i-1})$$

$$P(X; \Pi) = a_0 \prod_{i=1}^{L} e\pi_i(xi) a_{\pi_i \pi_{i+1}}$$

where $a_{0\pi_1} = P(\pi_1)$ and $a_{\pi_i \pi_{i+1}} = P(\pi_i = \pi_i | \pi_{i-1} = \pi_{i-1})$

Figure 6.3 Graphical model for a standard hidden Markov model.

probability $e_k(b) = P(x_i = b | \pi_i = k)$: (now viewed as an "emission" outcome of the hidden state, not a transition process in itself, see Figure 6.3):

6.2.2 Viterbi Path

In the Viterbi algorithm, a recursive variable is defined: $v_{kn} = v_k(n) = v_k(b_n) =$ "the most probable path ending in state $\lambda_n = k$ with observation b_n." The recursive definition of $v_k(n)$ is then: $v_l(n + 1) = e_l(b_{n+1}) \max_k [v_k(n) a_{kl}]$, where $e_l(b_{n+1})$ is the "emission" probability for the observed b_{n+1} when in state $\lambda_{n+1} = l$, and a_{kl} is the transition probability from state $\lambda_n = k$ to state $\lambda_{n+1} = l$. The optimal path information is recovered according to the (recursive) trace-back:

1) $\Lambda* = \text{argmax}_\Lambda \, P(B, \Lambda) = (\lambda_0^*, \ldots, \lambda_{L-1}^*); \; \lambda_{n | \lambda_{n+1}^* = l}^* = \text{argmax}_k [v_k(n) a_{kl}]$,
   and where
2) $\lambda_{L-1}^* = \text{argmax}_k [v_k(L-1)]$, for length L sequence.

The recursive algorithm for the most likely state path given an observed sequence (the Viterbi algorithm) is expressed in terms of v_{ki} (the probability of the most probable path that ends with observation $b_n = i$, and state $\lambda_n = k$). The recursive relation is lifted directly from the underlying probability definition: $v_{ki} = \max_n \{ e_{ki} a_{nk} v_{n(i-1)} \}$, where the $\max_n \{...\}$ operation returns the maximum value of the argument over different values of index n, and the boundary condition on the recursion is $v_{k0} = e_{k0} p_k$. The emission probabilities are the main place where the data is brought into the HMM–EM algorithm. An inversion on the emission probability is possible in this setting because the states and emissions share the same alphabet of states/quantized-emissions. The Viterbi path labelings are, thus, recursively defined by $p(\lambda_i | \lambda_{(i+1)} = n) = \text{argmax}_k \{ v_{ki} a_{kn} \}$. The evaluation of sequence probability (and its Viterbi labeling) take the emission and transition probabilities as a given. Estimates on those emission and transition probabilities themselves can be obtained by an Expectation/Maximization (EM) algorithm that is known as the Baum–Welch algorithm in this context. The 50-state generic HMM described above is used extensively in [1, 3], and will be described further in the Emission Variance Amplification (EVA) and other methods that follow.

6.2.2.1 The Most Probable State Sequence

Given an observation sequence $X =$ "$x_1\ x_2\ x_3...\ x_L$", we often want to know the most probable hidden state sequence that might be associated with it (shown in Figure 6.4):

$$\Pi* = \underset{\Pi}{\mathrm{argmax}}\ P(X;P)$$

This can be solved recursively using a dynamic programming algorithm known as the Viterbi algorithm. Let $v_k(i)$ be the probability of the most probable path ending in state "k" with i'th observation ($x_i = c$ and $k = e$, for exon, shown in Figure 6.5):

6.2.3 Forward and Backward Probabilities

The Forward and Backward probabilities occur when evaluating $p(b_0...b_{L-1})$ by breaking the sequence probability $p(b_0...b_{L-1})$ into two pieces via use of a single hidden variable treated as a Bayesian parameter: $p(b_0...b_{L-1}) = \Sigma_k\ p(b_0...b_i, \lambda_i = k)$ $p(b_{i+1}...b_{L-1}, \lambda_i = k) = \Sigma_k\ \mathbf{f_{ki}b_{ki}}$, where $\mathbf{f_{ki}} = p(b_0...b_i, \lambda_i = k)$ and $\mathbf{b_{ki}} = p(b_{i+1}...b_{L-1}, \lambda_i = k)$. Given stationarity, the state transition probabilities and the state probabilities at the ith observation satisfy the trivial relation $p_{qi} = \Sigma_k a_{kq}p_{k(i-1)}$, where $p_{qi} = p(\lambda_i = q)$, and $p_{q0} = p(\lambda_0 = q)$, and the latter probabilities are the state priors. The trivial recursion relation that is implied can be thought of as an operator equation, with operation the product by a_{kq} followed by summation (contraction) on

aacgcgtagctagttgactctcgaaacgcgtagctagttgactctcgaacgcgtagctagttgactctcgaacgcgtagctagttgactctt

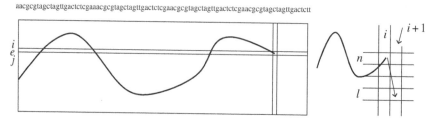

Figure 6.4 The most probable state sequence.

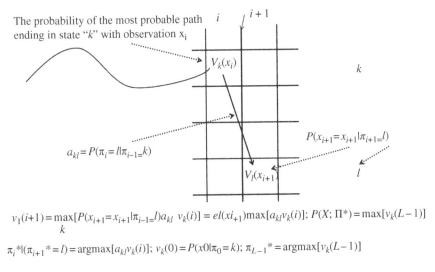

$$v_1(i+1)=\max_k[P(x_{i+1}=x_{i+1}|\pi_{i-1}=l)a_{kl}\;v_k(i)]=el(xi_{+1})\max[a_{kl}v_k(i)];\;P(X;\Pi^*)=\max[v_k(L-1)]$$

$$\pi_i^*|(\pi_{i+1}^*=l)=\operatorname{argmax}[a_{kl}v_k(i)];\;v_k(0)=P(x0|\pi_0=k);\;\pi_{L-1}^*=\operatorname{argmax}[v_k(L-1)]$$

Figure 6.5 Viterbi path. Optimal path identification.

the k index. The operator equation can be rewritten using an implied summation convention on repeated Greek-font indices (Einstein summation convention): $p_q = a_{\beta q}p_\beta$. Transition-probabilities in a similar operator role, but now taking into consideration local sequence information via the emission probabilities, are found in recursively defined expressions for the forward variables, $\mathbf{f_{ki}} = e_{ki}(a_{\beta k}f_{\beta(i-1)})$, and backward variables, $b_{ki} = a_{k\beta}e_{\beta(i+1)}b_{\beta(i+1)}$. The recursive definitions on forward and backward variables permit efficient computation of observed sequence probabilities using dynamic programming tables. It is at this critical juncture that side information must mesh well with the states (column components in the table), i.e. in a manner like the emission or transition probabilities. Length information, for example, can be incorporated via length-distribution-biased transition probabilities.

6.2.4 HMM: Maximum Likelihood discrimination

The maximum likelihood criterion is used to infer parameters, θ, for a model $\mathbf{M}$ from a data set $\mathbf{D}$ by simply taking the parameters that maximize $\mathbf{P(D|\theta,M)}$. Following the notation of [126] this is written:

$$\theta^{\mathrm{ML}} = \operatorname{argmax}_\theta \mathbf{P(D \mid \theta, M)}$$

(The terminology "likelihood" is taken to refer to the function of $\mathbf{y}$ described by $\mathbf{P(x|y)}$, while the function of $\mathbf{x}$ described by $\mathbf{P(x|y)}$ is interpreted as a probability.)

Along these same lines, the model itself could be the parameter by which to maximize. This was an actual implementation possibility: multiple HMM processes, separately trained to each individual class of current blockade signal, could be used to extend the HMMs feature extraction role to discrimination. After a few rounds of the EM filtering process, blockade probabilities could be obtained and ranked according to highest probability "template" match. The blockade signal could then be classified accordingly. As with maximum likelihood discrimination, however, there is a serious weakness with this approach given sparse data, and rejection is not as controlled as in full discriminatory frameworks like SVMs.

6.2.5 Expectation/Maximization (Baum–Welch)

EM is a general method to estimate the maximum likelihood when there is hidden or missing data. The method is guaranteed to find a maximum, but it may only be a local maximum, as is shown here (along the lines of [126]). For a statistical model with parameters θ, observed quantities $\mathbf{B}$, and hidden labels Λ, the EM goal is to maximize the log likelihood of the observed quantities with respect to θ: log $P(\mathbf{B}|\theta) = \log[\ \Sigma_\Lambda P(\mathbf{B},\Lambda|\theta)]$. At each iteration of the estimation process we would like the new log likelihood, $P(\mathbf{B}|\theta)$, to be greater than the old, $P(\mathbf{B}|\theta^*)$. The difference in log likelihoods can be written such that one part is a relative entropy, the positivity of which makes the EM algorithm work:

$$\log P(\mathbf{B} \mid \theta) - \log P(\mathbf{B} \mid \theta*) = Q(\theta \mid \theta*) - Q(\theta* \mid \theta*) + D\left[P(\Lambda \mid \mathbf{B}, \theta*)\middle\|P(\Lambda \mid \mathbf{B}, \theta)\right]$$

where $D[...\|...]$ is the Kullback–Leibler divergence, or relative entropy, and $Q(\theta|\theta*) = \Sigma_\Lambda P(\mathbf{B},\Lambda|\theta)$. Now a greater log likelihood results simply by maximizing $Q(\theta|\theta*)$ with respect to parameters θ. The EM iteration is comprised of two steps: (i) Estimation – calculate $Q(\theta|\theta*)$, and (ii) Maximization – maximize $Q(\theta|\theta*)$ w.r.t. parameters θ.

For an HMM the hidden labels Λ correspond to a path of states. Along path Λ the emission and transition parameters will be used to varying degrees. Along path Λ, denote usage counts on transition probability a_{kl} by $A_{kl}(\Lambda)$ and those on emission probabilities e_{kb} by $E_k(b,\Lambda)$ (following [126] conventions), $P(\mathbf{B},\Lambda|\theta)$ can then be written:

$$P(\mathbf{B}, \Lambda \mid \theta) = \Pi_{k=0}\Pi_b[e_{kb}]^\wedge E_k(b, \Lambda)\ \Pi_{k=0}\Pi_{l=1}[a_{kl}]^\wedge A_{kl}(\Lambda)$$

Using the above form for $P(\mathbf{B},\Lambda|\theta)$, A_{kl} for the expected value of $A_{kl}(\Lambda)$ on path Λ, and $E_k(b)$ for the expected value of $E_k(b,\Lambda)$ on path Λ, it is then possible to write $Q(\theta|\theta*)$ as:

$$Q(\theta \mid \theta*) = \Sigma_{k=1}\Sigma_b\ E_k(b)\ \log[e_{kb}] + \Sigma_{k=0}\ \Sigma_{l=1}\ A_{kl}\ \log[a_{kl}]$$

It then follows (relative entropy positivity argument again) that the maximum likelihood estimators (MLEs) for a_{kl} and e_{kb} are:

$$a_{kl} = A_{kl}/(\Sigma_l A_{kl}) \text{ and } e_{kb} = E_k(b)/(\Sigma_b E_k(b))$$

The latter estimation is for when the state sequence is known. For an HMM (with Baum–Welch algorithm) it completes the Q maximization step (M-step), which is obtained with the MLEs for a_{kl} and e_{kb}. The E-step requires that Q be calculated, for the HMM this requires that A_{kl} and $E_k(b)$ be calculated. This calculation is done using the forward/backward formalism with rescaling in the next section.

6.2.5.1 Emission and Transition Expectations with Rescaling

For an HMM, the probability that transition a_{kl} is used at position i in sequence **B** is:

$$p\big(\lambda_i = k, \lambda_{(i+1)} = l \mid X\big) = p\big(\lambda_i = k, \lambda_{(i+1)} = l, \boldsymbol{B}\big)/p(\boldsymbol{B}),$$

where

$$p\big(\lambda_i = k, \lambda_{(i+1)} = l, \boldsymbol{B}\big) = p(\boldsymbol{b}_0, ..., \boldsymbol{b}_i, \lambda_i = k)p\big(\lambda_{(i+1)} = 1 \mid \lambda_i = k\big)$$
$$p\big(\boldsymbol{b}_{i+1} \mid \lambda_{(i+1)} = l\big)p\big(\boldsymbol{b}_{i+2}, ..., \boldsymbol{b}_{L-1} \mid \lambda_{(i+1)} = l\big)$$

In terms of the previous notation with forward/backward variables:

$$p\big(\lambda_i = k, \lambda_{(i+1)} = l \mid X\big) = f_{ki}\, a_{kl}\, e_{l(i+1)}\, b_{l(i+1)}/p(\boldsymbol{B})$$

So the expected number of times a_{kl} is used, A_{kl}, simply sums over all positions i (except last with indexing):

$$A_{kl} = \Sigma_i f_{ki}\, a_{kl}\, e_{l(i+1)}\, b_{l(i+1)}/p(\boldsymbol{B})$$

Similarly, the probability that b is emitted by state k at position i in sequence **B**:

$$p(\boldsymbol{b}_i = b, \lambda_i = k \mid X) = [p(\boldsymbol{b}_0, ..., \boldsymbol{b}_i, \lambda_i = k)\, p(\boldsymbol{b}_{i+1}, ..., \boldsymbol{b}_{L-1} \mid \lambda_i = k)/p(\boldsymbol{B})]\, \delta\, (\boldsymbol{b}_i - b)$$

where a Kronecker delta function is used to enforce emission of b at position i. The expected number of times b is emitted by state k for sequence **B**:

$$E_k(b) = \Sigma_i f_{ki}\, b_{ki}/p(\boldsymbol{B})\, \delta\, (\boldsymbol{b}_i - b)$$

In practice, direct computation of the forward and backward variables can run into underflow errors. Rescaling variables at each step can control this problem. One rescaling approach is to rescale the forward variables such that $\Sigma_i F_{ki} = 1$, where F_{ki} is the rescaled forward variable, and B_{ki} is the rescaled backward variable: $F_{ki} = a_{\beta k} e_{ki} F_{\beta(i-1)}/s_i$, and $B_{ki} = a_{k\beta} e_{\beta(i+1)} B_{\beta(i+1)}/t_{i+1}$, where s_i and t_{i+1}, are the rescaling constants. The expectation on counts for the various emissions and transitions then reduce to:

$$A_{kl} = \Sigma_i F_{ki} \, a_{kl} \, e_{l(i+1)} \, B_{l(i+1)} / \left[\Sigma_k \Sigma_l F_{ki} \, a_{kl} \, e_{l(i+1)} \right] \text{ and } E_k(b) = \Sigma_i F_{ki} \, B_{ki} \, \delta(\boldsymbol{b}_i - \boldsymbol{b})$$

6.3   Standard HMM Weaknesses and their GHMM Fixes

A brief list of the typical weaknesses encountered with the standard HMM follows, with a description of the appropriate HMM generalization that eliminates those weaknesses (to be described in detail in later subsections):

1) Standard HMMs are at low Markov order in transitions (first) and in emissions (zeroth), and transitions are decoupled from emissions, which can miss critical structure in the model (e.g. state transition probabilities that are strongly sequence dependent). This weakness is eliminated if we generalize to the largest state-emission clique possible, fully interpolated on the data set, with use of a minimal state-length constraint to obtain an efficient implementation (see Figure 6.6).

    The generalized clique HMM (Figure 6.6) begins by enlarging the primitive hidden states associated with the individual base labels (such as with exon, intron, or junk in gene-structure identification) to substrings of primitive hidden states or footprint states. There is a key constraint, however, to keep the scaling of footprint states linear with footprint size: the footprint states are constrained to have self-transitions with a minimal length such that a footprint, and the overlapping "next" footprint, together can only have one primitive transition between states of different type (equivalent to constraining same-state transitions to have a minimal duration). The emissions are likewise expanded to higher order in the fundamental joint probability that is the basis of the generalized-clique, or "meta-State", HMM.

2) Standard HMMs do not properly model self-transition durations, imposing a "best-fit" geometric distribution on self-transition duration distributions instead. This weakness is eliminated if we generalize to a HMMD formalism, where direct modeling on self-transition duration distributions is incorporated. Standard HMMD methods are computationally expensive, however, when compared to Standard HMM. This weakness can be addressed, without loss of generality, via use of HMMBD representations [36].

3) The standard HMM approach lacks the means for directly incorporating side-information into the dynamic programming table based optimizations (used in the Viterbi and Baum–Welch algorithms, etc.). This is solved in [36], where HMM side-information is incorporated along with state duration information in a generalized HMMD implementation.

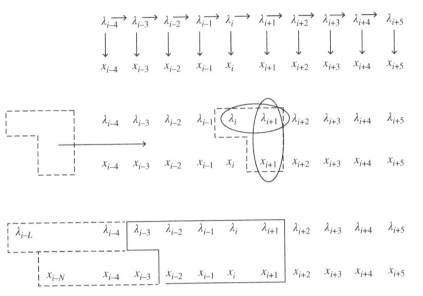

Figure 6.6 Comparison of standard HMM and the clique-generalized HMM. The upper graphical model is for the standard HMM and shows the "emission" observation sequence x_i, and the associated hidden label sequence λ_i, and the arrows denote the conditional probability approximations used in the model (for the transition and emission probabilities). Focusing at the level of the core joint-probability construct at instant "i" in the middle graph, the standard HMM is a subset of the joint probability construct $P(\lambda_i, \lambda_{i+1}, x_{i+1})$. The generalized-clique HMM is shown in the graphical model at the bottom for one particular clique generalization. The model can be exact on emission positionally, then extend via zone dependence and use of gIMM interpolation. The model can be exact to higher order in state (referred to as "footprint states", see [33]), and also extends modeling to have HMM with duration modeling. When doing the latter, zone-dependent and position dependent modeling can be incorporated via reference to the duration in the model, and can be directly incorporated into a generalized Viterbi algorithm (and other generalized HMM algorithms), as well as any other side-information of interest. *Source:* Based on Winters-Hilt [1]; Winters-Hilt [3]; Winters-Hilt et al. [34].

4) Standard HMM and HMMD methods suffer from a severe bottleneck if full table computation is used on a lengthy data sequence, where there is a need for a method for distributed processing, or "chunking," with overlaps sufficient for recovery. A method for distributed Viterbi and Baum–Welch will be described in what follows.

5) There is typically a need for a method for HMM Feature Extraction Selection, Compression, and Fusion. A modified form of Adaboost [110, 176] is used for this purpose in [55, 87].

6) The standard HMM has one "track" of hidden label information. There is a need for multitrack hidden Markov modeling in many applications but this

is not typically addressed in this direct way due to the significantly greater number of multitrack states indicated, and associated processing overhead. Multitrack hidden state constraints (and allowed transitions) are often present that can significantly limit model complexity, however, as already seen in the meta-HMM clique generalization with application in gene-finding in [33], and mentioned in item (1). If properly handled via a preliminary allowed state/transition analysis, significant multiple hidden-track model complexity can be accommodated. Chapter 7 and [1, 3, 18] preliminary statistical results are described that indicate that a two-track HMM alternative-splice gene-finder model is statistically well-supported for bootstrap learning with a wide range of eukaryotic genomes (*C. elegans* to *H. sapiens* genomes [18]).

7) There is a need for a standardized HMM method for handling power signal data, and this is accomplished by use of the EVA state projection preprocessing as described in Section 6.4, other uses for EVA are also indicated there as well, including identification of strong or weak signal stationarity.

8) There is a need for a standardized HMM usage in signal processing systems that draw upon the strengths of other Machine Learning methods. HMMs are very strong at extracting signal features from sequential data, for example, and at performing long-range structure identification along that sequential data. HMMs do not offer a scalable means to do classification when working with many classes, however, and HMMs are often a waste of computational resource) when an $O(L)$ complexity simple FSA "scan," for length L sequence, will often suffice for 95% of the data analysis (the popular BLAST [127] algorithm from Bioinformatics is an FSA/HMM hybrid algorithm for this same reason). The SSA Protocol is designed to handle this and other arrangements of signal processing methods.

In the SSA Protocol and the SCW Communications method (Chapter 13) the HMMD recognition of a signal's stationary statistics has benefits analogous to "time integration" heterodyning of a radio signal with a periodic carrier in classic electrical engineering, where longer observation time is leveraged into higher signal resolution. In order to enhance the "time integration," or longer observation, benefit in the signal recognition, one can introduce modulations (periodic, burst, or stationary stochastic) into the signal generator environment [1, 3].

In channel current state identification in a high noise background [1, 3], for example, modulations may be introduced such that some of the channel current state lifetimes have heavy-tailed, or multimodal, distributions. With these modifications, a state's signal could be recognizable in the presence of very high noise. The boost in sensitivity is mostly obtained by leveraging the SCW signal processing capabilities without further refinements to the channel monitoring device other than to, possibly, allow modulation. The SSA Protocol and SCW methods offer similar enhancement to signal processing capabilities in other devices as well. Any device generating a sequence of observations can be enhanced with use of SCW methods in a similar manner. Background on the SSA Protocol and its general-use is given in Chapter 12.

All of the HMM generalizations and feature extraction methods discussed in what follows can be optimized for speed with binned durations and thoroughly distributed table-chunking (and GPU-usage). This allows the limiting speed constraint on the core HMMBD component in the SSA protocol to be greatly reduced.

6.4 Generalized HMMs (GHMMs – "Gems"): Minor Viterbi Variants

6.4.1 The Generic HMM

An HMM that is designed to generate a specific signal type need only has a few states and transitions. In reverse, this HMM "template" can be used to detect signal with matching statistics. An HMM that is meant to generate a large family of signals, on the other hand, needs to have more states and associated transitions. The "Generic" HMM or "grayscale" HMM is an example of this in the case of the channel current analysis applications in [1, 3] and in many of the examples in this paper.

The generic or grayscale HMM used in [1, 3] is implemented with 50 states, corresponding to current blockades in 1% increments ranging from 20% residual current to 69% residual current. The HMM states, numbered 0–49, corresponded to the 50 different current blockade levels in the sequences that are processed. The state emission parameters of the HMM are initially set so that the state j, $0 < = j < = 49$ corresponding to level $L = j + 20$, can emit all possible levels, with the probability distribution over emitted levels set to a discretized Gaussian with mean L and unit variance. All transitions between states are possible, and initially are equally likely.

6.4.2 pMM/SVM

For start-of-coding recognition one can create a pMM based LLR classifier given by $\log[P_{start}/P_{non\text{-}start}] = \Sigma_i \log[P_{start}(x_i = b_i)/P_{non\text{-}start}(x_i = b_i)]$. Rather than a classification built on the sum of the independent log odds ratios, however, the sum of components could be replaced with a vectorization of components:

$$\Sigma_i \log\left[P_{start}(x_i = b_i)/P_{non-start}(x_i = b_i)\right]$$

$$\rightarrow \{..., \log\left[P_{start}(x_i = b_i)/P_{non-start}(x_i = b_i)\right], ...\}$$

These can be viewed as f.v.'s, and can be classified by use of an SVM. The SVM partially recovers linkages lost with the HMM's conditional independence approximations. For the 0th-order MM, for example, the positional probabilities are approximated as entirely independent – which is typically far from accurate. The SVM approach can recover statistical linkages between components in the f.v.'s in the SVM training process. Results along these lines are shown in [44].

There are generalizations for the MM sensor and its SVM f.v. implementation, and all are compatible with the SVM f.v. classification profiling. Markov Profiling with component-sum to component feature-vector mapping for SVM/MM profiling, thus, enhances the use of MMs, IMMs, gIMMs, hIMMs, and ghIMMs [1, 3, 60], with SVM usage via "vectorization" to SVM/MM, SVM/IMM, SVM/gIMM classification profiling.

6.4.3 EM and Feature Extraction via EVA Projection

EVA projection is used in the SSA Protocol to go from a power signal (or anything sampled from a continuum domain of possibilities) to a discrete, projected "EVA state", representation of the data. Once all states are discrete, higher order structure (or encoding) can be extracted by use of the meta-HMM generalization.

Using a standard implementation of a HMM with emissions probabilities parameterized by Gaussian distributions: emission_probabilities$[i][k] = \exp(-(k-i)*(k-i)/(2*variance))$, where "$i$" and "$k$" are each a state where $0 < = i, k < = 49$ in a 50 state system. To perform EVA, the variance is simply multiplied by a factor that essentially widens the gaussian distribution parameterized to best fit the emissions, and the equation simply becomes $\exp(-(k-i)*(k-i)/(2*variance*eva\_factor))$. For a sizable range of this parameter, HMM with EVA will remove the noise from the power signal while *strictly* maintaining the timing of the state transitions.

After EVA-projection, a simple FSA can easily extract level duration information (see Figure 6.7). Each level is identified by a simple threshold of blockade readings, typically one or two percent of baseline. When EVA boosts the variance of the distribution, for states near a dominant level in the blockade signal, the transitions are highly favored to points nearer than dominant level. This is a simple statistical effect having to do with the fact that far more points of departure are seen in the direction of the nearby dominant level than in the opposite direction. When in the local gaussian tail of sample distribution around the dominant level, the effect of transitions towards the dominant level over those away from the dominant level can be very strong. In short, a given point is much more likely to transition towards the dominant level than away from it, thereby arriving at a "focusing" on the levels, while preserving level transitions.

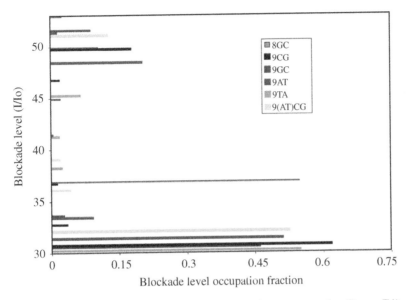

Figure 6.7 HMM/EM Viterbi-path level occupation feature extraction. Strong EVA projection is employed to project the data onto dominant levels, a Viterbi path Histogram then shows the barcode "fingerprints" of the different molecular species (the labels are for the DNA hairpins examined in [1, 3], and since then used as controls). *Source:* Based on Winters-Hilt [1]; Winters-Hilt [3].

EVA projection is used in the SSA Protocol to go from a power signal (or anything sampled from a continuum domain of possibilities) to a sparser, projected "EVA state," representation of the data. Quantization on the sparser representation can then provide a discrete representation. Once all states are discrete, higher order structure (or encoding) can be extracted by use of the meta-HMM generalization described in Section 7.1, and other methods. EVA makes use of EM, so that will be reviewed next before proceeding.

When EVA boosts the variance of the distribution, the states near a dominant level in the blockade signal are highly favored to transition to points nearer than dominant level. This is a simple statistical effect having to do with the fact that far more points of departure are seen in the direction of the nearby dominant level than in the opposite direction. When in the local Gaussian tail of sample distribution around the dominant level, the effect of transitions towards the dominant level over those away from the dominant level can be very strong. In short, a filtered datum is much more likely to transition towards the dominant level than away from it, thereby arriving at a "focusing" on the levels, while preserving level transitions.

When paired with HMMD modeling, EVA projection has additional synergy. EVA projects onto the dominant sub-levels, of which there can be many, all clearly separable after the projection. To the extent that they are not cleanly separable HMMD can greatly enhance performance (consider two sublevels that are close together, as a challenging case synthetic data is generated with such sublevels where their noise level standard deviance greatly exceed their sub-level separation (by a factor ranging from 4 to 50). In the "tight" two-level signal resolution studies in [1, 3], the performance difference is stark: the exact and adaptive HMMD decodings are 97.1% correct, while the HMM decoding is only correct 61% of the time (where random guessing would accomplish 50%, on average, in such a two-state system). Three parameterized distributions were examined in that study: geometric, Gaussian, and Poisson. Distributions that were segmented and "messy" were also examined. In all cases the HMMD performed robustly, similar to the above, and in all cases the adaptive binning HMMD optimization performed comparably to the more computationally expensive exact HMMD.

The EVA-projected/HMMD processing offers a hands-off (minimal tuning) method for extracting the mean dwell times for various blockade states (the core kinetic information on the blockading molecule's channel interactions). The results in [1, 3] clearly demonstrated the superior performance of the HMMD over the simpler standard HMM formulation on data with non-geometrically distributed same-state interval durations. In the stochastic carrier wave context, this describes a means to discern a stochastic stationary carrier with HMMD (while with HMM alone we are much weaker in this regard and cannot robustly discern carrier). With use of the EVA-projection method, this also affords a robust means to obtain kinetic-type (state duration) feature extraction. The HMM with duration enables accurate kinetic feature extraction when using EVA, thus the results in [1, 3] suggest that this problem can be elegantly solved with a pairing of the HMMD stabilization with EVA-projection.

6.4.4 Feature Extraction via Data Absorption (a.k.a. Emission Inversion)

A new form of "inverted" data injection is possible during HMM training when the states and quantized emission values share the same alphabet. This is typically the case in power signal analysis, such as the channel current cheminformatics problem described in what follows. Results from channel current signal classification consistently show approximately 5% improvement in accuracy (sensitivity + specificity) with the aforementioned data inversion upon SVM classification (and this holds true over wide ranges of SVM kernel parameters and collections of feature sets, see Figure 6.8). Transition and "absorption" statistical profiles are thought to work better than standard transition and emission profiles, in generalized classification

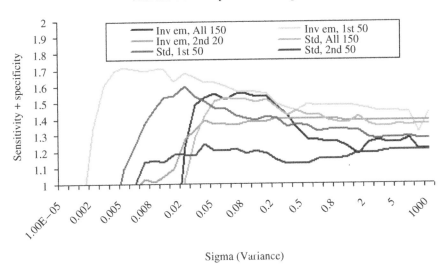

9AT vs 9CG binary classification performance

Figure 6.8 The binary classification performance using features extracted with HMM data inversion vs. HMM standard. Blockade data was extracted from channel measurements of 9AT and 9CG hairpins (both hairpins with nine base-pair stems), the data extraction involved either standard (std) emission data representations or inverted (inv) emission data, and was based on feature sets of the full 150 features, or the first 50, with the Viterbi-path level dwell-time percentages or the second 50, the emissions variances (much weaker features as expected). The inverted data offers consistently better discriminatory performance by the SVM classifier.

performance, due to regularization with an effective SRM (structural risk minimization [165]) constraint, via optimization with an added term that depends on the relative entropy between state prior probabilities and emission posterior probabilities.

By swapping $e_b(k)$ for $e_k(b)$ we introduce a multiplicative factor, the ratio of the priors on states to the frequencies on emissions: $e_k(b) = e_b(k)\,[P(b)/P(k)]$. This factor weights the computations in a manner that seems to track, and minimize, on the Kullback–Leibler divergence between the state prior distribution and the emission frequency distribution. This approximate notion follows from the evaluation of the extra terms that will occur on the maximum log-prob calculation for the Viterbi path. On the Viterbi solution, using the swapped emission probabilities, the sum (on log probabilities) at the end will differ by a sum of log ratios: log $[P(k_i)/P(b_i)] = -\log[P(b_i)/P(k_i)]$ normalized by length "L" over different k and b, this term is approximated by Diff Term $= -D(P(\mathbf{B})\|P(S))$, maximizing on this term is, thus, minimizing on the divergence, $D(P(\mathbf{B})\|P(S))$, between the priors and the emissions.

6.4.5 Modified AdaBoost for Feature Selection and Data Fusion

AdaBoost [110, 176] can take a collection of weak classifiers and boost them by forming a linear combination to have a single strong classifier. As a classification method, one of the main disadvantages of AdaBoost is that it is prone to overtraining. However, AdaBoost is a natural fit for feature selection. Here, overtraining is not a problem, as AdaBoost is only used to find diagnostic features, and those features are then passed on to a classifier that does not suffer from overtraining (such as an SVM). HMM features, and other features (from neural net, wavelet, or spike profiling, etc.), can be fused and selected via use of the Modified Adaboost selection algorithm [1, 3].

More specifically, AdaBoost learns from a collection of weak classifiers and then boosts them by a linear combination into a single strong classifier. The input to the algorithm is a training set $\{(x_1, y_1), ..., (x_N, y_N)\}$ where $y_i \in Y = \{-1, +1\}$ is the correct label of instance $x_i \in X$ and N is the number of training examples in the data set. A weak learning algorithm is repeatedly called in a series of rounds $t = 1, ..., T$ with different weight distributions D_t on the training data. This set of weights associated with the training data at each round t is denoted by $D_t(i)$. In general, sampling weights associated with each example are initially set equal, i.e. a uniform sampling distribution is assumed. For the tth iteration, a classifier is learned from the training examples and the classifier with error $\varepsilon_t \leq 0.5$ is selected. In each iteration, the weights of misclassified examples are increased which results in these examples getting more attention in subsequent iterations. AdaBoost is outlined below. It is interesting to note that α_t measures the importance assigned to the hypothesis h_t and it gets larger as the training error ε_t gets smaller. The final classification decision H of a test point x is a weighted majority vote of the weak hypotheses.

The AdaBoost algorithm

Input: $S = \langle (x_1, y_1), ..., (x_N, y_N) \rangle$ where $x_i \in X$ and $y_i \in Y = \{-1, +1\}$

Initialization: $D_1(i) = 1/N$, for all $i = 1, ..., N$
For $t = 1$ to T do
1. Train weak learners with respect to the weighted sample set $\{S, D_t\}$ and obtain hypothesis $h_t: X \rightarrow Y$.
2. Obtain the error rates ε_t of h_t over the distribution D_t such that $\varepsilon_t = P_{i \sim D_t} [h_t(x_i) \neq y_i]$.
3. Set $\alpha_t = \frac{1}{2} \ln(1 - \varepsilon_t / \varepsilon_t)$
4. Update the weights: $D_{t+1}(i) = (D_t(i)/Z_t) e^{-y_i h_t(x_i)\alpha_t}$, where Z_t is the normalizing factor such that $D_{t+1}(i)$ is a distribution.
5. Break if $\varepsilon_t = 0$ or $\varepsilon_t \geq \frac{1}{2}$.
end
Output: $H(x) = sign(\sum_{t=1}^{T} \alpha_t h_t(x_i))$

As has been shown in the tFSA spike analysis [55], careful selection of features plays a significant role in classification performance. However, adding non-characteristic or noisy features will hurt classification performance. The last set of 50 components from the standard HMM-based 150-component f.v. were chosen from compressed transition probabilities (where 50∗50 transitions are compressed to 50 features). A means of compression is necessary because many of these transitions are very unlikely and contribute noise to the f.v. (e.g. they offer weak generalization performance when passed to a classifier). Without compression, classification performance suffers, yet it is possible for diagnostic information to be discarded in such a feature compression. An automated approach is possible to solve the issue of feature selection.

6.4.5.1 The Modified Adaboost Algorithm for Feature Selection

In Modified AdaBoost [55] weights are given to the weak learners as well as the training data. The key modifications here are to give each column of features in a training set a weak learner and to update each weak learner every iteration, not just updates the weights on the data. In an example where there is a set of 150-component f.v.'s, 150 weak learners would be created. As previously mentioned, each weak learner corresponds to a single component and classifies a given f.v. solely based on that one component. Then, weights for these weak learners are introduced. In each iteration of this modified AdaBoost process, weights for both the input data and the weak learners are updated. The weights for the input data are updated as in the standard AdaBoost implementation, while weights on the individual weak learners are updated as if each were a complete hypothesis in the standard AdaBoost implementation. At the end of the iterative process, the weak learners with the highest weights, that is, the weak learners that represent the most diagnostic features, are selected and those features are passed to an SVM for classification (see [55] for more details). Thus, the benefits of both AdaBoost and SVMs are obtained.

6.4.5.2 Modified Adaboost in SSA Protocol

It is found that boosting from the set of 150 manually designed features worked better than from the 2600 naive Bayes, and boosting from the 50 features in the first group worked best (see Figures 6.9 and 6.10). This result is also consistent with the principal component analysis (PCA) filtering in [87] mostly reducing the 150 feature set to the first 50 features.

Classification improvement with Adaboost taking the best 50 from the inverted-emission 150 feature set is shown in Figure 6.11. An accuracy of 95% is possible for discriminating 9GC from 9TA hairpins with no data dropped with use of Adaboost. This demonstrates a significant robustness to what the SVM can "learn" in the presence of noise (some of the 2600 components have richer information, but even

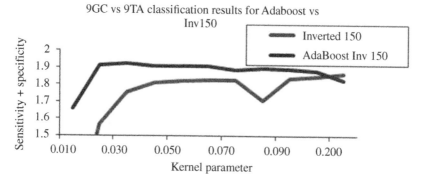

Figure 6.9 Adaboost feature selection strengthens the SVM performance of the Inverted HMM feature extraction set. Classification improvement with Adaboost taking the best 50 from the Inverted-emission 150 feature set. 95% accuracy is possible for discriminating 9GC from 9TA hairpins with no data dropped with use of Adaboost, without Adaboosting, the accuracy is approx. 91%.

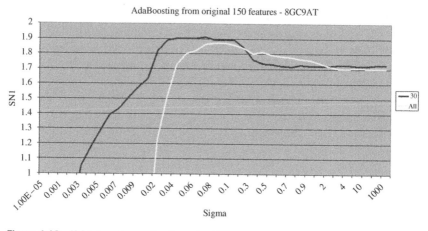

Figure 6.10 If Adaboost operates from the 150 component manual set, a reduced feature set of 30 is found to work best, and with notable improvement in kernel parameter stability in the region of interest.

more are noise contributors). This also validates the effectiveness with which the 150 parameter compression was able to describe the two-state dominant blockade data found for the nine base-pair hairpin and other types of "toggler" blockades, as well as the utility of the inverted features.

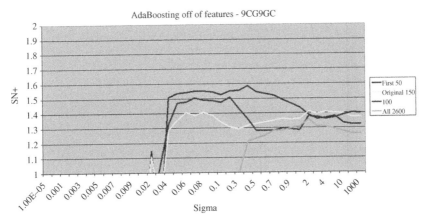

Figure 6.11 AdaBoosting to select 100 of the full set of 2600 features improves classification over just passing all 2600 components to the SVM. The best performance is still obtained when working with the Adaboosting from the manual set. (A principal component analysis (PCA) is done on the HMM projection data. 90% of the PCA information is contained in the first 50 principal components. The first 50 principal components are also listed as a feature set.)

6.5 HMM Implementation for Viterbi (in C and Perl)

```
int init_state_priors(int state_cardinality, double
*state_prior, int *data_quant,
      int total_pts, int stepsize, FILE *sigfile_hmm,int
      *state_count,
        int total_state_count, int j) {

   int print_quant_profile=0;
   int print_priors_profile=0;
   int prior_psuedocount=1;
   int k;
   double low_bound_cutoff = 0.05;
   double high_bound_cutoff = 0.05;
   double prior_test = 0.0;

   for (k=0;k<state_cardinality;k++) {
       state_count[k]=prior_psuedocount; // init
       psuedocounts
   }
```

```
    total_state_count =
    prior_psuedocount*state_cardinality; // init value
    for (k=0;k<total_pts/stepsize;k++) {
        state_count[data_quant[k]]++;
        total_state_count++;
    }
    for (k=0;k<state_cardinality;k++) {
        state_prior[k] = (double) state_count[k]/
        total_state_count;
        prior_test += state_prior[k];
    }

    // prior test
    if (prior_test>1.00001 || prior_test<0.99999) {
        printf("error: prior_test=%7.5f\n",prior_test);
        j++;
        return 1;
    }
    return 0;
}


void init_emissions(int state_cardinality, double
**emission_prob,  int *state_count){
    int i,k;
    double variance[state_cardinality];
    double em_prob_Z[state_cardinality];
    double rayleigh_var = 1.0;
    double emission_prob_test;
    for (i=0;i<state_cardinality;i++) {
        variance[i] = rayleigh_var;
    }
    for (i=0;i<state_cardinality;i++) {
        em_prob_Z[i] = 0.0;
        for (k=0;k<state_cardinality;k++) {
            emission_prob[i][k] = exp(-(k-i)*(k-i)/
            (2*variance[i]));
            em_prob_Z[i] += emission_prob[i][k];
        }
```

```c
    for (k=0;k<state_cardinality;k++) {
        emission_prob[i][k] = emission_prob[i][k]/
        em_prob_Z[i];
    }
    emission_prob_test = 0.0;
    for (k=0;k<state_cardinality;k++) {
        emission_prob_test += emission_prob[i][k];
    }
    if (emission_prob_test>1.01 ||
    emission_prob_test>0.99) {
        printf("error: emission_prob_test failure\n");
    }
    }
}
}

void init_transitions(int state_cardinality,
int *data_quant, double **trans_prob,
        int total_pts, int stepsize, double
        **emission_prob) {
    int i,k,l;
    int projection_init = 1;
    int trans_psuedocount = 3;
    int trans_count[state_cardinality][state_cardinality];
    int trans_count_total[state_cardinality];
    double trans_total[state_cardinality];
    double trans_prob_sum[state_cardinality];
    double test_total;
    double weight_total;
    double weight[state_cardinality][state_cardinality];
    int in_state,out_state;
    int shift = 10;
    int steps = 15; //by two
    int steplength = 2;
    int loop;
    int loop_max = 10;
    double weight_old[steps][steps];
    double weight_new[steps][steps];
```

```
    if (projection_init) {
        for (i=0;i<state_cardinality;i++) {
            for (k=0;k<state_cardinality;k++) {
                trans_prob[i][k] = 0.0;
            }
        }
        for (l=0;l<total_pts/stepsize-1;l++) {
            in_state = data_quant[l];
            out_state = data_quant[l+1];
            for (i=0;i<state_cardinality;i++) {
                trans_total[i] = 0.0;
                for (k=0;k<state_cardinality;k++) {
                    trans_prob[i][k] += emission_prob[i]
                    [in_state]
                                        *emission_prob[k]
                                        [out_state];
                    trans_total[i] += trans_prob[i][k];
                }
            }
        }
        for (i=0;i<state_cardinality;i++) {
            test_total = 0.0;
            for (k=0;k<state_cardinality;k++) {
                trans_prob[i][k] = trans_prob[i][k]/
                trans_total[i];
                test_total += trans_prob[i][k];
            }
            if (test_total>1.01 || test_total<0.99) {
                printf("error: trans_prob[%d][summed]=%6.4f
                \n",i,test_total);
            }
        }
    }
// omitted options ..........
}

void calculate_forward_backward(int state_cardinality,
double **forward,
```

```
          double **backward, int total_pts, int stepsize,
          double **emission_prob,
          double **trans_prob, int *data_quant, double
          *state_prior) {
    double scale[total_pts/stepsize];
    int i,k,l;
    double log_sig_prob;
    double fb_identity;

    //// calculate forward/backward variables (scaled
versions)
    scale[0] = 0.0;
    for (i=0;i<state_cardinality;i++) {
        forward[0][i] = emission_prob[data_quant[0]][i]
                                  *state_prior[i];
        scale[0] += forward[0][i];
    }
    // rescale forward vars
    for (i=0;i<state_cardinality;i++) {
        forward[0][i] = forward[0][i]/scale[0];
    }
    for (i=1;i<total_pts/stepsize;i++) {
        scale[i] = 0.0;
        for (k=0;k<state_cardinality;k++) {
            forward[i][k] = 0.0;
            for (l=0;l<state_cardinality;l++) {
               forward[i][k] += forward[i-1][l]*trans_prob
               [l][k];
            }
            forward[i][k] = forward[i][k]
                           *emission_prob[data_quant[i]][k];
            scale[i] += forward[i][k];
        }
        // rescale forward vars
        for (k=0;k<state_cardinality;k++) {
            forward[i][k] = forward[i][k]/scale[i];
        }
    }
}
```

```
    log_sig_prob = 0.0;
    for (i=0;i<total_pts/stepsize;i++) {
        log_sig_prob += log(scale[i]);
    }
    printf("-log_sig_prob=%9.7f\n",-log_sig_prob);
    // now have eval of forward variables

    for (i=0;i<state_cardinality;i++) {
      // without rescale, backward[total_pts-1]=1 is b.c.
        backward[total_pts/stepsize-1][i] = 1;
    }
    for (i=total_pts/stepsize-2;i>=0;i--) {
        for (k=0;k<state_cardinality;k++) {
            backward[i][k] = 0.0;
            for (l=0;l<state_cardinality;l++) {
                backward[i][k] += backward[i+1][l]
                *trans_prob[k][l]
                                *emission_prob[data_quant[i
                                +1]][l];
            }
            backward[i][k] = backward[i][k]/scale[i+1];
        }
    }
    for (i=0;i<total_pts/stepsize;i++) {
        fb_identity= 0.0;
        for (k=0;k<state_cardinality;k++) {
            fb_identity += forward[i][k]*backward[i][k];
        }
        if (fb_identity>1.0001 || fb_identity<0.9999) {
            printf("fb_identity failure\n");
            printf("i=%d:\t%6.4f\n",i,fb_identity);
            exit(0);
        }
    }
    // now have eval of backward variables
    // also passing identity check: forward and backward
variables good
}
```

```c
void get_expectated_values(int state_cardinality, int
total_pts, int stepsize,
            int *data_quant, double **emission_prob, double
            **trans_prob,
            double **forward, double **backward,double
            **expected_emission_count,
            double **expected_trans_count) {
    int i,k,l;
    double min_prob = 0.000000001;
    int in_state,out_state;
    double temp_trans_sum;
    double temp_trans[state_cardinality]
    [state_cardinality];

    for (i=0;i<state_cardinality;i++) {
        for (k=0;k<state_cardinality;k++) {
            expected_emission_count[k][i] = min_prob;
            expected_trans_count[k][i] = min_prob;
        }
    }

    // get expected counts on transitions
    for (i=0;i<total_pts/stepsize-1;i++) {
        in_state = data_quant[i];
        out_state = data_quant[i+1];
        temp_trans_sum = 0.0;
        for (k=0;k<state_cardinality;k++) {
            for (l=0;l<state_cardinality;l++) {
                // recall convention: emission[state][base]
                temp_trans[k][l] =
                    emission_prob[l][out_state]*trans_prob
                    [k][l]*forward[i][k];
                temp_trans_sum += temp_trans[k][l];
            }
        }
        for (k=0;k<state_cardinality;k++) {
            for (l=0;l<state_cardinality;l++) {
                expected_trans_count[k][l] +=
                    (temp_trans[k][l]/temp_trans_sum)
                    *backward[i+1][l];
            }
```

```
            }
        }
        // now have expected counts on transitions

        // get expected counts on emissions
        // recall convention: emission[state][base]
        for (i=0;i<total_pts/stepsize;i++) {
            in_state = data_quant[i];
            for (k=0;k<state_cardinality;k++) {
                expected_emission_count[k][in_state] +=
                    forward[i][k]*backward[i][k];
            }
        }

        for (i=0;i<state_cardinality;i++) {
            for (k=0;k<state_cardinality;k++) {
                if (expected_trans_count[i][k]<min_prob) {
                    printf("expected_trans_count error\n");
                    exit(0);
                }
                if (expected_emission_count[i][k]<min_prob) {
                    printf("expected_emission_count[%d][%d]=%
                    20.18f\n",
                            i,k,expected_emission_count[i][k]);
                    exit(0);
                }
            }
        }
    }
}

void obtain_viterbi_path(int state_cardinality,double
**emission_prob,
                    double **trans_prob,int total_pts,int
                    stepsize,
                        int *data_filtered,int *data_quant,
                            double *state_prior, int j)  {
    int i,k,l,penultima_index;
    double log_path_probt;
    int in_state,last_path_state;
    int min_ptr = 100;
    int min_loc = 0;
```

```c
int min_state = 0;
double max_path_value,path_value;
int back_ptr[total_pts/stepsize][state_cardinality];
double log_path_prob[total_pts/stepsize]
[state_cardinality];

in_state = data_quant[0];
for (k=0;k<state_cardinality;k++) {
    log_path_prob[0][k] = log(state_prior[k]) +
                    log(emission_prob[k][in_state]);
    back_ptr[0][k] = state_cardinality/2; // arbitrary
}

for (i=1;i<total_pts/stepsize;i++) {
    in_state = data_quant[i];
    for (k=0;k<state_cardinality;k++) {
        max_path_value = log_path_prob[i-1]
        [state_cardinality/2]
                        + log(trans_prob
[state_cardinality/2][k]);
        back_ptr[i][k] = state_cardinality/2;
        for (l=0;l<state_cardinality;l++) {
            path_value = log_path_prob[i-1][l] +
            log(trans_prob[l][k]);
            if (path_value > max_path_value) {
                max_path_value = path_value;
                back_ptr[i][k] = l;
            }
        }
        log_path_prob[i][k] = log(emission_prob[k]
        [in_state])
                            +max_path_value;
        if (back_ptr[i][k]<min_ptr) {
            min_ptr=back_ptr[i][k];
            min_loc=i;
            min_state=k;
        }
    }
}
```

```
        last_path_state = state_cardinality/2; //arbitrary,
        just not undef!
        log_path_probt = log_path_prob[total_pts/stepsize-1]
        [0];
        for (l=1;l<state_cardinality;l++) {
            if (log_path_prob[total_pts/stepsize-1][l]
            >log_path_probt) {
                log_path_probt = log_path_prob[total_pts/
                stepsize-1][l];
                last_path_state = l;
            }
        }
        data_filtered[total_pts/stepsize-1] = last_path_state;
        penultima_index = (total_pts/stepsize)-2;
        for (i=penultima_index;i>=0;i--) {
            data_filtered[i] = back_ptr[i+1][data_filtered[i
            +1]];
        }
    }
}
```

Code in Perl:

```
my $Init_State_Priors = sub {
    my ($level_one_ref,$HMM_state_cardinality,
    $state_shift,
        $pseudocount,$HMM_data_ref,$bin_size) = @_;
    my @HMM_data_array;
    my @state_count = ();
    my $total_instance_count=0;
    if (!$state_shift) { $state_shift = -20; }
    if (!$pseudocount) { $pseudocount = 1; }
    if (!$bin_size) { $bin_size = 1; }

    if ($HMM_data_ref) {
        @HMM_data_array = @{$HMM_data_ref};
        my $prior_training_data_instances = scalar
        (@HMM_data_array);
        my $training_index;
        for $training_index (0..
        $prior_training_data_instances-1) {
```

```perl
            my $state = int( ($HMM_data_array
            [$training_index]+$state_shift)/$bin_size );
            if ($state<0) { $state = 0; }
            $state_count[$state]++;
            $total_instance_count++;
        }
    }
    else { # get pseudocounts from here, if conditional
    defunct
        my $state;
        for $state (0..$HMM_state_cardinality-1) {
            $state_count[$state]=$pseudocount;
            $total_instance_count+=$pseudocount;
        }
    }

    my @lev1_array = @{$level_one_ref};
    my $training_data_instances = scalar(@lev1_array);
    my $training_index;
    for $training_index (0..$training_data_instances-1) {
        my $state = int( ($lev1_array[$training_index]+
        $state_shift)/$bin_size );
        if ($state<=0) {
            $state = 0;
#           print "zero state thresholding\n";
        }
        my $max_states = $HMM_state_cardinality;
        if ($state>=$max_states) {
            $state = $max_states-1;
#         print "upper_$max_states state thresholding\n";
        }
        $state_count[$state]++;
        $total_instance_count++;
    }

    my $state;
    my $prior_prob_test = 0;
    my @state_prior;
    for $state (0..$HMM_state_cardinality-1) {
        $state_prior[$state] = $state_count[$state]/
```

```perl
            $total_instance_count;
            $prior_prob_test += $state_prior[$state];
    }
    if ($prior_prob_test>1.00001 ||
    $prior_prob_test<0.99999) {
        print "error, prior_prob_test = $prior_prob_test
        \n";
    }
    my $prior_prob_ref = \@state_prior;
    return $prior_prob_ref;

};


my $Init_Emissions = sub {
    my ($HMM_state_cardinality,$esigma) = @_;

    my $state;
    my $emission_prob_test;
    my @emission_prob;
    my @emission_prob_Z;
    for $state (0..$HMM_state_cardinality-1) {
        $emission_prob_Z[$state] = 0.0;
        my $emission_state;
        for $emission_state (0..$HMM_state_cardinality-1) {
            $emission_prob[$state][$emission_state] =
            0.0000001;
            $emission_prob[$state][$emission_state] +=
                        exp(-(($emission_state-$state)
                        **2)/($esigma*2));
            $emission_prob_Z[$state] += $emission_prob
            [$state][$emission_state];
        }
        $emission_prob_test = 0;
        for $emission_state (0..$HMM_state_cardinality-1) {
            $emission_prob[$state][$emission_state] =
                $emission_prob[$state][$emission_state]/
                $emission_prob_Z[$state];
            $emission_prob_test += $emission_prob[$state]
            [$emission_state];
        }
```

```perl
        if ($emission_prob_test>1.01 ||
        $emission_prob_test<0.99) {
            print "error, emission_prob_test =
            $emission_prob_test\n";
        }
    }

    my $emission_prob_ref = \@emission_prob;
    return $emission_prob_ref;

};

my $Init_Transitions = sub {
    my ($level_one_ref,$HMM_state_cardinality,
    $state_shift,$HMM_data_ref,
        $emission_prob_ref,$bin_size,$decimation) = @_;
    if (!$bin_size) { $bin_size = 1; }
    if (!$decimation) { $decimation = 1; }

    my @emission_prob = @{$emission_prob_ref};
    my @HMM_data_array;

    my $in_state;
    my $out_state;
    my @transition_prob;
    my @transition_total;
    for $in_state (0..$HMM_state_cardinality-1) {
        for $out_state (0..$HMM_state_cardinality-1) {
            $transition_prob[$in_state][$out_state] = 0.0;
        }
    }

    my @lev1_array = @{$level_one_ref};
    my $training_data_instances = scalar(@lev1_array);
    my $training_index;

    for ($training_index=0; $training_index<
    $training_data_instances-1;
        $training_index+=$decimation) { # stops one short
        for out state +1 ref
        $in_state = int( ($lev1_array[$training_index]+
```

```
            $state_shift)/$bin_size );
            if ($in_state<0) { $in_state=0; }
            $out_state = int( ($lev1_array[$training_index+1]+
            $state_shift)/$bin_size );
            if ($out_state<0) { $out_state=0; }

        my $in_index;
        my $out_index;
        for $in_index (0..$HMM_state_cardinality-1) {
            $transition_total[$in_index] = 0.0;
            for $out_index (0..$HMM_state_cardinality-1) {
                $transition_prob[$in_index][$out_index] +=
                                $emission_prob[$in_index]
                                [$in_state]*
                                $emission_prob[$out_index]
                                [$out_state];
                $transition_total[$in_index] +=
                $transition_prob[$in_index][$out_index];
            }
        }
    }

my $in_index;
my $out_index;
my $test_total;
for $in_index (0..$HMM_state_cardinality-1) {
    $test_total = 0.0;
    for $out_index (0..$HMM_state_cardinality-1) {
        if ($transition_total[$in_index] == 0) {
            print "error\n";
        }
        else {
            $transition_prob[$in_index][$out_index] =
            $transition_prob[$in_index][$out_index]/
            $transition_total[$in_index];
        }
        $test_total += $transition_prob[$in_index]
        [$out_index];
    }
    if ($test_total>1.01 || $test_total<0.99) {
        print "error in trans_prob[$in_index][] not
```

```perl
                summing to unity\n";
        }
    }

    my $trans_prob_ref = \@transition_prob;
    return $trans_prob_ref;

};

my $Evaluate_Viterbi_Path = sub {
    my ($level_one_ref,$HMM_states_ref,$prior_prob_ref,
    $emission_prob_ref,
        $transition_prob_ref,$state_shift,$bin_size,
$subsample_size) = @_;
    if (!$bin_size) { $bin_size = 1; }
    my @HMM_states - @{$HMM_states_ref};
    my $HMM_state_cardinality = scalar(@HMM_states);
    my @lev1_array = @{$level_one_ref};
    my @state_prior = @{$prior_prob_ref};
    my @emission_prob = @{$emission_prob_ref};
    my @transition_prob = @{$transition_prob_ref};
    my $limit;
    my $training_data_instances = scalar(@lev1_array);
    if ($subsample_size) { $limit = $subsample_size-1; }
    else { $limit = $training_data_instances-1; }

    my @log_path_prob;
    my @back_ptr;
    my $in_state = int( ($lev1_array[0]+$state_shift)/
    $bin_size );
    my $in_index;
    for $in_index (0..$HMM_state_cardinality-1) {
        $log_path_prob[0][$in_index] = log($state_prior
        [$in_index]) +
                                    log($emission_prob
                                    [$in_index]
                                    [$in_state]);
        $back_ptr[0][$in_index] = int
        ($HMM_state_cardinality/2);
    # arbitrary, avoiding boundaries
    }
```

```perl
my $min_ptr;
my $min_loc;
my $min_state;


my $max_path_value;
my $training_index;
for $training_index (1..$limit) {
    $in_state = int( ($lev1_array[$training_index]+
    $state_shift)/$bin_size );
    my $arbitrary_state = int($HMM_state_cardinality/
    2);
    my $in_index;
    for $in_index (0..$HMM_state_cardinality-1) {
        $max_path_value = $log_path_prob
        [$training_index-1][$arbitrary_state]
                        + log($transition_prob
[$arbitrary_state][$in_index]);
        $back_ptr[$training_index][$in_index]=
        $arbitrary_state;
        my $out_index;
        for $out_index (0..$HMM_state_cardinality-1) {
            my $path_value = $log_path_prob
            [$training_index-1][$out_index]
                            + log($transition_prob
[$out_index][$in_index]);
            if ($path_value > $max_path_value) {
                $max_path_value = $path_value;
                $back_ptr[$training_index][$in_index]=
                $out_index;
            }
        }
        $log_path_prob[$training_index][$in_index]=
            log($emission_prob[$in_index][$in_state])
            + $max_path_value;
        if ($back_ptr[$training_index][$in_index]<
        $min_ptr) {
            $min_ptr = $back_ptr[$training_index]
            [$in_index];
```

```perl
            $min_loc = $training_index;
            $min_state = $in_index;
        }
    }
}

my $last_path_state = int($HMM_state_cardinality/2); #
arbitrary init
my $log_path_probt = $log_path_prob
[$training_data_instances-1][0];

my $out_index;
for $out_index (0..$HMM_state_cardinality-1) {
    if ($log_path_prob[$training_data_instances-1]
    [$out_index]>$log_path_probt) {
        $log_path_probt = $log_path_prob
        [$training_data_instances-1][$out_index];
        $last_path_state = $out_index;
    }
}

my @viterbi_path_data;
$viterbi_path_data[$limit] = $last_path_state;
my $penultima_index = $limit-1;
my $index;
for ($index = $penultima_index; $index >=0; $index--) {
    $viterbi_path_data[$index] = $back_ptr[$index+1]
    [$viterbi_path_data[$index+1]];
}
my $viterbi_score = $log_path_probt/
$training_data_instances;

return $viterbi_score;

};


my $Calculate_Forward_Backward = sub {
    my ($level_one_ref, $HMM_states_ref, $state_shift,
    $prior_prob_ref, $emission_prob_ref,
    $transition_prob_ref,$bin_size) = @_;
```

```perl
my @HMM_states = @{$HMM_states_ref};
my $HMM_state_cardinality = scalar(@HMM_states);
my @lev1_array = @{$level_one_ref};
my @prior_probs = @{$prior_prob_ref};
my @emission_probs = @{$emission_prob_ref};
my @transition_probs = @{$transition_prob_ref};
my $data_instances = scalar(@lev1_array);

my @rescale;
my $log_sig_prob;
my $fb_identity;
my @forward;
my @backward;

$rescale[0] = 0.0;
my $i;
my $k;
my $l;
for ($i=0; $i<$HMM_state_cardinality; $i++) {
    my $state = int( ($lev1_array[0] + $state_shift)/
    $bin_size );
    $forward[0][$i] = $emission_probs[$state][$i]*
    $prior_probs[$i];
    $rescale[0] += $forward[0][$i];
}

#rescale forward vars
for ($i=0; $i<$HMM_state_cardinality; $i++) {
    $forward[0][$i] = $forward[0][$i]/$rescale[0];
}

for ($i=1; $i<$data_instances; $i++) {
    $rescale[$i] = 0.0;
    for ($k=0; $k<$HMM_state_cardinality; $k++) {
        $forward[$i][$k] = 0.0;
        for ($l=0; $l<$HMM_state_cardinality; $l++) {
            $forward[$i][$k] += $forward[$i-1][$l]*
            $transition_probs[$l][$k];
        }
        my $state = $lev1_array[$i] + $state_shift;
```

```perl
            $forward[$i][$k] = $forward[$i][$k]*
            $emission_probs[$state][$k];
            $rescale[$i] += $forward[$i][$k];
        }
        #rescale Forward vars
        for ($k=0; $k<$HMM_state_cardinality; $k++) {
            $forward[$i][$k] = $forward[$i][$k]/$rescale
            [$i];
        }
    }
    $log_sig_prob = 0.0;
    for ($i=0; $i<$data_instances; $i++) {
        $log_sig_prob += log($rescale[$i]);
    }
    my $neglog = -$log_sig_prob;
    my $length = scalar(@lev1_array);
    my $renorm = int($neglog/$length);
    print "-log_sig_prob=$neglog\trenorm_log_prob=$renorm
    \n";
    # have now completed eval of forward vars

    for ($i=0; $i<$HMM_state_cardinality; $i++) {
        # without resclae, backward[$data_instances-1][]=1
is bc
        $backward[$data_instances-1][$i] = 1;
    }
    for ($i=$data_instances-2; $i>=0; $i--) {
        for ($k=0; $k<$HMM_state_cardinality; $k++) {
            $backward[$i][$k] = 0.0;
            for ($l=0; $l<$HMM_state_cardinality; $l++) {
                my $state = int( ($lev1_array[$i+1] +
                $state_shift)/$bin_size );
                $backward[$i][$k] += $backward[$i+1][$l]*
                $transition_probs[$k][$l]*
                                    $emission_probs[$state]
                                    [$l];
            }
            $backward[$i][$k] = $backward[$i][$k]/$rescale
            [$i+1];
        }
    }
```

```perl
    for ($i=0; $i<$data_instances; $i++) {
        $fb_identity = 0.0;
        for ($k=0; $k<$HMM_state_cardinality; $k++) {
            $fb_identity += $forward[$i][$k]*$backward[$i]
            [$k];
        }
        if ($fb_identity > 1.0001 || $fb_identity < 0.9999)
        {
            print "fb_identity_failure: oneval =
            $fb_identity\n";
        }
    }
    # now have backward vars, with fb_identity check

    my @ref_array = (\@forward,\@backward,$renorm);
    my $training_refs = \@ref_array;

    return $training_refs;
};

my $Get_Expected_Values = sub {
    my ($level_one_ref, $HMM_states_ref, $state_shift,
    $prior_prob_ref,
        $emission_prob_ref, $transition_prob_ref,
        $forward_ref, $backward_ref,
        $bin_size) = @_;

    my @HMM_states = @{$HMM_states_ref};
    my $HMM_state_cardinality = scalar(@HMM_states);
    my @lev1_array = @{$level_one_ref};
    my @prior_probs = @{$prior_prob_ref};
    my @emission_probs = @{$emission_prob_ref};
    my @transition_probs = @{$transition_prob_ref};
    my $data_instances = scalar(@lev1_array);
    my @forward = @{$forward_ref};
    my @backward = @{$backward_ref};

    my $i;
    my $k;
    my $l;
    my $min_prob = 0.000000001;
```

```perl
my $in_state;
my $out_state;
my $temp_trans_sum;
my @temp_trans;
my @expected_emission_count;
my @expected_transition_count;

for ($i=0; $i<$HMM_state_cardinality; $i++) {
    for ($k=0; $k<$HMM_state_cardinality; $k++) {
        $expected_emission_count[$k][$i] = $min_prob;
        $expected_transition_count[$k][$i] = $min_prob;
    }
}

# get expected counts on transitions
for ($i=0; $i<$data_instances; $i++) {
    $in_state = int( ($lev1_array[$i]+$state_shift)/
    $bin_size );
    $out_state = int( ($lev1_array[$i+1]+$state_shift)/
    $bin_size );
    $temp_trans_sum = 0.0;
    for ($k=0; $k<$HMM_state_cardinality; $k++) {
        for ($l=0; $l<$HMM_state_cardinality; $l++) {
            # convention: emission[state][base]
            $temp_trans[$k][$l] = $emission_probs[$l]
            [$out_state]*
                            $transition_probs[$k][$l]*
                            $forward[$i][$k];
            $temp_trans_sum += $temp_trans[$k][$l];
        }
    }
    for ($k=0; $k<$HMM_state_cardinality; $k++) {
        for ($l=0; $l<$HMM_state_cardinality; $l++) {
            $expected_transition_count[$k][$l] +=
            ($temp_trans[$k][$l]/$temp_trans_sum)*
            $backward[$i+1][$l];
        }
    }
}
# now have expected counts on transitions
```

```perl
        # get expected counts on emissions
        # convention: emission[state][base]
        for ($i=0; $i<$data_instances; $i++) {
            $in_state = int( ($lev1_array[$i]+$state_shift)/
            $bin_size );
            for ($k=0; $k<$HMM_state_cardinality; $k++) {
                $expected_emission_count[$k][$in_state] +=
                         $forward[$i][$k]*$backward[$i][$k];
            }
        }


    for ($i=0; $i<$HMM_state_cardinality; $i++) {
        for ($k=0; $k<$HMM_state_cardinality; $k++) {
            if ($expected_transition_count[$i][$k] <
            $min_prob) {
                print "expected_trans_count_error\n";
                die;
            }
            if ($expected_emission_count[$i][$k] <
            $min_prob) {
                print "expected_emission_count_error\n";
                die;
            }
        }
    }


    my $expected_emission_count_ref =
    \@expected_emission_count;
    my $expected_transition_count_ref =
    \@expected_transition_count;
    my @count_ref_array = ($expected_emission_count_ref,
    $expected_transition_count_ref );
    my $count_ref_array_ref = \@count_ref_array;
    return $count_ref_array_ref;
};

my $Eval_Maxlike_Estimators = sub {
    my ($HMM_states_ref, $emission_prob_ref,
    $transition_prob_ref,
        $expected_emission_count_ref,
        $expected_transition_count_ref, $em_loop) = @_;
```

```perl
if (!$em_loop) { $em_loop = 1; }

my @HMM_states = @{$HMM_states_ref};
my $HMM_state_cardinality = scalar(@HMM_states);
my @emission_prob = @{$emission_prob_ref};
my @transition_prob = @{$transition_prob_ref};
my @expected_emission_count = @
{$expected_emission_count_ref};
my @expected_transition_count = @
{$expected_transition_count_ref};
my @transition_probs = @{$transition_prob_ref};


my $k;
my $l;
my @expected_trans_count_sum;
my @expected_emission_count_sum;
my @trans_prob_test;
my $temp_trans_prob;
my $temp_emission_prob;
my $min_prob = 0.000000001;
my $emprob = 1.0;    # disables debug, have redundant
error check
my $transprob = 1.0;# disables debug, have redundant
error check

for ($k=0; $k<$HMM_state_cardinality; $k++) {
    $expected_trans_count_sum[$k] = 0.0;
    $trans_prob_test[$k] = 0.0;
    for ($l=0; $l<$HMM_state_cardinality; $l++) {
        $expected_trans_count_sum[$k] +=
        $expected_transition_count[$k][$l];
    }
    #new transition probabilities defined here
    for ($l=0; $l<$HMM_state_cardinality; $l++) {
        $temp_trans_prob = $expected_transition_count
        [$k][$l]/
                        $expected_trans_count_sum[$k];
        # error check
        if (abs($transition_prob[$k][$l]-
        $temp_trans_prob) > 1.0) {
            print "error: transition_prob[$k][$l] =
```

```
                        $transition_prob[$k][$l]\t
temp_trans_prob = $temp_trans_prob\n";
              }
              # lower bound to prevent underflow
              if ($transition_prob[$k][$l] < $min_prob) {
                  $transition_prob[$k][$l] = $min_prob;
              }
              # error and debug checks
              if (abs($transition_prob[$k][$l]-
              $temp_trans_prob) > $transprob) {
                  print "loop=$em_loop\t transition_prob[$k]
                  [$l]=$transition_prob[$k][$l]\t
temp_trans_prob=$temp_trans_prob\n";
              }
              $transition_prob[$k][$l] = $temp_trans_prob;
              $trans_prob_test[$k] += $temp_trans_prob;
          }
          # test on new trans_prob
        if ($trans_prob_test[$k] > 1.01 || $trans_prob_test
        [$k] < 0.99) {
            print "trans_prob_test[$k] = $trans_prob_test
            [$k]\n";
            die;
        }
    }


    for ($k=0; $k<$HMM_state_cardinality; $k++) {
        $expected_emission_count_sum[$k] = 0.0;
        for ($l=0; $l<$HMM_state_cardinality; $l++) {
            $expected_emission_count_sum[$k] +=
            $expected_emission_count[$k][$l];
        }
        # new emission probabilities defined here
        for ($l=0; $l<$HMM_state_cardinality; $l++) {
            $temp_emission_prob = $expected_emission_count
            [$k][$l]/
                            $expected_emission_count_sum
                            [$k];
            # error check
            if (abs($emission_prob[$k][$l]-
            $temp_emission_prob) > 1.0) {
                print "error: emission_prob[$k][$l] =
```

```perl
            $emission_prob[$k][$l]\t
            temp_emission_prob = $temp_emission_prob
            \n";
        }
        # lower bound to prevent underflow
        if ($emission_prob[$k][$l] < $min_prob) {
            $emission_prob[$k][$l] = $min_prob;
        }
        # error and debug checks
        if (abs($emission_prob[$k][$l] -
        $temp_emission_prob) > $emprob) {
          print "loop=$em_loop\t emission_prob[$k][$l]
          =$emission_prob[$k][$l]\t
          temp_emission_prob=$temp_emission_prob\n";
        }
        $emission_prob[$k][$l] = $temp_emission_prob;
    }
}
# test on new emission_prob
for ($k=0; $k<$HMM_state_cardinality; $k++) {
    $temp_emission_prob = 0.0;
    for ($l=0; $l<$HMM_state_cardinality; $l++) {
        $temp_emission_prob += $emission_prob[$k][$l];
    }


    if ($temp_emission_prob > 1.01 ||
    $temp_emission_prob < 0.99) {
        print "error: emission_prob[$k][summed] =
        $temp_emission_prob\n";
        die;
    }
}
# pass out re-est probs here........
my $emission_prob_ref = \@emission_prob;
my $transition_prob_ref = \@transition_prob;
my @new_HMMparam_ref_array =
($emission_prob_ref,$transition_prob_ref);
my $new_HMMparam_ref_array_ref =
\@new_HMMparam_ref_array;
return $new_HMMparam_ref_array_ref;
};
```

```perl
my $Relative_Entropy = sub {
    my ($new_prob_ref,$old_prob_ref) = @_;
    my @new_prob = @{$new_prob_ref};
    my @old_prob = @{$old_prob_ref};
    my $max_index = scalar(@new_prob)-1;
    my $index;
    my $relative_entropy = 0;
    for $index (0..$max_index) {
        if ($new_prob[$index]==0 || $old_prob[$index]==0) {
            print "error: new_prob[$index]=$new_prob
            [$index]\told_prob[$index]=$old_prob[$index]
            \n";
        }
        else {
            $relative_entropy += $new_prob[$index]*log
            ($new_prob[$index]/$old_prob[$index]);
        }
    }
    return $relative_entropy;
};


my $Do_Gaussian_Projection = sub {
    my ($HMM_states_ref,$emission_prob_ref) = @_;

    my @HMM_states = @{$HMM_states_ref};
    my $HMM_state_cardinality = scalar(@HMM_states);
    my @emission_prob = @{$emission_prob_ref};

    my $i;
    my $k;
    my $m;
    my @level_std_dev;
    my @mean;
    my @variance;
    my @em_prob_Z;
    my $eva = 1.0; # no eva projection

    for ($k=0; $k<$HMM_state_cardinality; $k++) {
        $level_std_dev[$k] = 0.0;
        $mean[$k] = 0.0;
```

```perl
    for ($m=0; $m<$HMM_state_cardinality; $m++) {
        $level_std_dev[$k] += $m*$m*$emission_prob[$k]
        [$m];
    }
    for ($m=0; $m<$HMM_state_cardinality; $m++) {
        $mean[$k] += $m*$emission_prob[$k][$m];
    }
    $level_std_dev[$k] -= $mean[$k]*$mean[$k];
    $variance[$k] = abs($level_std_dev[$k]);
    $level_std_dev[$k] = sqrt($variance[$k]);
}

for ($i=0; $i<$HMM_state_cardinality; $i++) {
    $em_prob_Z[$i] = 0.0;
    for ($k=0; $k<$HMM_state_cardinality; $k++) {
        $emission_prob[$i][$k] = exp(-($k-$i)*($k-$i)/
        (2*$eva*$variance[$k]));
        if ($emission_prob[$i][$k]<0.000000001) {
            $emission_prob[$i][$k]=0.000000001;
        }
        $em_prob_Z[$i] += $emission_prob[$i][$k];
    }
    for ($k=0; $k<$HMM_state_cardinality; $k++) {
        $emission_prob[$i][$k] = $emission_prob[$i]
        [$k]/$em_prob_Z[$i];
    }
    my $emission_prob_test = 0.0;
    for ($k=0; $k<$HMM_state_cardinality; $k++) {
        $emission_prob_test += $emission_prob[$i][$k];
    }
    if ($emission_prob_test > 1.01 ||
    $emission_prob_test < 0.99) {
        print "error: emission_prob_test \n";
    }
}
my $emission_ref = \@emission_prob;
return $emission_ref;
};
```

6.6 Exercises

6.1 Derive the recursive relation for the Viterbi algorithm shown in Section 6.2.1 (also see Figure 6.5).

6.2 Derive the recursive relation for the forward algorithm shown in Section 6.2.2.

6.3 Derive the recursive relation for the backward algorithm shown in Section 6.2.2.

6.4 Re-derive with more detail the Akl used in estimating akl.

6.5 Re-derive with more detail the $Ek(b)$ used in estimating ekb.

6.6 Implement Adaboost, use to do simple classification tests.

6.7 Implement modified adaboost, use to rank strongest features.

7

Generalized HMMs (GHMMs)

Major Viterbi Variants

Numerous prior book, journal, and patent publications by the author are drawn upon extensively throughout the text [1–68]. Almost all of the journal publications are open access. These publications can typically be found online at either the author's personal website (www.meta-logos.com) or with one of the following online publishers: www.m-hikari.com or bmcbioinformatics.biomedcentral.com.

For an overview of Hidden Markov Models (HMMs) and their (minor) Viterbi generalizations, see Chapter 6.

7.1 GHMMs: Maximal Clique for Viterbi and Baum–Welch

The generalized clique HMM begins by enlarging the primitive hidden states associated with individual base labeling (as exon, intron, or junk) to substrings of primitive hidden states or *footprint* states (details on the definitions of the base-label states and footprint are in what follows). In what follows, the transitions between primitive hidden states for coding {e} and noncoding {i,j}, {ei,ie,je,ej}, are referred to as "eij-transitions", and the self-transitions, {ee,ii,jj}, are referred to as "xx-transitions". The emissions are likewise expanded to higher order in the fundamental joint probability that is the basis of the generalized-clique, or "meta-state," HMM. In [33] we consider application to eukaryotic gene finding and show how a meta-state HMM improves the strength of eij-transition contributions to gene-structure identification. It is found that the meta-state eij-transition modeling can effectively "recapture" the exon and intron heavy tail distribution modeling capability as well as manage the exon-start "needle-in-the-haystack" problem [33].

The meta-state, clique-generalized, HMM entails a clique-level factorization rather than the standard HMM factorization (that describes the state transitions with no dependence on local sequence information). This is described in the

Informatics and Machine Learning: From Martingales to Metaheuristics, First Edition. Stephen Winters-Hilt.

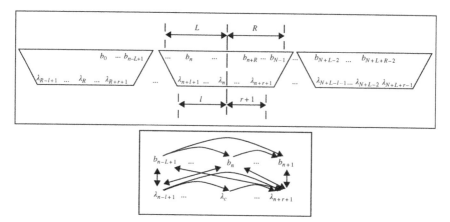

Figure 7.1 Top Panel. Sliding-window association (clique) of observations and hidden states in the meta-state hidden Markov model, where the clique-generalized HMM algorithm describes a left-to-right traversal (as is typical) of the HMM graphical model with the specified clique window. The first observation, b_0, is included at the leading edge of the clique overlap at the HMM's left boundary. For the last clique's window overlap we choose the trailing edge to include the last observation b_{N-1}. Bottom Panel. Graphical model of the clique-generalized HMM, where the interconnectedness on full joint dependencies is only partly drawn. The graphical model is significantly constrained, as well, in a manner not represented in the graphical model, in that state sequences are only allowed with at most one nonself transition.

general formalism to follow, where specific implementations are given for application to eukaryotic gene structure identification.

Observation and state dependencies in the generalized-clique HMM (see Figure 7.1) are parameterized according to the following:

1) Nonnegative integers L and R denoting left and right maximum extents of a substring, w_n, (with suitable truncation at the data boundaries, b_0 and b_{n-1}) are associated with the primitive observation, b_n, in the following way:

$$w_n = b_{n-L+1}, ..., b_n, ..., b_{n+R}$$
$$\widetilde{w}_n = b_{n-L+1}, ..., b_n, ..., b_{n+R-1}$$

2) Nonnegative integers l and r are used to denote the left and right extents of the extended (footprint) states, f. Here, we show the relationships among the primitive states λ, dimer states s, and footprint states f:

$$\delta_n = \lambda_n \lambda_{n+1} \qquad \text{(dimer state, length in } \lambda\text{'s} = 2)$$
$$f_n = \delta_{n-l+1}, ..., \delta_{n+r} \cong \lambda_{n-l+1}, ..., \lambda_n, ..., \lambda_{n+r+1}$$
$$\text{(footprint state, length in } \delta\text{'s} = l + r)$$

The probability of a sequence of observables $B = b_0, b_1,...,b_{L-1}$ being emitted by the sequence of hidden states $\Lambda = \lambda_0, \lambda_1,...,\lambda_{L-1}$ is solved by using $P(B, \Lambda) = P(B|\Lambda) P(\Lambda)$ in the standard factorization mentioned above, where the two terms in the factorization are described as the *observation model* and the *state model*, respectively. In the first-order HMM, the state model has the first-order Markov property and the observation model is such that the current observation, b_n, depends only on the current state, λ_n. Given (i) sequence of observations b_n, (ii) hidden labels λ_n, and (iii) stationary Markov statistics, one can calculate: (i) $p(B)$, or (ii) the most likely hidden labeling (path with largest contribution to $p(B; \Lambda)$), or (iii) the reestimation of emission and transmission probabilities such that $p(B; \Lambda)$ is maximized (using expectation/maximization).

As in the first-order HMM, the nth base observation b_n is aligned with the nth hidden state λ_n. Given the above, the clique-factorized HMM is as follows [33]:

$$P(B, \Lambda) = P(w_{-R}, f_{-R}) \left\{ \mathbf{\Pi}_{n=-R+1}^{N+L-2} [P(w_n, f_{n-1}, f_n)/P(\widetilde{w}_n, f_{n-1})] \right\}$$

The critical ratio of probabilities in the [...] term above retains the Martingale sequence properties on the generalized Viterbi path, as with the standard HMM/Viterbi implementation, and all of the elegant convergence and limit properties of Martingales are thereby inherited via the backward martingale convergence theorem (as discussed in [82]). The sliding-window clique overlap (see Figure 7.1) is much more significant than with the standard HMM, giving rise to many more table look-ups on *eij*-transition tables.

A generalization to the Viterbi algorithm can now be directly implemented, using the above form, to establish an efficient dynamic programming table construction. Generalized expressions for the Baum–Welch algorithm are also possible. For further details on the generalized Viterbi and Baum–Welch algorithms for the meta-state HMM see [33].

The gap and hash interpolating Markov Models (gIMM and hIMM) [34, 60] can be directly incorporated into meta-Hidden Markov Model with Binned Duration (HMMBD) gene-finding models as a further enhancement to the underlying Markov models, since they are already known to extract additional information that may prove useful, particularly in the zone-dependent emission regions (denoted "*zde*"s as in [34, 60]) where promoters and other gapped motifs might exist. Promoters and transcription factor binding sites often have lengthy overall gapped motif structure, and with the hash-interpolated Markov models it is also possible to capture the conserved higher order sequence information in the *zde* sample space. The hIMM and gIMM methods, thus, will not only strengthen the gene structure recognition, they can also provide the initial indications of anomalous motif structure in the regions identified by a gene-finder (in a post-genomic phase of the analysis) [34, 60].

By viewing state transitions, such as $e_0 e_1$ or $e_0 i_0$, as transition "dimer states", or as two-element "footprint" states, we begin to shift to a meta-HMM footing where we can model emissions more accurately. For the footprint states introduced in what follows a critical assumption is made – **at most one nonself transition is allowed per footprint transition**. This assumption is equivalent to a minimum length constraint on regions of self-transitions to be footprint size or greater. For genomic applications this is not a problematic constraint, and when a concern, different "gene-scans" can always be performed with different footprint sizes.

When encountered sequentially in the Viterbi algorithm, the sequence of (single) nonself state transition "dominated" *footprint* states would conceivably score highly when computed for the footprint-width number of footprint-states that overlap the non-self transition. In other words we can expect a *natural boosting* effect for the correct prediction at such nonself transitions (compared to the standard HMM). To describe bases in the irreducible joint probability we have: $w_n = b_{n-L+1}, ..., b_n, ..., b_{n+R}$, and $\widetilde{w}_n = b_{n-L+1}, ..., b_n, ..., b_{n+R-1}$ describes the base observations, while $s_n = \lambda_n \lambda_{n+1}$ (dimer states, length in λ's = 2), and $f_n = s_{n-l+1}, ..., s_{n+r} \cong \lambda_{n-l+1}, ..., \lambda_n, ..., \lambda_{n+r+1}$ (footprint state, length in s's = $l + r$), describes the associated labels. Given the above, the clique-factorized HMM is as shown in the previous equation.

The core term in the clique-factorization can also be written by introducing a Bayesian parameter, one that happens to provide a matching joint probability construct (to the extent possible) with the term in the numerator:

$$\rho = \frac{P(w_n, f_{n-1}, f_n)}{P(\widetilde{w}_n, f_{n-1})} = \frac{P(w_n, f_{n-1}, f_n)}{\sum\limits_{f'_{n(\text{allowed})}} P(\widetilde{w}_n, f_{n-1}, f'_n)}$$

$$= \frac{P(w_n \mid f_{n-1}, f_n) P(f_n \mid f_{n-1}) P(f_{n-1})}{\sum\limits_{f'_n} P(\widetilde{w}_n \mid f_{n-1}, f'_n) P(f'_n \mid f_{n-1}) P(f_{n-1})}$$

In the standard Markov model $R = 0$, $L = 1$, $r = -1$, $l = 0$: $f_n = \lambda_n$, $w_n = b_n$, $P(\widetilde{w}_n, f_{n-1}) = P(\lambda_n)$:

$$\left. \frac{P(w_n, f_{n-1}, f_n)}{P(\widetilde{w}_n, f_{n-1})} \right|_{\substack{\text{Standard Hidden} \\ \text{Markov Model}}} = P(b_n \mid \lambda_n) P(\lambda_n \mid \lambda_{n-1})$$

In the above we introduce the constraint notation with the vertical bar notation, where the expression on the left is the clique factorization term with the constraint that it approximate according to the standard HMM conditional probabilities.

We now examine specific cases of this equation to clarify the novel improvements that result. **In what follows we constrain our model to have a minimum length on regions (thus self-transitions) such that footprint states, and their transitions, can only have one transition between different states.**

Consider the case with the first footprint state being of *eij*-transition type, and the second footprint thereby constrained to be of the appropriate *xx*-type:

$$
\frac{P(w_n, f_{n-1}, f_n)}{P(\widetilde{w}_n, f_{n-1})}\Bigg|_{f_{n-1}\in eij} = \frac{P(w_n, f_{n-1}, f_n)}{\sum\limits_{f'_{n(allowed)}} P(\widetilde{w}_n, f_{n-1}, f'_n)}\Bigg|\quad \begin{array}{l} f_{n-1}\in eij \\[4pt] [\,f'_n \text{unique} \in xx\,] \end{array}
$$

$$
= P(b_{n+R}\mid \widetilde{w}_n, f_{n-1}, f_n)\Big|_{f_{n-1}\in eij}\qquad P(f_n\mid f_{n-1})\big|_{f_{n-1}\in eij}
$$

$$
= P(b_{n+R}\mid \widetilde{w}_n, f_{n-1})
$$

where use is made of the relation $P(f_n\mid f_{n-1})\big|_{f_{n-1}\in eij} = 1$ for the unique *xx*-footprint that follow the *eij*-transition given our minimum length constraint.

Consider, next, the case with the first footprint state being *xx*-type:

$$
\frac{P(w_n, f_{n-1}, f_n)}{P(\widetilde{w}_n, f_{n-1})}\Bigg|_{f_{n-1}\in xx} = \frac{P(w_n\mid f_{n-1}, f_n)\big|_{f_{n-1}\in xx} P(f_n\mid f_{n-1})}{\sum\limits_{f'_n} P(\widetilde{w}_n\mid f_{n-1}, f'_n)\big|_{f_{n-1}\in xx} P(f'_n\mid f_{n-1})}\Bigg|_{f_{n-1}\in xx}
$$

$$
= \frac{P(w_n\mid f_n)\,P(f_n\mid f_{n-1})}{\sum\limits_{f'_n} P(\widetilde{w}_n\mid f'_n)\,P(f'_n\mid f_{n-1})}\Bigg|_{f_{n-1}\in xx}
$$

If the second footprint is *eij*-transition type, then the equation has two sum terms in the denominator if the first transition is *ii* or *jj* transition, and a third sum contribution if the first transition is an *ee*-transition:

In what follows, dimer notation is used on footprints, since we are interested in the footprint-to-footprint transitions. Given their large overlap dependence, this notation and formalism directly generalizes to the same cases no matter the size of the footprint (due to the single major-transition in or between footprints constraint that is provided by a minimum length constraint).

If $f_{n-1}\in xx$ we have three cases: $xx \in \{ii, ee, jj\}$. For $f_{n-1} = ii$, we have two possible $f_n \in \{ii, ie\}$; for $f_{n-1} = jj$, we have two possible $f_n \in \{jj, je\}$; for $f_{n-1} = ee$, we have three possible $f_n \in \{ee, ej, ei\}$.

$$
\frac{P(w_n, f_{n-1}, f_n)}{P(\widetilde{w}_n, f_{n-1})}\Bigg|_{\substack{f_{n-1}=ii,\\ f_n=ie}} = \frac{P(w_n\mid ie)\,P(ie\mid ii)}{P(\widetilde{w}_n\mid ie)P(ie\mid ii) + P(\widetilde{w}_n\mid ii)P(ii\mid ii)}
$$

$$
= \frac{P(b_{n+R}\mid \widetilde{w}_n, ie)}{1 + \left(\dfrac{P(\widetilde{w}_n\mid ii)}{P(\widetilde{w}_n\mid ie)}\right)\left(\dfrac{P(ii\mid ii)}{P(ie\mid ii)}\right)}
$$

Where we have introduced the notation "*ii*" to denote the dimer state or the footprint state "*ii...iii*", and the notation "*ie*" to denote the dimer state or the footprint state "*ii...iie*".

Similarly, consider $f_{n-1} = jj$ and $f_n = je$:

$$\frac{P(w_n, f_{n-1}, f_n)}{P(\tilde{w}_n, f_{n-1})} \Bigg|_{\substack{f_{n-1} = jj \\ f_n = je}} = \frac{P(w_n \mid je) P(je \mid jj)}{P(\tilde{w}_n \mid je)P(je \mid jj) + P(\tilde{w}_n \mid jj)P(jj \mid jj)}$$

$$= \frac{P(b_{n+R} \mid \tilde{w}_n, je)}{1 + \left(\dfrac{P(\tilde{w}_n \mid jj)}{P(\tilde{w}_n \mid je)}\right) \left(\dfrac{P(jj \mid jj)}{P(je \mid jj)}\right)}$$

For the $f_{n-1} = ee$ and $f_n = ej$ we get a similar expression, but a third term in the sum due to the three possibilities allowed for f_n:

$$\frac{P(w_n, f_{n-1}, f_n)}{P(\tilde{w}_n, f_{n-1})} \Bigg|_{\substack{f_{n-1} = ee, \\ f_n = ej}}$$

$$= \frac{P(w_n \mid ej) P(ej \mid ee)}{P(\tilde{w}_n \mid ej)P(ej \mid ee) + P(\tilde{w}_n \mid ei)P(ei \mid ee) + P(\tilde{w}_n \mid ee)P(ee \mid ee)}$$

$$= \frac{P(b_{n+R} \mid \tilde{w}_n, ej)}{1 + \left(\dfrac{P(\tilde{w}_n \mid ei)}{P(\tilde{w}_n \mid ej)}\right) \left(\dfrac{P(ei \mid ee)}{P(ej \mid ee)}\right) + \left(\dfrac{P(\tilde{w}_n \mid ee)}{P(\tilde{w}_n \mid ej)}\right) \left(\dfrac{P(ee \mid ee)}{P(ej \mid ee)}\right)}$$

Likewise for the $f_{n-1} = ee$ and $f_n = ei$ we get a similar expression, but a third term in the sum:

$$\frac{P(w_n, f_{n-1}, f_n)}{P(\tilde{w}_n, f_{n-1})} \Bigg|_{\substack{f_{n-1} = ee, \\ f_n = ei}}$$

$$= \frac{P(w_n \mid ei) P(ei \mid ee)}{P(\tilde{w}_n \mid ei)P(ei \mid ee) + P(\tilde{w}_n \mid ej)P(ej \mid ee) + P(\tilde{w}_n \mid ee)P(ee \mid ee)}$$

$$= \frac{P(b_{n+R} \mid \tilde{w}_n, ei)}{1 + \left(\dfrac{P(\tilde{w}_n \mid ej)}{P(\tilde{w}_n \mid ei)}\right) \left(\dfrac{P(ej \mid ee)}{P(ei \mid ee)}\right) + \left(\dfrac{P(\tilde{w}_n \mid ee)}{P(\tilde{w}_n \mid ei)}\right) \left(\dfrac{P(ee \mid ee)}{P(ei \mid ee)}\right)}$$

Consider now the cases involving self-transitions: $f_{n-1} = xx$ and $f_n = xx$. The derivation parallels that above for $f_{n-1} = ii$ and $f_n = ii$:

$$\frac{P(w_n, f_{n-1}, f_n)}{P(\widetilde{w}_n, f_{n-1})}\bigg|_{\substack{f_{n-1} = ii, \\ f_n = ii}} = \frac{P(w_n \mid ii) \, P(ii \mid ii)}{P(\widetilde{w}_n \mid ie) P(ie \mid ii) + P(\widetilde{w}_n \mid ii) P(ii \mid ii)}$$

$$= \frac{P(b_{n+R} \mid \widetilde{w}_n, ii)}{1 + \left(\dfrac{P(\widetilde{w}_n \mid ie)}{P(\widetilde{w}_n \mid ii)}\right)\left(\dfrac{P(ie \mid ii)}{P(ii \mid ii)}\right)}$$

Similarly, consider $f_{n-1} = jj$ and $f_n = jj$:

$$\frac{P(w_n, f_{n-1}, f_n)}{P(\widetilde{w}_n, f_{n-1})}\bigg|_{\substack{f_{n-1} = jj, \\ f_n = jj}} = \frac{P(w_n \mid jj) \, P(jj \mid jj)}{P(\widetilde{w}_n \mid je) P(je \mid jj) + P(\widetilde{w}_n \mid jj) P(jj \mid jj)}$$

$$= \frac{P(b_{n+R} \mid \widetilde{w}_n, jj)}{1 + \left(\dfrac{P(\widetilde{w}_n \mid je)}{P(\widetilde{w}_n \mid jj)}\right)\left(\dfrac{P(je \mid jj)}{P(jj \mid jj)}\right)}$$

For the $f_{n-1} = ee$ and $f_n = ej$ we get the third term in the sum due to the three possibilities allowed for f_n:

$$\frac{P(w_n, f_{n-1}, f_n)}{P(\widetilde{w}_n, f_{n-1})}\bigg|_{\substack{f_{n-1} = ee \\ f_n = ee}}$$

$$= \frac{P(w_n \mid ee) \, P(ee \mid ee)}{P(\widetilde{w}_n \mid ej) P(ej \mid ee) + P(\widetilde{w}_n \mid ei) P(ei \mid ee) + P(\widetilde{w}_n \mid ee) P(ee \mid ee)}$$

$$= \frac{P(b_{n+R} \mid \widetilde{w}_n, ee)}{1 + \left(\dfrac{P(\widetilde{w}_n \mid ei)}{P(\widetilde{w}_n \mid ee)}\right)\left(\dfrac{P(ei \mid ee)}{P(ee \mid ee)}\right) + \left(\dfrac{P(\widetilde{w}_n \mid ej)}{P(\widetilde{w}_n \mid ee)}\right)\left(\dfrac{P(ej \mid ee)}{P(ee \mid ee)}\right)}$$

In the above expressions we clearly have sequence dependent transitions. For $f_{n-1} = ii$, and $f_n = ie$ for example, we have:

$$\rho|_{GCHMM} = \frac{P(w_n, f_{n-1}, f_n)}{P(\widetilde{w}_n, f_{n-1})}\bigg|_{\substack{f_{n-1} = ii, \\ f_n = ie}}$$

$$= \frac{P(w_n \mid ie) \, P(ie \mid ii)}{P(\widetilde{w}_n \mid ie) P(ie \mid ii) + P(\widetilde{w}_n \mid ii) P(ii \mid ii)} = \frac{P(b_{n+R} \mid \widetilde{w}_n, ie) P(ie \mid ii)}{P(ie \mid ii) + P(ii \mid ii)\left(\dfrac{P(\widetilde{w}_n \mid ii)}{P(\widetilde{w}_n \mid ie)}\right)}$$

While the standard HMM has this ratio with w_n a single element emission sequence, and $P(w_n, f_{n-1}, f_n) = P(w_n \mid f_n)\, P(f_n \mid f_{n-1})$, thus, for the standard HMM:

$$\rho\big|_{\text{std.HMM}} = \left. \frac{P(w_n, f_{n-1}, f_n)}{P(\widetilde{w}_n, f_{n-1})} \right|_{\substack{f_{n-1}=ii, \\ f_n=ie,}} = P(b_{n+R} \mid ie)\, P(ie \mid ii)$$

std.HMM

If we generalized the Std. HMM to higher order Markov models on emissions, to the same order as in the generalized clique, there is still the difference in the transition probability contributions:

$$\rho\big|_{\text{std.HMM}} = P(b_{n+R} \mid \widetilde{w}_n, ie)\, P(ie \mid ii),$$

HO EMs

as can be seen in the ratio of their contributions, and how it is sequence dependent (i.e., dependent on "$\widetilde{w}_i$"):

$$\frac{\rho\big|_{\substack{\text{Std.HMM} \\ \text{HO EMs}}}}{\rho\big|_{\text{GCHMM}}} = P(ie \mid ii) + P(ii \mid ii)\left(\frac{P(\widetilde{w}_i \mid ii)}{P(\widetilde{w}_i \mid ie)} \right).$$

Note that the sequence dependencies (in this and the other footprint transition choices) enter via likelihood ratio terms. These are precisely the type of terms examined in [44] in an effort to improve the HMM-based discriminatory ability via use of SVMs. The "discriminatory" aspect of the key new (sequence-dependent) contribution is most evident in forms like that above, where we have a likelihood ratio for the observed sequences given the different label "classifications" chosen. In the cases that follow we will examine the extreme cases of the likelihood-ratio discriminator strongly classifying one way or the other, or not strongly classifying either way with the given sequence information (making the contribution of knowing that sequence information negligible, which should then reduce to the std. HMM situation, as will be shown). Specifically, we will now examine the above equations in situations where the sequence-dependent likelihood-ratios strongly favor one state model over another, with particular attention as to whether there are sequence dependent scenarios offering recovery of the heavy-tail distribution in example one and recovery of contrast resolution in example two:

Example One:

For $f_{n-1} = ii$ and $f_n = ii$ we showed:

$$\rho = \left.\frac{P(w_n, f_{n-1}, f_n)}{P(\widetilde{w}_n, f_{n-1})}\right|_{\substack{f_{n-1} = ii, \\ f_n = ii}} = \frac{P(b_{n+R} \mid \widetilde{w}_n, ii)}{1 + \left(\dfrac{P(\widetilde{w}_n \mid ie)}{P(\widetilde{w}_n \mid ii)}\right)\left(\dfrac{P(ie \mid ii)}{P(ii \mid ii)}\right)}$$

Example One; Case 1: $P(\widetilde{w}_n \mid ie) \cong P(\widetilde{w}_n \mid ii)$ (likelihood ratio of probabilities is approximately one, leading to a weak (small) classification confidence if a confidence parameterized classifier, like an SVM, is referred to in place of the simple ratio)

$$\rho|_{ie \cong ii} \cong P(b_{n+R} \mid \widetilde{w}_n, ii)\, P(ii \mid ii)/[P(ii \mid ii) + P(ie \mid ii)]$$

$$= P(b_{n+R} \mid \widetilde{w}_n, ii)\, P(ii \mid ii)$$

Thus, in the "uninformed" case we recover regular first-order HMM theory, with geometric distribution on "ii". In this notation, $\rho|_{ie \cong ii}$ refers to the value of ρ when the observed sequence $\widetilde{w}_n$ has approximately the same probability regardless of the state being "ii" or "ie".

Example One; Case 2: $P(\widetilde{w}_n \mid ie) \gg P(\widetilde{w}_n \mid ii)$ (likelihood ratio of probabilities is very large, leading to a strong (large) classification confidence if a confidence parameterized classifier, like an SVM, is referred to in place of the simple ratio)

$$\rho|_{ie \gg ii} \cong P(b_{n+R} \mid \widetilde{w}_n, ii)\left[\frac{P(\widetilde{w}_n \mid ii)P(ii \mid ii)}{P(\widetilde{w}_n \mid ie)P(ie \mid ii)}\right]$$

In this case, we obtain contributions less than the regular first-order HMM counterpart, effectively shortening the geometric distribution on "ii" $\rightarrow$ e.g., it adaptively switches to a shorter, sharper, fall-off on the distribution in a sequence dependent manner.

Example One; Case 3: $P(\widetilde{w}_n \mid ie) \ll P(\widetilde{w}_n \mid ii)$ (likelihood ratio of probabilities is very small, leading to a strong (large) classification confidence if a confidence parameterized classifier, like an SVM, is referred to in place of the simple ratio)

$$\rho|_{ie \ll ii} \cong P(b_{n+R} \mid \widetilde{w}_n, ii)\, 1$$

In this case we obtain contributions greater than the regular first-order HMM theory. In particular, **we recover the heavy tail distribution in a sequence dependent manner**:

$$\left.\frac{P(w_n, f_{n-1}, f_n)}{P(\widetilde{w}_n, f_{n-1})}\right|_{\substack{f_{i-1} \in ie, \\ f_i \in ee}} = P(b_{n+R} \mid \widetilde{w}_n, f_{n-1})$$

Example Two:

One more example-case will be considered, that involving acceptor splice-site recognition. For $f_{n-1} = ii, f_n = ie$ we have:

$$\rho = \left. \frac{P(w_n, f_{n-1}, f_n)}{P(\widetilde{w}_n, f_{n-1})} \right|_{\substack{f_{n-1} = ii, \\ f_n = ie}} = \frac{P(b_{n+R} \mid \widetilde{w}_n, ie)}{1 + \left(\dfrac{P(\widetilde{w}_n \mid ii)}{P(\widetilde{w}_n \mid ie)} \right) \left(\dfrac{P(ii \mid ii)}{P(ie \mid ii)} \right)}$$

Example Two; Case 1: $P(\widetilde{w}_n \mid ie) \cong P(\widetilde{w}_n \mid ii)$

$$\rho|_{ie \cong ii} \cong P(b_{n+R} \mid \widetilde{w}_n, ie) \, P(ie \mid ii)$$

We recover regular HMM theory in the uninformed situation.

Example Two; Case 2: $P(\widetilde{w}_n \mid ie) \gg P(\widetilde{w}_n \mid ii)$

$$\rho|_{ie \gg ii} \cong P(b_{n+R} \mid \widetilde{w}_n, ie)$$

Greater than regular first-order HMM theory. Removes key penalty of $P(ie|ii)$ factor when sequence match overrides. **Resolves weak contrast resolution at first-order.**

Example Two; Case 3: $P(\widetilde{w}_n \mid ie) \ll P(\widetilde{w}_n \mid ii)$

$$\rho|_{ie \ll ii} \cong P(b_{n+R} \mid \widetilde{w}_n, ie) \left[\frac{P(ie \mid ii) P(\widetilde{w}_n \mid ie)}{P(ii \mid ii) P(\widetilde{w}_n \mid ii)} \right]$$

The result is less than that obtained with the regular first-order HMM. This effectively weakens the ie transition strength.

Use of the meta-HMM formalism resolves complications due to heavy-tail duration distributions and weak contrast. This is a new HMM modeling capability. The form of the clique factorization in [1, 3] also has LLR terms such as $P(\widetilde{w}_n \mid ie)/P(\widetilde{w}_n \mid ii)$ that allow for a simple switch from internal scalar-based state discriminant to a vector-based feature, allowing for a similar substitution of a discriminant based on a Support Vector Machine (SVM) as demonstrated for splice sites in [44], and described in the pMM/SVM subsection. These alternate representations do not introduce any significant increase in computational time complexity.

7.2 GHMMs: Full Duration Model

7.2.1 HMM with Duration (HMMD)

In the standard HMM, when a state i is entered, that state is occupied for a period of time, via self-transitions, until transitioning to another state j. If the state interval is given as d, the standard HMM description of the probability distribution on state intervals is implicitly given by:

$$p_i(d) = a_{ii}^{d-1}(1 - a_{ii}) \tag{7.1}$$

where a_{ii} is self-transition probability of state i. As mentioned previously, this geometric distribution is inappropriate in many cases. The standard HHMM-with-Duration (HMMD) replaces the equation above with a $p_i(d)$ that models the real duration distribution of state i. In this way explicit knowledge about the duration of states is incorporated into the HMM. When entered, state i will have a duration of d according to its duration density $p_i(d)$; it then transits to another state j according to the state transition probability a_{ij} (self-transitions, a_{ii}, are not permitted in this formalism). It is easy to see that the HMMD will turn into a HMM if $p_i(d)$ is set to the geometric distribution shown above. The first HMMD formulation was studied by Ferguson [138]. A detailed HMMD description was later given by [33]. There have been many efforts to improve the computational efficiency of the HMMD formulation given its fundamental utility in many endeavors in science and engineering. Notable amongst these are the variable transition HMM methods for implementing the Viterbi algorithm introduced in [139], and the hidden semi-Markov model (HSMM) implementations of the forward-backward algorithm [140].

In [1, 3, 34, 56] it is shown how to "lift" side information that is associated with a region, or transition between regions, by "piggybacking" that side information with the duration side information. We use, as example, HMM incorporation of duration itself as the guide in what follows. In doing so, we arrive at a HSMM formalism for a HMMD. An equivalent formulation of the HSMM was introduced in [139] for the Viterbi algorithm and in [140] for Baum–Welch. In these derivations, however, the maximum-interval constraint is still present (comparisons of these methods were subsequently detailed in [141]). Other HMM generalizations include Factorial HMMs [142] and hierarchical HMMs [143]. For the latter, inference computations scaled as $O(T^3)$ in the original description, and have since been improved to $O(T)$ by [144].

The HSMM formalism introduced here, however, is directly amenable to incorporation of side-information and to adaptive speedup (as described in [1, 3, 34, 36, 56]). For the state duration density $p_i(x = d)$, $1 \leq x \leq D$, we have:

$$p_i(x = d) = p_i(x \geq 1) \cdot \frac{p_i(x \geq 2)}{p_i(x \geq 1)} \cdot \frac{p_i(x \geq 3)}{p_i(x \geq 2)} \cdots \frac{p_i(x \geq d)}{p_i(x \geq d-1)} \cdot \frac{p_i(x = d)}{p_i(x \geq d)} \tag{7.2}$$

where $p_i(x = d)$ is abbreviated as $p_i(d)$ if there is no ambiguity. Define "self-transition" variable $s_i(d) =$ probability that next state is still $\lambda_t = i$, given that i has consecutively occurred d times up to now.

$$p_i(x = d) = \left[\prod_{j=1}^{d-1} s_i(j)\right](1 - s_i(d)), \text{where } s_i(d) = \begin{cases} \dfrac{p_i(x \geq d + 1)}{p_i(x \geq d)} & \text{if } 1 \leq d \leq D-1 \\ 0 & \text{if } d = D \end{cases}$$

$$(7.3)$$

We see with comparison of the equation for $p_i(d)$ above and $p_i(d) = (a_{ii})^{d-1}$ $(1 - a_{ii})$, that we now have similar form, there are "$d - 1$" factors of "s" instead of "a", with a "cap term" "$(1 - s)$" instead of "$(1 - a)$", where the "s" terms are not constant, but only depend on the state's duration probability distribution. In this way, each "s" can mesh with the HMM's dynamic programming table construction for the Viterbi algorithm at the column-level in the same manner that "a" does. Side-information about the local strength of EST matches or homology matches, etc., that can be put in similar form, can now be "lifted" into the HMM model on a proper, locally optimized Viterbi-path. The derivations of the Baum–Welch and Viterbi HSMM algorithms are in [1, 3, 34, 36, 63].

The memory complexity of this method is $O(TN)$. No forward table needs to be saved. The computation complexity is $O(TN^2 + TND)$. In an actual implementation, a scaling procedure may be needed to keep the forward–backward variables within a manageable numerical interval. One common method is to rescale the forward–backward variables at every time index t using the scaling factor $c_t = \Sigma_i f_t(i)$. Here we use a dynamic scaling approach. For this we need two versions of $\theta(k, i, d)$. Then at every time index, we test if the numerical values is too small, if so, we use the scaled version to push the numerical values up; if not, we keep using the unscaled version. In this way, no additional computation complexity is introduced by scaling.

As with Baum–Welch, the Viterbi algorithm for the HMMD is $O(TN^2+TND)$. Because logarithm scaling can be performed for Viterbi in advance, however, the Viterbi procedure consists only of additions to yield a very fast computation. For both the Baum–Welch and Viterbi algorithms, use of the HMMBD algorithm [36] can be employed (as in this work) to further reduce computational time complexity to $O(TN^2)$, thus obtaining the speed benefits of a simple HMM, with the improved modeling capabilities of the HMMD.

The standard HMMD replaces Eq. (7.1) with a $p_i(d)$ that models the real duration distribution of state i. In this way explicit knowledge about the duration of states is incorporated into the HMM. A general HMMD can be illustrated as shown in Figure 7.2.

When entered, state i will have a duration of d according to its duration density $p_i(d)$, it then transits to another state j according to the state transition probability a_{ij} (self-transitions, a_{ii}, are not permitted in this formalism). It is easy to see that the HMMD will turn into a HMM if $p_i(d)$ is set to the geometric distribution shown in Eq. (7.1). The first HMMD formulation was studied by Ferguson [138].

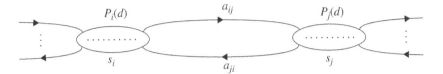

Figure 7.2 The transition schematic for the HMM with duration (HMMD).

A detailed HMMD description was later given by [36] (shown below). There have been many efforts to improve the computational efficiency of the HMMD formulation given its fundamental utility in many endeavors in science and engineering. Notable amongst these are the variable transition HMM methods for implementing the Viterbi algorithm introduced in [139], and the HSMM implementations of the forward–backward algorithm [140].

For an exact-HMMD formalism denote the following: N, the number of states; M, the number of distinct observations (where an observation sequence is denoted as: $B = b_1 b_2 \cdots b_T$); D, the maximum duration length; a_{ij}, the state transition probability; $e_i(k)$, the emission probability: probability of observing k in state i; π_i, the initial state probability: the probability of state i given the observation sequence B; $p_i(d)$, the state duration density: the probability of having exactly d consecutive state i observations after state i is entered. With these definitions the HMMD generalizations to the standard HMM resestimation formulae are as follows:

First define the forward–backward variables:

$f_t(i) = P(b_1...b_t, \lambda_i \text{ end at } t)$

$b_t(i) = P(b_{t+1}...b_T | \lambda_i \text{ end at } t)$

$f_t^*(i) = P(b_1...b_t, \lambda_i \text{ begins at } t + 1)$

$b_t^*(i) = P(b_{t+1}...b_T | \lambda_i \text{ begins at } t + 1)$

where $f_t(i)$ can be calculated by:

$$f_t(i) = \sum_{j=1}^{N} \sum_{d=1}^{D} f_{t-d}(j) a_{ji} p_i(d) \prod_{s=t-d+1}^{t} e_i(b_s) \tag{7.4}$$

Others can be calculated similarly. The relationships among $f, f*, b$, and $b*$ are:

$$f_t^*(i) = \sum_{j=1}^{N} f_t(j) a_{ji} \quad f_t(j) = \sum_{d=1}^{D} f_{t-d}^*(i) p_i(d) \prod_{s=t-d+1}^{t} e_i(b_s)$$

$$b_t(i) = \sum_{j=1}^{N} b_t^*(j) a_{ij} \quad b_t^*(i) = \sum_{d=1}^{D} b_{t+d}(i) p_i(d) \prod_{s=t+1}^{t+d} e_i(b_s)$$

Based on the above definitions and equations, we have the following maximum likelihood re-estimation formulas (the Baum–Welch algorithm) for HMMD [36]:

$$\pi_i^{\text{new}} = \frac{\pi_i b_0^*(i)}{P(B)} \tag{7.5}$$

$$a_{ij}^{\text{new}} = \frac{\sum\limits_{t=1}^{T} f_t(i)a_{ij}b_t^*(j)}{\sum\limits_{j=1}^{N}\sum\limits_{t=1}^{T} f_t(i)a_{ij}b_t^*(j)} \tag{7.6}$$

$$e_i^{\text{new}}(k) = \frac{\sum^T_{t=1}\; \underset{s.t.b_t = k}{\left\{\sum_{r<t}f_r^*(i)b_r^*(i) - \sum_{r>t}f_r(i)b_r(i)\right\}}}{\sum_{k=1}^{M}\sum^T_{t=1}\; \underset{s.t.b_t = k}{\left\{\sum_{r<t}f_r^*(i)b_r^*(i) - \sum_{r>t}f_r(i)b_r(i)\right\}}} \tag{7.7}$$

$$p_i^{\text{new}}(k) = \frac{\sum\limits_{t=1}^{T} f_t^*(i)p_i(d)b_{t+d}(i)\prod\limits_{s=t+1}^{t+d} e_i(b_s)}{\sum_{d=1}^{D}\sum\limits_{t=1}^{T} f_t^*(i)p_i(d)b_{t+d}(i)\prod\limits_{s=t+1}^{t+d} e_i(b_s)} \tag{7.8}$$

In the above equations we can see that $D^2/2$ times more computational cost is required than the standard HMM. We now introduce a more efficient implementation of the explicit duration HMM that uses HSMMs.

7.2.2 Hidden Semi-Markov Models (HSMM) with sid-information

It is shown in [34, 36, 56] how to "lift" side information that is associated with a region, or transition between regions, by "piggybacking" that side information along with the duration side information. We use, as example, HMM incorporation of duration itself as the guide. In doing so we arrive at a HSMM formalism. An equivalent formulation of the HSMM was introduced in [139] for the Viterbi algorithm and in [140] for Baum–Welch. In these derivations, however, the maximum-interval constraint is still present (comparisons of these methods were subsequently detailed in [141]). Other HMM generalizations include Factorial HMMs [142] and hierarchical HMMs [143]. For the latter, inference computations scaled as $O(T^3)$ in the original description, and has since been improved to $O(T)$ by [144].

The Baum–Welch algorithm in the martingale side-information HMMD formalism

We define the following three variables to simplify what follows:

$$\bar{s}_i(d) = \begin{cases} 1 - s_i(d+1) & \text{if } d = 0 \\ \dfrac{1 - s_i(d+1)}{1 - s_i(d)} \cdot s_i(d) & \text{if } 1 \leq d \leq D-1 \end{cases} \tag{7.9}$$

$$\theta(k, i, d) = e_i(k)\bar{s}_i(d) \qquad 0 \leq d \leq D-1 \tag{7.10}$$

$$\varepsilon(k, i, d) = e_i(k)s_i(d) \qquad 1 \leq d \leq D-1 \tag{7.11}$$

Define: $f'_t(i, d) = P(b_1 b_2 ... b_t, \lambda_t = i, \text{and i has cons.occ.} d \text{ times up to } t)$

$$f'_t(i, d) = \begin{cases} e_i(b_t) \displaystyle\sum_{j=1, j\neq i}^{N} F_{t-1}(j)a_{ji} & \text{if } d = 1 \\ f'_{t-1}(i, d-1)s_i(d-1)e_i(b_t) & \text{if } 2 \leq d \leq D \end{cases} \tag{7.12}$$

Define: $\bar{f}_t(i, d) = P(b_1 b_2 \cdots b_t, \lambda_t = i \text{ ends at } t \text{ with duration } d)$

$$= f'_t(i, d)(1 - s_i(d)) \qquad 1 \leq d \leq D$$

$$\bar{f}_t(i, d) = \begin{cases} \theta(b_t, i, d-1)F'_{t-1}(i) & \text{if } d = 1 \\ \theta(b_t, i, d-1)\bar{f}_{t-1}(i, d-1) & \text{if } 2 \leq d \leq D \end{cases} \tag{7.13}$$

where

$$F'_t(i) = \sum_{j=1, j\neq i}^{N} F_t(j) * a_{ji} \qquad F_t(i) = \sum_{d=1}^{D} f'_t(i, d)(1 - s_i(d)) \tag{7.14}$$

Define: $b'_t(i, d) = P(b_t b_{t+1} \cdots b_T, \lambda_t = i \text{ will have a duration of } d \text{ from } t)$

$$b'_t(i, d) = \begin{cases} \theta(b_t, i, d-1)B'_{t+1}(i) & \text{if } d = 1 \\ \theta(b_t, i, d-1)b'_{t+1}(i, d-1) & \text{if } 1 < d \leq D \end{cases} \tag{7.15}$$

where

$$B'_t(i) = \sum_{j=1, j\neq i}^{N} a_{ij}B_t(j) \qquad B_t(i) = \sum_{d=1}^{D} b'_t(i, d) \tag{7.16}$$

Now $f, f*, b$ and $b*$ can be expressed as:

$$f^*_t(i) = \frac{f'_{t+1}(i, 1)}{e_i(b_{t+1})}; \quad b^*_t(i) = B_{t+1}(i); \quad b_t(i) = B'_{t+1}(i); \quad f_t(i) = F_t(i)$$

Now define

$$\omega(t, i, d) = \bar{f}_t(i, d)B'_{t+1}(i) \tag{7.17}$$

$$\mu_t(i, j) = P(b_1 \cdots b_T, \lambda_t = i, \lambda_{t+1} = j) = F_t(i)a_{ij}B_{t+1}(j) \tag{7.18}$$

$$\varphi(i,j) = \sum_{t=1}^{T-1} \mu_t(i,j) \tag{7.19}$$

$$v_t(i) = P(b_1 \cdots b_T, \lambda_t = i) = \begin{cases} \pi(i)B_1(i) & \text{if } t = 1 \\ v_{t-1} + \sum_{j=1, j\neq i}^{N} (\mu_{t-1}(j,i) - \mu_{t-1}(i,j)) & \text{if } 2 \leq t \leq T \end{cases} \tag{7.20}$$

Using the above equations:

$$\pi_i^{\text{new}} = \frac{\pi_i b_1'(i,1)}{P(B)} \tag{7.21}$$

$$a_{ij}^{\text{new}} = \frac{\varphi(i,j)}{\sum_{j=1}^{N} \varphi(i,j)} \tag{7.22}$$

$$e_i^{\text{new}}(k) = \frac{\sum_{t=1 \text{ s.t.} O_t = k}^{T} v_t(i)}{\sum_{t=1}^{T} v_t(i)} \tag{7.23}$$

$$p_i(d) = \frac{\sum_{t=1}^{T} \omega(t,i,d)}{\sum_{d=1}^{D} \sum_{t=1}^{T} \omega(t,i,d)} \tag{7.24}$$

The Viterbi algorithm in the martingale side-information HMMD formalism.

Define $v_t(i, d)$ = the most probable path that consecutively occurred d times at state i at time t:

$$v_t(i,d) = \begin{cases} e_i(b_t) \max_{j=1, j\neq i}^{N} V_{t-1}(j)a_{ji} & \text{if } d = 1 \\ v_{t-1}(i, d-1)s_i(d-1)e_i(b_t) & \text{if } 2 \leq d \leq D \end{cases} \tag{7.25}$$

where

$$V_t(i) = \max_{d=1}^{D} v_t(i,d)(1 - s_i(d)) \tag{7.26}$$

The goal is to find: $\arg\max_{[i,d]} \{ \max_{i,d}^{N,D} v_t(i,d)(1 - s_i(d)) \}$

$$\text{Define: } \bar{s}_i(d) = \begin{cases} 1 - s_i(d+1) & \text{if } d = 0 \\ \dfrac{1 - s_i(d+1)}{1 - s_i(d)} \cdot s_i(d) & \text{if } 1 \le d \le D-1 \end{cases}$$

$$\theta(k, i, d) = \bar{s}_i(d-1) e_i(k) \qquad 1 \le d \le D \tag{7.27}$$

$$v_t'(i, d) = v_t(i, d)(1 - s_i(d)) \quad 1 \le d \le D = \begin{cases} \theta(b_t, i, d) \displaystyle\max_{j=1, j \ne i}^{N} V_{t-1}(j) a_{ji} & \text{if } d = 1 \\ v_{t-1}'(i, d-1)\theta(b_t, i, d) & \text{if } 2 \le d \le D \end{cases} \tag{7.28}$$

where

$$V_t(i) = \max_{d=1}^{D} v_t'(i, d) \tag{7.29}$$

The goal is now:

$$\arg\max_{[i,d]} \left\{ \max_{i,d}^{N,D} v_T'(i, d) \right\} \tag{7.30}$$

If we do a logarithm scaling on $\bar{s}$, a and e in advance, the final Viterbi path can be calculated by:

$$\theta'(k, i, d) = \log \theta(k, i, d) = \log \bar{s}_i(d-1) + \log e_i(k) \qquad 1 \le d \le D \tag{7.31}$$

$$v_t'(i, d) = \begin{cases} \theta'(b_t, i, d) + \displaystyle\max_{j=1, j \ne i}^{N} \left(V_{t-1}(j) + \log a_{ji} \right) & \text{if } d = 1 \\ v_{t-1}'(i, d-1) + \theta'(b_t, i, d) & \text{if } 2 \le d \le D \end{cases} \tag{7.32}$$

where the argmax goal above stays the same.

A summary of the Baum–Welch training algorithm is as follows:

1) Initialize elements (λ) of HMMD.
2) Calculate $b_t'(i,d)$ using Aqsa. (7.15) and (7.16) (save the two tables: $B_t(i)$ and $B_t'(i)$).
3) Calculate $\bar{f}_t(i, d)$ using Eqs. (7.13) and (7.14).
4) Reestimate elements (λ) of HMMD using Eqs. (7.17)–(7.24).
5) Terminate if stop condition is satisfied, else goto step 2.

The memory complexity of this method is O(TN). As shown above, the algorithm first does backward computing (step (2)), and saves two tables: one is $B_t(i)$, the other is $B_t'(i)$. Then at very time index t, the algorithm can group the computation of step (3) and (4) together. So no forward table needs to be saved. We can do a rough estimation of HMMD's computation cost by counting multiplications inside the loops of $\Sigma^T \Sigma^N$ (which corresponds to the standard HMM computational cost) and $\Sigma^T \Sigma^D$ (the additional computational cost incurred by the HMMD). The computation complexity is O(TN² + TND). In an actual implementation a scaling procedure may be needed to keep the forward–backward variables within a manageable numerical interval. One common method is to rescale the forward–backward variables at every time index t using the scaling factor $c_t = \Sigma_i f_t(i)$. Here we use a dynamic scaling approach. For this we need two versions of $\theta(k, i, d)$. Then at every time index, we test if the numerical values is too small, if so, we use the scaled version to push the numerical values up; if not, we keep using the unscaled version. In this way no additional computation complexity is introduced by scaling.

As with Baum–Welch, the Viterbi algorithm for the HMMD is O(TN² + TND). Because logarithm scaling can be performed for Viterbi in advance, however, the Viterbi procedure consists only of additions to yield a very fast computation. For both the Baum–Welch and Viterbi algorithms, use of the HMMBD algorithm [36] can be employed (as in this work) to further reduce computational time complexity to O(TN²), thus obtaining the speed benefits of a simple HMM, with the improved modeling capabilities of the HMMD.

7.2.3 HMM with Binned Duration (HMMBD)

The intuition guiding the HMMBD approach is that the standard HMM already does the desired duration modeling when the distribution modeled is geometric, suggesting that, with sufficient effort, a self-tuning explicit HMMD might be possible to achieve HMMD modeling capabilities at HMM computational complexity in an adaptive context.

The duration distribution of state i consists of rapidly changing probability regions (with small change in duration) and slowly changing probability regions. In the standard HMMD all regions share an equal computation resource (represented as D substates of a given state) – this can be very inefficient in practice. In this section, we describe a way to recover computational resources, during the training process, from the slowly changing probability regions. As a result, the computation complexity can be reduced to O(TN²+TND*), where D* is the number of "bins" used to represent the final, coarse-grained, probability distribution. A "bin" of a state is a group of substates with consecutive duration. For example, $f(i, d), f(i, d + 1), ...f(i, d + \delta d)$ can be grouped into one bin. The bin size is a measure of the granularity of the evolving length distribution approximation. A fine-granularity is retained in the active regions, perhaps with only one length state per bin, while a coarse-granularity is adopted in weakly changing regions, with possibly hundreds of length states per bin. An important generalization to the exact, standard, length-truncated, HMMD is suggested for handling long duration state intervals – a "tail bin". Such a bin is strongly indicated for good modeling on certain important distributions, such as the long-tailed distributions often found in nature, the exon and intron interval distributions found in gene-structure modeling in particular. In practice, the idea is to run the exact HMMD on a small portion, δT, of the training data, at $O(\delta TNN + \delta TND)$ cost, to get an initial estimate of the state interval distributions. Some preliminary course-graining is then performed, where strongly indicated, and the number of bins representing the length distribution is reduced from D to D'. The exact HMMD is then performed on the D' substate model for another small portion of the training data, at computational expense $O(\delta TNN + \delta TND')$. This is repeated until the number of bin states, $D*$, reduces no further, and the bulk of the training then commences with the D* bin-states length distribution model at expense $O(TN^2 + TND*)$. The key to this process is the retention of training information during the "freezing out" of length distribution states, and such that the D* bin state training process can be done at expense $O(TN^2 + TND*) \approx O(TN^2)$, which is the same complexity class as the standard HMM itself.

Starting from the above binning idea, for substates in the same bin, a reasonable approximation is applied:

$$\sum_{d'=d}^{d+\delta_d} f_t(i,d')\theta(b_t,i,d') \; = \theta(b_t,i,\overline{d}) \sum_{d'=d}^{d+\delta_d} f_t(i,d') \tag{7.33}$$

where $\overline{d}$ is the duration representative for all substates in this bin.

We begin in the first sub-section below with a description of the Baum–Welch algorithm in the adaptive HSMM formalism. This is followed by a sub-section with a description of the Viterbi algorithm in the adaptive HSMM formalism.

The Baum–Welch algorithm in the adaptive HMMD formalism
Define:

$$f\mathrm{prod}_t(i,n) \; = \prod_{t-\delta_d(i,n)}^{t} \theta(b_t,i,\overline{d}) \tag{7.34}$$

Based on the above approximation and equation, formulas (7.13) and (7.14) used by forward algorithm can be replaced by:

$$f\mathrm{bin}_t(i,n) = P(b_1...b_t, \lambda_t = i \text{ ends at } t \text{ with duration between } d \text{ and } d + \delta_d(i,n))$$

$$= \begin{cases} f\text{bin}_{t-1}(i,n)\theta(b_t,i,\overline{d}) - \text{pop}_t(i,n) + F'_{t-1}(i) & \text{if } n = 1 \\ f\text{bin}_{t-1}(i,n)\theta(b_t,i,\overline{d}) - \text{pop}_t(i,n) + \text{pop}_t(i,n-1) & \text{if } 1 < n < D^* \end{cases}$$

$$(7.35)$$

where

$$F_t(i) = \sum_{n=1}^{D^*} f\text{bin}_t(i,n) \quad F'_{t-1}(i) = \sum_{j=1;j\neq i}^{N} F_t(i)a_{ji} \tag{7.36}$$

$$\text{pop}_t(i,n) = \text{queue}(i,n) * f\text{prod}_t(i,n) \tag{7.37}$$

After the above calculations two updates are needed:

$$\text{queue}(i,n).\text{push}(\text{pop}_t(i,n-1)) \tag{7.38}$$

$$f\text{prod}_t(i,n) = f\text{prod}_t(i,n)/\theta(b_{t-\delta_d(i,n)},i,\overline{d}) \tag{7.39}$$

The explanation for push and pop operations, etc., begins with associating every bin with a queue *queue(i, n)*. The queue's size is equal to the number of substates grouped by this bin. At every time index, the oldest substate: $f(i, d + \delta_d(i, n))$ will be shifted out of its current bin and pushed into its next bin ("queue.push"), where *queue(i, n)* stores the original probability of each substates in that bin when they were pushed in. So when one substate becomes old enough to move to next bin, its current probability can be recovered by first popping out its original probability, then multiplied by its "gain". Then an update is applied. Similarly, define:

$$\text{bprod}_t(i,n) = \prod_{t}^{t+\delta_d(i,n)} \theta(b_t,i,\overline{d}) \tag{7.40}$$

Formulas (7.15) and (7.16) used by the backward algorithm can be replaced by

$b\text{bin}_t(i,n) = P(b_1 \ldots b_t, \lambda_t = i$ has remaining a duration between d and $d + \delta_d(i,n)$ at $t = b\text{bin}_{t+1}(i,n)\theta(b_t,i,\overline{d}) - \text{pop}_t(i,n) + B'_{t+1}(i)$
if $n = 1$ $b\text{bin}_{t+1}(i,n)\theta(b_t,i,\overline{d}) - \text{pop}_t(i,n) + \text{pop}_t(i,n+1)$ if $1 < n < D^*$

$$(7.41)$$

where

$$B_t(i) = \sum_{n=1}^{D^*} b\text{bin}_t(i,n) \quad B'_t(i) = \sum_{j=1;j\neq i}^{N} B_t(j)a_{ij} \tag{7.42}$$

$$\text{pop}_t(i,n) = \text{queue}(i,n).\text{pop} * \text{bprod}_t(i,n) \tag{7.43}$$

After the above calculation two updates are needed:

$$\text{queue}(i, n).\text{push}(\text{pop}_t(i, n + 1)) \tag{7.44}$$

$$\text{bprod}_t(i, n) = \text{bprod}_t(i, n)/\theta\left(b_{t + \delta_d(i,n)}, i, \overline{d}\right) \tag{7.45}$$

The reestimation formulas stay unchanged.

The Viterbi algorithm in the adaptive HMMD formalism

The idea is similar to the one for adaptive Baum–Welch training (with computation complexity also $O(TN^2 + TND*)$), where the following formulas are used:

$$\text{New}_t(i, n) = \begin{cases} \max_{j = 1, j \neq i}^{N}\left(m_{t-1}(j) + \log a_{ji}\right) & \text{if } n = 1 \\ \text{sum}_{t-1}(i, n) - \text{queue}(i, n - 1).\text{pop} & \text{if } 1 < n \leq D^* \end{cases} \tag{7.46}$$

$$\text{sum}_t(i, n) = \begin{cases} 0 & \text{if } t = 1 \\ \text{sum}_{t-1}(i, n) + \theta'\left(b_t, i, \overline{d}\right) & \text{if } 1 < t \leq T \end{cases} \tag{7.47}$$

$$D_t(i, n) = \text{sum}_t(i, n) - \text{new}_t(i, n) \tag{7.48}$$

$$\text{queue}(i, n).\text{push}(D_t(n, i)) \tag{7.49}$$

$$\text{sort}(i, n).\text{insert}(D_t(n, i)) \tag{7.50}$$

$$m_t(i, n) = \max\{m_t(i, n), D_t(n, i)\} \tag{7.51}$$

$$m_t(i) = \max_n^{D^*} m_t(i, n) \tag{7.52}$$

The usage of the above relations is described in [1, 36]. Note: there is non-trivial handling of many stack operations in order to attain the theoretically indicated $O(TND)$ to $O(TND^*)$ improvement in actual implementation, as described in detail in [205].

Adaptive null-state binning for O(TN) computation

During the HMM Viterbi table construction for each of T sequence data values there is a column entry, and for each of N states there is a row. At each column the HMM Viterbi algorithm must look to the past column entries as it populates the table from left to right, thus leading to an $O(TN^2)$ computation. If we establish an adaptive binning capability, reminiscent of what was done with the HMMBD method, then we can keep track of lists with respect to each state that correspond to prior column transitions to that state. If we, in particular, track those Viterbi most-probable-paths that arrive at our state cell with probability below some cutoff (with respect to the other probabilities arriving at that cell), we can ignore transitions from such cells in later column computations. What results is an initial $O(tN^2)$ ($t \ll T$) computation to learn the state lists for above cut-off transitions (suppose K on average), followed by the main body of the $O(TNK)$ computation (with $K \ll N$).

A method is also possible comprising use of a "fastViterbi" process where $O(TN^2) \rightarrow O(TmN)$ via learned, local, max-path ordering in a given column of the Viterbi computation for the highest "m" values. Subsequent columns first only examine the top "m" max-paths and if their ordering is retained, and their total probability advanced sufficiently, then the other states remain "frozen-out" with a large grouping (binning) on the probabilities on those states used to maintain their probability information (and correct normalization summing) when going forward column-by-column, with reset to full column evaluation on the individual state level when the m values fall out of their initially identified ordering.

A method is possible comprising use of a fastViterbi with null-binning process where $O(TN^2) \rightarrow O(Tmn) \rightarrow O(T)$ via learned global and local aspects of the data as indicated above. This approach offers significant utility as a purely HMM-based alignment algorithm that may outperform BLAST with comparable time complexity.

7.3 GHMMs: Linear Memory Baum–Welch Algorithm

Table chunking methods for the dynamic programming algorithms have been developed that involve only a single-pass computation analogous to the Viterbi algorithm (ignoring the O(L) traceback) [1, 3]. The Viterbi algorithm efficiently calculates the most probable state path. The Baum–Welch algorithm calculates the probability of having a state at a particular index, summing over all path probabilities that arrive at that state-instance, and is usually implemented as two passes, for the forward and backward parameters. In the Linear Memory HMM introduced in [42], however, the Baum–Welch implementation has a distinctive trait other than a linear memory implementation, it's also a "single-pass" implementation for the algorithm, which is needed for the Viterbi single-pass referenced, overlap-stitched, reconstituted signal in a distributed processing setting (described in Section 7.8 that follows). This can be used for brute force, and massively scalable, computational speed-up on all the HMM-based algorithms used in the SSA Protocol.

Following the notation used in [42], $t_{i,j}(t,m)$ is the weighted sum of probabilities of all possible state paths that emit subsequence $b_1,...,b_t$ and finish in state $\lambda_t = m$, taking an $\lambda_t = i \rightarrow \lambda_{t+1} = j$ ($i \rightarrow j$) transition at least once (for some t) where the weight of each state path is the number of $i \rightarrow j$ transitions that it takes. Processing of the entire $t_{i,j}(t,m)$ recurrence takes memory proportional to $O(NQ)$ and processor time $O(TNQQ_{max})$.

Initially, since no transitions have been made, we have $t_{i,j}(1,m) = 0$. After initialization we have the following recurrence steps

$$t_{i,j}(t, m) = f_i(t - 1) \, a_{im} \, em(b_t)\delta(m = j) + \sum_{n=1}^{N} t_{i,j}(t - 1, n) \, a_{nm} \, em(b_t)$$

$$(7.53)$$

The computation is in-step with the forward variable as a single-pass computation, where the delta function is defined as: $\delta(m = j) = \begin{cases} 1, & \text{if } m = j \\ 0, \text{otherwise} \end{cases}$. At a certain time moment t we need to score the evidence supporting transition between nodes i and j, which is the sum of probabilities of all possible state paths that emit subsequence $b_1,...,b_{t-1}$, and finish in state i (forward probability $f_{i(t-1)}$), multiplied by transition a_{ij} and emission $e_j(b_t)$ probabilities upon arrival to b_t. We extend the weighted paths containing evidence of $i{\rightarrow}j$ transitions made at previous time moments $1,...,t-1$ further down the trellis in the second part of the equation above. Finally, by the end of the recurrence, we marginalize the final state m out of probability $t_{i,j}(T,m)$ to get a weighted sum of state paths taking transition $i{\rightarrow}j$ at various time moments. Thus, we estimate transition utilization using

$$a_{ij} = \frac{\sum_{m=1}^{N} t_{i,j}(T, m)}{\sum_{j \in \text{out(state } i)} \sum_{m=1}^{N} t_{i,j}(T, m)} \qquad (7.54)$$

where out(state i) of nodes connected by edges from state i.

The following algorithm updates the "emission" parameters for the set of discrete symbol probability distributions $E = \{e_1(b),..., e_N(b)\}$ in O(NED) memory and O(TNEDQ$_{max}$) time. According to [42], $e_i(b,t,m)$ is the weighted sum of probabilities of all possible state paths that emit subsequence $b_1,...,b_t$ and finish in state m, for which state i emits observation b at least once where the weight of each state path is the number of b emissions that it makes from state i. Initialization step: $e_i(b,1,m) = f_{mi}\delta(i = m)\delta(b = b_1)$. After initialization we make the recurrence steps, where we correct emission recurrence presented in [145]:

$$e_i(b, t, m) = f_{mi}\delta(i = m)\delta(b = b_t) + \sum_{n=1}^{N} ei(b, t - 1, n) \, a_{nm} \, e_m(b_t) \qquad (7.55)$$

Finally, by the end of the recurrence, we marginalize the final state m out of $e_i(b,T,m)$ and estimate the emission parameters through normalization

$$ej(b) = \frac{\sum_{m=1}^{N} ei(b, T, m)}{\sum_{\gamma=1}^{D} \sum_{m=1}^{N} ei(b, T, m)} \qquad (7.56)$$

The forward sweep takes O(TNQ$_{max}$) time, where only the values of $f_{i(t-1)}$ for $1 \leq i \leq N$ are needed to evaluate f_{it}, thus rendering memory requirement to O(N) for the forward algorithm. Computing $e_i(b,t,m)$ takes O(NED) previous probabilities of $e_i(b, t-1, m)$ for $1 \leq m \leq N, 1 \leq i \leq E, 1 \leq b \leq D$. Recurrent updating of each $e_i(b, t, m)$ probability elements takes $O(Q_{max})$ summations, totaling $O(TNEDQ_{max})$.

7.4 GHMMs: Distributable Viterbi and Baum–Welch Algorithms

7.4.1 Distributed HMM processing via "Viterbi-overlap-chunking" with GPU speedup

In HMM signal processing latency becomes very prohibitive when attempting to increase device bandwidth or when input datasets are large. Described in what follows are results from performing HMM algorithms in a distributed manner by breaking the full HMM table computation into overlapping chunks and leveraging the Markovian assumption underlying the HMM to help arrive at a chunk to full table reconstruction. The pathological instances where the distributed merges can fail to exactly reproduce the non-distributed HMM calculation can be made as least likely as desired with sufficiently strict, but not computationally expensive, segment join conditions. In this way, the distributed HMM provides a feature extraction that is equivalent to that of the sequentially run, general definition HMM, and with a speedup factor approximately equal to the number of processes (threads) operating on the data. The Viterbi most probable path calculation and the expectation/maximization (EM) calculation can both be performed in this distributed processing context.

The linear memory implementation described previously (and in [145]) was optimized according to the observation that Viterbi traceback paths in the Viterbi procedure typically converge to the most likely state path and travel together to the beginning of the decoding table – the picture being much like a river with minor tributaries backtracking onto that river, and maybe those "tributaries" themselves have more minor state paths converging into them, etc. But the trait that is most notable in the convergence-durations to the "main-tributary," or what is to be the most likely (Viterbi) path, is that it is usually a modest number of columns for many data types. This backwards Markovian memory loss on a tributary with respect to its origin (said to occur when backtracked and mixed with the main, Viterbi, convergence path of the tributaries) is hypothesized to be an indicator of the span of overlap sequence needed to have Viterbi path probabilities in a given column that have settled into their properly ordered relative probabilities in that column. Further column processing refinement to bring the relative values of the Viterbi-path probabilities into better estimation is then possible. In distributed processing efforts, this "Viterbi relaxation time" is a key parameter that can be used to design an optimally overlapping chunking of the data sequence in a distributed speed-up on the sequence analysis. For further details see [1, 3, 12].

A distributed signal processing test of some basic chunk reconstruction heuristics was performed on 5 computers with 300 signals in the study in [1, 3, 12]. Each signal had 5000 samples. The resulting Viterbi paths matched between the distributed HMM and standard HMM on a 10-column segment. For the standard HMM, EM training (five loops) the Viterbi algorithm took 272 seconds. For distributed HMM with five CPU's, the computational time was reduced to 69 seconds. So using five computers, we had a speedup of 3.94. A perfect de-segmentation was performed with an $N = 10$ match window as indicated, initially, but it was found that a perfect re-stitching of segments was also possible simply with $N = 1$ (see Figure 7.3), due to the implicit stringency of the simultaneity condition (the overlap match, at the one position corresponding to $N = 1$, must globally index to the same observation data index for both segments). The multi-chunk re-stitching makes use of the Viterbi path and the entire set of Viterbi traceback pointers in a given overlap set of columns.

7.4.2 Relative Entropy and Viterbi Scoring

Sometimes tFSA methods can identify possible signal regions, but not all such regions contain true signals or good signals. This is simply because anomalous signal may come in multiple forms, some "good" in the sense that it is usable, some

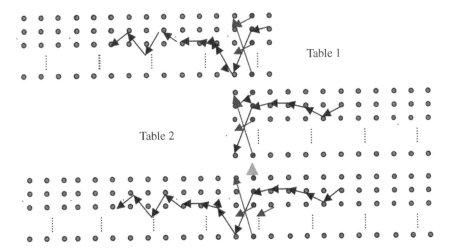

Figure 7.3 Viterbi column-pointer match de-segmentation rule. Table 1 and Table 2 (in figure) are overlapped. And their blue columns have the same pointers. Then the index of this blue column becomes the joint. The black pointers form the final Viterbi path.

"bad" in that the signals aren't sufficiently stable for use (and should be dropped from the analysis, like dropping "weak" confidence signals in SVM analysis [1, 3]). Having an objective, minimally informed, process for evaluating good or bad signals is needed. In this context relative entropy comparison on newly captured signal with known good signals can be used to determine if the unknown newly acquired signal is sufficiently useful. Relative entropy is the natural measure for comparison in this setting as the HMM priors, emissions, and transition probabilities, lead to a comparison between discrete probability based feature vectors (such as that used in [1, 3]). The proper measure for comparison of discrete probabilities is relative entropy [1, 3] (not Euclidean distance). The priors, transitions, and emissions obtained on a particular signal region can have strong characterization by use of the Baum–Welch algorithm to re-estimate these parameters in such a way that they very closely reflect the attributes of the signal acquisition in question. In addition to the HMM feature extraction in terms of emission and transition probabilities, and the priors from the transitions, another set of feature components is the Viterbi path state frequencies (similar to the priors).

It is possible to have a good-state confidence test simply in terms of the relative entropy between the acquired signals prior probabilities and that of the known "good" signal class. The weakness of this approach is that the priors entirely miss the transition structure, which might be highly distinctive, unless the prior states are actually dimer states or higher order footprint states thereby encapsulating the transition probabilities anyway [1, 3]. In the examples in [1, 3], the dimer and footprint states are introduced in analyzing genomic data for gene structure discovery, and the methods can be used, practically speaking, due to the low order of the genomic states. In channel current studies, and power signal analysis in general, there are typically too many states with the generic HMM. This problem can be overcome by use of EVA-projection on the acquired signal to project the noisy signal onto the dominant blockade states, however, resulting in a much smaller set of EVA-projected blockade states, often as few as four as with the genomic analysis, and in this setting the higher order states can be introduced and used in the higher-order-state prior comparisons.

7.5 Martingales and the Feasibility of Statistical Learning (further details in Appendix)

Martingale Definition [102]

A stochastic process $\{X_n; n = 0,1, ...\}$ is martingale if, for $n = 0,1, ...,$

1) $E[|X_n|] < \infty$
2) $E[X_{n+1}|X_0, ..., X_n] = X_n$

Def.: Let $\{X_n; n = 0,1, ...\}$ and $\{Y_n; n = 0,1, ...\}$ be stochastic processes. We say $\{X_n\}$ is martingale with respect to (w.r.t) $\{Y_n\}$ if, for $n = 0,1, ...$:

1) $E[|X_n|] < \infty$
2) $E[X_{n+1}|Y_0, ..., Y_n] = X_n$

Examples of Martingales:

a) Suns of independent random variables: $X_n = Y_1 + ... + Y_n$.
b) Variance of a Sum $X_n = \left(\sum_{k=1}^{n} Yk\right)^2 - n\sigma^2$
c) Have induced Martingales with Markov Chains! ....
d) For HMM learning, sequences of likelihood ratios are martingale....

The asymptotic equipartition theorem (AEP) and Hoeffding Inequalities (critical in Chapter 10) have both been generalized to Martingales.

Induced Martingales with Markov Chains

Let $\{Y_n; n = 0,1, ...\}$ be a Markov Chain (MC) process with transition probability matrix $P = ||P_{ij}||$. Let f be a bounded right regular sequence for P:

$f(i)$ is non-negative and $f(i) = \sum_{k=1}^{n} Pij f(j)$. Let $X_n = f(Y_n) \rightarrow E[|X_n|] < \infty$ (since f is bounded). Now we have:

$$E[X_{n+1} \mid Y_0, ..., Y_n]$$
$$= E[f(Y_{n+1}) \mid Y_0, ..., Y_n]$$
$$= E[f(Y_{n+1}) \mid Y_n] \text{ (due to MC)}$$
$$= \sum_{k=1}^{n} PYn, jf(j) \text{ (def.of } P_{ij} \text{ and} f)$$
$$= f(Y_n)$$
$$= X_n$$

In HMM learning have sequences of likelihood ratios, which is a martingale, proof:

Let $Y_0, Y_1, ...$ be iid rv.s and let f_0 and f_1 be probability density functions. A stochastic process of fundamental importance in the theory of testing statistical hypotheses is the sequence of likelihood ratios:

$$X_n = \frac{f1(Y0)f1(Y1)...f1(Yn)}{f0(Y0)f0(Y1)...f0(Yn)}, n = 0, 1, ...$$

Assume $f_0(y) > 0$ for all y:

$$E[Xn + 1 \mid Y0, ..., Yn] = E\left[Xn\left(\frac{f1(Y_{n+1})}{f0(Y_{n+1})}\right) \mid Y0, ..., Yn\right] = Xn E\left[\frac{f1(Y_{n+1})}{f0(Y_{n+1})}\right]$$

When the common distribution of the Y_k's (used in the "E" function) has f_0 as its probability density, have:

$$E\left[\frac{f1(Y_{n+1})}{f0(Y_{n+1})}\right] = 1$$

So, $E[X_{n+1}|Y_0, ..., Y_n] = X_n$. So likelihood ratios are martingale when the common distribution is f_0.

7.6 Exercises

7.1 Derive the $P(B,L)$ equation.

7.2 Re-do Example One with more detail or explanation, explain significance of overcoming bias towards geometric length distributions by the more informed situation in the example.

7.3 Re-do Example Two with more detail or explanation, explain significance of contrast resolution.

7.4 Rederive result that the standard HMM has geometric length distributions.

7.5 Re-derive 7.4–7.8.

7.6 Re-derive the linear memory recursion relations 7.53 and 7.54.

7.7 Re-derive the linear memory recursion relations 7.55 and 7.56.

7.8 Show O(TNEDQmax) for the linear memory recursion relations.

7.9 implement HMM code shown, apply to gene-finding using annotated genome ata (from genbank) to train test performance.

8

Neuromanifolds and the Uniqueness of Relative Entropy

8.1 Overview

This chapter goes into theoretical detail to show modern arguments for the choice of relative entropy as difference measure on distributions. Also shown is a fundamentally derived variant of Expectation Maximization (EM), referred to as "em." Following the derivation of Amari [113–115], geometric representations are considered for two types of statistical information: families of probability distributions and families of Neural Nets. Emphasis is placed on the dually flat formulation of "information geometry" by Amari, but the development is such that other formulations might be considered as well.

Why establish a geometric formulation? Here are some motivations:

1) to obtain a concise representation of the elements of families of probability distributions,
2) if a dynamical description is eventually sought, such as in adaptive algorithms that "learn," or optimize, then an established kinematical representation can offer much needed clarity,
3) if a variational formalism is provided, the spatial and "temporal" (inertial) kinematical structures can be clearly expressed, separate from the unknown dynamical structures.

Once the geometric formulation is established, we stand to gain substantially by systematic exploration of various algorithms, permitting the more efficient algorithms to be isolated. In part, the search for a more efficient algorithm is then simpler in that it may focus effort on the dynamical element of the algorithm, leaving the kinematics (geometry) to the underlying geometric formulation. Nonetheless, there are limitations to what can be done with a geometric representation. In

Informatics and Machine Learning: From Martingales to Metaheuristics, First Edition.
Stephen Winters-Hilt.
© 2022 John Wiley & Sons, Inc. Published 2022 by John Wiley & Sons, Inc.

particular, care must be taken to carry the "new orthodoxy" only as far as "natural" geometric interpretation allows – otherwise one begins to encode the information/ algorithm in a (geometric) representation that is no more informative than any other.

8.2 Review of Differential Geometry [206, 207]

The following short review draws from texts on differential geometry [206–209].

8.2.1 Differential Topology – Natural Manifold

First recall R^n, the n-dimensional space of vector algebra, with points the n-tuples $(x_1, x_2, ..., x_n)$ of real numbers. The concept of continuity in R^n is made precise by the study of its topology (where here topology is meant in the local or point-set sense, not the global or algebraic sense).

In the rigorous definition of manifold, to be described later, we find that manifolds have the property that they are "locally R^n." So it is possible that local structures on R^n might carry over to local structures on a manifold. One such structure turns out to be the notion of continuity, and the study of topology provides this and more.

A quick way to demonstrate that a topology is defined on R^n, i.e. that R^n is a topological space, is to introduce the Euclidean distance: $d(x,y)$. One can then define open sets according to unit radius neighborhoods, establish the Hausdorff (continuum) property, etc., and thereby obtain an induced topology from the definition and completeness of the open sets. Topology is more "primitive" then distance, however, non-"distance" measures such as $d(p\|q)$ and $d(q\|p)$ (the Kullback–Leibler measure) also induce a topology, and it is the same as the Euclidean $d(p,q)$, as can be seen from their local $(p \approx q)$ behavior.

Aside from familiarity with topological spaces, R^n being the primary example, an understanding of Manifolds requires an understanding of Mappings (Figure 8.1). A map from a space M to a space N, $f:M \rightarrow N$, is a rule that associates to an element x of M a *unique* element $f(x)$ of N, $x \rightarrow f(x)$:

a) A map is continuous at x in M if any open set of N containing $f(x)$ contains the image of an open set of M containing X. (Presupposes M and N are topological spaces.) Continuity on all of M is obtained iff the inverse image of every open set of N is open in M.

b) A map $f:S \rightarrow R^n$ that is continuous is also described as class C^k if all partial derivatives of $f(x)$ of order less than or equal to k exist.

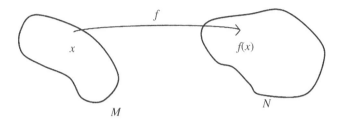

Figure 8.1 Map.

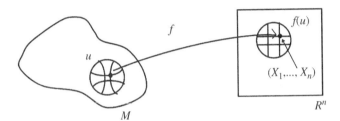

Figure 8.2 Chart.

c) If f is a one-to-one map of an open set M of R^n to N of R^n, then we can define the Jacobian, $J = \partial(f_1, ..., f_n)/\partial(x_1, ..., x_n)$. If the Jacobian is nonzero at a point, the *inverse function theorem* assures that the map f is one-to-one and onto in some neighborhood of x.

Definition for Manifold: A set (of "points") M is defined to be a manifold if each point of M has an open neighborhood which has a continuous one-to-one map onto an open set of R^n for some n. ("M is locally like R^n"). Dimension of M is n, local topology is that of R^n.

Since manifolds are "locally R^n" many of the tools of real analysis, defined on R^n, can be used on manifolds also.

Coordinates: by definition, $P \in M \rightarrow (x_1(P), ..., x_n(P))$, with $\{x_1, ..., x_n\}$ the coordinates of P under the map.

Chart: The pair consisting of a (bijective) neighborhood and its map (Figure 8.2): *Overlapping Charts* (coordinate transformations), shown in Figure 8.3:

The coordinates are C^k related if partial derivatives of order k or less are continuous.

Atlas: every point in M is in at least one Chart.

C^k *manifold:* Manifold with Atlas whose every Chart is C^k related on overlaps. ($\Rightarrow$ A Differentiable Manifold if at least C^1)

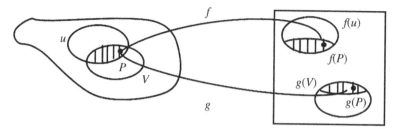

Figure 8.3 Overlapping charts.

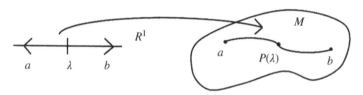

Figure 8.4 Curve. *Note:* The parameterization λ defines different curves, even if the image is the same.

Natural Structures on a Differentiable Manifold:

Curve: a differentiable mapping from an open set of R^1 into M (Figure 8.4):

Function: a function on M is a rule that assigns a real number to each point of M (Figure 8.5). If M maps differentiably to R^n, the function induces a function on R^n, which may be differentiable as well.

Vectors and Vector Fields: Consider curves and functions together. This allows one to obtain a differentiable function $g(\lambda)$ which gives f at a point with parameter value λ: $g(\lambda) = f(x^i(\lambda))$. Now have

i) The space of all tangent vectors at P and the space of all derivatives along curves at P are in one-to-one correspondence.

ii) Vectors lie, not in M, but in the tangent space to M at P, called "T_p."

iii) Vector field, a rule for defining a vector at each point of M.

Basis for Tangent Space T_p: Collection of n linearly independent vectors. Coordinate basis $\{x^i\}$ induces T_p basis $\{\partial/\partial x^i\}$.

Tangent bundle (fiber bundle): Consider manifold M combined with its tangent space T_p, call it TM (Figure 8.6).

Integral Curves exist for vector fields. Due to uniqueness of solutions the curves never cross except where $v^i = 0$. Aside from the $v^i = 0$ "caustics," the curves can be manifold filling, and if so you have a congruence (Hamiltonian vector fields).

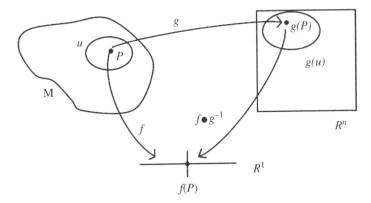

Figure 8.5 Function.

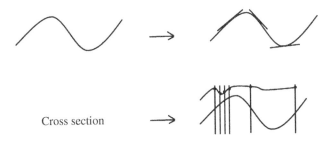

Cross section   ⟶

Figure 8.6 Tangent Bundle, where "cross-section" of TM (gives intuitive notion of "continuous" vector field). Note: even a simple manifold like S^2 has a nontrivial vector bundle.

1 Forms: linear real-valued functions of vectors. 1 Form fields can be described in terms of functions on TM (Figure 8.7):

Gradient can be interpreted as 1 form (Figure 8.8):

The gradient enables a picture of 1 form that is complementary to that of a vector:

Tensor and Tensor Field: A tensor at P is defined to be a linear function which takes as arguments N 1 forms and N' vectors and whose value is a real number.

i) A scalar is a tensor, such as $v^i w_i$. The v^1 component is not a scalar.
ii) Metric tensor field: A symmetric tensor field with inverse at every point of the manifold.
iii) Metric tensors reducible to diag $(-1, ..., -1, 1, ..., 1) \rightarrow O(n - k, k)$ symmetric group.

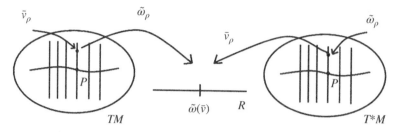

Figure 8.7 1 Forms and cotangent bundle. There is a duality between 1 forms and vectors, and the fiber bundle for 1 forms, $T*M$, is referred to as the cotangent bundle.

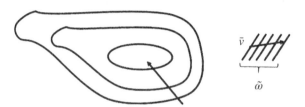

Figure 8.8 Gradient 1 form.

8.2.2 Differential Geometry – Natural Geometric Structures

The following short review is based on much lengthier analysis at [208].

Consider a tensor $T = T_\alpha^\beta e_\beta \otimes \omega^\alpha$ (where the "T_α^β" are the components and $\{e_\beta\}$ is an orthonormal basis in T_p, $\{\omega^\alpha\}$ on orthonormal basis in T_p^*). If we want to compare T at neighboring points in the manifold we must first move the T at one point to the other point. The intuition from vectors in Euclidean space is that we want to move a vector without rotation, i.e., we want to "parallel-transport" the vector. What that implies here is the need for an additional structure – a rule for parallel-transport (and one that generalizes to tensors).

The definition for (covariant) derivative then follows from the rule for parallel-transport (see Figure 8.9). Consider, for example, the "covariant derivative" $\nabla_u T$ of T along a curve $f(\lambda)$ whose tangent vector is $u = df/d\lambda$:

Consider the derivation of the parallel transport as indicated in Figure 8.9:

$$u = \frac{df}{d\lambda}\bigg|_{\lambda = 0}$$

$$\nabla_u T = \lim_{\varepsilon \to 0} \left\{ \frac{T(f(\varepsilon)) \mid \{parallel\ trans. to f(0)\} - T(f(0))}{\varepsilon} \right\}$$

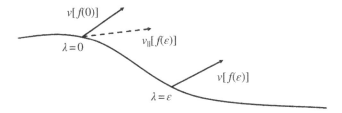

Figure 8.9 Parallel transport.

$$\overline{\nabla}T = \overline{\nabla}\left(T^{\beta}_{\alpha}e_{\beta}\otimes\omega^{\alpha}\right) = \overline{\nabla}\left(T^{\beta}_{\alpha}\right)e_{\beta}\otimes\omega^{\alpha} + T^{\beta}_{\alpha}\left(\overline{\nabla}e_{\beta}\right)\otimes\omega^{\alpha} + T^{\beta}_{\alpha}e_{\beta}\otimes\overline{\nabla}\omega^{\alpha}$$

Introduce standard connection coefficient notation:

$$\overline{\nabla}_{\lambda}e_{\beta} = \Gamma^{\alpha}_{\beta\lambda}e_{\alpha}$$

$$\overline{\nabla}_{\lambda}\omega^{\alpha} = -\Gamma^{\alpha}_{\beta\lambda}\omega^{\beta}$$

$$\overline{\nabla}_{\lambda}T = \left(T^{\beta}_{\alpha,\lambda} + T^{\mu}_{\alpha}\Gamma^{\beta}_{\mu\lambda} - T^{\beta}_{\mu}\Gamma^{\mu}_{\alpha\lambda}\right)e_{\beta}\otimes\omega^{\alpha}$$

$$T^{\beta}_{\alpha,\lambda} = T^{\beta}_{\alpha,\lambda} + T^{\mu}_{\alpha}\Gamma^{\beta}_{\mu\lambda} - T^{\beta}_{\mu}\Gamma^{\mu}_{\alpha\lambda}$$

Geodesic: A geodesic is a curve that is "straight" and uniformly parameterized as measured in each local R^n Chart. Thus, a geodesic is a curve that parallel-transports its tangent vector u along itself: $\nabla_u u = 0$. This is simple conceptually, and now for a concrete implementation. Introduce coordinate system x^{α} and describe u as $u(0) = dx^{\alpha}(\lambda)/d\lambda|_{\{\lambda=0\}}$:

$$\frac{d^2 x^{\alpha}}{d\lambda^2} + \Gamma^{\alpha}_{\mu\lambda}\frac{dx^{\mu}}{d\lambda}\frac{dx^{\lambda}}{d\lambda} = 0$$

Recap of Global Structures [209]

The review in this section largely draws from [209].

n-frame: a set of vector fields $V_{(1)}, V_{(2)}, ..., V_{(n)}$ defined on some subset of an n-dimensional manifold M such that they are linearly independent at each point of their domain of definition.

A Chart on $U \subset M$ with coordinates $x^1, x^2, ..., x^n$ determines an n-frame ∂/∂^x on U. In general there is no n-frame defined on all of M.

Parallelizable Manifold: A global n-frame exists. In general, however, it is difficult to find even a single vector field that is defined globally (hedgehog theorem: a smoothly combed hedgehog has at least one point of baldness → i.e., even dim sphere). Cartesian products of parallelizable manifolds are parallelizable. (In GR one usually has $M = R \times \sum_3$, and if a 3 dim. manifold is parallelizable it is orientable. Orientability is usually assumed, so parallelizability becomes common in GR).

Connections: The information inherent in the connection coefficients, or in parallel translation, can be associated with principle bundles over a manifold. In general, one considers a bundle space P, over base space M, with structural group G and projection map $\pi: P \rightarrow M$. It is then possible to establish a one-to-one correspondence between 1 forms on P (*Lie algebra valued*) and connections on P. These are referred to as connection 1 forms.

Affine Connection: If M denotes the Manifold and $B(M)$ its bundle of bases, then a connection on $B(M)$ is called an affine connection.

Difference Forms: Let ϕ and ψ be two connection 1 forms on $B(M)$. Define the difference from $\tau = \psi - \phi$.

Connections on parallelizable manifolds: Let α be an R^d valued 1 form on M which gives a parallelization of M. Define three connections associated with ρ as follows:

i) Direct connection: the one for which the vector fields $\rho^{-1}(x)$, for a fixed $x = R^d$, are parallel along every curve.
ii) Torsion zero connection: the one with the same geodesic as the direct connection but with torsion zero
iii) Opposite connection: the one with connection $\phi \, 2\tau$ where ϕ is the direct connection and $\phi + \tau$ is the torsion zero connection.

$$T_{XY} = [X, Y] - \nabla_X Y + \nabla_Y X \text{ (torsion)}$$

$$R_{XY} = \nabla_{[X,Y]} - \nabla_X \nabla_Y + \nabla_X \nabla_Y \text{ (failure of } \nabla \text{ to be a Lie algebra homomorphism)}$$

Direct: $T_{XY} = [X, Y]; R_{XY} = 0; \nabla_X Y = 0.$

Torsion zero: $T_{XY} = 0; R_{XY} = [[X, Y], Z]/4; \nabla_X Y = [X, Y]/2$

Opposite: $T_{XY} = -[X, Y]; R_{XY} = 0; \nabla_X Y = [X, Y]$

Let θ be a 2-form and X, Y, Z, vector fields, then:

$$(\nabla_X \theta)(Y, Z) = X\theta(Y, Z) - \theta(\nabla_X Y, Z) - \theta(Y, \nabla_X Z).$$

For metric connection:

$$X <Y, Z> \; = \; <\nabla_X Y, Z> \; + \; <Y, \nabla_X Z>.$$

For (g, ∇, ∇^*), with ∇, ∇^* dual connections:

$$X <Y, Z> \; = \; <\nabla_X Y, Z> \; + \; <Y, \nabla_X^* Z>.$$

With the above concepts reviewed, let's now proceed with the arguments of Amari.

8.3 Amari's Dually Flat Formulation [113–115]

The application of differential geometry methods to the study of statistical models traces back 1945, when it was noted that families of probability distributions could be described by a manifold and that the Fisher information matrix might be taken as a metric on that manifold.

Probability Distributions:

$$p(x) \geq 0 \ \forall x \in \chi \ \text{and} \ \sum_{\{x \in \chi\}} p(x) = 1.$$

Family of Probability Distributions:

$$S = \left\{ p_\theta = p(x; \theta) \mid \theta = \left[\theta^1, ..., \theta^n\right] \in \Phi \right\}$$

θ parametrizes an n-dimensional statistical model.

Fisher Information Matrix:

$$G(\theta) = \left[g_{ij}(\theta)\right]; g_{ij}(\theta) = E_\theta\left[\partial_i l_\theta \partial_j l_\theta\right], \quad \text{where} \ \partial_i = \frac{\partial}{\partial \theta^i} \ \text{and} \ l_\theta(x) = \log p(x; \theta)$$

Sufficient Statistic:

Consider

$$p(x; \theta) = p(x \mid y; \theta) q(y; \theta).$$

Suppose $p(x;\theta) = p(x|y;\theta) q(y;\theta)$, with associated spaces $S \equiv \{p(\cdot,\theta)\}$ and $S_F \equiv \{p(\cdot;\theta)\}$, and y is related to x by the constraint $y = F(x)$. If $\forall x \in \chi \ p(x|y;\theta)$ does not depend on θ, then F is a "sufficient statistic" w.r.t. S and there results $p(x;\theta) = p(x|y) q(y;\theta)$. Now, in order to estimate θ, it suffices to know y, and

$$\frac{\partial}{\partial \theta^i} \log p(x; \theta) = \frac{\partial}{\partial \theta^i} \log p(x; \theta) \Longrightarrow g_{ij} \text{ same for both } S \text{ and } S_F$$

$G(\theta)$ is symmetric and positive semi-definite (positive definite with linear independence in $\partial_1 p, ..., \partial_n p$). From this we can define inner product on the natural basis $[\theta^i]$ and obtain the "Fisher Metric." First revealed in an obscure paper [210] it was shown that the Fisher metric could be uniquely singled out based on invariance with respect to "sufficient statistics." *Also invariant w.r.t. sufficient statistics was a class of connections that Amari refers to as α-connections.*

"Invariance" w.r.t. sufficient statistics:

Here the invariance result of Chentsov is restated following the presentation of Amari. In essence, the Fisher metric and α-connections are uniquely characterized by invariance w.r.t. sufficient statistics on arbitrarily reduced probability spaces (reminiscent of Khinchin's reducibility axiom that entropy $H(p_1, p_2, ..., p_n, 0) = H(p_1, p_2, ..., p_n)$ in establishing the uniqueness of Shannon's entropy).

The formulation is done for a discrete set and then extended to (continuum) probability densities:

Consider events: $X_n \equiv \{0, 1, ..., n\}$; probabilities: $P_n \equiv P(X_n)$; and pairs (g_n, ∇_n) of metric and connection on $S \equiv \{P_n\}$. Consider $S_F \equiv \{P_m\}$ where $F: \chi_n \rightarrow \chi_m$ $(n \geq m$, and F is surjective). If F is sufficient w.r.t. S, then g_{ij} and Γ_{ijk} on S and S_F are the same. If sufficiency holds for all $n, m, s,$ and F, then g_n is the Fisher metric on P_n and ∇_n is an α-connection on P_n.

Recall Dual Connections:

$$X <Y, Z> = <\nabla_X Y, Z> + <Y, \nabla_X^* Z>.$$

$$\partial_k g_{ij} = \Gamma_{kij} + \Gamma_{kji}^*$$

$$\Gamma_{ijk}^{(\alpha)} = E_\theta \left[\left(\partial_i \partial_j l_\theta + \frac{1 - \alpha}{2} \partial_i l_\theta \partial_j l_\theta \right) (\partial_k l_\theta) \right], \text{ and } l_\theta = \log p(x; \theta)$$

Note: The α-connections and $(-\alpha)$ connection are dual w.r.t. the Fisher metric. **The α-connections are defined as follows:**

Consider the Exponential Family of distributions in this context:

$$p(x; \theta) = \exp \left[c(x) + \sum \theta^i F_i(x) - \psi(\theta) \right]$$

$$\partial_i l = F_i - \partial_j \psi$$

$$\partial_i \partial_j l = - \partial_i \partial_j \psi \longrightarrow \text{no } x \text{ dependence}$$

for $\alpha = 1$, we then have $\Gamma_{ijk}^{(1)} = - \partial_i \partial_j \psi E_\theta[\partial_k l_\theta] = 0$. So, $[\theta^i]$ is a $\nabla^{(1)}$-affine coordinate system, S is $\nabla^{(1)}$-flat. Amari refers to $\nabla^{(1)}$ as the exponential connection, or "e-connection."

Consider the Mixture Family of probability distributions in this context:

$$p(x; \theta) = \sum \theta^i p_i + \left(1 - \Sigma \theta^i \right) p_0(x) \partial_i l = \frac{p_i - p_0}{p}; \partial_i \partial_j l = - \frac{(p_i - p_0)(p_j - p_0)}{p^2}$$

$$\Rightarrow \partial_i \partial_j l + \partial_i l \partial_j l = 0 \Rightarrow \Gamma_{ijk}^{(-1)} = 0$$

So, $[\theta^i]$ is a $\nabla^{(-1)}$-affine coordinate system, S is $\nabla^{(-1)}$-flat. Amari refers to $\nabla^{(-1)}$ as the "mixture connection" or "m-connection."

Divergence: A triplet $(g, \nabla, \nabla*)$ can be defined locally from a "divergence" [211], where a divergence D is characterized by:

$$D(\cdot \| \cdot) : S \times S \rightarrow \Re, \text{ where } \forall p, \forall q \in S \times S: D(p \| q) \geq 0, \text{ and } D(p \| q) = 0 \text{ iff}$$

$$p = q.$$

Dually flat spaces: If ∇ and ∇^* are both symmetric then ∇ flat $\leftrightarrow \nabla^*$-flat. Since the α connections are symmetric, S is α-flat $\leftrightarrow$ S is $(-\alpha)$-flat, and, in particular, if $\nabla^{(1)}$ is flat then so is $\nabla^{(-1)}$.

If (S, g, ∇, ∇^*) is dually flat, then there exists ∇-affine coordinates $[\theta_i]$ and ∇^*-affine coordinates $[\eta_j]$. The inner product between the elements of the tangent spaces is a constant on S, from which it is possible to get the relations:

$$\frac{\partial \eta i}{\partial \theta^k} = g_{ik} \text{ and } \frac{\partial \theta^i}{\partial \eta j} = g^{ij}$$

Introduce Potentials: Suppose $\partial_i \psi = \eta_i$, then $\partial_i \partial_j \psi = g_{ij}$, and since g_{ij}, is a metric tensor the partial derivative must describe a positive definite matrix, which in turn implies that ψ is strictly convex. Likewise for $\partial^i \phi = \theta^i$. In terms of the potentials ψ and ϕ thus introduced, it is easy to show that the two coordinate systems $\{\theta^i, \eta_j\}$ can be related by Legendre transformation:

$$\phi = \theta^i \eta_i - \psi$$

Convexity in potentials that define a Legendre transformation leads to some interesting constructions. If we take our dually flat space (S, g, ∇, ∇^*) with coordinate systems $[[\theta^i], [\eta_i]]$ and their potentials $\{\psi, \phi\}$, then it can be shown that

$$\phi(q) = \lim_{p \epsilon S} \left\{ \theta^i(p)\eta_i(q) - \Psi(p) \right\}$$

and

$$\Psi(q) = \lim_{q \epsilon S} \left\{ \theta^i(p)\eta_i(q) - \phi(p) \right\}$$

Since we have:

$$\Psi(q) - \lim_{q \epsilon S} \left\{ \theta^i(p)\eta_i(q) - \phi(p) \right\} = 0$$

we have:

$$\Psi(q) - \left\{ \theta^i(p)\eta_i(q) - \phi(p) \right\} > 0$$

call this $D(p\|q)$:

$$D(p\|q) = \psi(p) + \phi(q) - \theta^i(p)\eta_i(q),$$

and it is easily shown that:

$$D(p\|q) \geq 0, \text{ and } D(p\|p) = 0 \Longleftrightarrow p = q$$

Thus, the Exponential and Mixture Families induce a dually-flat space which leads to the fundamental difference measure on distributions being a divergence. The "divergence" form of "distance" function is also indicated in the Link formalism (Chapter 9). Although the divergence family is singled out as fundamental at this point, the selection of a specific divergence (like Euclidean distance in the case of metrics) is yet to be determined. In Section 8.4 we will see that the "simplest" divergence, the Kullback–Leibler (relative entropy) divergence, is selected when maximizing log likelihood during learning, and that this is, fundamentally, because of the shortest path, or projection theorem, described in Section 8.3.2. *Implicit in this result is that the proper way to measure the difference between distributions is not the Euclidean distance between them but the Kullback–Leibler Divergence between them.*

8.3.1 Generalization of Pythagorean Theorem

Let p, q, and r be three points in S. Let γ_1 be the ∇-geodesic connecting p and q, and let γ_2 be the ∇_*-geodesic connecting q and r. If at the intersection q, the curves γ_1, and γ_2 are orthogonal (w.r.t. g), then

$$D(p||r) = D(p||q) + D(q||r)$$

To show this, first consider the γ_1 geodesic:

$$\theta_t^i = t\theta^i(p) + (1-t)\theta^i(q)\frac{d}{dt}\theta_t^j\partial_i = \left[\theta^i(p) - \theta^i(q)\right] * \partial_i \eta_{ti}$$

$$= t\eta_i(q) + (1-t)\eta_i(r)\frac{d}{dt}\eta_{ti}\partial^i = [\eta_i(q) - \eta_i(r)]\partial^i$$

So, prove relation with:

$$D(p||q) + D(q||r) - D(p||r)$$

$$= \Psi(p) + \phi(q) + \theta^i(p)[\eta_i(r) - \eta_i(q)] - \theta^i(q)\eta_i(r) = \left[\theta^i(p) - \theta^i(q)\right][\eta_i(r) - \eta_i(q)]$$

$$= \left\langle \left(\frac{d\gamma_1(t)}{dt}\right)_i, \left(\frac{d\gamma_2(t)}{dt}\right)^j \right\rangle = 0$$

8.3.2 Projection Theorem and Relation Between Divergence and Link Formalism

Suppose M, a, submanifold of S, is ∇^*-autoparallel, then

$$D(p||q) = \min_{\{r \in M\}} D(p||r) \text{ when the } \nabla\text{-geodesic connecting } p \text{ and } q$$
is orthogonal to M at q.

Relation to Link formalism, start with:

$$D(p||q) = \psi(p) + \phi(q) - \theta^i(p)\eta_i(q)$$

Use Legendre transformation:

$$\phi(p) = \theta^i(q)\eta_i(q) - \psi(q)$$

to get:

$$D(p||q) = \psi(\rho) - \psi(q) + \left(\theta^i(q) - \theta^i(p)\right)\eta_i(q), \quad \text{where } \eta_i = \frac{d\psi}{d\theta^i}$$

Thus, with shift in notation $\{f = \frac{\partial F}{\partial \omega}, \omega\} \to \{g, \theta\}$ we get the Link formalism used in Chapter 9:

$$D(\omega||\omega') = F(\omega) - F(\omega') - (\omega - \omega')\frac{\partial F}{\partial \omega}\bigg|_{\omega = \omega'}, \quad \text{where } \frac{\partial D}{\partial \omega} = f(\omega) - f(\omega').$$

The $\{f, \omega\} \leftrightarrow \{g, \theta\}$ duality corresponds precisely to the dually flat connection construction with potentials ϕ and ψ that are related via Legendre Transformation to the "coordinates" θ and ω. The link function $f(\omega) = \omega$, is associated with square loss and the gradient descent (GD) learning rule. The link function $f(\omega) = \ln(\omega)$, is associated with divergence loss and the exponentiated gradient descent (EG) learning rule. These will be explored in Chapter 9 in the link formalism context, along with an interpolating learning rule between GD and EG given by the link function $f(\omega) = \sinh^{-1}(\omega)$.

8.4 Neuromanifolds [113–115]

The information geometry methods described for families of probability distributions (parametrized by $\{\theta^i\}$) can just as easily be applied to neural networks; where now the parameters are the connection weights (further detailed description of neural nets is in Chapters 9 and 13). The statistical arguments also carry over and are applicable to stochastic neural nets, i.e., neural networks with noisy input or non-deterministic behavior. As Amari states, "even when a network is deterministic, it is sometimes effective to train it as if it were a stochastic network" [114]. With a stochastic network we then have probability distribution $p(x;\theta)$ and/or conditional probability distribution $p(y|x;\theta)$.

Complications associated with repeated observation:

For a single observation we have the distribution $p(x;\theta)$, an element of the family (space) S. For repeated independent observations there is the joint conditional distribution where θ generally changes from one observation to the next. The joint distribution, thus, is an element of a larger family (space)

$$S_T^* = S_1 \times S_2 \times \cdots \times S_T.$$

Multiple observations thus lead to a joint distribution whose manifold dimension (under direct product) increases with the number of observation, while the underlying parameterization is fixed – being the manifold dimension of the family of distributions considered for the individual distribution. This generally leads to a Curved Exponential family description on the joint distribution manifold (where the individual distribution was in an exponential family), i.e., a submanifold of the joint distribution manifold is specified. It is possible to describe repeated observations within the framework of the manifold S without referring to the product space S_T^*, but this holds only in the i.i.d. case. In general

$$\theta_T^* = (\theta_1, \theta_2, ..., \theta_T),$$

and

$$\theta_T^* = \theta(x_t, u_t), t = 1, ..., T.,$$

where x_t's are given and the u_t's are the only free parameters (the parameters of the underlying neural network). Suppose all x_t are subject to $p(x;\theta)$, $\theta_1 = \theta_2, = \cdots = \theta_T$, where $p(x;\theta)$ is the exponential family $p(x;\theta) = \exp(\theta \cdot x - \psi)$. The joint distribution is

$$p(x_1, ..., x_T; \theta) = \exp\{\Sigma x_t \cdot \theta - T\psi\}$$
$$p(\bar{x}; \theta) = \exp\{\bar{x} \cdot \theta - \psi\}$$

The maximum likelihood estimator (m.l.e.) $\hat{\theta}$ from the observed data $x_1, ..., x_T$ is given by maximizing $p(\bar{x};\theta)$:

$$\bar{x} = \frac{\partial}{\partial \theta}\psi(\theta)|_{\theta = \hat{\theta}} = \hat{\eta}$$

The observed data are then represented by the m.l.e $\hat{\eta}$ in S in the η-coordinate system.

Suppose that all x_t are subject to $p(x;\theta)$ where $p(x;\theta)$ is a curved exponential family. Again the observed data $x_1, ..., x_t$ are represented by $\bar{x} = \hat{\eta}$, now, however, we have that $\hat{\eta}$ does not necessarily belong to M. The m.l.e. $\hat{u}$, or corresponding distribution $\theta(\hat{u}) \in M$, is given by maximizing the log likelihood $x \cdot \theta(u) - \psi(\theta(u))$ w.r.t. u. Maximizing the log likelihood is equivalent to simply m-projecting $\hat{\theta}$ to M in the Amari differential geometry formalism, and this is equivalent to minimizing the Kullback–Leibler (KL) divergence $K(\hat{\theta}||\theta(\hat{u}))$ from $\hat{\theta}$ to $\theta(\hat{u}) \in M$. Thus, the KL Divergence is used in the Amari neuromanifold learning process.

EM Algorithm

The following EM/em discussion closely follows Amari [113–115].

Consider $M = \{p(r;\theta(u))\}$ a curved exponential family from which data is regenerated. Data r is observed. Consider $r = r(S_v, S_h)$, a sufficient statistic that includes hidden part S_h. Need to estimate unknown part of r information. Can do this based on observed S_v, and some candidate distribution u', via the conditional expectation:

E: $\hat{r}(u') = E[r \mid s_v; \theta(u')]$

Now estimate $\log p(\hat{r}; \theta(u))$ by conditional expectation also:

E: $LLH(s_v; \theta(u')) = E[\log p(\hat{r}; \theta(u)|s_v; \theta(u')] = \theta(u)*\hat{r}(u') - \psi(\theta(u))$

Now search for better candidate u by maximizing LLH, or equivalently, by minimizing the KL divergence $D(\hat{r}(u')\|\theta(u))$ from the guessed data point $\hat{\eta} = \hat{r}(u')$ to μ w.r.t. u.

The algorithm:

Step 0: Initialization step. Guess u_0, the initial guessed distribution $P_0 \in M$ is given by $\theta(u_0)$. Then repeat the following:

Step 1: E-step. Based on candidate probability distribution $P_i \in M$, calculate the conditional expectation of r. This gives the *i*th candidate for the observed point $Q_i \in D$, whose η-coordinate are $\eta_{(i)}$.

Step 2: M-step. Calculate the mle $u_{(i+1)}$ from $Q_i \in D$.

em Algorithm

Search for the pair of points $P \in M, Q \in D$ that minimizes the divergence between D and M, that is (originally proposed in 1984 [212]):

$$D(\hat{Q}\|\hat{P}) = \lim_{P \in M, Q \in D} D(Q\|P)$$

i) point $\hat{P} \in M$ that minimizes $D(Q\|P)$ is given by the m-projection of Q to M (i.e., by the m-geodesic connecting $\hat{P}$ and Q that is orthogonal to M at $\hat{P}$).

ii) point $\hat{Q} \in D$ that minimizes $D(Q\|P)$ is given by the e-projection of P to D (i.e., the e-geodesic connecting P and $\hat{Q}$ that is orthogonal to D at $\hat{Q}$).

The algorithm:

Step 0: Initialization Step. Guess $\hat{u}_0 \Rightarrow \hat{P}_0 \in M$. Then repeat:

Step 1: e-step. e-project $\hat{P}_i$ to D, gives $\hat{Q}_i$.

Step 2: m-step. m-project $\hat{Q}_i$ to M, gives $\hat{P}_{i+1}$.

Relation between EM and em:

1) M-step and m-step are the same
2) E-step and e-step differ depending on the conditional expectation of r_h given r_v:

Note: Divergence need not be symmetric or satisfy the triangle inequality, but given "orthogonal" learning steps, a Divergence satisfies a generalized Pythagorean theorem, leading to a the same critical Chapman–Kolmogorov-like propagation rule whether learning step assumes a Euclidean notion of distance (GD) or a KL divergence notion of distance (EG).

Which Is Better?

EM is more natural from statistical point of view, but the representation via a neural network does not correspond exactly to statistical inference, therefore em serves better as the approximator ideally suited to the neural network representation of the input–output relation. The two algorithms are asymptotically equivalent when T is large (in the framework of exponential families). If framework extended from exponential family to function space they are exactly equivalent. It may be that a hybrid of GD and EG, or of EM and em type learning, is best, and this will be explored further in Chapter 9 in an explicit loss bounds analysis on the learning process.

Amari's Dually Flat formulation is a more natural structure for representing exponential families and neural nets than it might appear at first sight. In particular, the Kullback–Leibler measure is naturally represented in terms of the dual coordinates and their relations via Legendre transformation (with the introduction of appropriate potentials). The potentials, in turn, provide a natural formulation that is precisely that exhibited when using the "link formalism." Similarly, the duality between the coordinates (via Legendre transformation) is precisely that exhibited in the neural net learning algorithms by [213] and [214] and described in Chapter 9.

While the e-projection and m-projection are natural geometric notions and certainly a strength of Amari's program, they also represent a weakness. This weakness is perhaps most clearly illustrated in the learning algorithm version where the projection is determined and then the actual update is interpolated (between old data and estimator and new, projected data and estimator). In this instance we must fall back on the gradient descent algorithm, or some such arguments, in order to have stable updates. One approach might be to use the SA link algorithm since it results in a formalism that interpolates between GD and EG learning according to the weight magnitude. Further details on this approach are given in Chapter 9.

8.5 Exercises

8.1 Re-derive the α-connections for $\alpha = 1$ and $\alpha = -1$ and show that they are "dually flat."

8.2 Introduce potentials on a dually flat space as described in Section 8.3, then show how their Legendre transformation induces a natural difference measure that is a divergence.

8.3 Prove the generalization of the Pythagorean theorem for dually flat spaces.

8.4 Show that em learning singles out the Kullback–Leibler Divergence as fundamental in dually flat space (this is akin to a Riemannian manifold giving rise to a fundamental locally flat spacetime reference and local Euclidean distance measure).

9

Neural Net Learning and Loss Bounds Analysis

In this chapter, neural learning loss bounds derivations are given for the gradient descent learning process (GD) as well as other learning processing indicated in the link formalism. The link formalism provides a unified variational formalism for learning or "information dynamics" (Section 9.2.1). To start, however, a detailed, but brief, description of neural nets will be given starting with the single neuron (Section 9.1.1). Section 9.1.2 describes neural nets and back-propagation (which is then largely explored in the TensorFlow implementation described in Chapter 13). Section 9.3 describes the motivation for the link function $f(\omega) = \sinh^{-1}(\omega)$. Section 9.4 provides a loss bounds analysis for the link function $f(\omega) = \sinh^{-1}(\omega)$.

In Machine Learning we have identified good families of "update rules" that can perform neural net learning. In examination of update rules that are "good" it is possible in some cases to explicitly derive the error rate or loss during learning and quantify loss-bounds on the overall learning process. In doing so there appears to be a learning phenomenology "regularization" for the better algorithms akin to the Newtonian Physics phenomenology of inertia in predicting motion. The Loss Bounds analysis reveals that statistical learning is feasible in these domains of application, i.e., they are effectively a proof of the feasibility of "statistical learning". However, the description and results also indicate that the choice of learning "update rules" and optimization constraints can greatly impact the performance of the algorithms (to be further explored with SVMs in Chapter 10). So the loss bounds method described is partly done to help understand what will likely be a good learning rule for developers optimizing a neural net or classifier for a particular input data environment.

Informatics and Machine Learning: From Martingales to Metaheuristics, First Edition.
Stephen Winters-Hilt.
© 2022 John Wiley & Sons, Inc. Published 2022 by John Wiley & Sons, Inc.

9.1 Brief Introduction to Neural Nets (NNs)

In Chapter 8 we saw brief mention of neural nets in an advanced mathematical setting where they are related to a manifold (a "neuromanifold") that can be used to perform learning (with dimensionality given by the number of weights in the neural net in the case of a neuromanifold). In Section 9.1.1 that follows, we start at the beginning of the computational neuron discussion, with a description of the Perceptron and Sigmoid neurons. In Section 9.1.2, the classic description of back-propagation is given, whereby a neural net can learn efficiently (akin to exponential to quadratic reduction in computational order in the Viterbi algorithm).

9.1.1 Single Neuron Discriminator

Shown in Figure 9.1 is a single neuron discriminator with a simple activation function (a step function) also known as the perceptron.

9.1.1.1 The Perceptron

The classic machine learning neural net classifier, the perceptron, will now be re-derived. The notation of [109] will be followed so that use can then be made of Problem 1.3 from [109] where the proof of the perceptron learning is done. Let $X = R^d$ be the input space. Let $Y = \{+1, -1\}$ be the output space (which denotes a yes/no decision typically). We will want to be able to adjust weight on the

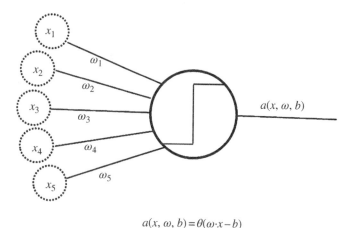

$$a(x, \omega, b) = \theta(\omega \cdot x - b)$$

Figure 9.1 **Single Neuron with step (threshold) activation function.** The inputs x_k are multiplied by the weights ω_k, and possibly have a bias applied, and the result is passed through a step function such that $a(x, \omega, b) = \theta(\omega \cdot x - b)$, where $\theta(z) = 1$ if $z > 0$ and $\theta(z) = -1$ if $z \le 0$.

importance of different features in the input space in arriving at a decision, and then have a simple threshold rule to make that decision:

Approve if

$$\sum_{i=1}^{d} \omega i x i > b$$

Deny if

$$\sum_{i=1}^{d} \omega i x i \leq b$$

Let us treat the threshold value as a weight $\omega_0 = -b$ and associate it with an added x coordinate, $x_0 = 1$. Let's also make use of the sign function: $\text{sign}(s) = 1$ if $s > 0$, and $\text{sign}(s) = -1$ if $s < 0$. The decision function $h(x)$ is $+1$ if approved and -1 if denied:

$$h(x) = \text{sign}(\omega \cdot x)$$

The classic perceptron learning algorithm then goes as follows (Figure 9.2). Suppose at the current iteration, t, of the learning algorithm, we have weight vector $\omega(t)$. If we have not already arrived at a solution then some of the $(x_1, y_1)...(x_N, y_N)$ will be misclassified, i.e., $h(x_k) \neq y_k$ for various k values, and one of these misclassifications will be used in the **update rule**:

$$\omega(t+1) = \omega(t) + y(t)x(t)$$

First note that: $y(t)[\omega(t) \cdot x(t)] < 0$ for each update (y has different sign than $\omega \cdot x$). Second, note: $y(t)[\omega(t+1) \oplus x(t)] > y(t)[\omega(t) \cdot x(t)]$ (so move toward positivity), where this last follows from $y(t)[\omega(t+1) \cdot x(t)] = y(t)[\omega(t) \cdot x(t)] + [y(t)]^2[x(t) \cdot x(t)]$. This move toward positivity is shown in Figure 9.2.

Within the infinite space of all weight vectors, the perceptron algorithm finds a weight vector that works. This means that an infinite hypothesis space will be searched in a finite number of steps, so finite time, to find a solution. This search

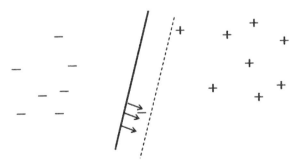

Figure 9.2 Perceptron update. The update rule shifts the position of the separating hyperplane such that classification errors eventually reduced to zero (assuming separable and that it does not get stuck in a local minimum of the minimization process).

of an infinite hypothesis space in finite time will be one of the hallmarks of the Feasibility of Learning Proof (the "First Law of Machine Learning").

To recap, the components of the Perceptron Learning Algorithm are:

1) There is an input x.
2) There is an unknown target function $f: X \rightarrow Y$.
3) There is a data set D of (input, output) examples $(x_1, y_1), ..., (x_N, y_N)$, where $y_n = f(x_n)$.
4) There is a learning algorithm that uses dataset D to pick a formula $g: X \rightarrow Y$ that approximates f.
5) The algorithm chooses g from a set of candidate formulas under consideration, which we call the hypothesis set H. In the case of the perceptron, $h(x) = \text{sign}(\omega \cdot x)$, $h \in H$.

The proof that the Perceptron Learning Algorithm converges to a solution follows the steps outlined in Problem 1.3 from [109]. Assuming the data is separable, there is a set of weights, $\omega*$, that separates the data:

1) Have $\rho = \min\limits_{n} y_n(\omega^* \cdot x_n) > 0$ since separable means all classified correct, so signs of y_n and $\omega^* \cdot x_n$ always match, thus their product always positive.
2) Have $\cdot \omega(t) \, \omega^* \geq \omega(t-1) \cdot \omega^* + \rho$, then with induction have $\omega(t) \cdot \omega^* \geq t\rho$.
3) Have $\|\omega(t)\|^2 \leq \|\omega(t-1)\|^2 + \|x(t-1)\|^2$,
4) By induction have $\|\omega(t)\|^2 \leq tR^2$, where $R = \max\limits_{n} \|x_n\|$.
5) Can put together to now have: $t \leq R^2 \|\omega^*\|^2 / \rho^2$.

Thus, an infinite space of weight vectors is searched in less than $R^2 \|\omega^*\|^2 / \rho^2$ steps (e.g., a finite number of steps).

9.1.1.2 Sigmoid Neurons

Let us now consider connecting up a collection of neurons to make a neural net. If we try to work with the perceptrons, their step function activation is detrimental in that it fires all-or-nothing, leading to an unstable learning process when networked. Almost as a rule, we want our learning process, here parametrizing a neural net, to be such that a small input change leads to a small output change. This is automatically the case if the activation function and formalism overall is differentiable, then a chain rule relating change in input to change in output can be written (and small enough input will give small enough output). Thus, the shift to the sigmoid activation function neuron (Figure 9.3) that is smooth and differentiable ("S shaped") in the next section. This differentiability is then critical to establishing learning via backpropagation, where the chain rule is used, and where backward propagation through the network is done because it allows a recursion relation to be written with loss in one layer only dependent on loss in its output layer

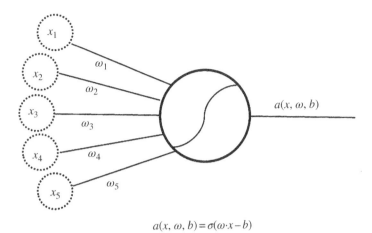

$$a(x, \omega, b) = \sigma(\omega \cdot x - b)$$

Figure 9.3 Single Neuron, Sigma activation function: the inputs x_k are multiplied by the weights ω_k, and possibly have a bias applied, and the result is passed through a function that is monotonically increasing, differentiable, with asymptotes (–1 or 0, and 1, typically): $a(x, \omega, b) = \sigma(\omega \cdot x - b)$.

(akin to Viterbi recursion with one column only dependent on its prior column). This reduces the computational task to polynomial time from exponential time (as with the Viterbi algorithm in the HMM chapters), so it is a critical methodology for practical learning implementations.

9.1.1.3 The Loss Function and Gradient Descent

Consider the simple activation function for the neuron: $a(x) = \omega \cdot x$ (inner product between vector of weights and vector of instance feature components). The notation in this section will assume any bias value, from the standard notation $\omega \cdot x - b$, is absorbed into the "$\omega \cdot x$" notationally as an extra $\omega^{n+1} = -b$ component with associated $x^{n+1} = 1$ value. If the observed value is $y(x)$, then our error, or loss, or cost associated with the neuron with its current learned ω values is:

$$L\left(y, \vec{\omega} \cdot \vec{x}\right) = \frac{1}{2}|y(x) - \omega \cdot x|^2$$

$$L\left(y, \vec{\omega} \cdot \vec{x}\right) = \sum_k \frac{\partial L}{\partial \omega_k} \omega_k = \nabla L \cdot \Delta \omega$$

Let us choose a change in weights given by the negative gradient on the Loss function (Gradient Descent or GD), with magnitude modulated by η, a "learning rate":

$$\Delta \omega = -\eta \nabla L.$$

This guarantees that the Loss will decrease with each learning step since:

$$L\left(y, \vec{\omega} \cdot \vec{x}\right) = -\eta |\nabla L|^2 \leq 0.$$

From the gradient descent update rule we get:

$$\omega_{t+1,i} = \omega_{t,i} - 2\eta(\omega \cdot x - y)x_i.$$

The description above can be thought of as describing information learning dynamics in terms of loss minimization subject to a learning rate, later (Section 9.2) we will see this can be fully encapsulated in a variational formulation in what follows. Most of this chapter explores the learning capabilities of a *single* neuron further, eventually crossing over to related SVM-based learning in Chapter 10. But first, let us briefly discuss how this will relate to multiple neuron systems in the form of neural nets, which will then be developed further in Chapter 13.

9.1.2 Neural Net with Back-Propagation

Neural Nets with backpropagation learning are implemented in Chapter 13 using the Google TensorFlow package. In this section is shown the core theory of the backpropagation learning methodology up to Deep Learning (which is left to Chapter 13). In what follows we see that the core model/theory for neural networks is remarkably simple, yet their performance, if they work at all, is generally very good. What makes the difference all (good) or nothing? It's a matter of having sufficient data to train the NN. If the neurons in the NN have a good activation function and a good learning rule, the amount of data needed is reduced. If there is sufficient data, however, any simple sigma activation and learning with gradient descent on the loss function (discussed next), will arrive at good results. Furthermore, the computation can be done efficiently (a must if success is premised on having immense amounts of data to process). We will see that the learning task can be distributed (multi and multicore CPUs) and vectorized (GPU), for multiple magnitude speedup in processing time.

9.1.2.1 The Loss Function – General Activation in a General Neural Net

Consider the neural net shown in Figure 9.4.

Notes on the backpropagation derivation follow. The key idea is that we make use of the chain rule, but do so in a the form of a recursively defined expression to have an efficient computational implementation.

The activation function is now referenced by layer and neuron index in that layer (following the notational convention on weights indicated in Figure 9.3). In component notation this is:

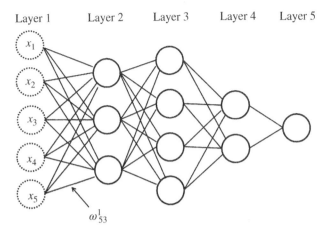

Layer 1   Layer 2   Layer 3   Layer 4   Layer 5

ω^{1}_{53}

Figure 9.4 Neural Net, layered topology, fully connected between layers ("dense").
The weight label shown indicates the labeling scheme, with superscript denoting
the layer, left subscript the neuron index coming from the layer indicated, to the
neuron index shown in the right subscript in the next layer. Shown is the weight labeled
for the fifth (lowest) input weight that connects into the third (lowest) neuron in the
next layer.

$$a^{l}_{j} = \sigma\left(\sum_{k}\omega^{l}_{jk}a^{l-1}_{k}\right) = \sigma\left(z^{l}_{j}\right).$$

If we work with neural nets with one node for final layer, then we can write the
Loss in terms of the inputs and activation at that last neuron (layer "L," related
recursively to earlier layers of neurons):

$$L\left(y, \vec{\omega}\cdot\vec{x}\right) = \frac{1}{2}|y(x) - a^{L}|^{2}$$

In the variational analysis that follows we can think of L a function of ω and a or
of $z = \omega \cdot a$, where the latter variation works out simplest, so it will be used. The
change in loss δ with change in "z" input can then be defined at the final layer and
then related recursively to earlier layers:

$$\delta^{L}_{j} = \frac{\partial L}{\partial z^{L}_{j}} = \sum_{k}\frac{\partial L}{\partial a^{L}_{k}}\frac{\partial a^{L}_{k}}{\partial z^{L}_{j}}.$$

Recall that $a^{l}_{j} = \sigma\left(z^{l}_{j}\right)$, so $\dfrac{\partial a^{L}_{k}}{\partial z^{L}_{j}} = \sigma'\left(z^{l}_{j}\right)$ if $k = j$, zero otherwise, so:

$$\delta_j^l = \frac{\partial L}{\partial a_j^L} \sigma'\left(z_j^l\right),$$

which established the boundary term, we now need a recursive relation to complete the derivation:

$$\delta_j^l = \frac{\partial L}{\partial z_j^l} = \sum_k \frac{\partial L}{\partial z_k^{l+1}} \frac{\partial z_k^{l+1}}{\partial z_j^l}.$$

Since $z_j^{l+1} = \sum_k \omega_{jk}^{l+1} a_k^l = \sum_k \omega_{jk}^{l+1} \sigma\left(z_k^l\right)$:

$$\frac{\partial z_k^{l+1}}{\partial z_j^l} = \omega_{kj}^{l+1} \sigma'\left(z_j^l\right)$$

So,

$$\delta_j^l = \frac{\partial L}{\partial z_j^l} = \sum_k \delta_k^{l+1} \omega_{kj}^{l+1} \sigma'\left(z_j^l\right).$$

Further elaboration on the above derivation, with more discussion of the critical role of sigmoid (differentiable) neurons can be found at neuralnetworksanddeeplearning.com.

As with the single neuron, depending on how the learning on the weights is regularized for stability, we arrive at a learning, or update, rule on the weights. For the typical square weight regularizer, we get gradient descent. Variations on this will be examined in Section 9.2. Critical to this learning process for neural nets is the existence of the above chain rule and a differentiable (stable) activation function. In Chapter 11, we will see that one of the simplest search heuristics is gradient descent, so this suggests that the neural net might be improved by use of one of the other, non-differentiable, stochastic sampling heuristics described there. If this is used to efficiently approximate the gradient itself, used in gradient descent, we arrive at stochastic gradient descent neural net learning. But that is as far as it goes. To extend to the variety of other more stochastic search methodologies to be described in Chapter 11, the "differentiability" of that search outcome is a much less stable, and when compounded via the chain rule we see a very unstable learning process. This will not work in large multilayer neural nets trained used immense amounts of data – the instability will act as noise to weaken the overall net performance. This is why the neurons used in the neural net described in Chapter 13 tend to be sigmoid (differentiable) with simple regularization.

9.2 Variational Learning Formalism and Use in Loss Bounds Analysis

9.2.1 Variational Basis for Update Rule

In the online learning process, it is proposed that the weights remain unchanged unless forced to change by a "learning" process. (A "first law" of information dynamics analogous to Newton's first law.) Instantaneous change, however, is not realizable physically and the same is assumed for information (whether or not a physical incarnation for the information is established). So when describing the response of the weights to a learning process it is also desirable to introduce an inertial property ("resistance" or "memory"). The resulting formalism achieves a balance between a tendency for the weights to remain unchanged referred to as the memory conservativeness, or inertia according to context and a change in the weights due to learning process (also referred to as correctness, or perhaps, the learning "force"). I'll refer to the latter principle as the second law of information dynamics due to its parallels with Newton's second law of dynamics, and, like its analogue, there is a variational formulation.

It is only in its variational formulation that an analysis of 'information dynamics" really begins to go its own way – separate from the physical Newtonian analogue. In large part this is due to the easy entry of a new conceptual element into the formalism: Entropy. The definition of Entropy is best understood in a variational context, and it is often defined in such terms. Sometimes the variational aspect of entropy is implicit, such as in thermodynamics, where the fundamental equilibrium states are ones of maximum entropy. Sometimes, on the other hand, the variational aspect is made quite clear, such as in the Jaynesian school of thought, where the links to thermodynamics are forged by the Maximum Entropy Principle [112].

When the second law of information dynamics is initially put into a variational formulation the measure of comparison between weight vectors is naturally defined by the Euclidean concept of "distance":

$$U\left(\vec{\omega}; \vec{s}, \vec{x}, y\right) = d\left(\vec{\omega}, \vec{s}\right) + \eta L\left(y, \vec{\omega} \cdot \vec{x}\right)$$

$$0 = \frac{\partial \mu}{\partial \omega_i}\bigg|_{\omega_t} = \left(\frac{\partial d\left(\vec{\omega}, \vec{s}\right)}{\partial \omega_i} + \eta \frac{\partial L\left(y, \vec{w} \cdot \vec{x}\right)}{\partial \omega_i}\right)_{\substack{\vec{\omega} = \vec{\omega}_t \\ \vec{s} = \vec{w}_{t+1}}}.$$

Suppose $\dfrac{\partial d(\omega, s)}{\partial \omega_i} = f(S_i) - f(\omega_i)$, then

$$f(\omega_{t+1,i}) = f(\omega_{t,i}) - \eta \frac{\partial L}{\partial \omega_i} \text{ (} f \text{ referred to as link function)}.$$

For $\delta(\omega, s) = \frac{1}{2} \|\omega - s\|_2^2$, Euclidean distance, and $L = \left(y - \vec{\omega} * \vec{x} \right)^2$, the familiar square loss, we obtain:

$$\omega_{t+1,i} = \omega_{t,i} - 2\eta(\omega \cdot x - y)x_i,$$

which is the Gradient Descent (GD) algorithm.

Other distance metrics lead to updates other than that of the classical GD algorithm (where the loss function is taken to be the usual square loss in all of the functions considered here). Furthermore it soon becomes apparent that the introduction of information entropy (as explored by Shannon), allows for an altogether different measure of comparison – one that need not satisfy the usual metric properties implied by the usage of the term "distance." The Kullback–Leibler measure indicated from Chapter 8 being an example of such, and one that leads to link function $f(\omega) = \ln\omega$.

The existence of loss bounds for the updates expressible in terms of the "laws" of information dynamics might be taken as an indication of future successes of an information dynamics perspective.

9.2.2 Review and Generalization of GD Loss Bounds Analysis [213, 214]

The proof of the worst-cast Loss Bounds centers on establishing the inequality

$$a(y_t - \omega_t \cdot x_t)^2 - b(y_t - u \cdot x_t)^2 \le d(u, \omega_t) - d(u, \omega_{t+1})$$

For constraints a and b as general as possible, where x_t is bounded by a 2-norm or ∞-norm, and the weights are bounded also, if necessary.

Begin with a review (and generalization) of the GD Loss Bounds analysis, where the variational basis for update rule is given as:

$$u(\omega) = d'(\omega, s) + \eta L(y, \omega \cdot x)$$

$$0 = \frac{\partial u}{\partial \omega_i}\bigg|_{\omega_{update}} = \left(\frac{\partial d(\omega, s)}{\partial \omega_i} + \eta \frac{\partial L(y, \omega \cdot x)}{\partial \omega_i} \right)_{\omega_{update}}$$

Properties of $d(\omega, s)$:

$d(\omega, s) \ge 0$; equality only for $\omega = s$.

$\frac{\partial d}{\partial \omega} = f(s) - f(\omega); f$ referred to as link function

$f(\omega)$ monotonically increasing and invertible

The update:

$$f(\omega_{t+1,i}) = f(\omega_{t,i}) - \eta \frac{\partial L}{\partial \omega}$$

The Loss, $L(y, \omega \cdot x)$, is assumed to be the square loss throughout:

$$\frac{\partial L}{\partial \omega_i} = 2(\omega \cdot x - y)x_i$$

Switching to the "p-q" notation:

$$p_t = y_t - \omega_t \cdot x_t, \quad q_t = y_t - u \cdot x$$

The update rule is reexpressed as:

$$f(\omega_{t+1,i}) = f(\omega_{t,i}) + 2\eta p_t x_i$$

$$\omega_{t+1,i} = f^{-1}[f(\omega_{t,i}) + 2\eta p_t x_i]$$

Consider the Taylor Expansion of $f^{-1}[\ldots]$ about $f^{-1}[f]$:

$$f^{-1}[f(\omega_{t,i}) + 2\eta p_t x_i] = f^{-1}[f(\omega_{t,i})] + \frac{\partial}{\partial \gamma} f^{-1}[f(\omega_{t,i}) + \gamma]$$

$$\left| \begin{matrix} \\ 2\eta p_t x_i + \frac{1}{2}\frac{\partial^2}{\partial \gamma^2} f^{-1}[f + \gamma] \\ \gamma = 0 \end{matrix} \right| \begin{matrix} (2\eta p_t x_i)^2 \\ \gamma = \tilde{\gamma} \end{matrix}$$

Thus,

$$\omega_{t+1,i} = \omega_{t,i} + \frac{\partial}{\partial \gamma} f^{-1}(f + \gamma) \left|_{\gamma = 0} \quad 2\eta p_t x_i + \frac{1}{2}\frac{\partial^2}{\partial \gamma^2} f^{-1}(f + \gamma) \right|_{\gamma = \tilde{\gamma}} (2\eta p_t x_i)^2$$

$$\omega_{t+1,i} - \omega_{t,i} = \alpha_1 (2\eta p_t x_i) + \alpha_2 (2\eta p_t x_i)^2$$

where $\quad \alpha_1 = \frac{\partial}{\partial \gamma} f^{-1}(f + \gamma) \left|_{\gamma = 0} \quad$ and $\alpha_2 = \frac{1}{2}\frac{\partial^2}{\partial \gamma^2} f^{-1}(f + \gamma) \right|_{\gamma = \tilde{\gamma}}$

($\alpha_1 > 0$ since f and f^{-1} are monotonically increasing.)

Taylor Expansion of distance function (existence of this expansion is the primary restriction on the update rule):

$$d(u, \omega_t, i) = d(u, \omega_t) + \frac{\partial d(u, \omega)}{\partial \omega_i} \left|_{\omega = \omega_t} (\omega_{t+1} - \omega_t)_i + \frac{1}{2}\frac{\partial^2 d(u, \omega)}{\partial \omega_i^2} \right|_{(\omega = \tilde{\omega})} (\omega_{t+1} - \omega_t)^2$$

$$d(u, \omega_t) - d(u, \omega_{t+1}) = -\frac{\partial d(u, \omega_t)}{\partial \omega_i}\left[\alpha_1(2\eta p_t x_i) + \alpha_2(2\eta p_t x_i)^2\right] - \frac{1}{2}\frac{\partial^2 d(u, \tilde{\omega})}{\partial \omega_i^2}[\ldots]^2$$

For GD: $f(\omega) = \omega$, $f' = 1$, $f'' = 0$ $\left(d(u, \omega) = \frac{1}{2}\|u - \omega\|_2^2\right)$

$$\alpha_1 = \left.\frac{\partial}{\partial \gamma} f^{-1}(f(\omega) + \gamma)\right|_{\gamma = 0} \qquad \frac{\partial}{\partial \gamma}(f(\omega) + \gamma) = 1$$

$$\alpha_2 = \left.\frac{1}{2}\frac{\partial^2}{\partial \gamma^2} f^{-1}(f + \gamma)\right|_{\gamma = \tilde{\gamma}} = 0$$

$$\left.\frac{\partial d(u, \omega)}{\partial \omega_i}\right|_{\omega = \omega_t} = (u_i - \omega_{t,i}) \qquad (= f(u) - f(\omega))$$

$$\left.\frac{\partial^2 d(u, \omega)}{\partial \omega_i^2}\right|_{\omega = \tilde{\omega}} = -1 \quad (= -f')$$

So,

$$d(u, \omega_t) - d(u, \omega_{t+1}) = (u - \omega_t)_i 2\eta p_t x_i - \frac{1}{2}(2\eta p_t x_i)^2$$
$$= ((y_t - \omega_t \cdot x) - (y_t - u \cdot x))2\eta p_t - 2\eta^2 p_t^2 x_i^2$$
$$\geq -2\eta p_t(q_t - p_t) - 2\eta^2 p_t^2 X^2, \|x\|_2 < X$$

In turn, $d(u, \omega_t) - d(u, \omega_{t+1}) \geq a p_t^2 - b q_t^2$ as long as $F(q, p, \eta) \leq 0$:

$$F(q, p, \eta) = a p_t^2 - b q_t^2 + 2\eta p_t(q_t - p_t) + 2\eta^2 p_t^2 X^2$$

Max F: $0 = \frac{\partial F}{\partial q} = 2\eta p - 2bq \Rightarrow p = \frac{b}{q}q$

$$\frac{\partial^2 F}{\partial q^2} = -2b, \text{ max for } b > 0$$

So, Max F at $\left(q = \frac{\eta p}{b}\right)$:

$$G(p, \eta) = F\left(q_{max} = \frac{\eta p}{b}\right) = 2\eta p_t\left(\frac{\eta p}{b} - p\right) + 2\eta^2 X^2 p^2 + a p^2 - b\left(\frac{\eta p}{b}\right)^2$$
$$= \frac{2\eta^2}{b}p^2 - 2\eta p^2 + 2\eta^2 X^2 p^2 + a p^2 - \frac{\eta^2 p^2}{b}$$

$$= p^2\left(\left(2X^2 + \frac{1}{b}\right)\eta^2 - 2\eta + a\right)$$

$G(p,\eta)$ is minimized for given p when $\dfrac{\partial G}{\partial \eta} = 0$:

$$\left(2X^2 + \frac{1}{b}\right)2\eta - 2 = 0$$

$$\Rightarrow \eta = 1/\left(2X^2 + 1/b\right)$$

Or

$$\eta = \frac{b}{1 + 2X^2 b}$$

where the latter choice η gives the broadcast range of possible $a's$. Suppose we take $\eta = b/(1 + 2X^2 b)$ always:

$$G(P,\eta_{min}) = p^2\left(-b + a + 2X^2 ab\right)$$

$$G(P,\eta_{min}) < 0 \text{ when } -b + a + 2X^2 ab < 0 \text{ or } \left(1 + 2X^2 b\right)a < b$$

Thus, we arrive at:

$$a \le \frac{b}{(1 + 2X^2 b)} = \eta \rightarrow \text{Condition for GD lemma 5.1 [214]}$$

Thus, for GD with learning rate η above and $a \le \dfrac{b}{(1 + 2X^2 b)}$ we satisfy:

$$a(y_t - \omega_t \cdot x_t)^2 - b(y_t - u \cdot x_t)^2 \le d(u, \omega_t) - d(u, \omega_{t+1}).$$

Once we know the constraints on a and b in the equation above, for the particular algorithm (GD in preceding), we can simply sum over a trial sequence to get the Worst Case Loss Bounds:

For GD Worst Case Loss Bounds, let $a = \dfrac{b}{(1 + 2X^2 b)} = \eta$:

$$(y_t - \omega_t \cdot x_t)^2 - \left[\frac{1}{(1 - 2X^2\eta)}\right](y_t - u \cdot x_t)^2 \le \left(\frac{1}{\eta}\right)[d(u, \omega_t) - d(u, \omega_{t+1})].$$

Now sum on instances from $t = 0$ to $t = T - 1$, for GD:

$$\text{Loss}(GD,\eta) \le \left[\frac{1}{(1 - 2X^2\eta)}\right]\text{Loss}(u) + \left(\frac{1}{\eta}\right)[d(u, \omega_0) - d(u, \omega_{T-1})],$$

Thus

$$\text{Loss }(GD,\eta) \le \left[\frac{1}{(1 - 2X^2\eta)}\right]\text{Loss}(u) + \left(\frac{1}{\eta}\right)d(u, \omega_0),$$

This result is equivalent to Lemma 5.2 in [213].

9.2.3 Review of the EG Loss Bounds Analysis

$$d_{reu}(u, \omega) = \sum_{i=1}^{N} \left(\omega_i - u_i + u_i \ln \frac{u_i}{\omega_i} \right), \text{where} \left(\beta = e^{-2\eta(\hat{y}_t - y_t)} \right)$$

$$d_{reu}(u, \omega_{t+1}) - d_{reu}(u, \omega_t) = \sum_{i=1}^{N} \left(\omega_{t+1,i} - \omega_{t,i} + u_i \ln \frac{\omega_{t,i}}{\omega_{t+1,i}} \right)$$

$$(\text{write } \omega_{t+1} = \omega_{t,i} \, \beta^{x_{t,i}})$$

$$= \sum_{i=1}^{N} (\omega_{t,i}(\beta^{x_{t,i}} - 1) + u_i \cdot x_{t,i} \ln \beta)$$

have $\beta > 0$, if $0 \le x_{t,i} \le X \Rightarrow 0 \le \frac{x_{t,i}}{X} \le 1$, call $\frac{x_{t,i}}{X} = z$

$$(\beta^{x_{t,i}} - 1) \le \frac{x_{t,i}}{X} (\beta^X - 1)$$

$$\alpha^z \le 1 - z(1 - \alpha) \text{ when } \alpha > 0, 0 \le z \le 1$$

$$d_{reu}(u, \omega_{t+1}) - d_{reu}(u, \omega_t) \le \omega \cdot x \frac{\beta^X - 1}{X} - u \cdot x \ln \beta$$

Here we have polynomial groupings in $\omega \cdot x$ and $u \cdot x$, but not $(\omega \cdot x - u \cdot x)$, so the analysis cannot proceed as in the key GD step where this was possible (allowing for polynomial expression in q's and p's and no ω or u references. Because of this, EGU will need more involved bounds on $\omega \cdot x \le y$, as will be apparent when comparison is made in the following $\sinh^{-1}(x)$ link analysis.

9.3 The "$\sinh^{-1}(\omega)$" link algorithm (SA)

9.3.1 Motivation for "$\sinh^{-1}(\omega)$" link algorithm (SA)

Given a set of instances and outcomes, the initial goal in using Loss Bounds analysis to guide algorithm development was to consider an algorithm that settles (quickly) on predications from the "best expert," where the experts will consist of the established neural net learning algorithms – Gradient Descent (GD) and un-normalized Exponentiated Gradient Descent (EGU), where the implementation is done in the frame-work of a neural net (NN) that learns via back-propagation. To do this, initially, an algorithm was proposed that alternated between the GD update and the EGU update with latency in each update dependent on the successes of that algorithm. In other words, if the EGU update provided smaller losses than the last GD update – then the EGU update would again be used. In order to prevent pathological domination of one update over another, it was

Figure 9.5 Link functions for GD (linear), EGU$^{+/-}$ (left and right curves), and SA (sigmoid).

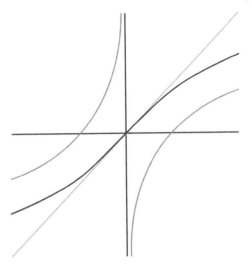

further proposed that the choice of next update have a stochastic element as well. The problem with this model was in the weight domains for GD vs EGU, GD having R for domain while EGU has R^+ or R^-. To remedy this an EGU-like update was sought with domain R, which suggested that the update with link function $\sinh^{-1}(w)$ be used. The $\sinh^{-1}(w)$ algorithm (SA) is asymptotically like EGU on large R^+, and like negative EGU on large R^-, and crosses the origin as an odd function – i.e. it behaves like GD for weights near zero (near the origin). So in seeking an algorithm like EGU on R, the SA algorithm was obtained that also has regions of behavior akin to the GD algorithm. Since SA behaves like both GD and EGU depending on the magnitude of the weight it will be examined in an exact study of loss bounds for SA and this is the focus of what follows.

Link functions for GD, EGU$^{+/-}$, and SA are shown in Figure 9.5:

Asymptotic properties of $f(\omega) = \sinh^{-1}(\omega)$: for $|\omega| \gg 1$: $f(\omega) \cong \ln(\omega) + \ln(2)$ for $(\omega > 0)$, so ~EGU$^+$, and $f(\omega) \cong -\ln(\omega) - \ln(2)$ for $(\omega < 0)$, so ~EGU$^-$, and for $-1 \ll \omega \ll 1$: $f(\omega) \cong \omega$, so ~GD.

Since $\sinh^{-1}(\omega) = \ln(\omega + \sqrt{\omega^2 + 1})$ this suggests the parameterized generalization via $f(\omega) = \ln(\omega + \sqrt{\omega^2 + \gamma})$. General (positive) γ leads to a GD-like update for small ω, but with a factor affecting the learning rate.

A statement of the learning algorithm for the sinh-link algorithm (SA) is:

Parameters:

L: a loss function from $R \times R$ to $[0,\infty)$

S: a start vector in R^N, and

η: a learning rate in $[0,\infty)$

Initialization: Before first trial, set $\omega_1 = S$

Prediction: upon receiving tth instance x_t, give prediction $\hat{y}_t = \omega_t * X_t$

Update: upon receiving the tth outcome y_t, update the weights according to the rule:

$$\sinh^{-1}(\omega_{t+1,i}) = \sinh^{-1}(\omega_{t,i}) - \eta x_{t,i} \left(\frac{\partial L(y_t, z)}{\partial z} \right) \bigg|_{z = \hat{y}_t}$$

9.3.2 Relation of sinh Link Algorithm to the Binary Exponentiated Gradient Algorithm

The SA link function is similar to the Binary Exponentiated Gradient algorithm (BEG) [216] link function (Figure 9.6).

A transformation on the weights allows BEG to be reexpressed as SA. This is accomplished by simplifying the following:

$$Z = \frac{\omega_{BEG} - L}{M - \omega_{BEG}} = \omega_s + \sqrt{\omega^2 + 1}$$

where $L < \omega_{BEG} < M \implies 0 < Z < \infty \implies -\infty < \omega_s < \infty$. Solving (call $\omega_{BEG} = y$ for this):

$$Z + \frac{L}{M} = \frac{(y-L) + \frac{L}{M}(M-y)}{(M-y)} = \frac{\left(1 - \frac{L}{M}\right)y}{M-y} = \frac{y}{M} \frac{(M-L)}{(M-y)}$$

$$Z + 1 = \frac{(y-L) + (M-y)}{(M-y)} = \frac{(M-L)}{(M-y)}$$

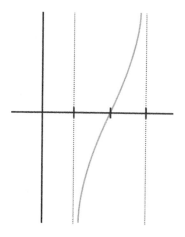

Figure 9.6 Link functions for BEG (sigmoid curve), where $f(\omega) = \ln\left(\frac{\omega - L}{M - \omega}\right)$, $L < \omega < M$.

$$\omega_{BEG} = \frac{M\left(Z + \dfrac{L}{M}\right)}{(Z + 1)} = \frac{MZ + L}{Z + 1} = \frac{M\left(\omega_s + \sqrt{\omega_s^2 + 1}\right) + L}{\left(\omega_s + \sqrt{\omega_s^2 + 1}\right) + 1}$$

The indicated (nonlinear) transformation on weights:

$$\omega_{BEG} = \frac{M\left(\omega_s + \sqrt{\omega_s^2 + 1}\right) + L}{\left(\omega_s + \sqrt{\omega_s^2 + 1}\right) + 1}$$

9.4 The Loss Bounds Analysis for sinh$^{-1}(\omega)$

The loss Bounds analysis for the $f(\omega) = \sinh^{-1}(\omega)$ link function is addressed next. The derivation follows the method of Kivinen and Warmuth and Jagota and Warmuth [213, 214] as summarized in Bylander [215]. The derivation in what follows was done in collaboration with Warmuth and Jagota.

The expression $d(u, \omega_{t+1}) - d(u, \omega_t)$ is now evaluated:

$$d(u, \omega_{t+1}) - d(u, \omega_t) = \sum_{i=1}^{N}\left(\sqrt{\omega_{t+1,i}^2 + 1} - \sqrt{\omega_{t,i}^2 + 1}\right.$$

$$\left. + u_i \ln\left[\frac{\omega_{t,i} + \sqrt{\omega_{t,i}^2 + 1}}{\omega_{t+1,i} + \sqrt{\omega_{t+1,i}^2 + 1}}\right]\right)$$

Since the update rule is:

$$\sinh^{-1}(\omega_{t+1,i}) = \sinh^{-1}(\omega_{t,i}) - \eta\left.\frac{\partial L(y, z = \omega_t \cdot x_t)}{\partial z}\right|_{z = \hat{y}_t} x_t$$

And only the square loss case considered, we have

$$L = (\omega x - y)^2; \qquad \left.\frac{\partial L}{\partial z}\right|_{z = \hat{y}_t} = 2(\hat{y} - y)$$

Thus, $\sinh^{-1}(\omega_{t+1,i}) = \sinh^{-1}(\omega_{t,i}) - 2\eta(\hat{y} - y)x_t$ and following [213], the substitution $\beta_t = \exp(-2\eta[\hat{y} - y])$ is made. Together with using the logarithmic form for the sinh^{-1}s this becomes:

$$\ln\left[\frac{\omega_{t+1,i} + \sqrt{\omega_{t+1,i}^2 + 1}}{\omega_{t,i} + \sqrt{\omega_{t,i}^2 + 1}}\right] = x_t \ln \beta_t$$

Now have:

$$d(u, \omega_{t+1}) - d(u, \omega_t) = \sum_{i=1}^{N} \left(\sqrt{\omega_{t+1,i}^2 + 1} - \sqrt{\omega_{t,i}^2 + 1} \right) - ux \ln \beta_t$$

This deceptively simple form must now be reexpressed with all ω_{t+1} dependence eliminated from the RHS. To get a relation for ω_{t+1} in terms of ω_t and $(x \ln \beta)$, it proves easiest to use both forms of the update rule:

$$\sinh^{-1}(\omega_{t+1,i} = \sinh^{-1}(\omega_{t,i}) + x_t \ln \beta_t$$

$$\Rightarrow (\omega_{t+1,i}) = \sinh\left[\sinh^{-1}(\omega_{t,i}) + x_t \ln \beta_t \right]$$

$$= \omega_{t,i} \cosh (x_t \ln \beta_t) + \sqrt{\omega_{t,i}^2 + 1} \sinh (x_t \ln \beta_t)$$

While the ln form for $\sinh^{-1}$ yields:

$$\ln \left(\omega_{t+1,i} + \sqrt{\omega_{t+1,i}^2 + 1} \right) = \ln \left(\omega_{t,i} + \sqrt{\omega_{t,i}^2 + 1} \right) + x_t \ln \beta_t$$

$$\Rightarrow \left(\omega_{t+1,i} + \sqrt{\omega_{t+1,i}^2 + 1} \right) = \left(\omega_{t,i} + \sqrt{\omega_{t,i}^2 + 1} \right) \beta^{x_t,i}$$

Now Grouping:

$$\sqrt{\omega_{t+1,i}^2 + 1} = \left(\omega_{t,i} + \sqrt{\omega_{t,i}^2 + 1} \right) \beta^{x_t,i} - \omega_{t+1,i}$$

$$= \left(\omega_{t,i} + \sqrt{\omega_{t,i}^2 + 1} \right) \beta^{x_t,i} - \omega_t \frac{1}{2} \left(\beta^{x_t,i} + \beta^{-x_t,i} \right)$$

$$- \sqrt{\omega_{t,i}^2 + 1} \frac{1}{2} \left(\beta^{x_t,i} + \beta^{-x_t,i} \right)$$

$$= \omega_{t,i} \sinh (x_t \ln \beta_t) + \sqrt{\omega_{t,i}^2 + 1} \cosh (x_t \ln \beta_t)$$

To get:

$$d(u, \omega_{t+1}) - d(u, \omega_t) = \sum_{i=1}^{N} \left[\omega_{t,i} \sinh (x_t \ln \beta_t) + \sqrt{\omega_{t,i}^2 + 1} (\cosh (x_t \ln \beta_t) - 1) \right]$$

$$- u \cdot x \ln \beta_t$$

The loss Bound statement is now expressed in terms of function F, $F(\omega_t, x_t, \omega_t \cdot x_t, y_t, u_t \cdot x_t, \eta) \leq 0$, where:

$$F = \sum_{i=1}^{N} \omega_{t,i} \sinh (x_t \ln \beta_t) + \sqrt{\omega_{t,i}^2 + 1} (\cosh (x_t \ln \beta_t) - 1)$$

$$- u \cdot x \ln \beta_t + a(y - \hat{y})^2 - b(y - u \cdot x)^2$$

First consider max F for some choice of $\cdot x = \bar{y}$:

$$\frac{\partial F}{\partial y} = -\ln \beta_t + 2b(y - \bar{y}) \quad \frac{\partial^2 F}{\partial y^2} = -2b$$

So, for positive b, we can establish the maximum value of F for any choice of $u \cdot x$, the Loss bound to be established accordingly. Substituting for $\frac{\partial F}{\partial y}\Big|_{\bar{y}} = 0$ leads to $\bar{y}_o = y - \frac{1}{2b} \ln \beta_t$, and

$$F(\omega_t, x_t, \omega \cdot x, y_t, u \cdot x = \bar{y}_o, \eta) = G(\omega, x, \hat{y}, y, \eta), \text{ and}$$

$$-b(y - \bar{y}_o)^2 = -b\left(y - \left[y - \frac{1}{2b} \ln \beta_t\right]\right)^2 = -\frac{1}{4b}(\ln \beta_t)^2$$

It is also simpler, at this point, to substitute for β_t:

$$\ln \beta_t = -2\eta(y - \bar{y}), \text{ but only in terms not in summation}$$

The resulting expression for G is:

$$G = \sum_{i=1}^{N}(\ldots\text{stuff}\ldots) - \left(y - \frac{1}{2b} \ln \beta_t\right) \ln \beta_t + a(y - \hat{y})^2 - \frac{1}{4b}(\ln \beta_t)^2$$

$$= \sum_{i=1}^{N}(\ldots) - y[-2\eta(y - \hat{y})] + \frac{1}{4b}(-2\eta(y - \hat{y}))^2 + a(y - \hat{y})^2$$

$$= \sum_{i=1}^{N}(\ldots) - 2\eta y(y - \hat{y}) + \left(a + \frac{\eta^2}{b}\right)(y - \hat{y})^2$$

$$= \sum_{i=1}^{N}\left[\omega_{t,i} \sinh(x_t \ln \beta_t) + \sqrt{\omega_{t,i}^2 + 1}(\cosh(x_t \ln \beta_t) - 1)\right]$$

$$- 2\eta y(y - \hat{y}) + \left(a + \frac{\eta^2}{b}\right)(y - \hat{y})^2$$

When $(y - \hat{y})$, $\ln \beta_t = 0$, and $G = 0$. So, following along the lines of the [215] calculation, if

$$\frac{\partial G}{\partial y}\Big|_{y = \hat{y}} = 0 \text{ and } \frac{\partial^2 G}{\partial y^2} < 0$$

can be established for some choice of a, b, η (and for x bounded in some sense), then a key lemma can be proved:

$$G = \sum_{i=1}^{N}\left[\omega_{t,i}\sinh\left(x_t 2\eta(y-\hat{y})\right) + \sqrt{\omega_{t,\;i}{}^2 + 1}\left(\cosh\left(x_t 2\eta(y-\hat{y})\right) - 1\right)\right]$$

$$-2\eta y(y-\hat{y}) + \left(a + \frac{\eta^2}{b}\right)(y-\hat{y})^2$$

$$\frac{\partial G}{\partial y} = \sum_{i=1}^{N}\left[2\eta\omega_{t,i}x_i\cosh\left(x_t 2\eta(y-\hat{y})\right) + 2\eta x_{t,i}\sqrt{\omega_{t,i}^2 + 1}\sinh\left(x_t 2\eta(y-\hat{y})\right)\right]$$

$$-2\eta(y-\hat{y}) - 2\eta y + 2\left(a + \frac{\eta^2}{b}\right)(y-\hat{y})$$

$$\frac{\partial G}{\partial y}\Big|_{y=\hat{y}} = 2\eta(\hat{y}-y)\Big|_{y=\hat{y}} = 0$$

$$\frac{\partial^2 G}{\partial y^2} = \sum_{i=1}^{N}\Big[(2\eta)^2 x_{t,i}(\omega_{t,i}x_i)\sinh\left(x_t 2\eta(y-\hat{y}) + (2\eta)^2 x_{t,i}^2 \sqrt{\omega_{t,i}^2 + 1}\right.$$

$$\cosh\left(x_t 2\eta(y-\hat{y})\right] - 4\eta + 2\left(a + \frac{\eta^2}{b}\right)$$

$$= 4\eta^2 \sum_{i=1}^{N} x_i^2\left[\omega_t\sinh\left(x_t \ln\beta_t\right) + \sqrt{\omega_t^2 + 1}\cdot\sqrt{\sin h^2(x_t\ln\beta_t) + 1}\right]$$

$$-4\eta + 2\left(a + \frac{\eta^2}{b}\right) = 4\eta^2\sum_{i=1}^{N} x_i^2\sqrt{\omega_{t+1,i}^2 + 1} - 4\eta + 2\left(a + \frac{\eta^2}{b}\right)$$

$$= 2\eta^2 R - 4\eta + 2\left(a + \frac{\eta^2}{b}\right), \text{ where } R = 2\sum x_i^2\sqrt{\omega_{t+1,i}^2 + 1}$$

So,

$$\frac{\partial^2 G}{\partial y^2} = 2\left[(-\eta + a) + \eta^2\left(-\frac{1}{\eta} + R + \frac{1}{b}\right)\right] \text{ and } \frac{\partial^2 G}{\partial y^2} \leq 0$$

$$\text{if } a \leq \eta \text{ and } R + \frac{1}{b} \leq \frac{1}{\eta} \text{ or } \eta \leq \frac{b}{1 + Rb}$$

The tightest restriction in the second inequality is then found by obtaining the upper bound on R:

$$R = 2\sum_{i=1}^{N} x_i^2\sqrt{\omega_{t+1,i}^2 + 1} = 2\sum_{i=1}^{N} x_i^2\left[\omega_t\sinh\left(x_t\ln\beta_t\right) + \sqrt{\omega_t^2 + 1}\cosh\left(x_t\ln\beta_t\right)\right]$$

In order to establish the inequality:

$$a(y_t - \omega_t\cdot x_t)^2 - b(y_t - u_t\cdot x_t)^2 \leq d(u,\omega_t) - d(u,\omega_{t+1})$$

$$\left(\text{where } 0 \leq a \leq \eta, \eta \leq \frac{b}{1 + Rb}, R = 2\sum_{i=1}^{N} x_i^2\sqrt{\omega_{t+1}^2 + 1}\right)$$

it is necessary to establish a bound on R. A bound on R then implies a bound on x_i and $\omega_{t+1,i}$ (if these variables are taken to be independent).

$$X_\infty \geq \max_i |x_{t,i}| W \geq \sum_{i=1}^{N} \sqrt{\omega_{t+1,i}^2 + 1}$$

Thus,

$$R \leq 2WX_\infty{}^2$$

It is interesting to note that unlike BEG, here there is no similar elimination of need for a bound on $\omega_{t+1,i}$ although there is the alternative:

$$W_\infty \geq \max_i \sqrt{\omega_{t+1,i}^2 + 1} X_2 \geq \sqrt{\sum_{i=1}^{N} x_{t,i}^2}$$

For which

$$R \leq 2W_\infty X_2^2.$$

Take $R = 2WX_\infty^2, 0 \leq a \leq \eta, \eta \leq \dfrac{b}{1 + Rb}$, then

$$aL(SA, S) \leq bL(u, s) + \sum_t (d(u, \omega_t) - d(u, \omega_{t+1}))L(SA, S) \leq \frac{b}{a}L(u, s)$$

$$+ \frac{1}{a}(d(u, \omega_0) - d(u, \omega_T)) \leq (1 + Rb)L(u, s) + \left(\frac{1}{b} + R\right)(d(u, \omega_0))$$

$$- d(u, \omega_T)) \leq (1 + c)L(u, s) + \left(1 + \frac{1}{c}\right)2W^2X_\infty^2$$

As with BEG, if $L(u, s) \leq K$, then with $\eta = \dfrac{1}{X_\infty \sqrt{2KW} + 2WX_\infty^2}$

If $L(u, s) = 0, \eta = \dfrac{1}{R}$ leads to

$$L(SA, s) \leq 2W^2X_\infty^2$$

9.4.1 Loss Bounds Analysis Using the Taylor Series Approach

Def: $\Delta_f(\omega^*, \omega) = \sum_i \int_{\omega[i]}^{\omega^*[i]} (f_i(r) - f_i(\omega[i]))dr$

$$\Delta_f(\omega^*, \omega_{t+1}) - \Delta_f(\omega^*, \omega_t) - \Delta_f(\omega_t, \omega_{t+1})$$

$$= \sum_i \left(\int_{\omega_{t+1}}^{\omega^*} - \int_{\omega_t}^{\omega^*} - \int_{\omega_{t+1}}^{\omega_t}\right)\{f(r)dr\} - \sum_i \int_{\omega_{t+1}}^{\omega^*} f(\omega_{t+1})dr$$

$$-\int_{\omega_t}^{\omega^*} f(\omega_t)dr - \int_{\omega_{t+1}}^{\omega_t} f(\omega_{t+1})dr$$

$$= \sum_i - \int_{\omega_t}^{\omega^*} f(\omega_{t+1})dr + \int_{\omega_t}^{\omega^*} f(\omega_t)dr$$

Thus,

$$\Delta - \Delta - \Delta = -\sum_i (\omega^* - \omega_t)(f(\omega_t) - f(\omega_{t+1}))_i$$

Let's begin with a variational formulation as before beginning with the definition:

$$u(\omega, s, x, y) = d(\omega, s) + \eta L(y, \omega x)$$

from which we get:

$$\left.\frac{\partial u}{\partial \omega}\right|_{\substack{s = \omega_t \\ \omega = \omega_{t+1}}} = 0 \Rightarrow f(\omega_{t+1}, i) - f(\omega_{t,i}) = -\eta \frac{\partial u}{\partial \omega_i}$$

The Taylor series for the Lagrangian:

$$L_t(\omega^*) = L_t(\omega_t) + \Sigma(\omega^* - \omega_t)\left.\frac{\partial L_t(\omega)}{\partial \omega}\right|_{\omega = \omega_t} + \frac{1}{2}\Sigma \left.\frac{\partial^2 L_t(\omega)}{\partial \omega}\right|_{\omega = \omega_t}(\omega^* - \omega_t)^2$$

$$L_t(\omega^*) \geq L_t(\omega_t) + \sum_i (\omega^* - \omega_t)\left.\frac{\partial L_t(\omega)}{\partial \omega}\right|_{\omega = \omega_t}$$

$$\left(\text{with restriction } \frac{\partial^2 L_t(\omega)}{\partial \omega} > 0, \text{i.e.convexity}\right) L_t(\omega^*) \geq L_t(\omega_t)$$

$$+ \sum_i (\omega^* - \omega_t)_i \left(\frac{f(\omega_{t,i}) - f(\omega_{t+1,i})}{\eta}\right) L_t(\omega^*) \geq L_t(\omega_t)$$

$$+ \frac{1}{\eta}\left[\Delta_f(\omega^*, \omega_{t+1}) - \Delta_f(\omega^*, \omega_t) - \Delta_f(\omega_t, \omega_{t+1})\right] \sum_{t=0}^{t-1} L_t(\omega^*)$$

$$\geq \sum_{t=0}^{t-1} L_t(\omega_t) + \frac{1}{\eta}\sum_{t=0}^{t+1}[.....]$$

Thus,

$$\sum_{T=0}^{T-1} L_t(\omega_t) \le \sum_{T=0}^{T-1} L_t(\omega_t) + \frac{1}{\eta}\left(\Delta_f(\omega^*,\omega_0) - \Delta_f(\omega^*,\omega_T)\right) + \frac{1}{\eta}\sum_{T=0}^{T-1}\Delta_f(\omega_t,\omega_{t+1})$$

9.4.2 Loss Bounds Analysis Using Taylor Series for the sinh Link (SA) Algorithm

In order to consider the loss bounds for SA in the Taylor approximation approach it is necessary to solve for the distance function indicated by the link function (note: two notations have converged here where distance is denoted as d or Δ_f: $d(s,\omega) = \Delta_f(s,\omega)$):

$$d'(s,\omega) = \frac{\partial d(s,\omega)}{\partial \omega} = \sinh^{-1}(\omega) - \sinh^{-1}(s)$$

Begin by taking the integral:

$$\int d'(s,\omega)d\omega = \int \ln\left[\omega + \sqrt{\omega^2+1}\right]d\omega - \ln\left[s + \sqrt{s^2+1}\right]\omega + C$$

when $\omega \to$ large, $\ln\left[\omega + \sqrt{\omega^2+1}\right] \simeq \ln[2\omega] \Rightarrow \int \ln\omega = \omega\ln\omega - \omega,$ so

guess that $\int \ln\left[\omega + \sqrt{\omega^2+1}\right]d\omega = \omega\ln\left[\omega + \sqrt{\omega^2+1}\right] - \sqrt{\omega^2+1}$

$$\frac{\partial}{\partial\omega}\left(\omega\ln\left[\omega + \sqrt{\omega^2+1}\right] - \sqrt{\omega^2+1}\right) = \ln\left[\omega + \sqrt{\omega^2+1}\right]$$

$$+ \omega + \frac{\left(1 + \frac{1}{2}(2\omega)\frac{1}{\sqrt{\omega^2+1}}\right)}{\omega + \sqrt{\omega^2+1}} - \frac{\frac{1}{2}(2\omega)}{\sqrt{\omega^2+1}}$$

$$= \ln\left[\omega + \sqrt{\omega^2+1}\right] + \omega\left(\sqrt{\omega^2+1} + \omega\right)$$

$$- \omega\left(\frac{\omega + \sqrt{\omega^2+1}}{\left(\omega + \sqrt{\omega^2+1}\right)\sqrt{\omega^2+1}} = \ln\left[\omega + \sqrt{\omega^2+1}\right]\right.$$

and we get:

$$d(s,\omega) = \omega\ln\left[\omega + \sqrt{\omega^2+1}\right] - \sqrt{\omega^2+1} - \omega\ln\left[s + \sqrt{s^2+1}\right] + C$$

The ω_i – independent function C is chosen such that $d(\omega,w) = 0$. Indices are reintroduced and summed over at this step also, for the $N - 1$ other integrations on $N - 1$ other ω_k parameters. Thus,

$$d(\omega, s) = \sum_{i=1}^{N} \omega_i \ln \left(\frac{\omega_i + \sqrt{\omega_i^2 + 1}}{s_i + \sqrt{s_i^2 + 1}} \right) - \sqrt{\omega_i^2 + 1} + \sqrt{s_i^2 + 1}$$

which will lead to:

$$\sum_{t=0}^{t-1} \Delta_f(\omega_t, \omega_{t+1})$$

$$= \sum_{t=0}^{t-1} \left(\sum_{i=1}^{N} \omega_{t,i} \ln \left[\frac{\omega_{t,i} + \sqrt{\omega_{t,i}^2 + 1}}{\omega_{t+1,i} + \sqrt{\omega_{t+1,i}^2 + 1}} \right] - \sqrt{\omega_{t,i}^2 + 1} + \sqrt{\omega_{t+1,i}^2 + 1} \right)$$

As before with square loss $L_t(\omega_t) = (\omega_t \cdot x_t - y)^2$ the update is;

$$\sinh^{-1}(\omega_{t+1,i}) - \sinh^{-1}(\omega_{t,i}) = -2\eta x_{t,i}(\hat{y} - y) = x_t \ln \beta$$

$$\ln \frac{\omega_{t+1,i} + \sqrt{\omega_{t+1,i}^2 + 1}}{\omega_{t,i} + \sqrt{\omega_{t,i}^2 + 1}} = x_t \ln \beta_t$$

$$\sum_{T=0}^{T-1} L_t(\omega_t) \le \sum_{T=0}^{T-1} L_t(\omega^*) + \frac{1}{\eta} \left(\Delta_f(\omega^*, \omega_0) - \Delta_f(\omega^*, \omega_T) \right)$$

$$+ \frac{1}{\eta} \sum_{T=0}^{T-1} \Delta_f(\omega_t, \omega_{t+1}) \le \sum_{T=0}^{T-1} L_t(\omega^*)$$

$$+ \frac{1}{\eta} \left(\Delta_f(\omega^*, \omega_0) + \frac{1}{\eta} \left[2 \sum_{T=0}^{T-1} x_t \omega_t (y - x_t \cdot \omega_t) + \sum_{i=1}^{N} \left\{ \sqrt{\omega_{T,i}^2 + 1} - \sqrt{\omega_{0,i}^2 + 1} \right\} \right] \right)$$

$$\le \sum_{T=0}^{T-1} L_t(\omega^*) + \frac{1}{\eta} \left(\Delta_f(\omega^*, \omega_0) - 2 \sum_{T=0}^{T-1} x_t \omega_t (y - x_t \cdot \omega_t) \right)$$

$$+ \frac{1}{\eta} \sum_{i=1}^{N} \left\{ \sqrt{\omega_{T,i}^2 + 1} - \sqrt{\omega_{0,i}^2 + 1} \right\}$$

$$\sum_{T=0}^{T-1} \left[(y - x_t \cdot \omega_t)^2 + 2x_t \cdot \omega_t (y - x_t \cdot \omega_t) \right]$$

$$\le \sum_{T=0}^{T-1} L_t(\omega^*) + \frac{1}{\eta} \left[\left(\Delta_f(\omega^*, \omega_0) + \sum_{i=1}^{N} \left\{ \sqrt{\omega_{T,i}^2 + 1} - \sqrt{\omega_{0,i}^2 + 1} \right\} \right) \right]$$

$$\sum_{T=0}^{T-1} \left[(y - x_t \cdot \omega_t)(y - x_t \cdot \omega_t) \right] \le \sum_{T=0}^{T-1} L_t(\omega^*)$$

$$+ \frac{1}{\eta} \left[\left(\Delta_f(\omega^*, \omega_0) + \sum_{i=1}^{N} \left\{ \sqrt{\omega_{T,i}^2 + 1} - \sqrt{\omega_{0,i}^2 + 1} \right\} \right) \right]$$

$$\sum_{T=0}^{T-1} \left(y^2 - (x_t \cdot \omega_t)^2 \right) \le \sum_{T=0}^{T-1} L_t(\omega^*)$$

$$+ \frac{1}{\eta} \left[\left(\Delta_f(\omega^*, \omega_0) + \sum_{i=1}^{N} \left\{ \sqrt{\omega_{T,i}^2 + 1} - \sqrt{\omega_{0,i}^2 + 1} \right\} \right) \right]$$

So, like BEG, the $\sinh^{-1}\omega$ link algorithm has the strengths of both GD and EGU. The loss bound for $\sinh^{-1}\omega$ is:

$$\text{Loss}(\sinh, S) \le (1 + c)\, \text{Loss}(u, S) + \left(1 + \frac{1}{c}\right) 2W^2 X_\infty^2$$

$$X_\infty \ge \max x_{t,i} \left(\eta = \frac{c}{2W^2 X_\infty^2 (1 + c)} \right) W \ge \sum_{i=1}^{N} \sqrt{\omega_{t+1,i}^2 + 1}$$

If $\text{loss}(u, S) \le K$ then choose $\eta = \dfrac{1}{X_\infty \sqrt{2KW} + 2W^2 X_\infty^2}$ to get:

$$\text{Loss}(\sinh, s) \le \text{Loss}(u, S) + 2WX_\infty \sqrt{2K} + 2W^2 X_\infty^2$$

The method involving Taylor series approximation gives a slightly weaker bound, but are more universally applicable.

9.5 Exercises

9.1 Derive the Perceptron learning rate and show it obtains separability, if it exists, in a finite number of learning steps.

9.2 Derive the recursive formulation of the backpropagation algorithm.

9.3 Obtain Loss bounds for the GD algorithm without using the Taylor Expansion method.

9.4 Obtain Loss bounds for the GD algorithm using the Taylor Expansion method.

9.5 Obtain Loss bounds for the EGU algorithm without using the Taylor Expansion method.

9.6 Obtain Loss bounds for the EGU algorithm using the Taylor Expansion method.

9.7 Obtain Loss bounds for the BEG algorithm without using the Taylor Expansion method.

9.8 Obtain Loss bounds for the BEG algorithm using the Taylor Expansion method.

9.9 Choose your own link function, do the Loss bounds with or without using the Taylor Expansion method. (Using Taylor series approach probably easier, but probably not as good a loss bound.)

10

Classification and Clustering

Numerous prior book, journal, and patent publications by the author [1–68] are drawn upon extensively throughout this chapter. Almost all of the journal publications are open access. These publications can typically be found online at either the author's personal website (www.meta-logos.com) or with one of the following online publishers: www.m-hikari.com or bmcbioinformatics.biomed-central.com.

A classifier is typically a simple rule whereby a class determination can be made, such as a decision boundary. Learning the decision rule, or a sufficiently good decision rule, especially if simple (and elegant), is the implementation aspect of a classifier, and can be difficult and time consuming. Even so, this is usually manageable because at least you have data to "learn from," e.g. supervised learning, where you have instances and their classifications (or "labels"). Learning for classification can be done very effectively using Support Vector Machines (SVMs), as will be described in what follows. With clustering efforts, or unsupervised learning, on the other hand, we do not have the label information during training. In what follows SVMs will also be shown to be incredibly effective at clustering when used with metaheuristics to recover label information in a bootstrap learning process. Also shown will be implementation details for distributed SVM training, and other speedup optimizations, for practical deployment of the powerful SVM classification and clustering methods in real-time operational situations (as will be demonstrated with results on a nanopore detector experiment in Chapter 14).

SVMs are variational-calculus-based methods that are constrained to have structural risk minimization (SRM) (maximum margin optimization), unlike neural net classifiers or perceptrons, such that they provide noise tolerant solutions for pattern recognition [1, 3, 59, 159–170]. An SVM determines a hyperplane that optimally separates one class from another, while the SRM criterion manifests as the hyperplane having a thickness, or "margin," that is

Informatics and Machine Learning: From Martingales to Metaheuristics, First Edition.
Stephen Winters-Hilt.
© 2022 John Wiley & Sons, Inc. Published 2022 by John Wiley & Sons, Inc.

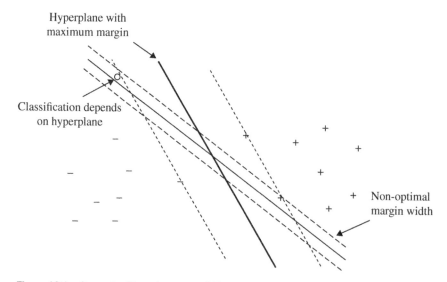

Hyperplane with
maximum margin

Classification depends
on hyperplane

Non-optimal
margin width

Figure 10.1 Supervised learning: separability and maximum margin. Shown in the figure is data separable by a line. With obvious generalization to a plane in 3D, and in higher dimensions, etc. We have thereby a (manifold) notion of separability on two datasets. Of all of the separating hyperplanes, it would appear to be optimal to choose one that could be made as "thick" as possible and still be a separating hyperplane. The thickness is known as the margin, so we are seeking a maximal margin hyperplane, as shown. Also shown is a nonoptimal margin hyperplane.

made as large as possible in the process of seeking a separating hyperplane (see Figure 10.1).

A benefit of using SRM is much less complication due to over-fitting. Once learned, the hyperplane allows data to be classified according to the side of the hyperplane in which it resides. The SVM approach encapsulates a significant amount of model-fitting information in its choice of kernel. The SVM kernel also provides a notion of distance in the neighborhood of the decision hyperplane. SVM binary discrimination outperforms other classification methods with or without dropping weak data. SVMs have a built parameter to assess confidence in a signal classification (related to the kernel distance from the separating hyperplane), thus have a built-in notion of weak data. Other classifier methods, if they have a notion of weak data, often introduce it as a separate evaluation that must itself be tuned and analyzed in order to be trusted. SVM multiclass discrimination and SVM-based clustering are also possible [1, 3, 13, 14, 54, 59, 167, 169]. In the stochastic sequential analysis (SSA) protocol SVMs play a central role in performing classification and clustering tasks.

Most SVM uses are restrictive in both training-set size and number of different classes, where most SVM applications involve datasets with fewer than 10 000 training instances and only two classes (the binary SVM). There are SVM implementations, however, that have no such limit on the number of training instances or the number of classes. Efficient new methods have been discovered for multi-class SVM, both internal to the optimization (multi-hyperplane) and external (decision tree and decision forest) [1, 3, 59]. In cases where the SVM training set is much larger than 10 000 instances, or when repeated training over the same training set is needed, significant SVM training computations are necessary. For this reason, distributed/GPU-optimized SVM training processes have been implemented [1, 3, 169].

There is a new approach to unsupervised learning that is based on use of supervised SVM classifiers. A fundamentally novel aspect of the proposed method is that it provides a nonparametric means for clustering (unsupervised learning) and partially supervised clustering. In preliminary work the SVM-based clustering method appears to offer prospects for inheriting the very strong performance of standard SVMs from the *supervised* classification setting. This offers a remarkable prospect for knowledge discovery and enhancing the scope of human cognition – the recognition of patterns and clusters without the limitations imposed by explicitly assuming a parametric model, where resolution of the identified clusters can be at an accuracy comparable to a supervised learning setting.

10.1 The SVM Classifier – An Overview

With an SVM "learning" process, once convergent to solution, what is learned, among other things, is the set of "support vectors" (SV)that define the thickness boundary of the separating hyperplane (see Figure 10.2).

So far we are implicitly describing a two-class problem, a "binary classifier," but what of multiple classes, can this be handled? Yes. Two broad categories of ways to handle this: (i) external refinement, such as via decisions trees (and forests) made from Binary classifier nodes (so more of what we have already got, to be discussed later); and (ii) internal refinement, such as via multiple hyperplanes (now inherently different, so discussed later).

So far we have been explicitly working with data that was separable. What if it is not? See Figure 10.3 Left for example where the data is not linearly separable. It is separable with a curve (or connected line segment as shown in Figure 10.3 Right).

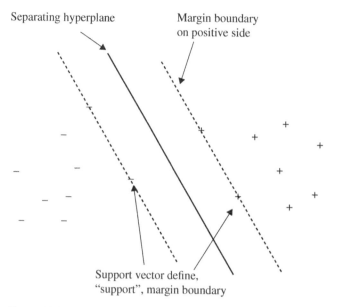

Figure 10.2 Support vectors on margin boundary.

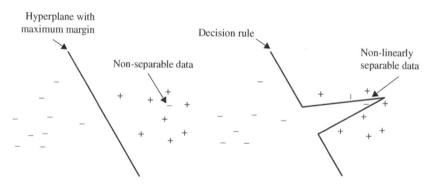

Figure 10.3 (Left) One of the former positives (central to the positive cluster) is now flipped to have a negative label. The data are now non-separable with a (single) straight line. (Right) The data are shown separable with a curve that happens to be a connected line segment.

10.2 Introduction to Classification and Clustering

In finding a separable solution (Figure 10.4, with maximum margin shown) we could also change the problem to separability on *most* of the data, with adjustments to account for those instances not consistent with the decision rule by

Hyperplane with
maximum margin

Margin width

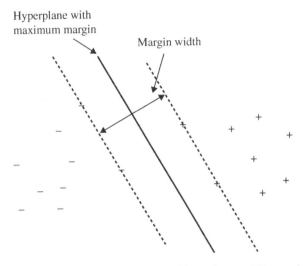

Figure 10.4 Separable solution with maximum width margin.

way of a penalty term (according to how much "wrong," perhaps). Alternatively, we could establish separability with a nonlinear discriminant by mapping the feature vector data to a higher dimensional space (e.g. introduction of the Kernel map and overall generalization), where linear separability would almost always be possible (almost provably always the case, due to hypersphere shattering in sufficiently high dimensions – something used to obtain an initial convergence in SVM clustering in Section 10.8)

Once training is done (have decision rule), we can then do classification (see Figure 10.5). Using the classifier (training data still shown), we simply classify according to which side of hyperplane (decision surface). Since we know the actual classes of the test data (used in the training/validation), we can score the performance of our classifier – this information, in turn, allows the classifier to be tuned for optimal performance.

So far we discussed the problem of learning to classify (with two class data: positive and negative), and we will find that we can solve any of the classification problems mentioned, whether extending to multiple classes or significantly non-separable. We have robust methods to do classification with SVMs and some other methods (carefully managed Neural Nets, for example). So what happens if we take away the label information and want to recover the identification of the positive and negative classes (assuming two classes)? Essentially, we have arrived at the classic clustering problem, where we want to identify clusters in the data. (Clustering is sometimes called unsupervised learning where the lack of label

Decision hyperplane

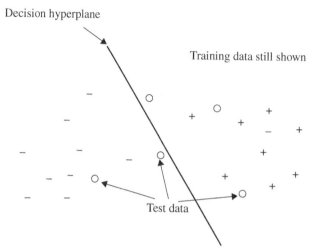

Figure 10.5 Post-training, have decision hyperplane. Now test data (circles in figure) can be introduced and classified according to which side of the hyperplane that they are on. Test data are usually held out from the original training data according to an N-fold cross-validation arrangement. So for analysis on one such fold, the test data used is data for which we "know" the answer ahead of time, but it presented without the knowledge as a test. If the test data are classified as positive and it was truly a positively labeled data, then it is a true positive (TP), and similarly for FP, TN, FN.

information is related to lack of "supervision.") It turns out that clustering is much more difficult than classification since we do not even have a definition for a "cluster."

What is a cluster? Vague Rule #1 is indicated in Figure 10.6, where intercluster distance is required to be greater than intra-cluster distance, and significantly so for good clustering (this fails for two parallel line of dots, for example).

Clustering is qualitative, thus lacks the same level of "well-definedness" as classification.

Having successfully obtained a clustering solution, you can then provide a labeling to the data, and revisit (bootstrap) in a supervised learning scenario to train a classifier.

Instead of bootstrapping from clustering into classification (with addition of label information), can we bootstrap from classification into clustering? By throwing any, random, label info, and having a learning process on the labels? That way we play to our strength, since classification as a learning process is on very form ground, both theoretically and in terms of efficient implementation. This would seem to indicate that a classification → clustering bootstrap solution could be done, but it would also require beginning the initial classification bootstrap run

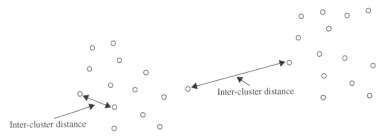

Figure 10.6 Two clusters, roughly, shown, with intra-cluster distance shown on right and intercluster distance shown between the edges of the two clusters.

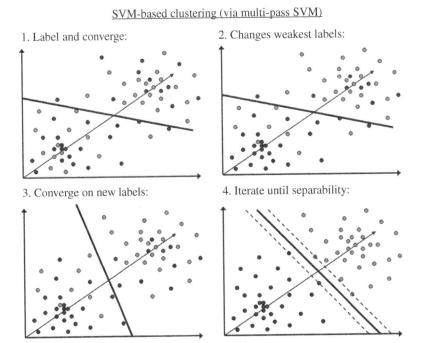

Figure 10.7 Schematic for SVM-based clustering, starting with randomly labeled data for which an SVM "learning" solution is sought (randomly labeled means extremely non-separable, typically). Remarkably, for certain kernels, convergence is possible.

with random data. Is this possible for any classifier to manage? The answer is yes, for the SVM, but only for certain choices of SVM kernel that are identified in Section 10.5. A schematic for the SVM-classifier-based clustering method is shown in Figure 10.7.

Suppose you have got a clustering solution. How can you do an objective measure of that cluster solution? One idea is to compute the total of the sum of the distances squared from the "center point" of the clusters.

10.2.1 Sum of Squared Error (SSE) Scoring

The sum of squared error (SSE) score for a cluster is the sum of the distances squared from the "center point" of the cluster (see Figure 10.8). The center point can be identified either as the most central data-point in the indicated cluster, or as the average data-point, i.e. centroid, that may not correspond with an actual data-point in the dataset. If you were hoping for more mathematical detail, that is covered in the subsection that follows on K-means clustering (one of the most popular clustering methods). It turns out that the central idea of the K-means clustering effort is an effort to minimize the SSE on the solution. For this reason, SSE scoring cannot be used (as effectively) for scoring K-means, since a different cluster evaluation would be needed to not be systematically blind to poor cluster solutions.

10.2.2 K-Means Clustering (Unsupervised Learning)

K-means is a simple algorithm for clustering a set of un-labeled feature vectors X: $\{x_1, \dots, x_n\}$ that are drawn independently from the mixture density $p(X|\theta)$ with a parameter set θ. At the heart of the K-means algorithm is optimization of the SSE function, J_i defined in definition SSE below.

Definition SSE (sum-of-squared-error): Given a cluster χ_i, the sum-of-squared, J_i is defined by:

$$J_i = \Sigma x \|x - m_i\|^2, x \in \chi_i$$

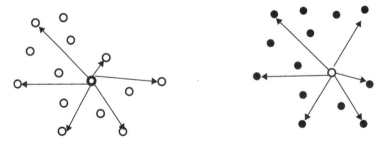

Figure 10.8 **Sum of squared error (SSE) scoring.** SSE, the total of the sum of the distances squared from the "center point" of the clusters.

and where m_i is the mean of the samples belonging to χ_i. The geometric interpretation of this criterion function is that for a given cluster χ_i the mean vector m_i is the centroid of the cluster by minimizing the length of the vector $x - m_i$. This can be shown by taking the variation of J_i with respect to the "centroid" m_i and setting it to zero,

$$\partial/\partial\ m_i J_i = \partial/\partial\ m_i \Sigma x (x - m_i) T (x - m_i) = 2\Sigma x (x - m_i) = 0$$

with minimum when equal to zero, where $x \in \chi_i$ in the sums, and solving for m_i:

$$m_i = 1/n_i \sum_x x$$

where $n_i = |\chi_i|$ is the number of feature vectors belonging to χ_i. The total SSE for all of the clusters, J_e is the sum of SSE for individual clusters. The value of J_e depends on the cluster membership of the data, (i.e. the shape of the clusters), and the number of clusters. The optimal clustering is the one that minimizes J_e for a given number of clusters, k, and K-means tries to do just that.

Kernel K-means [183] is a natural extension of K-means. Denote $M_{i\nu}$ to be the cluster assignment variables such that $M_{i\nu} = 1$ if and only if x_i belongs to cluster ν and 0 otherwise. As for K-means, the goal is to minimize the J_ν for all clusters, ν in feature space, by trying to find k means $\Phi(m_\nu)$ such that each observation in the data set when mapped using Φ is close to at least one of the means. Since the means lie in the span of $\Phi(x_1)$, ..., $\Phi(x_n)$, we can write them as:

$$\mu_\nu \equiv \Phi(m_\nu) = \Sigma_j \gamma_{\nu j} \Phi(x_j)$$

We can then substitute this in the J_i definition above to obtain,

$$J_\nu = \Sigma x \|\Phi(x) - \mu_\nu\|^2 = \Sigma x \|\Phi(x) - \Sigma_j \gamma_{\nu j} \Phi(x_j)\|^2$$
$$= K(x,x) - 2\Sigma_j \gamma_{\nu j} K(x, x_j) + \Sigma_{ij} \gamma_{\nu i} \gamma_{\nu j} K(x_i, x_j),$$

where $x \in \chi_i$ in Σx, $j \in 1...n$ in Σ_j, and $i, j \in 1...n$ in Σ_{ij}. We initially assign random feature vectors to means. Then Kernel K-means proceeds iteratively as follows: each new remaining feature vectors, x_{t+1}, is assigned to the closest mean μ_α:

$M_{t+1,\alpha} = 1$ if for all $\nu \neq \alpha$, $\|\Phi(x_{t+1}) - \mu_\alpha\|^2 < \|\Phi(x_{t+1}) - \mu_\nu\|^2$,
$M_{t+1,\alpha} = 0$ otherwise. Or, in terms of the kernel function,
$M_{t+1,\alpha} = 1$ if for all $\nu \neq \alpha$, $\Sigma_{ij} \gamma_{\alpha i} \gamma_{\alpha j} K_{ij} - \Sigma_j \gamma_{\alpha j} K_{t+1,j} < \Sigma_{ij} \gamma_{\nu i} \gamma_{\nu j} K_{ij} - \Sigma_j \gamma_{\nu j} K_{t+1,j}$
$M_{t+1,\alpha} = 0$ otherwise, where $K_{i,j} \equiv K_{ij} \equiv K(x_i, x_j)$. The update rule for the mean vector is then given by,

$$\mu_{t+1,\alpha} = \mu_{t,\alpha} + \Delta\left(\Phi(x_{t+1}) - \mu_{t,\alpha}\right), \text{where}, \Delta \equiv M_{t+1,\alpha}/\sum\nolimits_{1..t+1} M_{i\alpha}$$

A code sample in perl for performing K-means is shown below:

```perl
#!/usr/bin/perl

use strict;
use FileHandle;

my $train_fh = new FileHandle "9GC9TA.train";
my $test_fh = new FileHandle "9GC9TA.test";

my $index=0;
my @train_data;
my $comp_length;
while (<$train_fh>) {
    my ($label, @comps) = split;
    $comp_length = scalar(@comps);
#    @{$train_data[$index]} = ($label,@comps);
    @{$train_data[$index]} = @comps;
    $index++;
}
my $train_count = $index;

my ($K) = @ARGV;
if (!$K) { $K = 2; }

print "K=$K\n";

# initialize via random choice of data instances
my %ran_index_hash;
my @centroid;
my $index=0;
FOR: for $index (0..$K-1) {
    my $ran_index = int($train_count*rand() );
    print "$ran_index\n";
    if ($ran_index_hash{$ran_index}) {
        print "collision\n";
        $index-;
        next FOR;
    }
```

```perl
    else {
        $ran_index_hash{$ran_index}=1;
    }

    @centroid[$index] = @train_data[$ran_index];
}

if (0) { #### if0
my $index=0;
for $index (0..$train_count-1) {
    my $comp_index=0;
    for $comp_index (0..$comp_length-1) {
        print "$train_data[$index][$comp_index]\t";
    }
    print "\n\n";
}
} # endif0

if (0) { #### if0
my $index=0;
for $index (0..$K-1) {
    my $comp_index=0;
    for $comp_index (0..$comp_length-1) {
        print "$centroid[$index][$comp_index]\t";
    }
    print "\n\n";
}
} # endif0

my $old_first_diff=100;
my $first_diff=100;
# iterator loop
while (1) {

# have centroids, now evaluate distances of instances to
centroids
my @distances;
my @min_distance_labels;
my @min_distance_vals;
my $index=0;
for $index (0..$train_count-1) {
```

```perl
    my $j=0;
    my $min_dist_val=1000000;
    for $j (0..$K-1) {
        $distances[$j][$index]=0;
        my $comp_index=0;
        for $comp_index (0..$comp_length-1) {
            $distances[$j][$index] += ($centroid[$j]
            [$comp_index]-$train_data[$index]
            [$comp_index])**2;
        }

        if ($distances[$j][$index]<=$min_dist_val) {
            $min_dist_val=$distances[$j][$index];
            $min_distance_labels[$index]=$j;
#           print "$j loop: min_distance_labels[$index]=
            $j\n";
        }

#       print "index=$index\t j=$j\t distances[$j][$index]=
        $distances[$j][$index]\n";
    }
    my $mdl = $min_distance_labels[$index];
#   print "index=$index\t mdl=$mdl\t distances[$mdl]
    [$index]=$distances[$mdl][$index]\n";
}

my @old_centroid;
#copy and reset centroid array
my $j=0;
for $j (0..$K-1) {
    my $comp_index=0;
    for $comp_index (0..$comp_length-1) {
        $old_centroid[$j][$comp_index] = $centroid[$j]
        [$comp_index];
        $centroid[$j][$comp_index] = 0;
    }
}

# calculate new centroids
my @cent_count;
my $j=0;
```

```perl
for $j (0..$K-1) {
    $cent_count[$j]=0;
}

my $index=0;
for $index (0..$train_count-1) {
    my $comp_index=0;
    for $comp_index (0..$comp_length-1) {
        $centroid[$min_distance_labels[$index]]
        [$comp_index] += $train_data[$index][$comp_index];
    }
    $cent_count[$min_distance_labels[$index]]++;
}

my $j=0;
for $j (0..$K-1) {
    my $comp_index=0;
    for $comp_index (0..$comp_length-1) {
        $centroid[$j][$comp_index] /= $cent_count[$j];
    }
}

# if (0) { # if0
# print new centroids
my $index=0;
for $index (0..$K-1) {
    my $comp_index=0;
    for $comp_index (0..$comp_length-1) {
        print "$centroid[$index][$comp_index]\t";
    }
    print "\n\n";
}
# } # if0

$old_first_diff = $first_diff;

# test for movement of first centroid:
$first_diff=0;
my $comp_index=0;
```

```perl
for $comp_index (0..$comp_length-1) {
    $first_diff += ($centroid[0][$comp_index]-
$old_centroid[0][$comp_index])**2;
}
print "old_first_diff = $old_first_diff\n";
print "first_diff=$first_diff\n";
my $df = abs($old_first_diff - $first_diff);
print "$df\n";
my $epsilon = 0.000000001;

if ($df<$epsilon) {
    print "exiting with no change in centroid 1 position
\n";
    my $index=0;
    for $index (0..$train_count-1) {
        print "$index\t$min_distance_labels[$index]\n";
    }
    exit;
}

print "end of loop, with new centroid evaluation, starting
next loop\n";
} # no indent
```

10.2.3   k-Nearest Neighbors Classification (Supervised Learning)

Classification is generally easier than clustering, and that is often reflected in a simpler classification algorithm. Case in point is the classic k-nearest neighbors (k-NN) classification algorithm, where, as the name suggest, classification is based on the k-NN to instances in some given reference "training" set (where the data are labeled). To keep this simple, k is often chosen to be an odd number, such that a simple majority vote of the neighbors according to their class will then determine the class of the unlabeled instance presented.

   A code sample in perl for performing k-NN is shown below:

```perl
#!/usr/bin/perl
use strict;
use FileHandle;
my ($K) = @ARGV; # the K in K-nn
if (!$K) { $K = 5; }
# hardcoded data-file entries
```

```perl
my $train_data = "9GC9TA.train";
my $test_data = "9GC9TA.test";
my $train_fh = new FileHandle "$train_data";
my $test_fh = new FileHandle "$test_data";
my $index=0;
my @train_data;
while (<$train_fh>) {
    my ($label, @comps) = split;
    @{$train_data[$index]} = ($label,@comps);
    $index++;
}
my $train_count = $index;
my $index=0;
my @test_data;
while (<$test_fh>) {
    my ($label, @comps) = split;
    @{$test_data[$index]} = ($label,@comps);
    $index++;
}
my $test_count = $index;

my ($TP,$TN,$FP,$FN) = (0,0,0,0);
my $test_index;
for $test_index (0..$test_count-1) {
    my ($test_label,@test_comps) = @{$test_data
    [$test_index]};

    my @min;
    my @ind;
    my @lab;
    my $k_loop;
    for $k_loop (0..$K-1) {
        $min[$k_loop] = 100;
        $ind[$k_loop] = 0;
        $lab[$k_loop] = 0;
    }

    my $train_index;
    for $train_index (0..$train_count-1) {
        my ($train_label,@train_comps) = @{$train_data
```

```perl
            [$train_index]};
        my $comp_count = scalar(@train_comps);
        my $square_distance=0;
        my $comp_index;
        for $comp_index (0..$comp_count-1) {
            $square_distance += ($test_comps[$comp_index]-
            $train_comps[$comp_index])**2;
        }
        KFOR: for $k_loop (0..$K-1) {
            if ($square_distance<$min[$k_loop]) {
                my @min_shift = @min[$k_loop..($K-2)];
                my @ind_shift = @ind[$k_loop..($K-2)];
                my @lab_shift = @lab[$k_loop..($K-2)];

                $min[$k_loop] = $square_distance;
                $ind[$k_loop] = $train_index;
                $lab[$k_loop] = $train_label;

                @min = (@min[0..$k_loop],@min_shift);
                @ind = (@ind[0..$k_loop],@ind_shift);
                @lab = (@lab[0..$k_loop],@lab_shift);

                last KFOR;
            }
        }
    }
    # get majority vote of k nearest neighbors
    my $vote=0;

    for $k_loop (0..$K-1) {
        $vote += $lab[$k_loop];
    }

    my $test_predict;
    if ($vote>0) { $test_predict = 1; }
    elsif ($vote<0) { $test_predict = -1; }
    else { print "whoaaa!\n"; }

    if ($test_predict == $test_label && $test_predict == 1) {
        $TP++;
    }
```

```
elsif ($test_predict == $test_label && $test_predict ==
-1) {
     $TN++;
}
elsif ($test_predict != $test_label && $test_predict ==
1) {
     $FP++;
}
elsif ($test_predict != $test_label && $test_predict ==
-1) {
     $FN++;
}


}
print "K-nn with K=$K:\tTN=$TN\t TP=$TP\t FN=$FN\t FP=$FP
\t\t";
print "traindata=$train_data\ttestdata=$test_data\n";
```

10.2.4 The Perceptron Recap (See Chapter 9 for Details)

Let $X = R^d$ be the input space. Let $Y = \{+1, -1\}$ be the output space. We will want to be able to adjust weight on the importance of different features in the input space in arriving at a credit decision, and then have a simple threshold rule to make that decision:

$$\text{Approve if: } \sum_{i=1}^{d} w_i x_i > \theta$$
$$\text{Deny if: } \sum_{i=1}^{d} w_i x_i \leq \theta$$

Let us treat the threshold value as a weight $w_0 = b$ and associate it with an added X coordinate, $x_0 = 1$. Let us also make use of the sign function: $\text{sign}(s) = 1$ if $s > 0$, and $\text{sign}(s) = -1$ if $s < 0$. The decision function $h(x)$ is 1 if approved and -1 if denied:

$$h(x) = \text{sign}(w \cdot x)$$

The classic perceptron learning algorithm then goes as follows (Figure 10.9):

Suppose at the current iteration, t, of the learning algorithm, we have weight vector $w(t)$. If we have not already arrived at a solution then some of the $(x_1, y_1)...(x_N, y_N)$ will be misclassified, i.e. $h(x_k) \neq y_k$ for various k values, and one of these misclassifications will be used in the **update rule**:

$$w(t+1) = w(t) + y(t)x(t)$$

Figure 10.9 Perceptron update. The update rule shifts the position of the separating hyperplane such that classification errors eventually reduced to zero (assuming separable and that it does not get stuck in a local minimum of the minimization process).

Note: $y(t)[w(t) \cdot x(t)] < 0$ for each update (y has different sign than $w \cdot x$)

Note: $y(t)[w(t+1) \cdot x(t)] > y(t)[w(t) \cdot x(t)]$ (so move toward positivity)

Last follows from $y(t)[w(t+1) \cdot x(t)] = y(t)[w(t) \cdot x(t)] + [y(t)]^2[x(t) \cdot x(t)]$

Within the infinite space of all weight vectors, the perceptron algorithm finds a weight vector that works. This means that an infinite hypothesis space has been searched in a finite number of steps, so finite time, to find a solution. This search of an infinite hypothesis space in finite time will be one of the hallmarks of the Feasibility of Learning Proof (the "First Law of Machine Learning").

To recap, the components of the Perceptron Learning Problem are:

1) There is an input x.
2) There is an unknown target function $f: X \rightarrow Y$ (the ideal formula for credit approval in the credit problem).
3) There is a data set D of (input, output) examples $(x_1, y_1), ..., (x_N, y_N)$, where $y_n = f(x_n)$.
4) There is a learning algorithm that uses dataset D to pick a formula $g: X \rightarrow Y$ that approximates f.
5) The algorithm chooses g from a set of candidate formulas under consideration, which we call the hypothesis set H.

In the case of the perceptron, $h(x) = \text{sign}(w \cdot x), h \in H$.

10.3 Lagrangian Optimization and Structural Risk Minimization (SRM)

10.3.1 Decision Boundary and SRM Construction Using Lagrangian

In Figure 10.10, a decision boundary is shown as the solid line for a 2D domain, where separating hyperplane generalizations to planes in 3D and hyperplanes in higher dimensional domains are also possible (with in any orientable

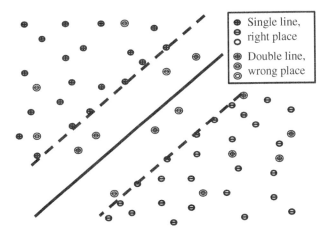

Figure 10.10 Decision boundary (solid line); with margin (region between dotted lines). Instances in the "wrong" place (double circles) are allowed but incur a penalty.

hyperplane manifold). The notion of a separating hyperplane is not unique to the SVM approach, but it is with use of further SRM constraints via maximizing a margin around the decision hyperplane. The margin is shown as the region around the decision boundary in Figure 10.10 that is between the dotted lines on either side of the decision hypersurface. Generalizations to compact (or circular) or multiple decision surfaces (e.g. multiple lines in 2D case) are also possible.

Obviously the decision boundary will rarely be a straight line in the feature space of the observed instances. Kernels can be employed to map to higher dimensional spaces, however, where hyperplane boundaries do work well, and are part of the standard SVM implementation for this purpose. In effect, on the lower dimensional feature space, this endows us with curve modeling capabilities that can reach any conceivable deformation of the decision-boundary line in 2D (similarly in higher dimension), where not even differentiability need be preserved in those deformations (as will be seen with the Kernel generalizations). In other words, we can model any boundary that simply has the topology of a line (or hyperplane in higher dimensions). The question naturally arises, if the infinite line (or hyperplane) decision surfaces work so well, what about a compact decision surface, such as a circle? And what does it mean? Given that the decision boundary implicitly provides a binary classifier, another possibility is to use multiple separating hyperplanes and majority vote to arrive at a multiclass classifier.

In the case of the circle we have a method for clustering, where we try to fit everything in the cluster circle, while at the same time minimizing the size of the

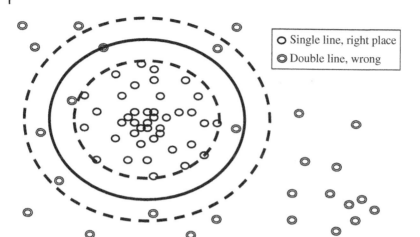

Figure 10.11 The solid circle is the decision boundary. In the implementation in [185], described as "SVM-internal" later, and the circle "collapse" under variational optimization is prevented by the condition that there be as few penalties as possible in a balance between the cluster boundary focusing on a dense "core" (high specificity automatically) or an outer fringe of outliers (high specificity automatically), or somewhere, optimally, in-between. The enclosing boundary balance approach may still lack the full SRM benefits of a true margin implementation that might be possible in a two-label system with use of indicated margin, labels obtained via label-flipping algorithms as described in the linear discriminant (the binary SVM) starting configuration for the SVM process. *Source:* Based on Ben-Hur et al. [185].

circle (see Figure 10.11). SVM clustering based on a circle topology decision boundary can be effective in many situations, but is burdened with a significant tuning task.

Other Lagrangian formulations for clustering might have the form shown in Figure 10.12, but the cluster decision boundary radii and the two-cluster center-separation (small dotted line in Figure 10.12) must all be explicitly modeled as separate tuning parameters, which can be a time-consuming approach, and there are entirely different approaches for SVM clustering, as will be discussed, that will be more effective.

The SVM method builds from a Lagrangian formulation that will be unfamiliar and very theoretical to people with a computer science background. Lagrangians play a prominent role in physics and control theory, so would be very familiar to those with that background. As the derivations get sorted out in what follows, however, one of the benefits of the SVM approach is that it directly relates to a small set of easily encoded tests and actions, and, in the end, a lot can be done without understanding the theoretical derivation in-depth by simply picking up with the use of the Karush–Kuhn–Tucker (KKT) relations. The theoretical

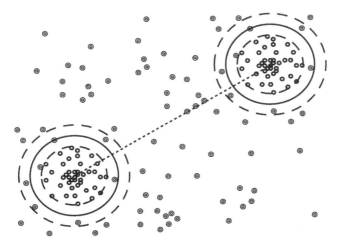

Figure 10.12 A possible modeling configuration for two-cluster explicit situation. In the case of the enclosing boundary method the hyperspherical clustering can map back to topologically disconnected components, such as that shown as the two cluster groupings shown in the feature space, but the model freedom is restricted and difficult to manage, so, again, separate approaches may be more useful to consider, to some extent for arriving at a computationally efficient solution.

derivations, however, do provide a greater understanding that is helpful in tuning.

It is conceivable to have a properly coded SVM but to initiate training with model parameters, such as the kernel or kernel parameter, that are so far out of the operational regime that no convergence is obtained in training. So training must be repeated, with tuning on SVM parameters, to optimize. For some feature vectors, such as probability vectors, this can partly be done automatically with choice of kernel. Overall, in many situations the SVM tuning can be done quickly, manually, and to some extent automatically with simple range testing, where only small, separated, subsets of the training data are used in the tuning tests, before performing SVM training on the full dataset minus the tuning data. Sometimes, however, and most certainly for the SVM applications in clustering that will be described, more automated tuning procedures are needed. Tuning is a form of optimization, and excellent metaheuristics are known for identifying optimal solutions when a scoring function (a fitness function) can be identified (such as for the SVM sensitivity and specificity score). Metaheuristic optimization includes genetic algorithms (GAs), simulated annealing, swarm intelligence, Ant colony optimization (ACO), steepest ascent hill-climbing, among others (as described in Chapter 12). Applications of many of these methods are shown in the results involving SVM-externalclustering in Section 10.8.1.

In setting up an SVM classifier you need to have training data in the form of feature vectors, where all of the feature vectors are the same length. You then need to specify a choice of kernel and kernel parameter (and possibly other parameters), and therein lies the rub. The SVM may not converge with your specification. SVMs have a surprising amount of practical functionality, as will be shown, so it turns out to be fairly easy to tune them, in many cases, by simply using a default set of kernel's and parameter ranges. In the SVM applications to bootstrap a Clustering solution, on the other hand, there is more sensitivity to kernel and kernel parameter, and more sophisticated tuning methods will clearly be helpful, as will be described further in Sections 10.8.2 and 10.8.3.

The ability to do fast SVM training (with distributed chunking and, possibly, GPU enhancements) means that *online* SVM learning can be managed in a brute-force fashion, with retuning on kernels periodically, and directly re-training on a moving window of data.

We describe a new form of clustering, via use of the SVM convergence process. Single-convergence initialized clustering methods, involving label-flipping between SVM convergence training runs, have been studied and will be detailed first. The single-convergence methods outperform other methods on the test sets considered, but in examining the clustering failures (albeit fewer than with parameterized methods), there appears to be room for improvement. Efforts to handle this with more sophisticated tuning have met with initial success, as will be related in what follows. But before proceeding further with efforts along these lines there is one other form of initialization to mention that may be adopted and modify the overall approach, probably to its betterment and reduction in complexity. That different approach is to initialize with information from *multiple* SVM convergences, with selection on training data based on algorithmic methods that leverage the clustering groupings indicated (from the "steepest ascent" multiple convergence data) to more effectively cluster, or simply cluster with less sophistication needed in the overall tuning requirements, and this will be discussed in Section 10.8.4.

In application to channel current signal analysis there is generally an abundance of experimental data available, if not, the experimenter can usually just take more samples and make it so. In this situation it is appropriate to seek a method good at both classifying data and evaluating a confidence in the classifications given. In this way, data that is low confidence can simply be dropped. The SRM at the heart of the SVM method's robustness *also provides a strong confidence measure*. For this reason, SVM's are the classification method of choice for channel current analysis, as they have excellent performance at 0% data drop, and as weak data is allowed to be dropped, the SVM-based approaches far exceed the performance of most other methods known as shown in what follows.

The applications of the SVM methods not only include classification and clustering, but also impacts feature extraction and identification in Hidden Markov model (HMM)-based methods, using an HMM/SVM vectorization/classification boost [44] (see Chapter 7).

10.3.2 The Theory of Classification

The formal description of the problem of classification is to find a general rule to match a set of objects, or observations, to their appropriate classes. In one form the binary classifier's task is to estimate a function $f: R^N \rightarrow \{\pm 1\}$, given examples and their $\{\pm 1\}$ classifications:

$$(x_1, y_1), ..., (x_n, y_n) \in R^N \times Y, Y = \{ \pm 1 \},$$

where (x, y) are assumed to be independent and identically distributed training data drawn from (unknown) probability distribution $P(x, y) \cdot f$ perfectly classifies if $y = +1$ when $f(x) \geq 0$, with $y = -1$ otherwise, where this holds for all of the n training instances.

In the loss-function formalism, the optimal f is obtained by minimizing the **expected risk** function (expected error) [162]:

$$R[f] = \int l(f(x), y) dP(x, y)$$

where l is a suitable loss function. For instance, in the case of "0/1 loss"

$$l(f(x), y) = \Theta(-yf(x))$$

where Θ is the Heaviside function ($\Theta(z) = 0$ for $z < 0$ and $\Theta(z) = 1$ otherwise; and where the argument chosen relates to the KKT relations in what follows) In most realistic cases $P(x, y)$ is unknown and therefore the risk function above cannot be used to find the optimum function f. To overcome this fundamental limitation one has to use the information hidden in the limited training examples and the properties of the function class F to approximate this function. Hence, instead of minimizing the expected risk, one minimizes the **empirical risk**

$$R_{emp}[f] = 1/n \sum l(f(x_i), y), \text{ with sum on } i \in 1...n.$$

The learning machine can ensure that for $n \rightarrow \infty$ the empirical risk will asymptotically converge to expected risk, but for a small training set the deviations are often large. This leads to a phenomenon called "over-fitting," where a small generalization error cannot be obtained by simply minimizing the training error. One way to avoid the over-fitting dilemma is to restrict the complexity of the function class [160]. The intuition, which will be formalized in the following, is that a "simple" (e.g. linear) function that explains most of the data is preferable to a complex one (i.e. an application of Occam's razor). This is often introduced via a regularization term that limits the complexity of the function class used by the learning machine [171].

A specific way of controlling the complexity of a function class is described by the Vapnik–Chervonenkis (VC) theory and the SRM principle [160, 161]. Here, the concept of complexity is captured by the VC dimension **h** of the function class **F** from which the estimate *f* is chosen. The following set of definitions indicate the role of SRM – in the SVM construction that follows SRM is implemented via maximum margin separation (and clustering).

Definition 1 (Shattering). A learning machine *f* can shatter a set of points x_1, $x_2, \dots, x_n$ if and only if for every possible training set of the form $(x_1, y_1), \dots, (x_n, y_n)$ there exists some parameter set that gets zero training error.

Definition 2 (VC Dimension). Given a learning machine *f*, the VC-dimension **h** is the maximum number of points that can be arranged so that *f* shatters them. Roughly speaking, the VC dimension measures how many (training) points can be shattered (i.e. separated) for all possible labelings using functions of the class. Constructing a nested family of function classes $F_1 \subset \cdots \subset F_k$ with nondecreasing VC dimension the SRM principle proceeds as follows:

Definition 3 (SRM Principle). Let $f_1, \dots, f_k$ be the solutions of the empirical risk minimization in the function classes F_i. SRM chooses the function class F_i (and the function f_i) such that an upper bound on the generalization error is minimized which can be computed making use of theorems such as the following one.

Theorem 4 (Expected Risk Upper bound). *Let **h** denote the VC dimension of the function class **F** and let R_{emp} be defined by using the "0/1 loss." For all delta > 0 and $f \in F$ the inequality bounding the risk*

$$R[f] \leq R_{\mathrm{emp}}[f] + [(h/n)(\ln(2n/h) + 1) - (1/n)\ln(\delta/4)]^{1/2}$$

holds with probability of at least $1-\delta$ for $n > h$ [160, 161]. For a derivation of this relation see Section 10.3.3. Note: this bound is only an example and similar formulations are available for other loss functions [161] and other complexity measures [172].

Thus, in the effort to minimize the generalization error $R[f]$ two extremes can arise: (i) a very small function class (like F_1) yields a vanishing square root term, but a large training error might remain, while (ii) a huge function class (like F_k) may give a vanishing empirical error but a large square root term. The best class is usually in between, as one would like to obtain a function that explains the data quite well and to have a small risk in obtaining that function. This is very much in analogy to the bias-variance dilemma scenario described for neural networks (see, e.g. [173]).

What these bounds universally indicate is that the minimized generalization error is bounded by a balance between training error and size of function class (i.e. structural risk). The standard SVM formulation (described in Section 10.4) directly implements such an optimization problem by balancing such terms using a Lagrangian formalism, and allowing for kernel selection model fitting via choice of kernel parameter selected according to minimal generalization error.

10.3.3 The Mathematics of the Feasibility of Learning

The question of whether D tells us anything outside of D that we did not know before has two different answers: D cannot tell us something outside of D with certainty, but D can tell us something *likely* about f outside of D.

It is in a probabilistic view (not deterministic) that we can show the feasibility of learning. The only assumption in the probabilistic framework is that the examples in D are generated independently (or, minimally, exchangeably).

10.3.3.1 The Hoeffding Inequality
Here is a form of the Hoeffding Inequality used in what follows:

$$P\left(\mid \overline{X} - E\left[\overline{X}\right] \mid \; \geq k\right) \leq 2e^{-2nk^2} \tag{10.1}$$

This can be understood in relation to the classic problem of drawing N samples from a Bin containing red and green marbles (Figure 10.13).

10.3.3.2 Hoeffding Inequality is Related to Chebyshev Inequality
1853/1867 Bienayme/Chebyshev: $P(\mid X - E(X) \mid > k) \leq \mathrm{Var}(X)/k^2$.

1963 Wassily Hoeffding: $P\left(\mid \overline{X} - E\left[\overline{X}\right] \mid \; \geq k\right) \leq 2e^{-2nk^2}$

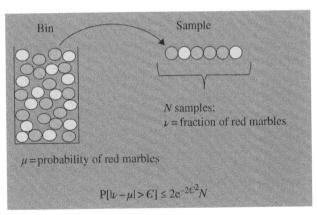

Bin **Sample**

N samples;
ν = fraction of red marbles

μ = probability of red marbles

$$P[\mid \nu - \mu \mid > \epsilon] \leq 2e^{-2\epsilon^2 N}$$

Figure 10.13 The Hoeffding inequality can be reduced to $P[\mid \nu - \mu \mid > \epsilon] \leq 2e^{-2\epsilon N}$, where this is the version of the Hoeffding inequality with range [0,1], and values {0,1}, where red is "1." The bin can be large, small, finite, or infinite.

More formally: Let $X_1,..., X_n$ be independent random variables. Assume that the X_i are almost surely bounded: $P(X_i\varepsilon[a_i, b_i]) = 1$. Define the empirical mean of the sequence of variables as: $\overline{X} = \frac{1}{n}(X_1 + \cdots + X_n)$.

Hoeffding [104] proves the following (relations critical to the theory underpinning SVMs, the entire field of classifiers, statistical learning (below), and is used in the proof of the feasibility of learning – further details in Appendix C):

$$P(\overline{X} - E[\overline{X}] \ge k) \le \exp\left(-\frac{2n^2k^2}{\sum_{i=1}^{n}(bi - ai)^2}\right)$$

$$P(|\overline{X} - E[\overline{X}]| \ge k) \le 2\exp\left(-\frac{2n^2k^2}{\sum_{i=1}^{n}(bi - ai)^2}\right)$$

For each X almost surely bounded have another relation if $E(X) = 0$ known as the Hoeffding Lemma:

$$E[e^{\lambda X}] \le \exp\left(\frac{\lambda^2(b - a)^2}{8}\right)$$

10.3.3.3 Sample Error

In sample error ("v"): $E_{in}(h) =$ (fraction of D where f and h disagree)

$$= \frac{1}{N}\sum_{n=1}^{N}[h(x_n) \ne f(x_n)]$$

where, "[...]" $= 1$ if true, else $= 0$. Out-of-sample error ("μ"): $E_{out}(h) = P[h(x) \ne f(x)]$. So, from Hoeffding:

$$P(|E_{in}(h) - E_{out}(h)| \ge \epsilon) \le 2e^{-2N\epsilon^2} \text{ for any } \epsilon > 0. \tag{10.2}$$

This result is for one hypothesis h from $H = \{h_1, ..., h_M\}$. Instead of Hoeffding in terms of generalization error on some h in H, want in terms of error on g, the final hypothesis based on D, where g has to be one of the h_M. Using the union-bound property we can obtain an error bound for g:

$$|E_{in}(g) - E_{out}(g)| > \epsilon \Rightarrow |E_{in}(h_1) - E_{out}(h_1)| > \epsilon \text{ OR.....OR}$$
$$|E_{in}(h_M) - E_{out}(h_M)| > \epsilon$$

Note: If $X \Rightarrow Y$ ("X implies Y"), then $P[X] \le P[Y]$.

Union bound rule: $P(X \text{ or } Y \text{ or...or } Z) \le P[X] + P[Y] + \cdots + P[Z]$ (weak bound since it does not count overlapping). Thus:

$$P(|E_{in}(g) - E_{out}(g)| > \epsilon) \le 2Me^{-2N\epsilon^2} \tag{10.3}$$

where, M can be thought of as a measure of the complexity of the hypothesis set H.

10.3.3.4 The Generalization Bound (Establishes First ML Law for |*H*| < ∞)

If M is a measure of the complexity of the hypothesis space, then the relation:

$$P(\,|E_{in}(g) - E_{out}(g)| > \epsilon) \le 2Me^{-2N\epsilon^2} \qquad (10.4)$$

tells us about a hypothesis-complexity/generalization-error-minimization trade-off in the hypothesis space selection: the more complex the H, the better (smaller) the $Ein(g)$, but the greater its complexity M. So seek the lowest complexity hypothesis space that still has small $E_{in}(g)$, to obtain a small $E_{out}(g)$. This is, in this context, basically a restatement of Occam's Razor (*lex parsimoniae*, ~ 1200 AD, the law of parsimony, also traces to Ptolemy and Aristotle).

The probabilistic form of the relation is not clear, so let us restate it as: With probability at least $(1 - 2M\ e^{-2N\epsilon^2}) \equiv (1 - \delta)$ have $|E_{in}(g) - E_{out}(g)| \le \epsilon$, and in terms of δ:

$$E_{out}(g) \le E_{in}(g) + \sqrt{\frac{1}{2N} \ln \frac{2M}{\delta}} \qquad (10.5)$$

This bound on the generalization error is often referred to as the "generalization bound". Now the trade-off is clearer, to have $E_{out}(g)$ small we simultaneously need $E_{in}(g)$ to be small and M to be not so huge that its log cannot be eliminated with a "reasonable-sized" sample N. For most cases of interest, however, |*H*| = ∞, even the simple perceptron the problem is the union bound has grossly overestimated by assuming no overlap (e.g. complete independence) between the hypotheses in the hypothesis set (an infinitesimal shift in the perceptron's decision hyperplane is counted as an entirely nonoverlapping hypothesis).

10.3.3.5 The VC Generalization Bound (Establishes First ML Law for |*H*| = ∞)

When accounting for overlapping (nonindependent) hypothesis, does the value "*M*" still go to infinity? Typically it does not. To account for the overlaps with the binary classifier consider the following notation:

Define the "growth function" for the binary classifier ($h(x) \rightarrow \{\pm 1\}$):
$m_H(N) = \max_{x_1, \ldots, x_N \in x} |H(x_1, \ldots, x_N)|$, where the "|...|" operation is set cardinality.

So, have $m_H(N) \le 2^N$. The question becomes, does the growth function continue to grow exponentially as $N \rightarrow \infty$? Here is a promising step in that direction. It is found that any growth function with a break point (k in what follows) is bounded by a polynomial:

If $m_H(k) < 2^k$ for some value k, then

$$m_H(N) \le \sum_{i=1}^{k-1} \binom{N}{i}, \forall N,$$

where N chooses I in this context and has degree $k - 1$.

Definition: The Vapnik–Chervonenkis dimension of a hypothesis set H, denoted by $d_{VC}(H)$, or just d, is the largest value of N for which $m_H(N) = 2^N$. Since $k = d + 1$, can simplify $\sum_{i=1}^{d} \binom{N}{i}$ to $N^d + 1$, and can then write:

$$m_H(N) \le N^d + 1 \tag{10.6}$$

The VC Dimension and Generalization

The VC dimension of the perceptron is only 3, because only up to three points in the plane, with arbitrary labels, will always be separable by a line, e.g. with four or more points can have non-separable situations. So can we just use $m_H(N)$ in place of M now that we know it is generally polynomial bound? It turns out there are some subtleties that lead to some of the constants involved changing, but otherwise, YES (for exact derivation see the Appendix in the book by Abu-Mostafa et al. [109]). We thus arrive at the following:

VC Generalization Bound

For any tolerance $\delta > 0$ have

$$E_{out}(g) \le E_{in}(g) + \sqrt{\frac{8}{N} \ln \frac{4m_H(2N)}{\delta}}$$

where $m_H(N)$ is polynomially bound, so $\frac{1}{N} \ln(\text{poly } N) \to 0$ as $N \to \infty$, have feasibility of learning with infinitely large hypothesis spaces H.

10.3.4 Lagrangian Optimization

The margin width can be easily determined by simply using a point on H_+ (the positive support vector boundary – e.g. the positive face of the hyperplane) as reference and taking the nearest point on H_- (similarly for negative face) (Figure 10.15 will have further details). Consider $x_i^{(+)}$ a feature vector from x_i that has $y_i = +1$, and that resides on ("supports") the H_+ boundary (e.g. a support vector). Likewise for the $x_i^{(-)}$, that is orthogonally positioned, will have $(x_i^{(+)} - x_i^{(-)}) \cdot \omega = 2$.

Since $(x_i^{(+)} - x_i^{(-)})$ is perpendicular to the hyperplane the distance between the hyperplanes is given by $d = \|(x_i^{(+)} - x_i^{(-)}) = 2/\|\omega\|$.

A variational derivation of the result $d = 2/\|\omega\|$, the distance between hyperplanes H_+ and H_- will be instructive for those not familiar with Lagrangians, and will be shown next. The variational derivation will provide a refresher on methods to be used in what follows. In Figure 10.14 we show a line in 2D space.

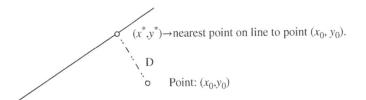

Figure 10.14 Line $Ax + By + C = 0$ and a nearby point. $D = \sqrt{(x^* - x_0)^2 + (y^* - y_0)^2}$.

Want to minimize D subject to constraint $Ax^* + By^* + C = 0$ (i.e. that the nearest point to (x_0, y_0) reside on the line. This suggests the following Lagrangian formulation:

$$L(x^*, y^*, \alpha) = D(x^*, y^*) + \alpha\left[Ax^* + By^* + C\right] \tag{10.7}$$

The Lagrangian solution is obtained by minimizing L on choice of $\{x^*, y^*\}$, i.e. minimize D (x^*, y^*), but subject to the constraint $Ax^* + By^* + C = 0$ (encapsulated in the term with the Lagrange multiplier):

$\dfrac{\partial L}{\partial \alpha} = [Ax^* + By^* + C]$, requiring $\dfrac{\partial L}{\partial \alpha} = 0$ then restores constraint, other variations yield:

$$0 = \frac{\partial L}{\partial \alpha^*} = \frac{(x^* - x_0)}{D} + \propto A; 0 = \frac{\partial L}{\partial y^*} = \frac{(y^* - y_0)}{D} + \propto B$$

$$\frac{(x^* - x_0)^2}{D^2} + \propto^2 A^2; \frac{(y^* - y_0)^2}{D^2} = \alpha^2 B^2 \Rightarrow 1 = \alpha^2\left(A^2 + B^2\right)$$

And, with other set of equations from grouping extremal equations multiplied by A (or B) instead of squaring, it is then possible to solve to get:

$$D = \frac{Ax_0 + By_0 + C}{\sqrt{A^2 + B^2}} \tag{10.8}$$

Generalization from 2D space to m-D space is to $|Ax_0 + By_0 + C| \rightarrow |\omega \cdot x - b|$ which will be $|\omega \cdot x - b| = 1$ for the geometries we have described, so simplifies significantly. For the "ω" parameters in H_0 (same as in H_+) have:

$$\sqrt{A^2 + B^2} \rightarrow \sqrt{\sum_k w_k^2} = \|\omega\|$$

So, $D = 1/\|\omega\|$ from H_+ to H_0, and thus twice that for H_+ to H_-; and, thus, $d = 2/\|\omega\|$ as before.

10.3.5 The Support Vector Machine (SVM) – Lagrangian with SRM

A SVM is a machine learning classification method with robust learning and minimal overtraining concerns. SVMs can also be used for bootstrap clustering. A fundamental constraint on SVM learning is the management of the training set. This is because the order of computations during the learning process typically goes as the square of the size of the training set. In Chapter 14 we examine experimental data involving a nanopore detector, where 150-component feature data is gathered on individual molecules (that are drawn into the nanopore). SVM training data can be produced by the nanopore detector in prodigious amounts, to arrive at a set of 150-component feature vectors that number from 10 000 to 100 000, depending on the number of molecular classes being examined for a particular application. For the 150-component feature data examined here, training sets of 1000 (500 positives and 500 negatives, for example) can be managed on a PC without hard-drive I/O thrashing. Training sets of 10 000 or more, however, cannot be managed with a single PC-based resource. For this reason most SVM implementations must contend with some kind of *chunking* process to learn parts of the data at a time. In later sections, results show that chunk aliasing and outlier accumulation may pose problems for distributed SVM learning. The results also present new methods and how they offer a stable learning solution to these problems at minimal cost. One of those methods extends the learning process with modified alpha-selection heuristics that enable a support-vector reduction phase. What is not as commonly discussed about *distributed* SVM learning are the details of the distributed, or approximately parallel, chunk processing methods. The distributed SVM described here was implemented using Java RMI, and was developed to run on a network of multi-core computers.

As mentioned previously SVMs are discriminators that use SRM to find a decision hyperplane with a maximum margin between separate groupings of feature vectors [160]. When SVMs were first implemented in 1995, a quadratic programming algorithm was used [159]. This was slow and only small datasets could be run with them. In 1998, Platt implemented sequential minimal optimization (SMO), which is an algorithm that uses incremental (minimal) learning steps in a Lagrangian implementation to bypass having to use a quadratic algorithm (to be implemented in Section 10.4.1) [163, 164]. The SMO SVM iterates through the dataset comparing and updating the Lagrange multipliers two at a time. SMO typically provides a significant increase in the speed of learning (once trained, SVM classification is at the speed of computation for an inner-product calculation on the features vectors, here amounting to 150 multiplications). If a GPU is used this can be reduced further, to the time of computing one multiplication (where the inner product multiplications are done in parallel). Although the SMO implementation did much to advance the feasibility and ease-of-use of SVM classifiers, there has still been the key constraint of computing the kernel matrix, which is quadratic in the size of the training data. In the sections that follow we introduce (i) the standard binary SVM; kernel variants; (iii) alpha-selection-variants (including simple chunking); chunking methods; and a synopsis of previous work with a multiclass SVM formulation. The work with the multiclass formulation provides a clear example of the importance of managing and tracking SV's (and other feature vector categories) during the learning process. The importance of tracking the SV's during the learning process will be revisited in the distributed learning results presented in the Results section.

10.3.5.1 Kernel Modeling and Other Tuning

The so-called curse of dimensionality from statistics says that the difficulty of an estimation problem increases drastically with the dimension N of the space, since in principle as N increases, the number of required patterns to sample grows exponentially. This statement may cast doubts on using higher dimensional feature vectors as input to learning machines. This must be balanced with results from statistical learning theory [174], however, that shows that the likelihood of data separability by linear learning machines is proportional to their dimensionality.

Thus, instead of working in the R^N, one can design algorithms to work in feature space, F, where the data has much higher dimension (but with sufficiently small function class). This can be described via the following mapping

$$\Phi : R^N \to F; x \to \Phi(x)$$

Consider the prior training description with data $x_1, ..., x_n \in R^N$ is mapped into a potentially much higher dimensional feature space F. For a given learning algorithm one now considers the same algorithm in F instead of R^N. Hence, the learning machine works with the following:

$$(\Phi(x_1), y_1), ..., (\Phi(x_n), y_n) \in F \times Y, Y = \{\pm 1\}$$

It is important to note that this mapping is also implicitly done for (one hidden layer) neural networks, radial basis networks [175] and boosting algorithms [110, 176] where the input data is mapped to some representation given by the hidden layer, the radial basis function (RBF) bumps or the hypotheses space, respectively.

As mentioned above, the dimensionality of the data does not detract us from finding a good solution, but it is rather the complexity of the function class F that contributes the most to the complexity of the problem. Similarly, in practice, one need never know the mapping function Φ. Therefore, the complexity and intractability of computing the actual mapping is also irrelevant to the complexity of the problem of classification. To this end, algorithms are transformed via a kernel generalization to take advantage of this aspect of the method.

10.3.6 Kernel Construction Using Polarization

The kernels used in the analysis are based on a family of previously developed kernels [59, 67], here referred to as "Occam"s Razor," or "Razor" kernels. As will be seen, the Gaussian kernel is included in the family of Razor kernels. All of the Razor kernels examined perform strongly on the channel current data analyzed, with some regularly outperforming the Gaussian Kernel itself. The kernels fall into two classes: regularized distance (squared) kernels; and regularized information divergence kernels. The first set of kernels strongly models data with classic, geometric, attributes or interpretation. The second set of kernels is constrained to operate on $(R^+)^N$, the feature space of positive, nonzero, real-valued feature vector components. The space of the latter kernels is often also restricted to feature vectors obeying an L_1-norm $= 1$ constraint, i.e. the feature vector is a discrete probability vector.

Given any metric space (x, d) one can build a positive-definite kernel of the form $e^{-\lambda d^2}$. Conversely, any positive definite kernel with form $e^{-\lambda d^2}$ must have a "d" that is a metric (this is Mercer's condition in another form). This suggests that the "simplest" kernel is the Gaussian kernel, since the "simplest" distance, the Euclidean distance, is used. Functional variations on the Gaussian kernel are described in what follows (see [59] for further details), including variations that are no-longer represented as distances (nonmetric), but that operate on a constrained domain and provide a positive definite kernel (or close enough). In what follows a quick synopsis is given of the novel kernels that are used – two of these kernels regularly outperformed the Gaussian kernel (and all other kernels), as will be shown in the following.

Tuning is needed to optimize the choice of kernel and kernel-parameter used by the SVM. This is often handled simply by ranging over a collection of roughly 10 kernel types and each at roughly 10 kernel parameter setting (where each is single-parameter kernel), and to do this only on smaller test sets in the training data, where the time complexity of the SVM training is directly tied to the training-kernel computation, which is quadratic in the number of training instances. Although caching can modify the assumptions on time-complexity, there is generally an approximately quadratic time-complexity in the size of the training instances regardless. Chunking must be used to break past this, or more extensive use of GPU capabilities (still need to eventually do chunking). Chunking algorithms will be shown to be effective, but susceptible to training-failure pathologies if certain safeguards are not observed, as will be discussed.

Once the small test set is done on the initial kernel screening indicated above, a sub-set of kernels will emerge as best, and these are considered again with larger training sets, eventually allowing selection of a good choice of kernel and kernel parameter. More directed tuning paradigms typically involve simulated annealing in this setting (to be shown later). Algorithmic and implementation parameters can also be considered in the tuning, which means we now have a collage of different parameter types in a coupled optimization task. For this type of generalization, GAs have been applied with amazing success (but not shown in what follows). These more sophisticated tuning methods may not always be necessary in the SVM classification applications but will allow for successful classifications in some situations where simple methods do not, and in the SVM-based clustering methods to be described in what follows these tuning methods generally play an important role. Further details on the tuning methods themselves are described in Chapter 12.

In [67], novel, information-theoretic, kernels were introduced for notably better performance over standard kernels – especially when discrete probability distributions or other constrained feature vectors were used as the feature vector data. The use of probability vectors, and L_1-norm feature vectors in general, turns out to be a very general formulation, wherein feature extraction makes use of signal decomposition into a complete set of separable states that can be interpreted or represented as a probability vector (or normalized collection of such, or concatenation, then normalization, etc.). A probability vector formulation also provides a straightforward hand-off to the SVM classifiers since all feature vectors have the same length with such an approach. What this means for the SVM, however, is that geometric notions of distance are no longer the best measure for comparing feature vectors. For probability vectors (i.e. discrete distributions), the best measures of similarity are the various information-theoretic divergences: Kullback–Leibler, Renyi, etc. By symmetrizing over the arguments of those divergences a rich source of kernels is obtained that works well with the types of probabilistic data obtained, as shown in [1, 3].

The SVM Lagrangian formulation and kernel incorporation are described in what follows. The SVM discriminators are trained by solving their KKT relations using the SMO procedure [33, 164], shown in what follows. Multi-class SVM training is also explored and involves thousands of blockade signatures for each signal class. A chunking variant of SMO (similar to [179, 180]) also is employed to manage the large SVM training task. Data rejection heuristics are also explored.

For the binary SVM and "single-class" SVM, SRM can be simply implemented via a maximal margin on the boundaries indicated, for the multiclass SVM there is the additional constraint used in the multiclass method in Section 10.7 that the decision hypersurface distances to a common origin be minimized (this remarkable and practical constraint allows the optimization equations to decouple, and the *multiclass* SVM to be easily implemented on a computer). This can be implemented with a variety of variations, including simply requiring a square term to be minimized that will do the necessary decoupling.

10.3.7 SVM Binary Classifier Derivation

SVMs use a Lagrangian formulation to construct a separating hyperplane (see Figure 10.15), surrounded by the thickest margin, using a set of training data represented as feature vectors. Prediction is made according to some measure of the "distance" between the test data and the hyperplane. Complex pattern-classification problems can be transformed into a new feature space where the patterns are more likely to be linearly separable, provided that the transformation itself is nonlinear and that the feature space is in a high enough dimension [178]. The SVM construction described in what follows achieves such a transformation by choosing a qualified kernel.

Linear Binary SVM: N training data "points" (feature vectors with binary labels), are denoted:

$$\left\{ \left(\vec{x_1}, y_1 \right), \left(\vec{x_2}, y_2 \right), ..., \left(\vec{x_n}, y_n \right) \right\}, \; \vec{x_i} \in R^m, y_i = \pm 1$$

One possible assumption to proceed: The positive and negative labeled data ($y_i = \pm 1$) is sufficiently separable and grouped as positives and negatives, that notions such as positive and negative data clusters, and a hyperplane separating them, are meaningful. This is accomplished by finding a separating hyperplane between positives and negatives, $y_i = \pm 1$ (see Figure 10.15), where we assume full separability possible with choice of feature vector (*f. v.*) components:

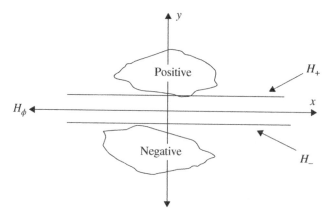

Figure 10.15 Hyperplane separability. $H_\varphi: y = \vec{\omega} \cdot \vec{x} - b = \mathbf{0}$.

All horizontal hyperplanes in Figure 10.15 are parallel, $H_\varphi \| H_+ \| H_-$ so all have hyperplanes proportional to $\vec{\omega}$. So, have without loss of generality:

$$H_+ : y = \vec{\omega} \cdot \vec{x} - b_+ = 0$$
$$H_- : y = \vec{\omega} \cdot \vec{x} - b = 0 \tag{10.9}$$

Again, rescaling, to bring into the form for fully separable binary data $(\vec{x}_i, y_i)$:

$$\vec{\omega} \cdot \vec{x} - b \geq +1 \text{ for } y_i = +1$$
$$\vec{\omega} \cdot \vec{x} - b \leq -1 \text{ for } y_i = -1 \tag{10.10}$$

Or, unifying:

$$y_i \left(\vec{\omega} \cdot \vec{x} - b \right) - 1 \geq 0 \, \forall_i \tag{10.11}$$

This is the constraint that must be satisfied for separable (binary) data (in this formulation).

In the binary SVM implementation described in what follows we follow the notation and conventions used previously [59]. Feature vectors are denoted by x_{ik}, where index i labels the feature vectors ($1 \leq i \leq M$) and index k labels the N feature vector components ($1 \leq k \leq N$). For the binary SVM, labeling of training data is done using label variable $y_i = \pm 1$ (with sign according to whether the training instance was from the positive or negative class). For hyperplane separability, elements of the training set must satisfy the following conditions: $w_\beta x_{i\beta} - b \geq +1$ for i such that $y_i = +1$, and $w_\beta x_{i\beta} - b \leq -1$ for $y_i = -1$, for some values of the coefficients $w_1, \ldots, w_N$, and b (*using the convention of implied sum on repeated Greek indices*). This can be written more concisely as: $y_i(w_\beta x_{i\beta} - b) - 1 \geq 0$. Data points that satisfy the equality in the above are known as "SV" (or "active constraints").

Once training is complete, discrimination is based solely on position relative to the discriminating hyperplane: $w_\beta x_{i\beta} - b = 0$. The boundary hyperplanes on the two classes of data are separated by a distance $2/w$, known as the "margin," where $w^2 = w_\beta w_\beta$. By increasing the margin between the separated data as much as possible, the optimal separating hyperplane is obtained. In the usual SVM formulation, the goal to maximize w^{-1} is restated as the goal to minimize w^2. The Lagrangian variational formulation then selects an optimum defined at a saddle point of

$$L(w, b; \, \alpha) = \frac{W_\beta W_\beta}{2} - \alpha_\gamma y_\gamma \left(w_\beta w_{\gamma\beta} - b \right) - \alpha_0 \tag{10.12}$$

where, $\alpha_0 = \sum_\gamma \alpha_\gamma, \alpha_\gamma \geq 0 \, (1 \leq \gamma \leq M)$

The saddle point is obtained by minimizing with respect to $\{w_1, \ldots, w_N, b\}$ and maximizing with respect to $\{\alpha_1, \ldots, \alpha_M\}$. If $y_i(w_\beta x_{i\beta} - b) - 1 \geq 0$, then maximization on α_i is achieved for $\alpha_i = 0$. If $y_i(w_\beta x_{i\beta} - b) - 1 = 0$, then there is no constraint on α_i. If $y_i(w_\beta x_{i\beta} - b) - 1 < 0$, there is a constraint violation, and $\alpha_i \to \infty$. These three relations are known as the KKT, relations. If absolute separability is possible, the last case will eventually be eliminated for all α_i, otherwise it is natural to limit the size of α_i by some constant upper bound, i.e. $\max(\alpha_i) = C$, for all i. This is equivalent to another set of inequality constraints with $\alpha_i \le C$. Introducing sets of Lagrange multipliers, ξ_γ and $\mu_\gamma (1 \le \gamma \le M)$, to achieve this, the Lagrangian becomes:

$$L(w, b : a, \xi, u) = \frac{W_\beta W_\beta}{2} - \alpha_\gamma \left[y_\gamma \left(w_\beta x_{\gamma\beta} - b \right) + \xi_\gamma \right] + \alpha_0 + \xi_0 C - \mu_\gamma \xi_\gamma$$

$$(10.13)$$

where, $\xi_0 = \sum_\gamma \alpha_\gamma$ and $\alpha_\gamma \ge 0$ and $\xi \ge 0$ $(1 \le \gamma \le M)$

At the variational minimum on the $\{w_1, ..., w_N, b\}$ variables, $w_\beta = \alpha_\gamma y_\gamma x_{\gamma\beta}$, and the Lagrangian simplifies to:

$$L(\alpha) = \alpha_0 - \frac{\alpha_\delta y_\delta x_{\delta\beta} \alpha_\gamma y_\gamma x_{\gamma\beta}}{2}$$

$$(10.14)$$

With $0 \le \alpha_\gamma \le C$ $(1 \le \gamma \le M)$ and $\alpha_\gamma y_\gamma = 0$, where only the variations that maximize in terms of the α_γ remain (known as the Wolfe Transformation).

Thus, the Wolfe Dual Calculations, with or without slack variable, have the form (with sums explicit):

$$\tilde{L}(\propto) = \sum_i \alpha_i - \frac{1}{2} \sum_{i,j} \alpha_i \alpha_j y_i y_j \vec{x}_i \vec{x}_j, \alpha_j \ge 0$$

$$(10.15)$$

where, we want to find the α's that maximize $L(\alpha)$. Similarly, for the L_0 Dual:

$$\tilde{L}_0(\propto) = \sum_i \alpha_i - \frac{1}{2} \sum_{i,j} \alpha_i \alpha_j y_i y_j \vec{x}_i \vec{x}_j, C \ge \alpha_i \ge 0$$

$$(10.16)$$

So, the duals are the same, with or without start variable, aside from the $C \ge \alpha_i$ $(\max(\alpha \le C))$ constraint.

In the Wolfe Dual form the computational task can be greatly simplified. By introducing an expression for the discriminating hyperplane: $f_i = w_\beta x_{i\beta} - b = \alpha_\gamma y_\gamma x_{\gamma\beta} x_{i\beta} - b$, the variational solution for $L(\alpha)$ reduces to the following set of KKT relations:

i) $\alpha_i = 0, y_i f_i \ge 1$
ii) $0 < \alpha_i < C, y_i f_i = 1$ (10.17)
iii) $\alpha_i = C, y_i f_i \le 1$

When the KKT relations are satisfied for all of the α_γ (with $\alpha_\gamma y_\gamma = 0$ maintained) the solution is achieved. The constraint $\alpha_\gamma y_\gamma = 0$ is satisfied for the initial choice of multipliers by setting the α's associated with the positive training instances to $1/N$ $(+)$ and the α's associated with the negatives to $1/N(-)$, where $N(+)$ is the number of positives and $N(-)$ is the number of negatives. Once the Wolfe transformation is performed it is apparent that the training data (SV in particular, KKT class (ii) above) enter into the Lagrangian solely via the inner product $x_{i\beta}x_{j\beta}$. Likewise, the discriminator f_i, and KKT relations, are also dependent on the data solely via the $x_{i\beta}x_{j\beta}$ inner product.

Generalization of the SVM formulation to data-dependent inner products other than $x_{i\beta}x_{j\beta}$ are possible and are usually formulated in terms of the family of symmetric positive definite functions (reproducing kernels) satisfying Mercer's conditions [25].

To see the origin of the KKT relations more clearly, consider the SVM Lagrangian with multipliers for the collection of separability constraints on solutions: $y_i(\omega \cdot \boldsymbol{x} - b) - 1 \geq 0 \, \forall \, i$. For SRM we then need to maximize $d = 2/\|\omega\|$, or minimize $\|\omega\|^2$, which is chosen due to simplifications in the formalism that follows (i.e. if we max $2/\|\omega\|$ by min on $\|\omega\|$, it could just as well be done with min on $\|\omega\|^2$). The Lagrangian formulation then should have one multiplier constraint for each training instance, where $y_i(\omega \cdot \boldsymbol{x} - b) - 1 \geq 0$, and minimize on $\|\omega\|^2$ overall, so:

$$L\left(\vec{\omega}, b, \vec{a}\right) = \frac{1}{2}\|\omega\|^2 - \sum_i \alpha_i[y_i(\omega x_i - b) - 1], \alpha_i \geq 0 \qquad (10.18)$$

As with the practice Lagrangian described earlier, we seek to minimize L on $\{\vec{\omega}, b\}$ and to extremize (maximize in this case) L on $\{\vec{a}\}$, i.e. what results is a minimization – maximization saddle-point optimization for the solution. Note how the inequality constraints above differ from the distance-to-line problem. In the latter case, the constant was an exact equality $(Ax* + By* + C = 0)$ (as with the common holonomic constraints in theoretical physics), and the term entering the Lagrangian was: "$\alpha[Ax^* + By^* + C]$", and the recovery of the constraint from $\partial L/\partial \alpha = 0$ was straightforward. Now, however, the Lagrange multipliers are no longer free to be negative (recall the $\alpha = \pm (A^2 + B^2)^{-1}$ solution before). Now the α's are restricted to be positive, and the term entering the Lagrangian has an overall negative in front: "$-\alpha_i[y_i(\omega_i x_i - b) - 1]$." To understand the inequality constraint recovery, and with it the KKT relations, we must consider the overall influence on the Lagrangian in its saddle-point approximation when three possibilities are considered:

- If $[y_i(\omega \, x_i - b) - 1] > 0$ (constraint satisfied), then maximization (on α_i's) for $\sum_i \alpha_i[y_i(\omega x_i - b) - 1]$ is achieved for $\alpha_i \to 0$ (since $\alpha_i \geq 0$ constraint).

- If $[y_i(\omega x_i - b) - 1] = 0$ (constraint satisfied, a support vector), then there is no constraint on α_i.

- If $[y_i(\omega x_i - b) - 1] < 0$, then $\alpha_i \to \infty$.

The last case, $\left[y_i \left(\omega \cdot \vec{x}_i - b \right) - 1 \right] < 0$, is an example of where the constraint is not satisfied. For completely separable data this case will not occur in the solution, but may occur when incrementally optimizing to achieve that selection. As we shall see, non-separable (perfectly) data can have constraint violations in the solution. How is this managed if the Lagrangian optimization will drive the associated Lagrange multipliers to larger and larger positive values ($\alpha_i \to \infty$)? The answer is to establish a max α cut off: max $(\alpha_i) = C$. In practice, the max α cutoff is usually imposed whether or not you have separable or non-separable data, and thus we have:

$$L = \frac{1}{2}\|\omega\|^2 - \sum_i \alpha_i [y_i(\omega x_i - b) - 1], \alpha_i \ge 0, \alpha_i \le C \ (C - \alpha_i \ge 0) \qquad (10.19)$$

The $(C - \alpha_i \ge 0)$ constraint can itself be absorbed into the Lagrangian:

$$L = \frac{1}{2}\|\omega\|^2 - \sum_i \alpha_i [y_i(\omega x_i - b) - 1] + \sum \sigma(C - \alpha_i), \sigma_i \ge 0, \alpha_i \ge 0 \ (C - \alpha_i \ge 0)$$

$$(10.20)$$

This can be rewritten where the role of the new Lagrange multiplier is much more apparent:

$$= \frac{1}{2}\|\omega\|^2 - \sum_i \alpha_i [y_i(\omega x_i - b) - 1 + \sigma_i] + C \sum \sigma_i, \sigma_i \ge 0, \alpha_i \ge 0 \qquad (10.21)$$

The revised form relates back to an initial formulation with constraints:

$$\omega \cdot x_i - b \ge 1 - \sigma_i \text{ for } y_i = +1$$

$$\omega \cdot x_i - b \le -1 + \sigma_i \text{ for } y_i = -1 \qquad (10.22)$$

So, the Lagrange multiplier σ_i, introduced to deal with the max(α_i) = C) constraint, can be interpreted as a "slack variable" (see Figure 10.16):

With the $\alpha's$ we have more control with $[y_i(\omega x_i - b) - 1]$, where previously $\alpha_i \to \infty$ resulted. Can now avoid with $C < \alpha_i$ condition by establishing initial conditions without $C < \alpha_i$ and maintaining those conditions as the Lagrangian optimization goes forward.

To recap, the Wolfe Dual Calculations, with or without slack variable, have the same form:

$$L = \frac{1}{2}\|\omega\|^2 - \sum_i \alpha_i [y_i(\omega x_i - b) - 1], \alpha_i \ge 0 \qquad (10.23)$$

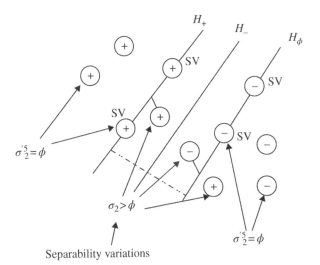

Separability variations

Figure 10.16 IF $C > \alpha_i$, $\sigma_i \to \emptyset$; If $C = \alpha_i$, σ_i free($\geq \emptyset$); If $C < \alpha_i$, $\sigma_i \to \infty$.

$$L_\sigma = \frac{1}{2}\|\omega\|^2 - \sum_i \alpha_i[y_i(\omega x_i - b) - 1 + \sigma_i] + C\sum \sigma_i, \sigma_i \geq 0, \alpha_i \geq 0 \quad (10.24)$$

$$0 = \frac{\partial L}{\partial \omega_j} = \omega_j - \sum_i \alpha_i y_i (x_i)_j \forall_j \Rightarrow \vec{\omega} = \sum_i \alpha_i y_i \vec{x}_i$$

$$0 = \frac{\partial L}{\partial \omega_j} = \sum_i \alpha_i y_i \Rightarrow \sum_i \alpha_i y_i = 0$$

$$\tilde{L} = \frac{1}{2}\sum_j \left(\sum_i \alpha_i y_i (x_i)_j\right)^2 - \sum_i \alpha_i y_i \left(\sum_j \left(\sum_i \alpha_i y_i (x_i)_j\right)(x_i)_j\right) + \sum_i \alpha_i$$

$$\tilde{L}(\alpha) = \sum_i \alpha_i - \frac{1}{2}\sum_{i,j} \alpha_i \alpha_j y_i y_j \vec{x}_i \cdot \vec{x}_j, \alpha_i \geq 0 \quad (10.25)$$

where, we want to find the $\alpha's$ that maximize $L(\alpha)$. Similarly, for the L_σ Dual:

$$L_\sigma \cdot \tilde{L}(\sigma, \alpha) = \sum_i \alpha_i - \frac{1}{2}\sum_{i,j} \alpha_i \alpha_j y_i y_j \vec{x}_i \cdot \vec{x}_i - \sum_\sigma (\alpha_i - C), \sigma_i \geq 0, \alpha_i \geq 0$$

$$\tilde{L}_\sigma(\alpha) = \sum_i \alpha_i - \frac{1}{2}\sum_{i,j} \alpha_i \alpha_j y_i y_j \vec{x}_i \cdot \vec{x}_j, C \geq \alpha_i \geq 0 \quad (10.26)$$

So, the duals are the same aside from the $C \geq \alpha_i(\max(\alpha) > C)$ constraint.

10.4 SVM Binary Classifier Implementation

10.4.1 Sequential Minimal Optimization (SMO)

The SVM discriminators are trained by solving their KKT relations using the SMO procedure of [163, 164]. The method described here follows the description of [163, 164] and begins by selecting a pair of Lagrange multipliers, $\{\alpha_1, \alpha_2\}$, where at least one of the multipliers has a violation of its associated KKT relations. For simplicity it is assumed in what follows that the multipliers selected are those associated with the first and second feature vectors: $\{x_1, x_2\}$. The SMO procedure then "freezes" variations in all but the two selected Lagrange multipliers, permitting much of the computation to be circumvented by use of analytical reductions:

$$\tilde{L}_\sigma\left(\vec{\alpha}\right) = \tilde{L}_\sigma(\alpha_1, \alpha_2; \alpha_3...\alpha_n) = \left(\alpha_1 + \alpha_2 + \sum_{i \geq 3} \alpha_i\right)$$

$$-\frac{1}{2}\left(\alpha_1^2 k_{11} + \alpha_2^2 k_{22} + 2\alpha_1\alpha_2 s k_{12}\right) \tag{10.27}$$

$$-\frac{1}{2}\left(2\alpha_1 y_1 \sum_{j \geq 3} \alpha_j y_j k_{1j} + 2\alpha_2 y_2\right) - \frac{1}{2} \sum_{j \geq 3, i \geq 3} \alpha_i \alpha_j y_i y_j k_{ij}$$

where, $s = y_1 y_2$.

Let $v_i = \sum_{j \geq 3} \alpha_j y_j k_{ij} = \vec{\omega} \cdot \vec{x}_i - \alpha_1 y_1 k_{ij} = \vec{\omega} \cdot \vec{x}_i - \alpha_1 y_1 k_{i1} - \alpha_2 y_2 k_{il} - \alpha_2 y_2 k_{i2}$

$$\tilde{L}_\sigma\left(\vec{\alpha}\right) = \alpha_1 + \alpha_2 - \frac{1}{2}\left(\alpha_1^2 k_{11} + \alpha_2^{22} k_{22} + 2\alpha_1\alpha_2 s k_{12}\right) - \alpha_1 y_1 v_1 - \alpha_2 y_2 v_2$$

$$+ \sum_{i \geq 3} \alpha_i - \frac{1}{2}\alpha_i \alpha_j y_i y_j K_{ij} \tag{10.28}$$

Or, in shorthand notation:

$$L\left(\alpha_1, \alpha_2; \alpha_{\beta \geq 3}\right) = \alpha_1 + \alpha_2 - \frac{\left(\alpha_1^2 K_{11} + 2\alpha_1\alpha_2 y_1 y_2 K_{12}\right)}{2} - \alpha_1 y_1 y_2 - \alpha_2 y_2 v_2$$

$$+ \alpha_{\beta'} U_{\beta'} - \frac{\alpha_{\beta'} \alpha_y y_{\beta'} y_y K_{\beta' y}}{2} \tag{10.29}$$

with $\beta', \gamma' \geq 3$, and where $K_{ij} \equiv K(x_i, x_j)$, and $v_i \equiv \alpha_{\beta'} y_{\beta'} K_{i\beta'}$ with $\beta' \geq 3$. U projects the sum on $\alpha_{\beta'}$ for $\beta' \geq 3$. Due to the constraint $\alpha_\beta y_\beta = 0$, we have the relation: $\alpha_1 + s\alpha_2 = -\gamma$, where $\gamma \equiv y_1 \alpha_{\beta'} y_{\beta'}$ with $\beta' \geq 3$ and $s \equiv y_1 y_2$.

Now consider variational parameters other than $\{\alpha_1, \alpha_2\}$ to be fixed in the $\{\alpha_1, \alpha_2\}$ variational optimization. Furthermore:

$$\sum_i y_i \alpha_j = 0 \Rightarrow y_1 \alpha_1 + y_2 \alpha_2 = -\sum_{i \geq 3} y_i \alpha_i; (\alpha_1 + S\alpha_2) = \gamma$$

$$\gamma = -y \sum_{i \geq 3} y_1 \alpha_i; \alpha_i = \gamma - s\alpha_2$$

$$L(\alpha_2; \alpha_3, \dots \alpha_m) = (\gamma - s\alpha_2) + \alpha_2 - \frac{1}{2}\left((\gamma - s\alpha_2)^2 k_{11} + 2\alpha_2(\gamma - s\alpha_2)sk_{12}\right)$$

$$- (\gamma - s\alpha_2)y_1 v_1 - \alpha_2 y_2 v_2 + [\text{terms independent of } \{\alpha_1, \alpha_2\}] \quad (10.30)$$

$$0 = \frac{\partial L}{\partial \alpha_2} = (1 - s) - \alpha_2(k_{11} + k_{22} - 2k_{12}) + s\gamma k_{12} + sy_1 v_1 - y_2 v_2$$

Let: $-\eta = k_{11} + k_{22} - 2k_{12}$

α_2^*: *new* $\propto$, the optimization solution:

$$\alpha_2^* = \alpha_2 - \frac{y_2\left[\vec{\omega} \cdot x_1 - y_1 - \left(\vec{\omega} \cdot x_2 - y_2\right)\right]}{\eta} \quad (10.31)$$

Rewriting, and explicitly relating the new alpha to the old:

$$\alpha_2^{\text{new}} = \alpha_2^{\text{old}} - \frac{y_2\left((w_\beta x_{1\beta} - y_1) - (w_\beta x_{2\beta} - y_2)\right)}{\eta} \quad (10.32)$$

Once α_2^{new} is obtained, the constraint $\alpha_2^{\text{new}} \leq C$ must be re-verified in conjunction with the $\alpha_\beta y_\beta = 0$ constraint. If the $L(\alpha_2; \alpha_\beta \geq 3)$ maximization leads to a α_2 new that grows too large, the new α_2 must be "clipped" to the maximum value satisfying the constraints. For example, if $y_1 \neq y_2$, then increases in α_2 are matched by increases in α_1. So, depending on whether α_2 or α_1 is nearer its maximum of C, we have max $(\alpha_2) = \text{argmin}\{\alpha_2 + (C - \alpha_2); \alpha_2 + (C - \alpha_1)\}$. Similar arguments provide the following boundary conditions:

(i) if $s = -1$, $\max(\alpha_2) = \text{argmin}\{\alpha_2 ; C + \alpha_2 - \alpha_1\}$, and $\min(\alpha_2) = \text{argmax}\{0; \alpha_2 - \alpha_1\}$, and (ii) if $s = +1$, $\max(\alpha_2) = \text{argmin}\{C; \alpha_2 + \alpha_1\}$, and $\min(\alpha_2) = \text{argmax}\{0; \alpha_2 + \alpha_1 - C\}$.

In terms of the new $\alpha_2^{\text{new,clipped}}$, clipped as indicated above if necessary, the new α_1 becomes:

$$\alpha_1^{\text{new}} = \alpha_1^{\text{old}} + s\left(\alpha_2^{\text{old}} - \alpha_2^{\text{new,clipped}}\right), \quad (10.33)$$

where, $s \equiv y_1 y_2$ as before. After the new α_1 and α_2 values are obtained there still remains the task of obtaining the new b value. If the new α_1 is not "clipped" then the update must satisfy the non-boundary KKT relation: $y_1 f(x_1) = 1$, i.e. $f^{\text{new}}(x_1) - y_1 = 0$. By relating f^{new} to f^{old} the following update on b is obtained:

$$b_1^{\text{new}} = b - \left(f^{\text{new}}(x_1) - y_1\left(\alpha_1^{\text{new}} - \alpha_1^{\text{old}}\right)K_{11} - y_2\left(\alpha_2^{\text{new,clipped}} - \alpha_2^{\text{old}}\right)K_{12} \quad (10.34)$$

If α_1 is clipped but α_2 is not, the above argument holds for the α_2 multiplier and the new b is:

$$b_2^{\text{new}} = b - \left(f^{\text{new}}(x_2) - y_2\left(\alpha_2^{\text{new}} - \alpha_2^{\text{old}}\right)K_{22} - y_1\left(\alpha_1^{\text{new,clipped}} - \alpha_1^{\text{old}}\right)K_{12}\right)$$

$$(10.35)$$

If both α_1 and α_2 values are clipped then we do not have a unique solution for b. The Platt convention was to take:

$$b^{\text{new}} = \frac{b_1^{\text{new}} + b_2^{\text{new}}}{2} \tag{10.36}$$

and this works well much of the time. Alternatively, Keerthi [184] has devised an alternate formulation without this lacuna, as have Crammer and Singer [181], with the latter described in the multiclass SVM section. Perhaps just as good as any exact solution for "b" in the double-clipped scenario is to manage this special case by rejecting the update and picking a new pair of alphas to update (in this way only unique "b" updates are made). Alpha-selection variants are briefly discussed next.

10.4.2 Alpha-Selection Variants

In the standard Platt SMO algorithm, $-\eta = k_{11} + k_{22} - 2k_{12}$, and speedup variations are described to avoid calculation of this value entirely. A middle ground is sought with the following definition "$\eta = 2k_{12} - 2$; if $(\eta >= 0)$ $\{\eta = -1;\}$" (in [59], where underflow handling and other details differ slightly, and nonstandard kernels are explored).

A comparison of some of the SVM Kernels of interest, with "regularized" distances or divergences, where they are regularized if in the form of an exponential with argument the negative of some distance-measure squared ($d2(x, y)$) or symmetrized divergence measure ($D(x, y)$), the former if using a geometric heuristic for comparison of feature vectors, the latter if using a distributional heuristic. Results are shown in Figure 10.17 for the Gaussian Kernel: $d2(x, y) = \Sigma k(xk - yk)2$; for the Absdiff Kernel $d2(x, y) = (\Sigma k \,|\, xk - yk|)1/2$; and for the Symmetrized Relative Entropy Kernel $D(x, y) = D(x\|y) + D(y\|x)$, where $D(x\|y)$ is the standard relative entropy.

10.4.3 Chunking on Large Datasets: $O(N^2) \rightarrow n\; O(N^2/n^2) = O(N^2)/n$

SVM chunking provides an alternative method to running a typical SVM on a dataset by breaking up the training data and running the SVM on smaller chunks of data. In the chunking process feature vectors associated with strong data points are retained from chunk to chunk, while weak data points are discarded.

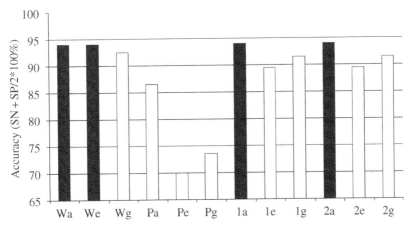

Figure 10.17 Comparative results are shown on performance of Kernels and algorithmic variants. The classification is between two DNA hairpins and uses thousands of samples (in terms of features from the blockade signals they produce when occluding ion flow through a nanometer-scale channel). Implementations: WH SMO (W); Platt SMO (P); Keerthi1 (1); and Keerthi2 (2). Kernels: Absdiff (a); Entropic (e); and Gaussian (g). The best algorithm/ kernel on this and other channel blockade data studied has consistently been the WH SMO variant and the Absdiff and Entropic Kernels. Another benefit of the WH SMO variant is its significant speedup over the other methods (about half the time of Platt SMO and one-fourth the time of Keerthi 1 or 2).

The variable projection method (VPM) is developed in [216] for training SVMs in parallel. This method is based off of the SVM light decomposition techniques [180] which delve further into the inner workings of the SMO algorithm [163, 164]. In the latter, the feature vector indices are divided into two categories, the free and fixed sets based upon their "alphas" (Lagrange multipliers). The free set represents the KKT violators which need to be further optimized while the fixed set is comprised of the alphas that already fulfill the KKT equations. An alpha from each set is used to solve each quadratic sub problem in order to optimize the free set alphas until convergence. VPM provides a parallel solution to computing the kernel matrix which is the most memory intensive part of the SVM. The kernel calculations are spread among several processing elements and the rows of the matrix are spread and usually duplicated across the memory of those processing elements. Since the rows are duplicated, they must be synchronized after each local computation. VPM is implemented using standard C and MPI communication routines.

In [166] is shown the Cascade SVM method to parallelize SVMs. This method begins by breaking the large dataset into chunks. The SVM is run on each separate chunk in the first layer. When the SVMs have all converged, new chunks are created from the resulting SV from the pairs of first layer chunks which make up the

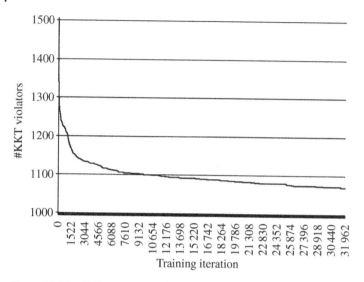

Figure 10.18 SVM convergence failure seen with 100% SV passing on distributed learning topologies. SVM training dataset reduction with 100% SVs passed on a distributed learning topology.

second layer of chunks. This occurs until a final chunk is reached. The final set of SV is then fed back into each first layer chunk. If further optimization is possible and needed, the entire process is rerun until the global optimum is met. This method seems intuitive, but after testing we have found that passing 100% of SV down to the next set of chunks, without also passing non-SV or using the SVR method, typically results in systematic convergence failure (with the various 150-component deoxyribonucleic acid (DNA) feature datasets examined). The data run never finishes, in other words, since it cannot sufficiently reduce the SV to converge (see Figure 10.18). The weakness of the method, not apparent at first sight, is not simply that the SVs from different chunks might be sufficiently different to pose complications. The added subtlety is to prevent the accumulation of outliers during the distributed learning/merging of the SVM with chunking.

Chunking becomes a necessity when classifying large datasets. The number and size of the chunks depends on the size of the dataset to be trained. In the Java implementation used here, the user specifies the size of each chunk and the chunks are broken up accordingly. If the chunks do not divide evenly, which is the case most of the time, the few remaining feature vectors are added to the last chunk. When training on the chunk is complete, the resulting trained feature vectors split into distinct sets (SV, polarization set, penalty set, and KKT violator). If the SVM learning is done well, the largest set consists of the support and polarization feature vectors. The polarization set consists of the feature vectors that have been properly classified. These feature vectors pass the KKT relations and have an alpha coefficient equal to zero. The penalty set consists of the feature vectors which pass the KKT relations and have alpha coefficients equal to C (the max value). The KKT violators make up another set consisting of feature vectors that violate one of the KKT relations. (The KKT violator set is usually zero at the end of the training process, unless some minimal number of violators is allowed upon learning completion.). These sets give the user different categories of feature vectors that they can pass to the next chunk(s). To keep the SVM converging to a better solution on the next chunk run, however, SV (and sometimes some of the polarization set) are passed to the next chunk(s). The optimal pass-percentages of each feature vector set depend on which kernel is used, the dataset, and the manner of merging information from different chunks.

There are different methods of extracting the feature vectors from the different sets. The specified percentages of feature vectors are typically randomly chosen from each of the sets, except for one. *The SVs extraction method differs since it extracts the SV that are nearest to the decision hyperplane.* In the results we choose feature vectors whose scores are closer to the hyperplane in order to pass a tighter hyperplane on to the next chunk(s), and to manage accumulation of outliers.

The chunk learning topology used in our distributed approach is slightly different from the Binary Tree splitting described in the Cascade SVM presented in [166]. As discussed above, the large dataset is broken into smaller chunks and the SVM is run on each separate chunk. Instead of bringing the results of paired chunks together, all chunk results are brought together and re-chunked as occurred in the first layer. This process occurs until the final chunk is calculated which gives the trained result. At each training stage, the user has the option to tune the percentage of SV and non-SV to pass to the next set of chunks. Additionally, passed SV can be chosen to satisfy some max value (approx. $C/10$ in cases examined) to produce a tighter hyperplane to better distinguish the polarization sets and eliminate outliers. We also incorporate SVR post-processing in some of the dataruns (method below), where SVR runs as part of the core SVM learning task on each chunk. It uses a user-defined alpha cutoff value for further tuning and can significantly reduce the number of SV passed to the next set of chunks (with bias towards elimination of outliers and the large non-boundary alphas). These additional steps reduce the size of the chunks, thus making the algorithm run faster without loss of accuracy. The SVR post-processing also appears to offer similar immunity to the convergence pathology (noted previously for 100% SV passing on distributed learning topologies).

Four sets of results are now described. The first two sets concern the optimal performance/learning-rate configurations for feature vector passing ("pass-tuning"), where the SVM training is performed using chunking with different learning topologies (sequential, partially sequential distributed). The experiments are: (Section 10.4.3.1) SV/non-SV pass-tuning on binary subsets of {9AT, 9TA, 9CG, 9GC, 8GC}; and, (Section 10.4.3.2) SV/non-SV pass-tuning for binary classification of (9AT, 9TA) vs. (9CG, 9GC).

There appear to be instabilities when learning on distributed topologies, and there are a couple of new approaches that appear to be robust in addressing these instabilities. In turn, this allows the hopes of a distributed speedup to be directly realized. The Support Vector Reduction (SVR) Method described in Section 10.4.4 uses a post-processing phase: after all KKT violators have been eliminated, the SV alphas near the boundaries are coerced to their boundary sets (i.e. to the polarization set at alpha = 0 or the penalty set at alpha = C). Results on this method are shown in the third section (Section 10.4.4.1). The fourth and last section of the Results shows the combined operation of various methods for comparative purposes, and shows how robust distributed SVM learning may be possible (Section 10.4.4.2): Distributed SVM with pass-tuning and SVR.

10.4.3.1 Distributed SVM Processing (Chunking)

There are a variety of ways to avoid the pure SV training-set pathology [13, 59]. Since we are interested in training set reduction overall, we consider the possibility of simply reducing the SV set. This appears to work in preliminary tests on well-studied datasets of interest (see Table 10.1), where the SV's nearest to the decision

Table 10.1 Performance comparison table for the different SVM methods.

SVM method	Sensitivity	Specificity	(SN + SP)/2	Time (ms)
SMO (non-chunked)	0.87	0.84	0.86	47 708
Sequential chunking	0.84	0.86	0.85	27 515
Multi-threaded chunking	0.88	0.78	0.83	7855
SMO (non-chunked) with SV reduction	0.91	0.81	0.86	43 662
Sequential chunking with SV reduction	0.90	0.82	0.86	18 479
Multi-threaded chunking with SV reduction	0.85	0.83	0.84	5232
Multi-threaded dist. chunking with SVR	0.85	0.83	0.84	5973

The distributed chunking used three identical networked machines. Dataset = 9GC9CG_9AT9TA (1600 feature vectors). SVM parameters: Absdiff kernel (with sigma = 0.5, C = 10, Epsilon = 0.001, Tolerance = 0.001). For chunking methods: Pass 90% of support vectors, starting chunk size = 400, maxChunks = 2. For SVR methods: Alpha cutoff value = 0.15.

hyperplane (most supporting the hyperplane) are retained. For the channel current data examined in, with 150-component feature vectors, we find that 30% SV passing is optimal on distributed learning topologies. The low SV-passing percentage that is found to work in *distributed* chunking might fundamentally be an issue of outlier control during distributed learning. Further reduction of SV passed is possible with dropping SV's with confidence values at the other extreme, near zero (i.e. those nearest and most strongly supporting the hyperplane). This entails a additional SVR process that is run right after the SVM learning step is complete, where we further reduce the support vector set according to some confidence cut-off (actually imposed via cut-off on associated Lagrange multiplier in the SVM/ SMO implementation). By reducing the number of SV propagated into the next round, we further accelerate the chunked processing. In this way, a strongly performing distributed chunk-training process is possible, with speedup by ~10 in the example shown in the table shown in Table 10.1 (with no significant loss in accuracy). It appears possible to automate the tuning and selection procedures. To achieve this, it is necessary to examine the stability of the algorithmic parameters such as the pass percentages on the different types of learned data (e.g. see Table 10.1 for pass percentages indicated). Further results on distributed SVM learning is given in [13, 59].

To further enhance processing speed, one can not only perform distributed processing as indicated, but can also boost thread-processing speed on a given computer via use of GPU processing. This has already been undertaken, where distributed chunks of SVM training data were processed using a CPU/GPU that, at marginal added cost (a graphics card), provided as much as a 32-fold speedup on the channel current blockade classification [170]. Similar GPU speed enhancements to the other machine learning algorithms are possible as well.

10.4.3.2 SV/Non-SV Pass-Tuning on Train Subsets: An Outlier-Management Heuristic

For DNA hairpin feature vector datasets, our observations have shown the best kernels to be the Gaussian, Absdiff, and Sentropic kernels. Of the three kernels indicated, Absdiff and Sentropic produce similar results when measuring accuracy as the average of the Sensitivity (SN) and Specificity (SP) typically significantly outperforms the third best kernel, Gaussian. The Gaussian kernel, on the other hand, is found to be the best performing of the three at keeping the growth in chunk-size as small as possible.

Sequential chunking is a simple form of chunking which is not multi-threaded. This method runs the SVM on the first chunk, and then sends the support feature vectors (SVs) and sometimes non-SVs to be added onto the training data for the next chunk. This continues until the final chunk has been run. When using sequential chunking, feature vector passing can be difficult since passing too many

Table 10.2 Sequential chunking using different DNA hairpin datasets.

Sequential chunked SMO			Chunk size 200 of 800 total feature vectors			
Data	Iterations	# of SVs	SN	SP	(SN + SP)/2	Elapsed time (ms)
8GC9AT	100	554	0.96	0.95	0.955	12 610
8GC9CG	114	557	0.92	0.92	0.92	16 901
8GC9GC	58	524	0.94	0.97	0.955	8914
8GC9TA	68	542	0.97	0.95	0.96	10 000
9AT9CG	37	727	0.83	0.8	0.815	10 936
9AT9GC	23	727	0.83	0.83	0.83	9757
9AT9TA	9	661	0.93	0.93	0.93	7563
9CG9GC	15	751	0.78	0.77	0.775	9218
9CG9TA	41	597	0.92	0.89	0.905	10 267
9GC9TA	51	567	0.95	0.92	0.935	9695
Mean	52	621	0.903	0.893	0.898	10 586

This table shows the different sequential chunking data runs performed on datasets deriving from pairs of DNA hairpin data. The last line of the table shows the mean of the data runs. The SVM parameters used: Absdiff kernel with sigma = 0.5, $C = 10$, epsilon = 0.001, tolerance = 0.001, and passing 100% of support vectors.

features on to the next chunk can result in training datasets that are too large in the later chunks in the process. For the sequential chunking method, the accuracy with Absdiff kernel (0.898) is shown in Table 10.2, with kernel and chunking parameters: sigma = 0.5, $C = 10$, Epsilon = 0.001, Tolerance = 0.001, and passing 100% of SV. Using the Sentropic kernel we obtain a similar accuracy of 0.891, where the Sentropic kernel and chunking parameters were: sigma = 0.5, $C = 10$, Epsilon = 0.001, Tolerance = 0.001, and passing 100% of SV. Using the Gaussian kernel provides accuracy 0.864, where the Gaussian kernel and chunking parameters were: sigma = 0.05, $C = 10$, Epsilon = 0.001, Tolerance = 0.001, and passing 100% of SV. As noted, in these data runs, 100% of the support vector set was passed to the next set of chunks. The chunking parameters indicated for the table represent the best accuracy for the given chunking method, and all of the parameter selections are verified for stability (via tests confirming that minor changes of tuned parameter setting do not strongly alter classifier accuracy). In each SVM classifier test the positive class is named first, so "8GC9AT" is a SVM training with 8GC data used for positives, and 9AT data used for negatives. Since the 8 and 9 refer different lengths in the molecules, it is not surprising that the best scoring classifications are for telling the 8GC molecule apart from one of the nine base-pair

Table 10.3 Multi-threaded chunking using different DNA hairpin datasets

	Distributed chunked SMO		Chunk size 200 of 800 total feature vectors			
Data	Iterations	# of SVs	SN	SP	(SN + SP)/2	Elapsed time (ms)
8GC9AT	14	221	0.97	0.89	0.93	2667
8GC9CG	30	202	0.91	0.9	0.905	1993
8GC9GC	27	208	0.91	0.93	0.92	2003
8GC9TA	38	208	0.95	0.88	0.915	2017
9AT9CG	8	232	0.79	0.72	0.755	2531
9AT9GC	21	237	0.71	0.8	0.755	2121
9AT9TA	8	234	0.85	0.87	0.86	2318
9CG9GC	9	237	0.74	0.69	0.715	2132
9CG9TA	8	230	0.84	0.94	0.89	2003
9GC9TA	10	224	0.94	0.87	0.905	1945
Mean	17.3	223.3	0.86	0.849	0.855	2173

This table shows the different multi-threaded chunking data runs performed on assortments of DNA hairpin pairs. The last line of the table presents the mean of the data runs. SVM Parameters: **Sentropic kernel** with sigma = 0.5, $C = 10$, Epsilon = 0.001, Tolerance = 0.001. Passing 30% of support vectors.

molecules. The most challenging case was for discriminating between 9GC and 9CG, where an accuracy of only 77.5% was observed (dropping weak data can boost this classification to 99.9% [1, 3, 13, 59]).

For the multi-threaded chunking method, the average accuracy of Sentropic is best (0.855) (see Table 10.3). Absdiff (0.854) is very similar in performance, and Gaussian has average accuracy 0.833. In these data runs, 30% of the support vector set was passed to the next set of chunks, with kernel parameters unchanged. If 100% SV-passing is attempted there is typically failure to converge. As with the sequential Results, these chunking parameters chosen represent the best accuracy for the given chunking method, and all of the parameter selections are verified for stability.

For the multithreaded chunking, the SVs in the final distributed chunk with Gaussian kernel have an average 78% reduction from the original dataset to final chunk SV decision-set, while the Sentropic kernel has a 72% reduction. The SV number in the final sequential chunk had a 22.5% reduction for the Absdiff kernel in the sequential setting, compared with a 44.3% reduction for the Gaussian kernel. So the improved accuracy of the Absdiff and Sentropic kernels, over the standard Gaussian kernels, comes at a minor cost in computational time in the distributed-chunking setting, while it can involve significantly more time in the sequential-chunking setting.

From tuning over the number of SVs to pass, we find that sequential learning topologies strongly benefit from 100% SV passing, whereas distributed learning topologies have a non-optimality at 100% SV passing (and is prone to non-convergence to a solution – see Figure 10.18), while 30% SV-passing performs as well and with greater stability. There are a variety of ways to deal with the distributed learning instabilities found with passing "base" SV's, including the solution of pipelining the learning process to always have SV's merge into an untrained chunk to avoid outlier accumulation (and gridlock) in the learning process. In the Discussion we suggest that the low SV-passing percentage that is found to work in *distributed* chunking might fundamentally be an issue of outlier control during distributed learning.

10.4.3.3 SV/Non-SV Pass-Tuning on (9AT,9TA) vs. (9CG,9GC)

For the DNA hairpin datasets considered in the previous section, and considered here on a larger dataset, we find that the ideal chunking parameter for sequential chunking is 100% of the support vector set. The Absdiff kernel produced the best accuracy (0.855) with stable conditions (see Figure 10.19). Table 10.4 displays a

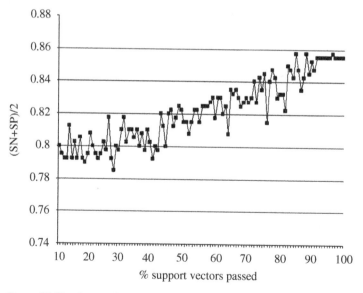

Figure 10.19 Sequential learning topology SV pass-tuning. Dataset = 9GC9CG_9AT9TA (1600 feature vectors). SVM parameters: Absdiff kernel with sigma = 0.5, C = 10, Epsilon = 0.001, Tolerance = 0.001.

Table 10.4 Sequential chunking with the Absdiff kernel.

	Chunk 1	Chunk 2	Chunk 3	Chunk 4
Total chunk size	400	787	1143	1472
Support vectors	373	700	1002	1320
Polarization set	27	86	140	152
Penalty set	0	0	0	0
Violator set	0	1	1	0
Support vectors passed	373	700	1002	
Polarization set passed	14	43	70	
Total passed set	387	743	1072	

Dataset = 9GC9CG_9AT9TA (1600 feature vectors). SVM parameters: Absdiff kernel with sigma = 0.5, C = 10, Epsilon = 0.001, Tolerance = 0.001. Pass 100% of support vectors and 50% of polarization set. Final chunk performance: {SN, SP} = {0.87, 0.84}. A breakdown of each feature vector set is displayed to show how the percentage parameters are used to pass portions of each set to the next chunk.

sample run using Absdiff and the size of each chunk as the algorithm progresses through the chunks. Table 10.4 also shows the feature vector set composition of each chunk. Similar results were obtained when using the Sentropic and Gaussian kernels (not shown).

The Figure 10.19 results show a clear trend for *sequential* chunking when using different support vector and polarization set percentage parameters. (During the tuning operation, every variation of multiples of ten up to 100 was used for each of the two sets. For example, when the SV % parameter was 10, the polarization set % parameter would vary from 10 to 100 in steps of 10.) For most of the data run, especially the more stable part at 100% SVs, the variation of the small polarization set did not seem to have much effect on the outcome. Table 10.4 shows results for Absdiff kernel passing 100% of SV and 50% of polarization set. Final Chunk Performance: {SN, SP} = {0.87, 0.84}. For the Sentropic kernel with sigma = 0.5, C = 10, Epsilon = 0.001, Tolerance = 0.001, passing 100% of SV and 50% of polarization set the final Chunk Performance: {SN, SP} = {0.875, 0.82} (table not shown). For the Gaussian kernel with sigma = 0.05, C = 10, Epsilon = 0.001, Tolerance = 0.001, passing 100% of SV and 50% of polarization set, the final Chunk Performance: {SN, SP} = {0.715, 0.85}.

For the DNA hairpin datasets considered in the previous section, and considered here on a larger dataset, results have shown that the ideal chunking parameter for *distributed* chunking can be as low as 30% of the support vector set. This produced

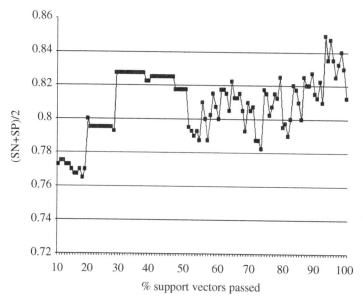

Figure 10.20 Distributed learning topology SV pass-tuning. Dataset = 9GC9CG_9AT9TA (1600 feature vectors). SVM parameters: Absdiff kernel with sigma = 0.5, *C* = 10, Epsilon = 0.001, Tolerance = 0.001. This shows the trend for multi-threaded chunking when using different support vector and polarization set percentage parameters. (Every variation of multiples of ten up to 100 was used for each of the two sets. For example, when the SV % parameter was 10, the polarization set % parameter would vary from 0 to 100 in multiples of 10.) For most of the data run, especially the more stable part around 30% SV-passing, the variation of the polarization set did not have much effect on the outcome.

the best accuracy (0.83) with stable conditions (see Figure 10.20 for results with the Absdiff kernel).

For the *sequential* chunking method, the training process shown in Table 10.4 is optimal when passing 100% of the support vector set and 50% of the polarization set. For *multithreaded* chunking, with the Absdiff kernel, shown in Table 10.5, on the other hand, good performance was found when passing a wide range of parameters, with results shown for a training run with passing on 80% of the support vector set and 60% of the polarization set. Similar results are found for the Sentropic and Gaussian kernels (Tables not shown). Note that only 30% SV-passing was needed for the best performing multi-threaded learning. With only 30% SV passing there was no weakening of performance or convergence instabilities, and the results shown in Figure 10.20 reveal this to be a more stable chunk learning parameter for the datasets examined (for discussion on how to implement an SVM with auto-tuning on the kernel and chunking parameters).

Table 10.5 Multi-threaded chunking with the Absdiff kernel.

	C1	C2	C3	C4	C5	C6	C7	C8	C9	C10
Total chunk size	400	400	400	400	423	423	425	504	504	791
Support vectors	373	377	378	388	402	402	403	466	460	699
Polarization set	27	23	22	12	21	21	22	38	43	92
Penalty set	0	0	0	0	0	0	0	0	0	0
Violator set	0	0	0	0	0	0	0	0	1	0
Support vectors passed	1218	—	—	—	968	—	—	742	—	—
Polarization set passed	53	—	—	—	40	—	—	49	—	—
Total passed set	1271	—	—	—	1008	—	—	791	—	—

Dataset = 9GC9CG_9AT9TA (1600 feature vectors). SVM parameters: Absdiff kernel with sigma = 0.5, C = 10, Epsilon = 0.001, Tolerance = 0.001. Pass 80% of support vectors and 60% of polarization set. Final chunk performance: {SN, SP} = {0.855, 0.795}. A breakdown of each feature vector set is displayed to show how the percentage parameters are used to pass portions of each set to the next set of chunks.

For Sentropic kernel with sigma = 0.5, C = 10, Epsilon = 0.001, Tolerance = 0.001, passing 80% of SV and 60% of polarization set. Final chunk performance: {SN, SP} = {0.845, 0.755} (Table not shown). For Gaussian kernel with sigma = 0.05, C = 10, Epsilon = 0.001, Tolerance = 0.001, passing 80% of SV and 60% of polarization set. Final Chunk Performance: {SN, SP} = {0.85, 0.83} (Table not shown).

10.4.4 Support Vector Reduction (SVR)

SVR is a process that is run right after the SVM learning step is complete. Instead of going on to testing data against the training results to get accuracy, we further reduce the support vector set. One way to do this is to coerce some alphas to zero which means they would now fall into the polarization set. Converting the smaller alphas to zeros makes the most sense since a larger alpha indicates that the data point is stronger towards its grouping (polarized sign). This is done using a user-defined alpha cutoff value. All alpha values that are under the cutoff are pushed to zero. It is not entirely trivial since certain mathematical constraints must be met. The constraint that must be met for this method is the linear equality constraint [1, 3, 59]:

$$\sum_i y_i \alpha_i = 0$$

Therefore, the alpha values not meeting the cutoff cannot just be forced to zero unless the value is retained somewhere else in the set. This is done by first sorting the alpha values of the SV. Then for each alpha that does not meet the cutoff value, the small left over value is added to the largest alpha of the same polarity. Since the list is sorted it can loop through and evenly distribute the left over values through the larger alphas starting with the largest. The reduction process can cut the number of SV significantly, while not significantly diminishing the accuracy. Other observations have shown that the easier the dataset to classify, the larger the reduction via this process.

SVR is a process that is run right after the SVM learning step is complete. Instead of going on to merge subsets of feature vectors or to test data against known results, the idea is to further reduce the support vector set. One way to do this is to coerce some alphas to zero which means they would now fall into the polarization set. This process is described further in the Methods.

Figure 10.21 shows the results of the SVR method on the non-chunking SMO SVM. For this dataset, 0.19 was found to be the best cutoff value since it retains accuracy while reducing the SV. For the 9GC9CG vs. 9AT9TA dataset, 140 SV (10.5% of total) were dropped without affecting the accuracy.

SVR-enabled data runs using sequential chunking methods (Figure 10.22) and multi-threaded chunking methods (Figure 10.23) show similar results. The chunking results tend to be noisier since the SVM algorithm makes some approximations, thus the hyperplane will not be exactly the same for every data run and this behavior is amplified in the chunking methods. Nonetheless, the SVR method cuts down on SV and decreases testing time. For sequential chunking

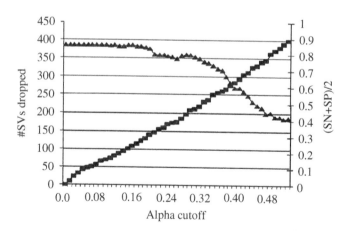

Figure 10.21 SMO (non-chunking) support vector reduction. Dataset: 9GC9CG_9AT9TA (1600 feature vectors). SVM parameters: Absdiff kernel with sigma = 0.5, C = 10, Epsilon = 0.001, Tolerance = 0.001. This graph shows the rate of support vectors reduced as the alpha cutoff value is increased. The alpha cutoff value 0.19 is chosen as the best since it is the last value before accuracy begins to degrade. This chosen value reduces 140 support vectors.

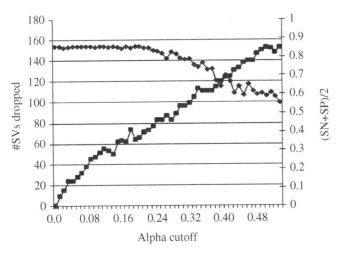

Figure 10.22 **Sequential chunking support vector reduction**. Dataset: 9GC9CG_9AT9TA (1600 feature vectors), starting chunk size = 400. SVM parameters: Absdiff kernel with sigma = 0.5, *C* = 10, Epsilon = 0.001, Tolerance = 0.001. Passing 100% of support vectors. This graph shows the rate of support vectors reduced as the alpha cutoff value is increased. The alpha cutoff value 0.25 is chosen as the best since it is the last value before accuracy begins to degrade. This chosen value reduces 87 support vectors.

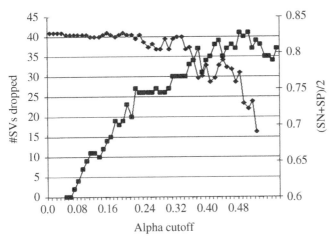

Figure 10.23 **Multi-threaded chunking support vector reduction**. Dataset: 9GC9CG_9AT9TA (1600 feature vectors), starting chunk size = 400. SVM parameters: Absdiff kernel with sigma = 0.5, *C* = 10, Epsilon = 0.001, Tolerance = 0.001. Passing 30% of support vectors. This graph shows the rate of support vectors reduced as the alpha cutoff value is increased. The alpha cutoff value 0.22 is chosen as the best since it is the last value before accuracy begins to degrade. This chosen value reduces 26 support vectors.

(Figure 10.22), an alpha cutoff value of 0.25 caused 87 SV (7.2%) to be dropped without affecting accuracy. For multi-threaded chunking (Figure 10.23), an alpha cutoff value of 0.22 dropped 26 SV (6.2%) while retaining the same accuracy.

10.4.4.1 Multi-Threaded Chunking with SVR

The multi-threaded chunking method simultaneously runs the chunks using multiple threads. Once all of the threaded chunks are finished training, the chunk results are collected. The same user defined percentages of feature vector sets are used here except this time those percentages of feature vectors are extracted from each chunk. All of the chosen feature vectors to be passed are stored together then re-chunked if the current data set is large enough to be chunked again. Re-chunking occurs when the data set is greater than or equal to twice the specified chunk size. If this is not the case, the final chunk is run alone to get the final result. The main use of the multi-threaded chunking method is with a single computer with multiple processors/cores. Results are shown in Table 10.6.

10.4.4.2 Multi-Threaded Distributed Chunking with SVR

The multi-threaded *distributed* chunking implementation is a multi-server/multi-CPU (core) version of the previous multi-threaded chunking method. Java RMI is used to handle the remote calls between the client and servers. The client program runs multi-threaded remote calls to a user specified set of servers (round robin). Each server and the client machine have an SVM Server listening. When the client program runs, a chunk is passed to each available processor/core in the network

Table 10.6 Performance comparison of the different SVM methods.

SVM method	Sensitivity	Specificity	(SN + SP)/2	Time (ms)
SMO (non-chunked)	0.87	0.84	0.86	47 708
Sequential chunking	0.84	0.86	0.85	27 515
Multi-threaded chunking	0.88	0.78	0.83	7855
SMO (non-chunked) with SVR	0.91	0.81	0.86	43 662
Sequential chunking with SVR	0.90	0.82	0.86	18 479
Multi-threaded chunking w/ SVR	0.85	0.83	0.84	5232
Multi-threaded distributed chunking w/ SVR	0.85	0.83	0.84	5973

The distributed chunking used three identical networked machines. Dataset = 9GC9CG_9AT9TA (1600 feature vectors). SVM parameters: Absdiff kernel with sigma = 0.5, $C = 10$, Epsilon = 0.001, Tolerance = 0.001. For chunking methods: Pass 90% of support vectors, Starting chunk size = 400, maxChunks = 2. For SV reduction methods: Alpha cutoff value = 0.15.

until all or as many as possible are training simultaneously. As the chunks finish, the results are passed back to the client. Each "chunk level" may take multiple batches depending on the chunk size and amount of processors/cores available. The final chunk is largest so the client program should be processed on the machine with the most computing power. This not only speeds up the final chunk but allowing larger chunks should produce better final results. The main benefit of this method is a significant decrease in run time for large datasets. As shown in Table 10.6, multi-threaded distributed chunking performs almost as well as non-chunked learning. Network overhead causes it to be slightly slower than the single-server multi-threaded chunking method. With extremely large datasets (i.e. 60 000 feature vectors and larger), the multi-threaded and distributed method is thus shown to work.

SVMs are extremely useful for classifying data. Since the main weakness of SVMs is the long training time when running large datasets, it is only natural to develop multi-threaded distributed SVM training methods, especially since multiple cores/processors are becoming commonplace.

An overall comparison of the SVM with chunking methods is shown in Table 10.6. It is found that sequential chunking has the benefit of holding onto accuracy when compared to running the straight SVM (SMO) but the run times can be inefficient since the method does not run in parallel. Multi-threaded chunking has a significant run time performance improvement, which is further improved when employing the SVR method. The multi-threaded aspect allows training of extremely large datasets which may not be possible using sequential chunking. Additionally, using the multi-threaded distributed method allows users to add machines to make the algorithm train even faster. An instability of the multithreaded approach is found to sometimes occur when the percentage of information passed to later chunks involves 100% of the SV, possibly due to outlier accumulation in the chunk-carry set of feature vectors. This instability is eliminated by passing less than 100% of the SV, favoring those nearest to the decision surface (thereby excluding outliers), and by passing some of the non-support-vector set of data as well. A stable distributed platform for training SVM thus appears possible. With the size of the dataset no longer a significant limitation in SVM training, many practical applications of SVM methods become accessible in Big Data classification tasks, and in SVM-intensive applications, such as SVM-based clustering mentioned in the Methods.

10.4.5 Code Examples (in OO Perl)

Portions of two OO-PERL programs are shown that perform the SVM computation indicated above. The first is part of a PERL control script, the second is part of a PERL module file.

```perl
#!/usr/bin/perl

# Perl program Binary_Classifier.pl

use File::Basename;
use lib dirname($0);
use FileHandle;
use SVM_Discriminator;
use strict;
my ($train_file,$label_file,$test_file,$testlabel_file,
$output_file,
    $kernel_type,$kernel_param,$bvalue,$Cvalue,$epsilon,
$skip_train_loop,
    $class_file,$perm_index,$N,$feature_count) = @ARGV;
my $SVM_Process = new SVM_Discriminator
($train_file,$label_file,$output_file,
    $kernel_type,$kernel_param,$bvalue,$Cvalue);
my $signal_count = $SVM_Process->Count_Features();
print "Signal Count = $signal_count\n";
my $badKKT=0;
my $tolerance=0;
if {!$ skip_train_loop) { ### no indent follows
print "Performing Load_Features on Segment $chunk_index\n";
$SVM_Process->Load_Features($feature_count);
print "Performing Eval_Kernel_Matrix\n";
$SVM_Process->Eval_Kernel_Matrix();
print "Performing Initialize_Alphas\n";
$SVM_Process->Initialize_Alphas();
print "Performing Initialize_bvalue\n";
$SVM_Process->Initialize_bvalue($bvalue);
print "Performing Initialize_Disc_Function_Cache\n";
$SVM_Process->Initialize_Disc_Function_Cache();
my @signs = @{$SVM_Process->{signs_ref}};
my @alphas = @{$SVM_Process->{alphas_ref}};
my @kernel = @{$SVM_Process->{kernel_ref}};
my $max_index=scalar(@signs)-1;
my @disc_function = @{$SVM_Process->{disc_function_ref}};

my $datacount = scalar(@signs);
my $looplimit = $datacount*5;
print "Performing Evaluate_KKT to initialize violator
lists\n";
```

```perl
my $KKTsatisfied=$SVM_Process->Evaluate_KKT($epsilon); #
to init violator lists
print "KKTsatisfied = $KKTsatisfied\n";
my $loopcount=0;
print "Performing Alpha Selections\n";
WHILE: while (!$KKTsatisfied) {
    $loopcount++;
    my $skip_index_default=-1;
    my $nonboundKKT_index =
            $SVM_Process->Select_KKT_nonboundViolator
            ($skip_index_default);
    my $boundKKT_index =
            $SVM_Process->Select_KKT_BoundViolator
($skip_index_default);
    my $first_index=-1;
    if ($nonboundKKT_index!=-1) {
        $first_index = $nonboundKKT_index;
    }
    elsif ($boundKKT_index!=-1) {
        $first_index = $boundKKT_index;
    }
    else {
        print "zero nonBound violators or alpha-one
selection failure\n";
        $KKTsatisfied=1;
        last;
    }

    my $second_index = $SVM_Process->Select_SecondKKT
($first_index);
    print "Performing Update_Cache\n";
    $SVM_Process->Update_Cache
($first_index,$second_index);
    print "Performing Evaluate_KKT via KKTsatisfied update
    \n";

    if ($loopcount>=$looplimit && $loopcount<2*$looplimit) {
        $tolerance = 1+int($datacount/100);  # approx 1%
        violators
        if ($loopcount==$looplimit) {
```

```perl
                print "KKTlooplimit=$looplimit exceeded, ";
                print "allowing up to $tolerance violators.\n";
            }
            $KKTsatisfied=$SVM_Process->Evaluate_KKT($epsilon,
$tolerance);
        }
        elsif ($loopcount>=2*$looplimit) {
            print "Bad KKT convergence, exiting.\n";
            $badKKT=1;
            last WHILE;
        }
        else {
            $KKTsatisfied=$SVM_Process->Evaluate_KKT
($epsilon);
        }
    }
    print "KKT loop count: $loopcount\n";

    if (!$badKKT) {
        $SVM_Process->Save($class_file,$perm_index,$N);
        $SVM_Process->Call($test_file,$testlabel_file,
$tolerance);
    }


#############
} # for skip on train loop
else {
    $SVM_Process->Call($test_file,$testlabel_file,
                    $tolerance,$class_file,$perm_index,$N);
}
```

The Perl module referred to, SVM_Discriminator.pm, is partly given below

```perl
#!/usr/bin/perl
#
# Perl module SVM_Discriminator.pm

package SVM_Discriminator;
use strict;
```

```perl
sub new {
    my ($class,$data_file,$label_file,$output_file,
        $kernel_type,$kernel_param,$bvalue,$Cvalue) = @_;
    my $self = bless {}, $class;
    $self->{data_file}=$data_file;
    $self->{output_file}=$output_file;
    $self->{label_file}=$label_file;
    $self->{kernel_type}=$kernel_type;
    $self->{kernel_param}=$kernel_param;
    $self->{bvalue}=$bvalue;
    $self->{Cvalue}=$Cvalue;

    if (!$kernel_type) {
        $self->{kernel_type}="gaussian";
    }
    if (!$kernel_param) {
        $self->{kernel_param}="1";
    }
    if (!$bvalue) {
        $self->{bvalue}=0;
    }
    if (!$Cvalue) {
        $self->{Cvalue}=100;
    }
    return $self;
}

sub Eval_Kernel_Matrix {
    my ($self) = @_;
    my $kernel_type = $self->{kernel_type};

    my $feature_count = $self->{feature_count};
    print "feature_count = $feature_count\n";
    my $sigma = $self->{kernel_param};
    my @kernel;
    my @data = @{$self->{data_ref}};
    my $index;
    my $max_index=scalar(@data)-1;
    my $count = $max_index+1;
    print "training data count = $count\n";
```

```perl
my $other;
my $sqdistance;
for $index (0..$max_index) {
    if ($index%10==0) {
        print "eval on matrix row $index\n";
    }
    my $start_index;
    $start_index = $index;
    for $other ($start_index..$max_index) {
        $sqdistance=0;
        my $feat;
        for $feat (0..$feature_count-1) {
            my $p = $data[$index][$feat];
            my $q = $data[$other][$feat];
            if ($kernel_type =~ "poly") {
                $sqdistance += $p*$q;
            }
            elsif ($kernel_type =~ "power") {
              my ($power) = ($kernel_type =~ /power(\d+
              \.\d+)/);
                if (!$power) { $power=1; }
              $sqdistance += ($p**$power)*($q**$power);
            }
            elsif ($kernel_type =~ "gamma") {
              my ($gamma) = ($kernel_type =~ /gamma(\d+
              \.\d+)/);
                if (!$gamma) { $gamma=1; }
              $sqdistance += ($p**$gamma)*($q**$gamma);
            }
            elsif ($kernel_type =~ "gaussian") {
              my ($power) = ($kernel_type =~ /gaussian
              (\d+\.\d+)/);
                if (!$power) { $power=1; }
                $sqdistance += ($p**$power-$q**$power)
                **2;
            }
            elsif ($kernel_type =~ "absdiff") {
                $sqdistance += abs($p-$q);
            }
            elsif ($kernel_type eq "squareexp") {
                $sqdistance += (exp($p)-exp($q))**2;
```

```perl
        }
        elsif ($kernel_type eq "squarelog") {
            $sqdistance += (log($p)-log($q))**2;
        }
        elsif ($kernel_type eq "sentropic") {
            my $minprob = 0.0000001;
            if ($p<$minprob) { $p = $minprob; }
            if ($q<$minprob) { $q = $minprob; }
          $sqdistance += ($p-$q)*(log($p)-log($q));
        }
    }
    if ($kernel_type =~ "poly") {
        my ($poly) = ($kernel_type =~ /poly(\d+\.\d
        +)/);
        if (!$poly) { $poly=1; }
        $kernel[$index][$other]=$sqdistance**$poly/
        $sigma;
    }
    elsif ($kernel_type =~ "power") {
        $kernel[$index][$other]=$sqdistance/$sigma;
    }
    elsif ($kernel_type =~ "gamma") {
      my ($gamma) = ($kernel_type =~ /gamma(\d+\.\d
      +)/);
        if (!$gamma) { $gamma=1; }
        $kernel[$index][$other]=$sqdistance**(1/
        $gamma)/$sigma;
    }
    elsif ($kernel_type =~ "gaussian" ||
            $kernel_type eq "sentropic" {
      $kernel[$index][$other]=exp(-$sqdistance/(2*
      $sigma));
    }
    elsif ($kernel_type =~ "absdiff") {
      my ($power) = ($kernel_type =~ /absdiff(\d+.
      \d+)/);
        if (!$power) { $power=1; }
        $kernel[$index][$other]=exp(-$sqdistance**
        $power/(2*$sigma));
    }
}
```

```perl
        }
        $self->{kernel_ref}=\@kernel;
    }

sub Initialize_Alphas {
    my ($self) = @_;
    my @alphas;

    my @signs = @{$self->{signs_ref}};
    my $plus_count=$self->{plus_count};
    my $minus_count=$self->{minus_count};

    my $plus_alpha_init;
    if ($plus_count==0) {
        die "plus_alpha_init=0\n";
    }
    $plus_alpha_init = 1/$plus_count;
    my $minus_alpha_init;
    if ($minus_count==0) {
        die "minus_alpha_init=0\n";
    }
    $minus_alpha_init = 1/$minus_count;

    my $max_index=scalar(@signs)-1;
    my $index;
    for $index (0..$max_index) {
        if ($signs[$index]==1) {
            $alphas[$index]=$plus_alpha_init;
        }
        elsif ($signs[$index]==-1) {
            $alphas[$index]=$minus_alpha_init;
        }
        else {
            die "signs[index] error\n";
        }
    }
    $self->{alphas_ref}=\@alphas;
}

sub Initialize_bvalue {
    my ($self,$bvalue) = @_;
```

```perl
        $self->{bvalue} = $bvalue;
        if (!$bvalue) {
            $self->{bvalue} = 0;
        }
    }

sub Initialize_Disc_Function_Cache {
    my ($self ) = @_;
    my @labels = @{$self->{labels_ref}};
    my @signs = @{$self->{signs_ref}};
    my @data = @{$self->{data_ref}};
    my $plus_count=$self->{plus_count};
    my $minus_count=$self->{minus_count};
    my @alphas = @{$self->{alphas_ref}};
    my @kernel = @{$self->{kernel_ref}};
    my $max_index=scalar(@signs)-1;

    my @disc_function;
    my $index;
    my $other_index;
    for $index (0..$max_index) {
        for $other_index (0..$max_index) {
            $disc_function[$index] += $alphas[$other_index]
                                     *$signs[$other_index]
                                      *$kernel[$index]
[$other_index];
        }
        $disc_function[$index]+=$self->{bvalue};
    }
    $self->{disc_function_ref} = \@disc_function;
}

sub Evaluate_KKT {
    my ($self,$epsilon,$tolerance) = @_;
    if (!$epsilon) {
        $epsilon = 0.001;
    }
    if (!$tolerance) {
        $tolerance = 0;
    }
    my @KKT_Status;
```

```perl
my @KKT_violators;
my @Bound;
my @Bound_violators;
my @Nonbound_violators;
my @Nonbound;
my @KKT_non_violators;
my @KKT_supportvectors;
my @signs = @{$self->{signs_ref}};
my @alphas = @{$self->{alphas_ref}};
my $max_index=scalar(@signs)-1;
my $C = $self->{Cvalue};
my @disc_function = @{$self->{disc_function_ref}};
my $index;
for $index (0..$max_index) {
    $KKT_Status[$index]=0;
   my $score = $signs[$index]*$disc_function[$index];
    if ($alphas[$index]==0 && $score>=1+$epsilon) {
        $KKT_Status[$index]=1;
    }
   elsif ($alphas[$index]>=0 && $alphas[$index]<=$C &&
            $score<1+$epsilon && $score>1-$epsilon) {
        $KKT_Status[$index]=1;
      my $tightness = 5;   # tighten for support vector
      passing
        my $new_epsilon = $epsilon/$tightness;
        if ($score<1+$new_epsilon && $score>1-
        $new_epsilon) {
            my $vindex = scalar(@KKT_supportvectors);
            $KKT_supportvectors[$vindex]=$index;
        }
    }
   elsif ($alphas[$index]==$C && $score<=1-$epsilon) {
        $KKT_Status[$index]=1;
    }

    if ($alphas[$index]==0 || $alphas[$index]==$C) {
        my $bound_index = scalar(@Bound);
        $Bound[$bound_index]=$index;
    }
    else {
        my $nonbound_index = scalar(@Nonbound);
        $Nonbound[$nonbound_index]=$index;
    }
```

```perl
      if ($KKT_Status[$index]!=1) { # KKT violators
          my $vindex = scalar(@KKT_violators);
          $KKT_violators[$vindex]=$index;
        if ($alphas[$index]!=0 && $alphas[$index]!=$C) {
              my $nonbound_index = scalar
              (@Nonbound_violators);
            $Nonbound_violators[$nonbound_index]=$index;
          }
          else {
            my $bound_index = scalar(@Bound_violators);
              $Bound_violators[$bound_index]=$index;
          }
      }
      else {
          my $non_index = scalar(@KKT_non_violators);
          $KKT_non_violators[$non_index]=$index;
      }
  }
  $self->{KKT_Status_ref} = \@KKT_Status;
  $self->{KKT_violators} = \@KKT_violators;
  $self->{Bound} = \@Bound;
  $self->{Bound_violators} = \@Bound_violators;
  $self->{Nonbound} = \@Nonbound;
  $self->{Nonbound_violators} = \@Nonbound_violators;
  $self->{KKT_non_violators} = \@KKT_non_violators;
  $self->{KKT_supportvectors} = \@KKT_supportvectors;

  my $violator_count = scalar(@KKT_violators);
  if ($violator_count%10==0) {
      print "violator_count = $violator_count\n";
  }
  print "KKT_violators @KKT_violators\n";
  if ($violator_count>$tolerance) {
      return 0;
  }
  else {
      print "KKT pass obtained\n";
      return 1;
  }
}
```

10.5 Kernel Selection and Tuning Metaheuristics

10.5.1 The "Stability" Kernels

The SVM Kernels of interest are "regularized" distances or divergences, where they are regularized if in the form of an exponential with argument the negative of some distance-measure squared ($d2(x, y)$) or symmetrized divergence measure ($D(x, y)$), the former if using a geometric heuristic for comparison of feature vectors, the latter if using a distributional heuristic. The Gaussian and Absdiff kernels are regularized distances in the form of an exponential distance measure ($d^2(x, y)$). The Gaussian kernel ($d^2(x, y) = \Sigma_k(x_k - y_k)^2$) is common since it tends to produce good results when used with a wide variety of datasets. The Absdiff ($d^2(x, y) = \Sigma_k(|x_k - y_k|)^{1/2}$) and Sentropic ($D(x, y) = [D(x\|y) + D(y\|x)]/2$) Kernels [1, 3, 59] tend to work better with all of the datasets considered here and in other tests not shown. The Sentropic kernel is based on a regularized information divergence ($D(x, y)$) instead of a geometric distance.

The so-called curse of dimensionality from statistics says that the difficulty of an estimation problem increases drastically with the dimension N of the estimate configuration space, since in principle as N increases, the number of required patterns to sample grows exponentially. This statement may cast doubts on using higher dimensional feature vectors as input to learning machines. This must be balanced with results from statistical learning theory [161], however, that show that the likelihood of data separability by linear learning machines is proportional to (and improves with) their dimensionality. The feature space may grow in dimensionality, but does so under constraint to remain in a small (manageable) function class.

The kernels used in the analysis fall into two classes: regularized distance (squared) kernels; and regularized information divergence kernels. The first set of kernels strongly models data with classic, geometric, attributes or interpretation. The second set of kernels is constrained to operate on $(R^+)^N$, the feature space of positive, nonzero, real-valued feature vector components. The space of the latter kernels is often also restricted to feature vectors obeying an L_1-norm $= 1$ constraint, i.e. the feature vector is a discrete probability vector.

Given any metric space (x, d) one can build a positive-definite kernel of the form $e - \lambda d^2$ Conversely, any positive definite kernel with such form must have a "d" that is a metric (this is Mercer's condition in another form). (The metric appears as a squared entity in the argument, which is the mathematical grouping that satisfies the triangle inequality. It so happens that there is another set of entities, other than metrics squared, that have a similar relation to the triangle inequality (related by a Lagrange transformation), and these relate to the Bregman divergences, including the information divergences such as relative entropy, in particular, which is the fundamental (simple) divergence, just as Euclidean distance is a fundamental (simple) distance. This suggests that the "simplest" distance-based kernel is the Gaussian kernel, since the "simplest" distance, the Euclidean distance, is used. Likewise, this suggests that the simplest divergence-based kernel would be the aforementioned entropic kernel.

The use of probability vectors, and L_1-norm feature vectors in general (often in conjunction with the entropic kernel), turns out to provide a very general formulation, wherein feature extraction makes use of signal decomposition into a complete set of separable states that can be interpreted or represented as a probability vector (or normalized collection of such, or concatenation, then normalization, etc.). A probability vector formulation also provides a straightforward hand-off to the SVM classifiers since all feature vectors have the same length with such an approach. What this means for the SVM, however, is that geometric notions of distance are no longer the best measure for comparing feature vectors. For probability vectors (i.e. discrete distributions), the best measures of similarity are the various information-theoretic divergences: Kullback–Leibler, Renyi, etc. By symmetrizing over the arguments of those divergences, the entropic kernels are obtained, where the (symmetrized) Kullbach–Leibler Diveregence is used in the entropic kernel in [1, 3, 59, 217] and what follows.

Notice how in the Dual reduction the dependence on the training data only appears in the $\vec{x}_i \cdot \vec{x}_j$ inner product term:

$$\sum_{i,j} \alpha_i \alpha_j y_i y_j \vec{x}_i \cdot \vec{x}_j$$

We can generalize from the simple inner product term in a number of ways, and in doing so arrive at the SVM Kernel generalization. First, however, let us consider remapping the feature vectors into some higher dimensional space that is hyperspherically bounded. The volume of a thin shell at the boundary hypersphere dominates at higher dimensions, allowing for the data in the mapping to become approximately unit hyperspherical when the bounded hypersphere has its hyperspherical bound dilated to be unity by going to higher dimension. I will proceed then with a new ith feature vector, where the same feature vector label, $\vec{x}_i$, will be used.

If we have $x_i \, x_j = \|x_i\| \|x_j\| \cos \theta_{ij}$ and $\|x_i\| = 1 \; \forall_i$ (unit hyperspherical data), then $x_i \, x_j = \cos \theta_{ij}$. On unit hypersphere, the spherical arc angle between points x_i and x_j on its surface, for small angle:

$$\cos \theta = 1 - \frac{1}{2}\theta^2 + \dots \text{ and } x_i \cdot x_j = 1 - \frac{1}{2}\theta_{ij}^2 + \dots$$

In the Lagrangian, the constant term does not matter due to the linear equality constraint:

$$\sum_i y_i \alpha_i = 0$$

So, for unit hyperspherical, proximate, feature vectors, arc-length is approximately Euclidean distance, and we have:

$$x_i \cdot x_j = 1 - \frac{1}{2}\theta_{ij}^2 + \ldots = 1 - \left|x_i - x_j\right|^2 + \ldots$$

Since there is a conformal angle-preserving mapping (triangle-inequality, so distance square preserving, and divergence preserving) with Lorentz transformation on observation sphere (with generalization to higher dimension, where we can ignore red-shift information, physically, for 4D (3D spatial), this occurs in special relativity, the near-lightspeed observer causes more and more to map into the forward view). The point is that we can do another mapping to achieve the small angle approximation on the entire dataset, where the compound mapping preserves the triangle inequality relation that is thought to be important. This specific mapping example is meant to indicate what might be argued as existing, however, while in actual implementation the focus is on the kernel function and the mapping need never be known.

For unit hyperspherical data, $x_i \cdot x_j = 1 - |x_i - x_j|^2 + \ldots$ can be thought of as measuring a distance that has been regularized in some manner when the distance grows large (i.e. as angular coordinate limitation, or in exponentiation to be seen in what follows), such that it is impossible or harshly penalized to have significantly large distances (equivalent to the original inner product term being close to zero, its contribution in the Lagrangian optimization made irrelevant), while data that is "near" has its contribution in the Lagrangian optimization set close to one according to $1 - |x_i - x_j|^2$).

What this demonstrates, thus far, is that we can manipulate our feature vectors by some mapping (with inverse) such that $\bar{x}_i \rightarrow \Phi(\bar{x}_i)$, and such that the inner product can become $\Phi(\bar{x}_i)\Phi(\bar{x}_j) \approx \emptyset\left(-\left|\vec{x}_i - \vec{x}_j\right|^2\right)$, where $\emptyset\left(-\left|\vec{x}_i - \vec{x}_j\right|^2\right)$ is an exponentially regularized distance squared, e.g. the Gaussian Kernel:

$$\emptyset_G\left(\vec{x}_i, \vec{x}_j\right) = \exp\left(\frac{\left|\vec{x}_i - \vec{x}_j\right|^2}{2\sigma^2}\right), \text{ and for } \bar{x}_i \approx \bar{x}_j, \emptyset_G \approx 1 - \frac{1}{2\sigma^2}\left|\vec{x}_i - \vec{x}_j\right|^2,$$

which relates to the earlier derivation for kernel parameter (the variance in the case of the Gaussian Kernel) equals one half. It is therefore possible to generalize using $\Phi: \vec{x}_i \mapsto \Phi\left(\vec{x}_j\right)$ to obtain:

$$\tilde{L}_\sigma\left(\vec{a}\right) = \sum_i \alpha_i - \frac{1}{2}\alpha_i \alpha_j y_i y_j K_{ij}, 0 \le \alpha_i \le C, \text{where } K_{ij} = \Phi(\bar{x}_i) \cdot \Phi(\bar{x}_j)$$

$$(10.37)$$

Kernel functions expressible in this way must have a positive semi-definite kernel, typically tested using what is known as Mercer's condition. Not all kernels satisfy Mercer's condition, and are therefore are not describable in terms of a mapping Φ on feature vectors. Although all kernels examined appear to satisfy Mercer's condition, this will not be taken as a critical limitation, especially if working with algorithmic developments that make use of the KKT relations separate from the originating derivation.

10.5.1.1 The Mercer Test
The kernel is positive definite if and only if:

$$\sum_{ij} k\left(\vec{x}_i, \vec{x}_j\right) C_i C_j \ge 0 \forall \vec{C} \in R^m \text{ (for positive semidefinite K)},$$

where C-vectors are randomly generated and tested against the kernel obtained on the training data.

10.5.1.2 The Positive Principal Minors Test
The kernel is positive definite if and only if the determinants of all of the principal minors are positive.

10.5.2 Derivation of "Stability" Kernels

For the Gaussian kernel the stability property is exhibited when the log Kernel variation on feature vector components is calculated:

$$\partial \ln \left(K_G(\mathbf{x}_i, \mathbf{x}_j)\right)/\partial x_i^k = \left(x_j^k - x_i^k\right)/\sigma^2$$

where "x_i^k" is the kth component of the ith feature vector and "stability" is indicated by the sign of the difference term $(x_j^k - x_i^k)$, e.g. for

$$K_G\left(\vec{y}, \vec{z}\right) = \exp\left(-\frac{\left\|\vec{y}-\vec{z}\right\|^2}{2\sigma^2}\right); \frac{\partial \ln K_G\left(\vec{y}, \vec{z}\right)}{\partial y_K} = (y_K - z_K)/\sigma^2 \quad (10.38)$$

Clearly, the sign is important, as is a notion of difference. Suppose we generalize on this basis to decouple the sign convention from the "notion of distance," here providing a new kernel expression, for the "variational kernel" by way of an integration factor:

$$\frac{\partial \ln K_G\left(\vec{y}, \vec{z}\right)}{\partial y_k} = \frac{-1}{2\sigma^2} \frac{\text{sign}(y_k - z_k)}{\sqrt{\sum_k |y_k - z_k|}} \tag{10.39}$$

$$K_v\left(\vec{y}, \vec{z}\right) = \exp\left(-\frac{\sqrt{\sum_k |y_k - z_k|}}{2\sigma^2}\right) \tag{10.40}$$

The subscript "v" in K_v is meant to denote "variational" kernel (sometimes referred to as "indicator" kernel or "Absdiff" kernel). For suitable choice of tuning parameter σ, the variational kernel offers the best performance on the data sets considered. The regularized distance in K_v is the square root of the "Variational" distance: $V(\mathbf{x}_i \| \mathbf{x}_j) = \sum_k |x_j^k - x_i^k|$. It is found that the variational kernel is usually the best performing kernel on the L_1 normed data considered in the channel current analysis (L_1 norm: $|x| = \sum_k |x_k|$, a discrete prob. dist if $x_k > 0$ also). The argument of the exponential in the variational kernel is a distance squared (which satisfies the triangle inequality, etc.), with $K_v = \exp\left(-\frac{d_v^2}{2\sigma^2}\right)$, thus the variational kernel automatically satisfies Mercer's conditions.

Consider now the case where the notion of difference is not arithmetic but multiplicative, i.e. based on $(1 - z_k/y_k)$ rather than $(y_k - z_k)$ (for the Gaussian). In doing so, we must restrict to $y_k \neq 0$ of course. As before, the sign of $(y_k - z_k)$ is information preserved in $(1 - z_k/y_k)$, but the latter is not integrable. However, $\ln(y_k/z_k)$ also provides sign info – positive when $y_k > z_k$, etc., as before, and also includes a ratio. Which to go with? A combination seems best as this is integrable:

$$\frac{\partial \ln K_g(y_k - z_k)}{\partial y_k} = \left(\frac{-1}{2\sigma^2}\right)\left[\left(1 - \frac{z_k}{y_k}\right) + \ln\left(\frac{z_k}{y_k}\right)\right] \tag{10.41}$$

$$k_\sigma\left(\vec{y}, \vec{z}\right) = \exp\left(-\left[D(y\|z) + D(z\|y)/2\sigma^2\right]\right) \tag{10.42}$$

This is usually a close second to the K_v kernel, sometimes out performing. This kernel relates feature vectors via relative entropy terms:

$$D\left(y\|z\right) = \sum_k y_k \ln\frac{y_k}{z_k} \tag{10.43}$$

The doubly novel aspect of the entropic kernel is that it would be the very first guess if one wanted to generalize from kernels based on exponentially regularized, square distances, to exponentially regularized, symmetrized, divergences (beginning with the most fundamental, symmetrized "relative entropy" also known as then Kullback–Leibler information divergence).

Note that we began with the supposition that sign was important, as was some well-behaved notion of difference (whether it is distance-based or divergence-based, etc.). Remarkably, the entropic kernel K_σ appears to satisfy Mercer's condition, when properly restricted to discrete probability distributions: $y_k > 0, \sum y_k = 1$. This is not established with precise mathematical proof, but tested through exhaustive numerical testing using the Mercer test and the positive principle minors test on test data.

Since the feature vectors can be interpreted as probabilities, and satisfy the probability relation $\sum_k (x_i^k) = 1$, it is, perhaps, not surprising that the symmetric-entropic kernel should be a good performer. There is also an interesting relationship between the Variational distance in the Variational Kernel and the Kullback–Leibler divergence in the Entropic Kernels, known as the Pinsker inequality:

$$D(\mathbf{x}_i||\mathbf{x}_j) \geq V(\mathbf{x}_i||\mathbf{x}_j)^2/2 \tag{10.44}$$

It may prove possible to generate other such inequalities by use of other integrating factors in the above kernel selection process.

10.5.3 Entropic and Gaussian Kernels Relate to Unique, Minimally Structured, Information Divergence and Geometric Distance Measures

Using the Shannon entropy measure it is possible to derive the classic probability distributions of statistical physics by maximizing the Shannon measure subject to appropriate linear momentum constraints. Constrained variational optimizations involving the Shannon entropy measure can, thus, provide a unified framework with which to describe all, or most, of statistical mechanics. The distributions derivable within the maximum entropy formalism include the Maxwell–Boltzmann, Bose–Einstein, Fermi–Dirac, and Intermediate distributions. The maximum entropy method for defining statistical mechanical systems has been extensively studied [112].

Both statistical estimation and maximum entropy estimation are concerned with drawing inferences from partial information. The maximum entropy approach estimates a probability density function when only a few moments are known (where there are an infinite number of higher moments). The statistical approach estimates the density function when only one random sample is available out of an infinity of possible samples. The maximum entropy estimation may be significantly more robust (against over-fitting, for example) in that it has an Occam's Razor argument that "cuts both ways" – use *all* of the information given and avoid using any information not given. This means that out of all of the probability distributions consistent with the set of constraints, choose the one that has maximum uncertainty, i.e. maximum entropy [112].

At the same time that Jaynes was doing his work, essentially an optimization principle based on Shannon entropy, Soloman Kullback was exploring optimizations involving a notion of probabilistic distance known as the Kullback–Leibler distance, referred to above as the relative entropy [60]. The resulting minimum relative entropy (MRE) formalism reduces to the maximum entropy formalism of Jaynes when the reference distribution is uniform. The information distance that Kullback and Leibler defined was an oriented measure of "distance" between two probability distributions. The MRE formalism can be understood to be an extension of Laplace's *Principle of Insufficient Reason* (e.g. if nothing is known assume the uniform distribution) in a manner like that employed by Khinchine in his uniqueness proof, but now incorporating constraints.

In their book *Entropy Optimization Principles with Applications* [111], Kapur and Kesavan argue for a generalized entropy optimization approach to the description of distributions. They believe every probability distribution, theoretical or observed, is an entropy optimization distribution, i.e. it can be obtained by maximizing an appropriate entropy measure, or by minimizing a relative entropy measure with respect to an appropriate *a priori* distribution. The primary objective in such a modeling procedure is to represent the problem as a simple combination of probabilistic entities that have a simple set of moment constraints. Generalized measures of distributional distance can also be explored along the lines of generalized measures of geometric distance. In physics, not every geometric distance is of interest, however, since the special theory of relativity tells us that spacetime is locally flat (Lorentzian, which is Euclidean on spatial slices), with metric generalization the Riemannian metrics. Likewise, perhaps not all distributional distance measures are created equally either. The locally flat equivalent in information geometry [113–115] is the Kullback-Leibler divergence (e.g. the symmetrized relative entropy used in the entropic kernel). We propose generalization to all exponentially regularized distance squared and information divergence kernels and further generalization to include the larger class of stability kernels and "triangle-inequality" kernels (if not a larger class, then the generalizations consist of metrical and divergence-measure generators).

A comparison of some of the SVM Kernels of interest is shown in Figure 10.16, with "regularized" distances or divergences, where they are regularized if in the form of an exponential with argument the negative of some distance-measure squared ($d2(x, y)$) or symmetrized divergence measure ($D(x, y)$), the former if using a geometric heuristic for comparison of feature vectors, the latter if using a information divergence heuristic. Results in Figure 10.16 are shown for the Gaussian Kernel: $d2(x, y) = \Sigma k(xk - yk)2$; for the Absdiff or Variational Kernel $d2(x, y) = (\Sigma k \,|\, xk - yk|)1/2$; and for the Symmetrized Relative Entropy Kernel $D(x, y) = D(x\|y) + D(y\|x)$, where $D(x\|y)$ is the standard relative entropy.

The SVM algorithm variants that have been explored are minimally detailed here: in the standard Platt SMO algorithm, $\eta = 2 * K_{12} - K_{11} - K_{22}$, while speedup variations are described to avoid calculation of this value entirely. A middle ground is obtained with the following definition $\eta = 2 * K_{12} - 2$; If $(\eta >= 0)$ $\{\eta \geq -1;\}$ (labeled WH SMO in Figure 10.16, with underflow handling and other details that differ slightly in the WH-SMO implementation as well).

The best algorithm/kernel in Figure 10.16, and in other channel blockade data studied, has consistently been the WH SMO variant and the Absdiff and Entropic Kernels. Another benefit of the WH SMO variant is its significant speedup over the other methods (about half the time of Platt SMO and one fourth the time of Keerthi 1 or 2). The alpha handling and other modifications in WH SMO [1, 3, 59, 217] relate to boundary support vector (BSV) handling (associated with handling on outliers), which is also critical to enhancements to a multiclass SVM solution described in the next section.

10.5.4 Automated Kernel Selection and Tuning

An automated tuning solution for SVM classifiers is described. This is done by implementing a simulated annealing with perturbation tuning procedure on the SVM's kernel and algorithmic parameters. The SVM performance on training data is used to define a fitness function and the tuning results obtained were as good as or better than those obtained manually. SVM methods are described for data classification, data clustering, and signal analysis using experimental data obtained from a nanopore detector. The SVM implementations described involve SVM algorithmic variants, SVM kernel variants, and SVM chunking variants, where tuning metaheuristics offer an automated way to obtain a powerful SVM classifier for a given dataset. The SVM discussion is interwoven with application to data involving channel current analysis, with application to the signal processing associated with the nanopore transduction detector (NTD) described in Chapter 14.

A classifier is typically a simple rule whereby a class determination can be made, such as a decision boundary. Figure 10.23 shows labeled training data and a decision boundary with a margin region. Learning the decision rule, or a sufficiently good decision rule, especially if simple and elegant, is the implementation aspect of a classifier, and can be difficult and time consuming. Even so, this is often manageable because at least there is data to "learn from," e.g. supervised learning, with instances and their classifications (or "labels"). Learning for classification can be done very effectively using generalized SVMs. With clustering efforts, or unsupervised learning, on the other hand, we do not have the label information during training. In what follows SVMs will also be shown to be incredibly effective at clustering when used with metaheuristics to recover label information in a bootstrap

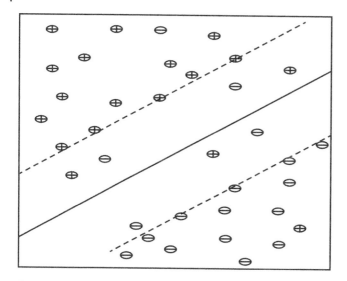

Figure 10.24 **Decision boundary (solid line); with margin** (region between dotted lines). Instances are indicated as positive class (+) or negative class (−), where misclassified data on wrong side of hyperplane (or penalized if in margin), is allowed by incurring a penalty in the optimization process. The misclassified or partly penalized margin-region instances are shown with double walled circles.

learning process. Also shown will be implementation details for distributed SVM training [13], and other speedup optimizations, allowing practical deployment, with the auto-tuning methods described here, of the generalized SVM classification and clustering methods in real-time operational situations (as demonstrated in applications in nanopore detector experiments) (Figure 10.24).

The SVM-based clustering method that will be described in what follows makes use of the SVM-classifier convergence process. Single-convergence initialized clustering methods, involving label-flipping between SVM convergence training runs, have been studied previously and will be described in detail. The single-convergence methods outperform other methods on the test sets considered, but in examining the clustering failures (albeit fewer than with parameterized methods), there is room for improvement. Efforts to handle this with more sophisticated tuning have met with initial success, as will be related in the following, where multiple convergence processes are examined and scored according to a post-processing SSE criterion, where a minimal SSE is sought, to reliably obtain very well-tuned strong SVM performance.

In setting up an SVM Classifier, one must have training data in the form of feature vectors, where all of the feature vectors are the same length. One typically needs to specify a choice of kernel and kernel parameter (and possibly other parameters), and therein lies the rub. The SVM may not converge with your specification. SVMs have a surprising amount of practical functionality, however, as will be shown. It is fairly easy to tune SVMs, in many cases, by simply using a default set of kernel's and parameter ranges. There is more robust performance, however, with more sophisticated tuning. In the SVM applications that attempt to bootstrap a clustering solution, there is more sensitivity to kernel and kernel parameter overall, and more sophisticated tuning methods are clearly needed for the more challenging SVM applications.

Use of SVMs for clustering (unsupervised learning) is possible in a number of different ways. As with the multiclass SVM discriminator generalizations, the strong performance of the binary SVM enables SVM-External as well as SVM-Internal approaches to clustering. Nonparametric SVM-based clustering methods may allow for much improved performance over parametric approaches, particularly since they can apparently be designed to inherit the strengths of their supervised SVM counterparts as will be shown. The "external" SVM clustering algorithm, to be described in detail, clusters data vectors with no *a priori* knowledge of each vector's class.

It is possible to initiate SVM training with model parameters, such as the kernel or kernel parameter, that are so far out of the operational regime that no convergence is obtained in training. So training must be repeated with tuning on SVM parameters to optimize. For some feature vectors, such as probability vectors, this can partly be done automatically with choice of kernel. In many situations the SVM tuning can simply be done manually, or partly automatically, with simple range testing, where only small, separated, subsets of the training data are used in the tuning tests, before performing SVM training on the full dataset minus the tuning data. Sometimes more elaborate tuning procedures are needed, however, and thus necessary for performance guarantees, and also for the SVM applications in clustering that will be described next. Tuning is a form of optimization, and excellent metaheuristics are known for identifying optimal solutions when a scoring function (a fitness function) can be identified and such is provided by the SVM via sensitivity and specificity scores on training data). Metaheuristic optimizations attempted include GAs, simulated annealing, and steepest ascent hill-climbing, among others. Applications of many of these methods are shown in the results involving SVM-external clustering.

Our ability to assess a score with SVMs, and thereby assign a fitness, allows for a collection of metaheuristics that basically reduce to "look around and take the best way forward" via a series of tweaks. This is not possible for some problems, however, because the "looking around" part is not that informative, e.g. the fitness landscape has sections that are at a fixed level (with noise variations about that level, for example). This is the larger problem of the simple globalization algorithm, via random restart: if the fitness landscape or configuration space is too large random restart would not offer a solution (even if it can) in a reasonable amount of time. This is where more clever metaheuristics must be drawn upon to extend to a global optimization algorithm.

One of the weaknesses of the brute force random restart approach mentioned is that the parameter "tweak" involved is with a *bounded* perturbative change, which may *already* exclude the possibility of reaching the solution sought (given the computational resources and a reasonable amount of time). So one generalization is to allow for tweaks that are unbounded, but in some perturbatively stable way, such as with a Boltzmann factor for regularization, and in doing this we arrive at the Simulated Annealing approach (see Chapter 12 for details).

The global optimization metaheuristics mentioned thus far have worked well, as described in what follows, and suggest more sophisticated configuration selection may benefit further at the component level, on the one hand, and at the population level, on the other hand, especially as further refinements are made to the probabilistic simulated annealing approach. This is because the population and history aspects point to a general metaheuristics that operates on populations of configurations (or populations of "agents" that interact as intermediaries to determining a configuration selection). The notion of "history" can be incorporated in various ways, with conveyance of history or learned information internally (such as genome in GA approaches) or externally via "artifact" (such as via stygmergy in the ACO method). Note that the population based search metaheuristics allow a simple means for distributed computational speedup.

10.6   SVM Multiclass from Decision Tree with SVM Binary Classifiers

The SVM binary discriminator offers high performance and is very robust in the presence of noise. This allows a variety of reductionist multiclass approaches, where each reduction is a binary classification. The SVM Decision Tree is one such approach used extensively with the datasets examined in [1, 3, 59, 67], where a collection of SVM Decision Trees (a SVM Decision Forest) can be used to avoid problems with throughput biasing. Alternatively, the variational formalism can be modified to perform a multi-hyperplane optimization situation for a direct multiclass solution [1, 3, 59, 67], as is described next.

The SVM Decision Tree shown in Figure 10.25 obtained nearly perfect sensitivity and specificity, with a high data rejection rate, and a highly nonuniform class

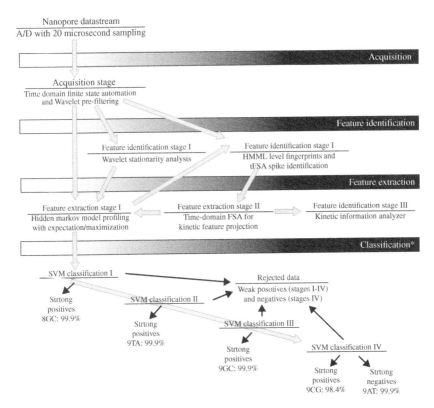

Figure 10.25 Nanopore detector signal analysis architecture, with use of an SVM decision tree for classification. *Source:* Modified from Winters-Hilt et al. [67].

signal-calling throughput. In Figure 10.26, the Percentage Data Rejection vs. SN + SP curves are shown for test data classification runs with a binary classifier with one molecule (the positive, given by label) vs. the rest (the negative). Since the signal calling was not passed through a Decision Tree, the way these curves were generated, they do not accurately reflect total throughput, and they do not benefit from the "shielding" shown in the Decision Tree in Figure 10.25 prototype. In the SVM Decision Tree implementation described in Figure 10.25 [67], this is managed more comprehensively, to arrive at a five-way signal-calling throughput at the furthest node of 16% (in Figure 10.25, 9CG and 9AT have to pass to the furthest node to be classified), while the best throughput, for signal calling on the 8GC molecules, is 75%.

The SVM Decision Tree classifier's high, nonuniform, rejection can be managed by generalizing to a collection of Decision Trees (with different species at the

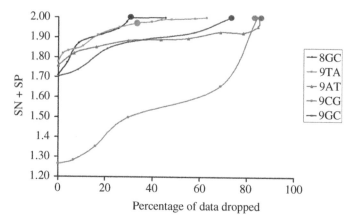

Figure 10.26 The percentage data rejection vs. SN + SP curves are shown for test data classification runs with a binary classifier with one molecule (the positive, given by label) vs. the rest (the negative). Since the signal calling was not passed through a Decision Tree, it does not accurately reflect total throughput, and they do not benefit from the "shielding" shown in the Decision Tree in Figure 10.25 prototype. The relative entropy kernel is shown because it provided the best results (over Gaussian and Absdiff).

furthest node). The problem is that tuning and optimizing a single decision tree is already a large task, even for five species (as in [67]). With a collection of trees, this problem is seemingly compounded, but can actually be lessened in some ways in that now each individual tree need not be so well-tuned/optimized. Although more complicated to implement than an SVM-External method, the SVM-Internal multiclass methods are not similarly fraught with tuning/optimization complications. Figure 10.27 shows the Percentage Data Rejection vs. SN + SP curves on the same train/test data splits as used for Figure 10.26, except now the drop curves are to be understood as *simultaneous* curves (not sequential application of such curves as in Figure 10.26). Thus, comparable, or better, performance is obtained with the multiclass-internal approach and with far less effort, since there is no managing and tuning of Decision Trees. Another surprising, and even stronger argument for the SVM-Internal approach to the problem, for many situations, is that a natural drop zone is indicated by the margin.

Suppose we define the criteria for dropping weak data as the margin: For any data point x_i; let $\max_m\{f_m(x_i)\} = f_{yi}$, and Let $f_m = \max_m\{f_m(x_i)\}$ for all $m \neq yi$, then we define the margin as: $(f_{yi} - f_m)$, hence data point x_i is dropped if $(f_{yi} - f_m) \leq$ Confidence Parameter. (For this data set using Gaussian, AbsDiff and Sentropic kernel, a confidence parameter of at least $(0.000\,01)*C$ was required to achieve 100% accuracy.) Using the margin drop approach, there is even less tuning, and there is improved throughput (approximately 75% for *all* species) [1, 3, 59].

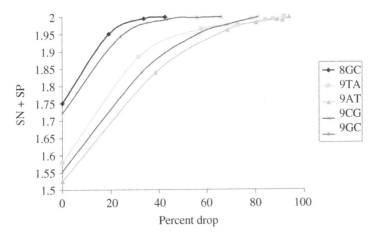

Figure 10.27 The percentage data rejection vs. SN + SP curves are shown for test data classification runs with a *multiclass* discriminator. The following criterion is used for dropping weak data: for any data point x_i; if $\max_m\{f_m(x_i)\} \leq$ confidence parameter, then the data point x_i is dropped. For this dataset using AbsDiff kernel ($\sigma^2 = 0.2$) performed best, and a confidence parameter of 0.8 achieve 100% accuracy.

10.7 SVM Multiclass Classifier Derivation (Multiple Decision Surface)

In [59] we make use of a variant of a formulation by Crammer and Singer [177]. In the variant [59], the formulation is modified at the Lagrangian level to allow for an analytic decoupling on the constraint equations, permitting a partly analytic solution to the multiclass problem, with significant reduction in algorithmic complexity at the implementation (coding) stage. Using the notation in [59, 177]: there are "k" classes and hence "k" linear decision functions. For a given input "x", the output vector corresponds to the output from each of these decision functions. The class of the largest element of the output vector gives the class of "x." Each decision function is given by: $f_m(x) = w_m \cdot x + b_m$ for all $m = (1, 2, ..., k)$. If y_i is the class of the input x_i, then for each input data point, the misclassification error is defined as follows: $\max_m\{f_m(x_i) + 1 - \delta_i^m\} - f_{yi}(x_i)$, where δ_i^m is 1 if $m = y_i$ and 0 if $m \neq y_i$. We add the slack variable ζ_i where $\zeta_i \geq 0$ for all i that is proportional to the misclassification error: $\max_m\{f_m(x_i) + 1 - \delta_i^m\} - f_{yi}(x_i) = \zeta_i$, hence $f_{yi}(x_i) - f_m(x_i) + \delta_i^m \geq 1 - \zeta_i$ for all i, m. To minimize this classification error and maximize the distance between the hyperplanes (SRM) we have the following formulation:

Minimize: $\sum_i \zeta_i + \beta(1/2)\sum_m w_m^T w_m + (1/2)\sum_m b_m^2$,
where $\beta > 0$ is defined as a regularization constant.
Constraint: $w_{yi} \cdot x_i + b_{yi} - w_m \cdot x_i - b_m - 1 + \zeta_i + \delta_i^m \geq 0$ for all i, m

Note: the term $(1/2)\sum_m b_m^2$ is added for de-coupling, $1/\beta = C$, and $m = y_i$ in the above constraint is consistent with $\zeta_i \geq 0$, and that periods are being used to denote inner products. The Lagrangian is:

$$L(w, b, \zeta) = \sum_i \zeta_i + \beta(1/2)\sum_m w_m^T w_m + (1/2)\sum_m b_m^2$$
$$- \sum_i \sum_m \alpha_i^m \left(w_{yi}.x_i + b_{yi} - w_m.x_i - b_m - 1 + \zeta_i + \delta_i^m \right)$$

(10.45)

Where all $\alpha_i^m s$ are positive Lagrange multipliers. Now taking partial derivatives of the Lagrangian and equating them to zero (Saddle Point solution): $\partial L/\partial \zeta_i = 1 - \sum_m \alpha_i^m = 0$. This implies that $\sum_m \alpha_i^m = 1$ for all i. $\partial L/\partial b_m = b_m + \sum_i \alpha_i^m - \sum_i \delta_i^m = 0$ for all m. Hence $b_m = \sum_i(\delta_i^m - \alpha_i^m)$. Similarly: $\partial L/\partial w_m = \beta w_m + \sum_i \alpha_i^m x_i - \sum_i \delta_i^m x_i = 0$ for all m. Hence $w_m = (1/\beta)[\sum_i(\delta_i^m - \alpha_i^m)x_i]$ Substituting the above equations into the Lagrangian and after simplification reduces into the dual formalism:

Maximize: $-\frac{1}{2}\sum_{i,j}\sum_m (\delta_i^m - \alpha_i^m)(\delta_j^m - \alpha_j^m)(K_{ij} + \beta) - \beta \sum_{i,m} \delta_i^m \alpha_i^m$

Constraint: $0 \leq \alpha_i^m, \sum_m \alpha_i^m = 1, i = 1...l; m = 1...k$

where $K_{ij} = x_i \cdot x_j$ is the Kernel generalization. In vector notation:

Maximize : $-\frac{1}{2}\sum_{i,j}(\Delta_{yi} - A_i)(\Delta_{yj} - A_j)(K_{ij} + \beta) - \beta \sum_i \Delta_{yi} A_i$

Constraint : $0 \leq A_i, A_i.1 = 1, i = 1...l$

Let $\tau_i = \Delta_{yi} - A_i$. Hence after ignoring the constant: $-\frac{1}{2}\sum_{i,j}\tau_i \cdot \tau_j(K_{ij} + \beta) + \beta\sum_i \Delta_{yi}\tau_i$, subject to: $\tau_i \leq \Delta_{yi}, \tau_i.1 = 0, i = 1...l$. The dual is solved (determine the optimum values of all the τs) using the decomposition method.

Minimize: $\frac{1}{2}\sum_{i,j}\tau_i^m.\tau_j^m(K_{ij} + \beta) - \beta\sum_{i,m}\delta_i^m \tau_i^m$
Constraint: $\tau_i \leq \Delta_{yi}, \tau_i.1 = 0, i = 1...l$

The Lagrangian of the dual is:

$$L = \frac{1}{2}\sum_{i,j,m}\tau_i^m.\tau_j^m(K_{ij} + \beta) - \beta\sum_{i,m}\delta_i^m \tau_i^m - \sum_{i,m}u_i^m(\delta_i^m - \tau_i^m)$$
$$- \sum_i v_i \sum_m \tau_i^m;$$

Subject to $u_i^m \geq 0$

(10.46)

We take the gradient of the Lagrangian with respect to τ_i^m:

$$\nabla_{\tau_i}^m[L] = \sum_i \tau_j^m(K_{ij} + \beta) - \beta\delta_i^m + u_i^m - v_i = 0$$

Introducing $f(\tau) = \sum_i \tau_j^m(K_{ij} + \beta) - \beta\delta_i^m + u_i^m - v_i = 0$ and $f_i^m = \sum_i \tau_j^m(K_{ij} + \beta)$ $- \beta\delta_i^m$, then $f(\tau) = f_i^m + u_i^m - v_i = 0$. By KKT conditions we get two more equations:

$$u_i^m(\delta_i^m - \tau_i^m) = 0 \text{ and } u_i^m \geq 0 \tag{10.47}$$

Case I: if $\delta_i^m = \tau_i^m$, then $u_i^m \geq 0$, hence $f_i^m \leq v_i$. Case II: if $\tau_i^m < \delta_i^m$, then $u_i^m = 0$, hence $f_i^m = v_i$. Note: There is at least one "m" for all i such that $\tau_i^m < \delta_i^m$ is satisfied. Therefore, combining Case I and II, we get:

$$\max_m\{f_i^m\} \leq v_i \leq \min_{m:\tau_i^m < \delta_i^m}\{f_i^m\}$$
$$\text{Or } \max_m\{f_i^m\} \leq \min_{m:\tau_i^m < \delta_i^m}\{f_i^m\} \tag{10.48}$$
$$\text{Or } \max_m\{f_i^m\} - \min_{m:\tau_i^m < \delta_i^m}\{f_i^m\} \leq \varepsilon$$

Note: $\tau_i^m < \delta_i^m$ implies that $\alpha_i^m > 0$. Since $\sum_m \alpha_i^m = 1$, for any i each α_i^m is treated as the probability that the data point belongs to class m. Hence we **define KKT violators as:**

$$\max_m\{f_i^m\} - \min_{m:\tau_i^m < \delta_i^m}\{f_i^m\} > \varepsilon \text{ for all } i. \tag{10.49}$$

10.7.1 Decomposition Method to Solve the Dual

To solve the Dual, maximize:

$$Q(\tau) = -\tfrac{1}{2}\sum_{i,j}\tau_i.\tau_j(K_{ij} + \beta) + \beta\sum_i \Delta_{yi}\tau_i; \text{Subject to:}$$
$$\tau_i \leq \Delta_{yi}, \tau_i.1 = 0, i = 1...l$$

Expanding in terms of a single "τ" vector:

$$Q_p(\tau_p) = -\tfrac{1}{2}A_p(\tau_p.\tau_p) - B_p.\tau_p + C_p$$

Where:

$$A_p = K_{pp} + \beta; B_p = -\beta\Delta_{yp} + \sum_{i\neq p}\tau_i(K_{ip} + \beta);$$
$$C_p = -\tfrac{1}{2}\sum_{i,j\neq p}\tau_i.\tau_j(K_{ij} + \beta) + \beta\sum_{i\neq p}\tau_i\Delta_{yi}$$

Therefore ignoring the constant term "C_p," we have to minimize:

$$Q_p(\tau_p) = \tfrac{1}{2}A_p(\tau_p.\tau_p) + B_p.\tau_p; \tau_p \leq \Delta_{yp} \text{ and } \tau_p.1 = 0$$

The above equation can also be written as:

$$Q_p(\tau_p) = \tfrac{1}{2}A_p(\tau_p + B_p/A_p).(\tau_p + B_p/A_p) - B_p.B_p/2A_p$$

Substitute $v = (\tau_p + B_p/A_p)$ and $D = (\Delta_{yp} + B_p/A_p)$ in the above equation. Hence, after ignoring the constant term $B_p.B_p/2A_p$ and the multiplicative factor "A_p" we have to minimize:

$$Q(v) = \tfrac{1}{2}v.v = \tfrac{1}{2}||v||^2; v \le D \text{ and } v.1 = D.1 - 1$$

The Lagrangian is given by:

$$L(v) = \tfrac{1}{2}||v||^2 - \sum_m \rho_m(D_m - v_m) - \sigma\left[\sum_m(v_m - D_m) + 1\right]; \rho_m \ge 0$$

Hence $\partial L/\partial v_m = v_m + \rho_m - \sigma = 0$. By KKT conditions we have: $\rho_m(D_m - v_m) = 0$ and $\rho_m \ge 0$, also $v_m \le D_m$. Hence by combining the above in-equalities, we have: $v_m = \text{Min}\{D_m, \sigma\}$, or $\sum_m v_m = \sum_m \text{Min}\{D_m, \sigma\} = \sum_m D_m - 1$. The above equation uniquely defines the "σ" that satisfies the above equation AND that "σ" is the optimal solution of the quadratic optimization problem. (Refer to [177] for a formal proof).

Solve for "σ": We have $\text{Min}\{D_m, \sigma\} + \text{Max}\{D_m, \sigma\} = D_m + \sigma$, hence $\sum_m[D_m + \sigma - \text{Max}\{D_m, \sigma\}] = \sum_m D_m - 1$, or $\sigma = 1/K[\sum_m \text{Max}\{D_m, \sigma\} - 1]$, hence we find σ (iteratively) that satisfies the equation: $|(\sigma_l - \sigma_{l+1})/\sigma_l| \le \text{tolerance}$. The initial value for "σ" is set to $\sigma_1 = 1/K[\sum_m D_m - 1]$.

Update rule for "τ": Once we have "σ," $\tau_{\text{new}}{}^m = v_m - B_p{}^m/(K_{pp} + \beta)$, or:

$$\tau_{\text{new}}{}^m = v_m - f_p{}^m/(K_{pp} + \beta) + \tau_{\text{old}}{}^m$$

10.7.2 SVM Speedup via Differentiating BSVs and SVs

If we track the status of SV according to whether they are Boundary (penalty) or not, and select accordingly, we can get speedup (even in the binary SVM, where choice of $C = 100$ is usually good, but $C \ge 10$ typically usually okay too). For the multiclass-internal SVM, on the other hand, the speedup with choice of C can be more significant, as shown in what follows, Figures 10.28 and 10.29, where $C \ge 100$ typically is needed. Figures 10.28 shows the percent increase in iterations-to-convergence against the "C" value. Figure 10.29 shows the number of bounded support vectors (BSV) as a function of "C" value.

Since the algorithm presented in [177] does not differentiate between SV and BSV, a lot of time is spent in trying to adjust the weights of the BSV i.e. weak data. The weight of a BSV may range from 0 to 0.5 in their algorithm. In our modification to the algorithm, shown below, as soon as we identify the BSV (as specified by Case III conditions), its weight is no longer adjusted. Hence faster convergence is achieved without sacrificing accuracy.

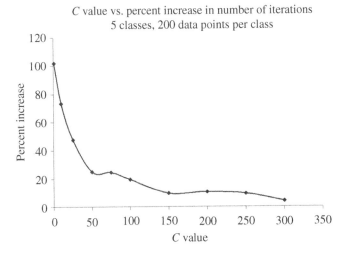

Figure 10.28 The percent increase in iterations-to-convergence against the "*C*" value. For very low values of "*C*" the gain is doubled while for very large values of "*C*" the gain is low (almost constant for *C* > 150). Thus we note the dependence of the gain on "*C*" value.

Figure 10.29 The number of bounded support vectors (BSV) as a function of "*C*" value. There are many BSVs for very low values of "*C*" and very few BSVs for large values of "*C*." Thus we can say that the number of BSVs plays a vital role in the speed of convergence of the algorithm.

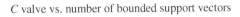

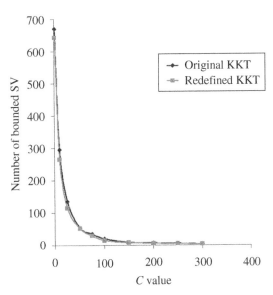

For the BSV/SV-tracking speedup, the KKT violators are redefined as:
For all $m \neq y_i$ we have:
$$\alpha_i^m \{f_{yi} - f_m - 1 + \zeta_i\} \geq 0$$
Subject to: $1 \geq \alpha_i^m \geq 0$; $\sum_m \alpha_i^m = 1$; $\zeta_i \geq 0$ for all i, m
Where $f_m = (1/\beta)[w_m \cdot x_i + b_m]$ for all m

Case I:
If $\alpha_i^m = 0$ for m S.T $f_m = f_m^{\max}$
Implies $\alpha_i^{yi} > 0$ and hence $\zeta_i = 0$
Hence $f_{yi} - f_m^{\max} - 1 \geq 0$

Case II:
If $1 > \alpha_i^m > 0$ for m S.T $f_m = f_m^{\max}$ and $\alpha_i^{yi} > \alpha_i^m$
Implies $\zeta_i = 0$
Hence $f_{yi} - f_m^{\max} - 1 = 0$

Case III:
If $1 \geq \alpha_i^m > 0$ for m S.T $f_m = f_m^{\max}$ and $\alpha_i^{yi} \leq \alpha_i^m$
Implies $\zeta_i > 0$
Hence $f_{yi} - f_m^{\max} - 1 + \zeta_i = 0$
Or $f_{yi} - f_m^{\max} - 1 < 0$

10.8 SVM Clustering

The goal of clustering analysis is to partition objects into groups, such that members of each group are more "similar" to each other than the members of other groups. Similarity, however, is determined subjectively as it does not have a universally agreed upon definition. In [174] the author suggests a formal perspective on the difficulty in finding such a unification, in the form of an impossibility theorem: for a set of three simple properties described in [174], there is no clustering function satisfying all three. Furthermore, the author demonstrates that relaxations of these properties expose some of the interesting (and unavoidable) trade-offs at work in well-studied clustering techniques such as single-linkage, sum-of-pairs, K-means, and k-median.

Ideally, one would like to solve the clustering problem given all the known and unknown objective functions. This is provably a NP-Hard problem [218]. This brings us to the work presented here, which seeks to provide a new perspective on clustering by introducing an algorithm that does not require an objective function. Hence, it does not inherit the limitations of an embedded objective function. We propose an algorithm that is capable of suggesting solutions that can be later evaluated using a variety of cluster validators.

10.8.1 SVM-External Clustering

SVMs provide a powerful method for supervised learning. Use of SVMs for clustering (unsupervised learning) is also possible in a number of different ways. As with the multiclass SVM discriminator generalizations, the strong performance of the binary SVM enables SVM-External as well as SVM-Internal approaches to clustering. Nonparametric SVM-based clustering methods may allow for much improved performance over parametric approaches, particularly since they can apparently be designed to inherit the strengths of their supervised SVM counterparts as will be shown. Our external-SVM clustering algorithm clusters data vectors with no *a priori* knowledge of each vector's class (Figure 10.30).

10.8.1.1 Single-Convergence Initialized SVM-Clustering: Exploration on Sensitivity to Tuning

The algorithmic variant works by first running a Binary SVM against a data set, with each vector in the set randomly labeled (usually half positives and half

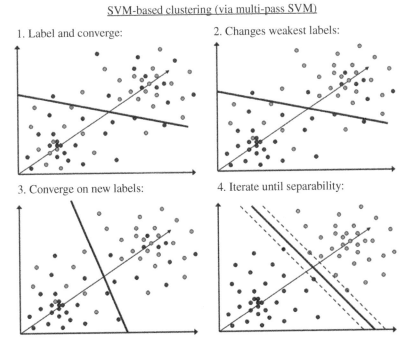

Figure 10.30 (enlarged version of Figure 10.7) SVM-external clustering method that uses label flipping with decision making based on the SVM's confidence parameter on classifications.

negatives), until the SVM converges (Figure 10.30). Choice of an appropriate kernel and an acceptable sigma value will affect convergence. After the *initial* convergence is achieved, the (sensitivity + specificity) will be low. The algorithm now improves this result by iteratively relabeling the worst misclassified vectors, which have confidence factor values beyond some threshold, followed by rerunning the SVM on the newly relabeled data set. This continues until no more progress can be made. Progress is determined by an increasing value of (sensitivity + specificity). With sub-cluster identification upon iterating the overall algorithm on the positive and negative clusters identified (until the clusters are no longer separable into sub-clusters), this method provides a way to cluster data sets without prior knowledge of the data's clustering characteristics, or the number of clusters (by iteration on clusters in the binary SVM or direct with merge of multiclass SVM with label flipping algorithms to directly model cluster number without embedded recursion).

Figures 10.31 and 10.32 show clustering runs on a dataset with a mixture of 8GC and 9GC DNA hairpin data. The set consists of 400 elements. Half of the elements belong to each class. The SVM uses a Gaussian Kernel and allows 3% KKT Violators. The algorithmic variant shown in Figure 10.32 works by first running a Binary SVM against a data set, with each vector in the set randomly labeled, until the SVM converges. In order to obtain convergence, an acceptable number of KKT violators must be found. This is done through running the SVM on the randomly labeled data with different numbers of allowed violators until the number of violators allowed is near the lower bound of violators needed for the SVM to converge on the particular data set. In practice, the initialization step, that arrives at the first SVM convergence, typically takes longer than all subsequent partial relabeling and SVM rerunning steps.

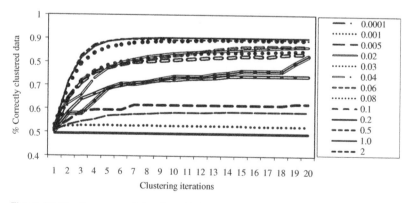

Figure 10.31 Summary of the degradation in clustering performance for less optimal selection of kernel and tuning parameter – with averages of the five test-runs are used as representative curves for that kernel/tuning selection in the above.

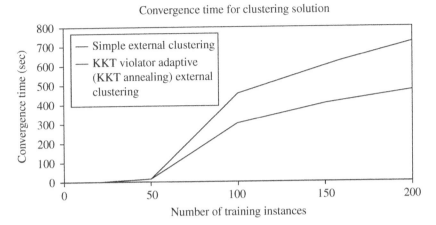

Convergence time for clustering solution

Figure 10.32 Efforts to use simulated annealing in the number of KKT violators tolerated on each iteration of the external clustering algorithm, to accelerate the convergence (clustering) process. In results shown, cluster time is approximately halved.

In Figure 10.33 the decision hyperplane is circular in the feature space. The clustering kernel used in Figure 10.33 was the polynomial kernel (the linear kernel failed in this case).

10.8.1.2 Single-Convergence SVM-Clustering: Hybrid Clustering

Although convergence is always achieved with the SVM-clustering method in the label-flippings, after the initial convergence, convergence to a *global* optimum is not guaranteed. Figure 10.34a and b shows the purity and entropy (with the RBF kernel) as a function of Number of Iterations, while Figure 10.34c, on the right, shows the SSE as a function of Number of Iterations. The stopping criteria used for the algorithm is based on the unsupervised (external) SSE measure. Comparison to fuzzy c-means and kernel K-means is shown on the same dataset (the solid lines in Figure 10.34a and b).

The ability to get stuck in local minima motivates the introduction of perturbations into the methods, as shown in Figures 10.35 and 10.36 that follow.

It is found that the result of the Re-labeler algorithm can be significantly improved by randomly perturbing a weak clustering solution and repeating the SVM-external label-swapping iterations as depicted in Figure 10.36. To explore this further, a hybrid SVM-external approach to the above problem is introduced to replace the initial random labeling step with K-means clustering or some other fast clustering algorithm. The initial SVM-external clustering must then be slightly and randomly perturbed to properly initialize the relabeling step;

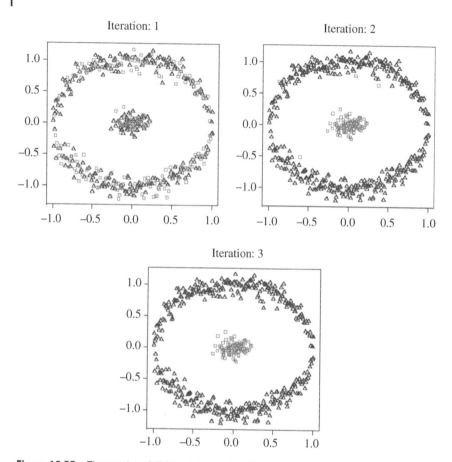

Figure 10.33 The results of SVM-relabeler algorithm using a third degree polynomial kernel.

otherwise the SVM clustering tends to return to the original K-means clustering solution. A complication is the unknown amount of perturbation of the K-means solution that is needed to initialize the SVM-clustering well – it is generally found that a weak clustering method does best for the initialization (or one weakened by a sufficient amount of perturbation).

10.8.2 Single-Convergence SVM-Clustering: Comparative Analysis

The single-convergence initialized SVM-based clustering algorithm begins by first running a binary SVM classifier against a data set with each vector in the set randomly labeled, this is repeated until an *initial convergence* occurs. The convergence

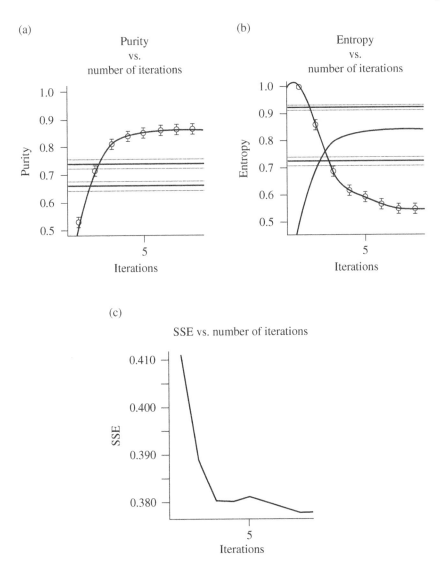

Figure 10.34 SVM-external clustering results. (a) and (b) show the boost in purity and entropy as a function of number of iterations of the SVM clustering algorithm. (c) Shows that SSE, as an unsupervised measure, provides a good indicator in that improvements in SSE correlate strongly with improvements in purity and entropy. The blue and black lines are the result of running fuzzy *c*-mean and kernel *k*-mean (respectively) on the same dataset.

(a)

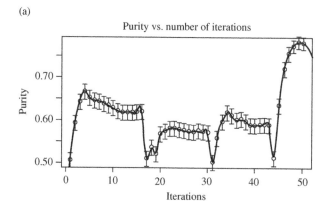

(b)

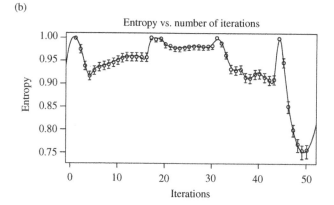

(c)

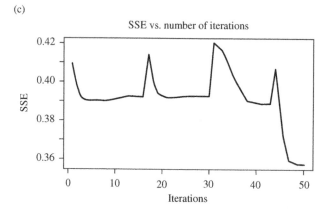

Figure 10.35 The result of relabeler algorithm with perturbation. The top plots demonstrate the various purity and entropy scores for each perturbed run. The spikes are drops followed by recovery in the validity of the clusters as a result of random perturbation. The bottom plot is a similar demonstration, by tracking the unsupervised quality of the clusters. Note that after four runs of perturbation best solution is recovered.

(a)

(b)

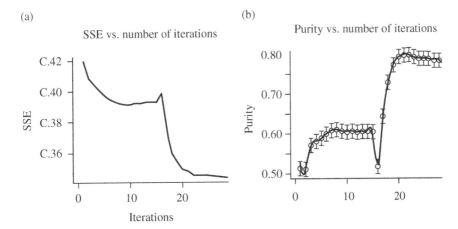

Figure 10.36 (a, b) represent the SSE and purity evaluation of hybrid Re-labeler with perturbation on the same dataset. Data is initially clustered using K-means to initialize the relabeler algorithm. The first segment of the plot (right before the spike at 16) is the result of relabeler after 10% perturbation, while the second segment is the result after 30% perturbation. Purity number of iterations.

sometimes has to be attempted several times (with different randomized initializations) before a SVM solution is obtained. Once an SVM solution is obtained, however, the strengths of the SVM classifier can be used to full advantage. SVMs are ideal in this effort as they not only classify, but offer a confidence parameter with their classification, and can do so in a generalized kernel space. Once a convergent solution is obtained label-flipping (from positive to negative) can be done for low-confidence labels in an iterative process, with SVM re-training after each round of weak-label changes. At each iteration we can potentially have unequal numbers of positives and negatives changing their labels, thus, asymmetrically sized clusters can be realized from a half-positive/half-negative initialization. This iterative process continues until there is no longer a low-confidence classification by the SVM, or until an external cluster validation, such as the SSE on each cluster, remains relatively unchanged. There are numerous tuning parameters in the SVM-classification process itself, as well as in the SVM-clustering halting specification, and even tuning choices in the SVM chunk-training (that may be necessary for larger data sets). As shown in Figure 10.37, SVM-based clustering often outperforms other methods.

The problem with the single-convergence initiated SVM clustering approach is that it can get stuck in a weak solution or occasionally fail more seriously. Stabilization could be done with numerous repeats of the SVM clustering process, but this is computationally over-kill and more efficient processes, including

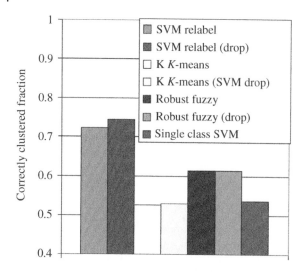

Figure 10.37 Clustering performance comparisons: SVM-external clustering compared with explicit objective function clustering methods. Nanopore detector blockade signal clustering resolution from a study of blockades due to individual molecular capture-events with 9AT and 9CG DNA hairpin molecules. The SVM-external clustering method consistently out-performs the other methods. The optimal drop percentage on weakly classified data differed for the different methods for the scores shown: Our SVM relabel clustering with drop: 14.8%; Kernel K-means with drop: 19.8%; Robust fuzzy with drop: 0% (no benefit); Vapnik's single-class SVM (internal) clustering: 36.1%.

distributed intelligence tuning (with GAs, for example) are sought in the label-flipping convergence process, and initializing in a more informed way, with more than one initial convergence required.

10.8.2.1 SVM "Internal" Clustering

The SVM-Internal approach to clustering was originally defined by [166]. Data points are mapped by means of a kernel to a high dimensional feature space where we search for the minimal enclosing sphere. The minimal enclosing sphere, when mapped back into the data space, can separate into several components; each enclosing a separate cluster of points. The width of the kernel (say Gaussian) controls the scale at which the data is probed while the soft margin constant helps to handle outliers and over-lapping clusters. The structure of a dataset is explored by varying these two parameters, maintaining a minimal number of SV to assure smooth cluster boundaries.

We have used the algorithm defined in [185] to identify the clusters, with methods adapted from [186, 187] for their handling. If the number of data points is "n", then we require $n(n-1)/2$ number of comparisons. We have made modifications to the algorithm such that we eliminate comparisons that do not have an impact on the cluster connectivity. Hence the number of comparisons required will be less than $n(n-1)/2$.

In each comparison we sub-divide the line segment connecting the two data points into 20 parts; hence we obtain 19 different points on this line segment. The two data points belong to the same cluster only if all the 19 points lie inside the cluster. Given the cost of evaluating utmost 19 points for every comparison, the need to eliminate comparisons that do not have an impact on the cluster connectivity becomes even more important. Finally we have used Depth First Search (DFS) algorithm for the cluster harvest. The approach to the solving the Dual problem is shown below.

Let $\{x_i\}$ be a data set of "N" points in R^d. Using a nonlinear transformation φ, we transform "x" to some high-dimensional Kernel space and look for the smallest enclosing sphere of radius "R". Hence we have: $||\varphi(x_j) - a||^2 \leq R^2$ for all $j = 1, ...,$ N; where "a" is the center of the sphere. Soft constraints are incorporated by adding slack variables "ζ_j":

$$||\varphi(x_j) - a||^2 \leq R^2 + \zeta_j \text{ for all } j = 1, ..., N; \zeta_j \geq 0$$

We introduce the Lagrangian as:

$$L = R^2 - \sum_j \beta_j \left(R^2 + \zeta_j - ||\varphi(x_j) - a||^2\right) - \sum_j \zeta_j \mu_j + C \sum_j \zeta_j; \beta_j \geq 0, \mu_j \geq 0$$

where C is the cost for outliers and hence $C\sum_j \zeta_j$ is a penalty term. Setting to zero the derivative of "L" w.r.t. R, a and ζ we have: $\sum_j \beta_j = 1$; $a = \sum_j \beta_j \varphi(x_j)$; and $\beta_j = C - \mu_j$.

Substituting the above equations into the Lagrangian, we have the dual formalism as:

$$W = 1 - \sum_{i,j} \beta_i \beta_j K_{ij} \text{ where } 0 \leq \beta_i \leq C;$$

with $K_{ij} = \exp(-||x_i - x_j||^2/2\sigma^2)$ typically used. Subject to: $\sum_i \beta_i = 1$

By KKT conditions we have: $\zeta_j \mu_j = 0$ and $\beta_j(R^2 + \zeta_j - ||\varphi(x_j) - a||^2) = 0$.

In the kernel space of a data point "x_j" if $\zeta_j > 0$, then $\beta_j = C$ and hence it lies outside of the sphere i.e. $R^2 < ||\varphi(x_j) - a||^2$. This point becomes a bounded support vector or BSV. Similarly if $\zeta_j = 0$, and $0 < \beta_j < C$, then it lies on the surface of the sphere i.e. $R^2 = ||\varphi(x_j) - a||^2$. This point becomes a support vector or SV. If $\zeta_j = 0$, and $\beta_j = 0$, then $R^2 > ||\varphi(x_j) - a||^2$ and hence this point is enclosed with-in the sphere.

10.8.2.2 Solving the Dual (Based on Keerthi's SMO [184])

The dual formalism is: $1 - \sum_{i,j} \beta_i \beta_j K_{ij}$ where $0 \leq \beta_i \leq C$; $K_{ij} = \exp(-||x_i - x_j||^2/2\sigma^2)$ is used, also $\sum_i \beta_i = 1$. For any data point "x_k", the distance of its image in kernel space from the center of the sphere is given by: $R^2(x_k) = 1 - 2\sum_i \beta_i K_{ik} + \sum_{i,j} \beta_i \beta_j K_{ij}$. The radius of the sphere is $R = \{R(x_k) \mid x_k \text{ is a Support Vectors}\}$, hence data points which are SV lie on cluster boundaries. Outliers are points that lie outside of the sphere and therefore they do not belong to any cluster i.e. they are BSV. All other points are enclosed by the sphere and therefore they lie inside their respective cluster. KKT Violators are given as: (i) If $0 < \beta_i < C$ and $R(x_i) \neq R$; (ii) If $\beta_i = 0$ and $R(x_i) > R$; and (iii) If $\beta_i = C$ and $R(x_i) < R$.

The Wolfe dual is: $f(\beta) = \text{Min}_\beta\{\sum_{i,j} \beta_i \beta_j K_{ij} - 1\}$. In the SMO decomposition, in each iteration we select β_i and β_j and change them such that $f(\beta)$ reduces. All other β's are kept constant for that iteration. Let us denote β_1 and β_2 as being modified in the current iteration. Also $\beta_1 + \beta_2 = (1 - \sum_{i=3} \beta_i) = s$, a constant. Let $\sum_{i=3} \beta_i K_{ik} = C_k$, then we obtain the SMO form: $f(\beta_1, \beta_2) = \beta^2_1 + \beta^2_2 + \sum_{i,j=3} \beta_i \beta_j K_{ij} + 2\beta_1 \beta_2 K_{12} + 2\beta_1 C_1 + 2\beta_2 C_2$. Eliminating $\beta_1 : f(\beta_2) = (s - \beta_2)^2 + \beta^2_2 + \sum_{i,j=3} \beta_i \beta_j K_{ij} + 2(s - \beta_2)\beta_2 K_{12} + 2(s - \beta_2)C_1 + 2\beta_2 C_2$. To minimize $f(\beta_2)$, we take the first derivative w.r.t. β_2 and equate it to zero, thus $f(\beta_2) = 0 = 2\beta_2(1 - K_{12}) - s(1 - K_{12}) - (C_1 - C_2)$, and we get the update rule: $\beta_2^{\text{new}} = [(C_1 - C_2)/2(1 - K_{12})] + s/2$. We also have an expression for "$C_1 - C_2$" from: $R(x_1^2) - R(x_2^2) = 2(\beta_2 - \beta_1)(1 - K_{12}) - 2(C_1 - C_2)$, thus $C_1 - C_2 = [R(x_2^2) - R(x_1^2)]/2 + (\beta_2 - \beta_1)(1 - K_{12})$, substituting, we have:

$$\beta_1^{\text{new}} = \beta_1^{\text{old}} - [R(x_2^2) - R(x_1^2)]/[4(1 - K_{12})]$$

10.8.2.3 Keerthi Algorithm

Compute "C": if percent outliers $= n$ and number data points $= N$, then: $C = 100/(N * n)$

Initialize β: Initialize $m = \text{int}(1/C) - 1$ number of randomly chosen indices to "C"

Initialize two different randomly chosen indices to values less than "C" such that $\sum_i \beta_i = 1$.

Compute $R^2(x_i)$ for all "i" based on the current value of β. Divide data into three sets: Set I if $0 < \beta_i < C$; Set II if $\beta_i = 0$; and Set III if $\beta_i = C$. Compute $R^2\_\text{low} = \text{Max} \{ R^2(x_i) \mid 0 \leq \beta_i < C\}$ and $R^2\_\text{up} = \text{Min} \{ R^2(x_i) \mid 0 < \beta_i \leq C\}$.

In every iteration, execute the following two paths alternatively until there are no KKT violators:

1) Loop through all examples (call examine example subroutine)
2) Keep count of number of KKT Violators.
3) Loop through examples belonging only to Set I (call Examine Example subroutine) until $R^2\_\text{low} - R^2\_\text{up} < 2 * \text{tol}$.

Examine example subroutine

a) Check for KKT Violation. An example is a KKT violator if:
b) Set II and $R^2(x_i) > R^2\_\text{up}$; choose R^2_up for joint optimization

c) Set III and $R^2(x_i) < R^2$_low; choose R^2_low for joint optimization

d) Set I and $R^2(x_i) > R^2$_up + 2 * tol OR $R^2(x_i) < R^2$_low − 2 * tol; choose R^2_low or R^2_up for joint optimization depending on which gives a worse KKT violator

e) Call the Joint Optimization subroutine

Joint optimization subroutine

a) Compute $\eta = 4(1 - K_{12})$ where K_{12} is the kernel evaluation of the pair chosen in Examine Example

b) Compute $D = [R^2(x_2) - R^2(x_1)]/\eta$

c) Compute Min$\{(C - \beta_2), \beta_1\} = L1$

d) Compute Min$\{(C - \beta_1), \beta_2\} = L2$

e) If $D > 0$; then $D =$ Min $\{D, L1\}$

f) Else $D =$ Max$\{D, -L2\}$

g) Update β_2 as: $\beta_2 = \beta_2 + D$

h) Update β_1 as: $\beta_1 = \beta_1 - D$

i) Recompute $R^2(x_i)$ for all "i" based on the changes in β_1 and β_2

j) Recompute R^2_low and R^2_up based on elements in Set I, $R^2(x_1)$ and $R^2(x_2)$

10.8.3 Stabilized, Single-Convergence Initialized, SVM-External Clustering

The External SVM Clustering data set is chosen to be an equal positives vs. negatives sample of 200 8GC blockade signals and 200 9GC blockade signals (see [1, 3, 59, 67] for details about these molecules). Each feature vector is 150 dimensional and normalized to satisfy the L_1 (norm = 1) constraint. Features from the eight and nine base-pair blockade signals were extracted using HMMs (for details, see [1, 3, 59, 67]). Although convergence was easily achieved with the External SVM Clustering algorithm (see the Methods), convergence to a global optimum was not guaranteed.

In [54], we see that a small value of Kernel-SSE (herein referred to as SSE) is shown to provide us with a reliable cluster validation measure. The Exteranl SVM Clustering (SVM-Relabeler) algorithm does not use an objective function, and the hope is that by running the algorithm in its purest form, the resulting clusters are reliable solutions. However, running this algorithm in this basic fashion does not consistently provide us with a satisfying clustering solution. In fact, the solution space can be divided into three sets: successful, local-optimum, and unsuccessful. Unsuccessful solutions and local optima solutions are undesirable and the objective is to find a method to eliminate their usage by simply re-clustering for objectively improved clustering (via SSE scoring, for example). Since, the solutions in the unsuccessful set are expected to be easily identified in any experiment that calculates the SSE of a randomly labeled dataset, they can be

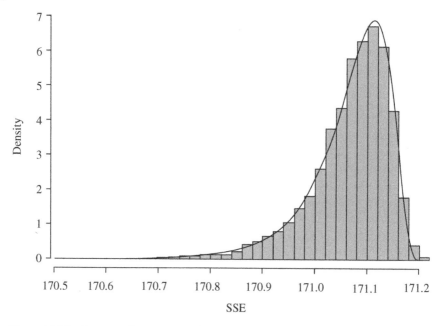

Figure 10.38 Nanopore feature vector data (in standard 150 component, L_1-norm, format) is randomly labeled 5000 times followed by evaluation of SSE values and production of a histogram of those values as shown. The resulting distribution has a good fit to Johnson's SB distribution with gamma = −5.5405, delta = 1.8197, lambda = 2.7483, epsilon = 168.46.

simply eliminated by post-processing. In a control experiment we have randomly labeled the dataset 5000 times and calculated the SSE distribution for the experiment. The resulting distribution has a good fit to Johnson's SB distribution and is illustrated in the histogram of Figure 10.38. Using a fitted distribution one can calculate the *p*-value of a given SSE. For a SSE threshold of 170.5 (accidentally very unlikely) we can directly eliminate the unsuccessful set.

To substantially reduce the local optimum solutions, however, thresholding does not scale well. One solution is to use a simple hill climbing algorithm which is to run the algorithm for a sufficiently long number of iterations to find the solution with the lowest SSE value. To do this, the clustering algorithm is run repeatedly and randomly initialized every time. A solution is accepted as the best solution if it has a lower SSE than the previously recorded value. This can be a very slow learning process, and is a familiar scenario in statistical learning, and one of the popular solutions in those situations works well here as well – simulated annealing.

It is observed that random perturbation by flipping each label at some probability, p_{pert}, is often sufficient to switch to another subspace where a better solution

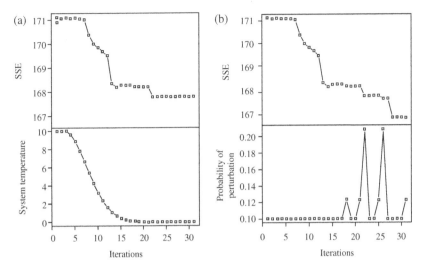

Figure 10.39 (a) Simulated annealing with constant perturbation, (b) simulated annealing with variable perturbation. As shown in left, top panel, simulated annealing with a 10% initial label-flipping results in a local-optimum solution. In the right panel this is avoided by boosting the perturbation function depending on the number of iterations of unchanged SSE (b, top panel). These results were produced using an exponential cooling function, $T_{k+1} = \beta^k T_k$, with $\beta = 0.96$ and $T_0 = 10$.

could be found. (Note that $p_{pert} = 0.50$ has the effect of random reinitialization and $p_{pert} = 1$ flips the entire labels.) The hope is that perturbation with $p_{pert} \leq 0.50$ results in a faster convergence. Reliability can be achieved by searching through the solution space. To do this efficiently, Monte Carlo Methods could be used by taking advantage of perturbation to evaluate the neighboring configuration. The procedure described next uses a modified version of Simulated Annealing to achieve this desired reliability.

As shown in Figure 10.39 left, top panel, constant perturbation with $p_{pert} = 0.10$ results in a local-optimum solution that could be otherwise avoided by using a perturbation function depending on the number of iterations of unchanged SSE (Figure 10.39 right, top panel). These results were produced using an exponential cooling function, $T_{k+1} = \beta^k T_k$, with $\beta = 0.96$ and $T_0 = 10$. The initial temperature, T_0 should be large enough to be comparable with the change of SSE, ΔSSE, and therefore increase the randomness by making the Boltzman factor $e^{-\Delta SSE/T} \approx e^0$, while $\beta(<1)$ should be large enough to speed up the cooling effect.

In the effort shown in Figures 10.39 and 10.40, it was found that random perturbation and hybridized methods (with more traditional clustering methods) could help stabilize the clustering method, but often at significant cost to its performance edge over other clustering methods (apparently due to getting stuck in local

(a)

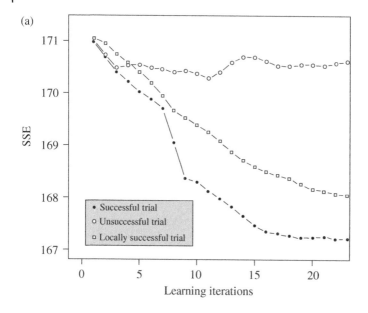

(b)

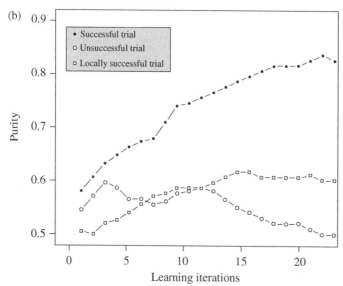

Figure 10.40 Multiple-convergence, SVM-external clustering. Three multiple clustering convergences (different trials) of SVM-Relabeler algorithm demonstrating the range of the possible solution space as measured by SSE and purity (a "successful", an unsuccessful, and a partly-successful trial). Choosing the good SSE external measure (a) typically provides a generalized clustering that has high purity (b). The improvements in purity and SSE with learning iterations are shown. Once learning slows (SSE unimproved), a restart for a new convergence clustering is done. Usually in the first two or three attempts a strongly performing convergence is seen as with the example shown here.

minima traps to which the other parametric clustering methods are susceptible). The "pure" SVM-external clustering method appears to offer very strong solutions about half the time – which allows for optimization simply by repeated clustering attempts and looking for the most tightly clustered (smallest SSE) solution. This suggested a simulated annealing approach for greater computational efficiency, as shown in Figure 10.39 (recent work with GAs, not shown, exhibit even stronger stability). Results of this effort (Figure 10.39) significantly improve and stabilize the SVM clustering process.

Given the wide variety of dissimilar tuning parameters in the SVM classification process alone, tests on SVM classification with GA-based tuning seems optimal. The very robust and rapid auto-tuning with the GA approach on SVM classification in initial tests strongly suggests that this, or any swarm intelligence search/tuning paradigms, offer important refinement to the SVM-classification efforts and critical refinement to the single-convergence initialization SVM-clustering efforts.

10.8.4 Stabilized, Multiple-Convergence, SVM-External Clustering

Re-examining the same dataset using the Absdiff kernel (gamma = 1.8) and computing SSE, but now iterating with multiple convergences instead of perturbations.

SVM methods are described for classification, clustering, as well as aiding with signal analysis and pattern recognition on stochastic sequential data. Analysis tools for stochastic sequential data, Markovian (or causal) data for example, have broad-ranging application in that almost any device producing a sequence of measurements can be made more sensitive, or "smarter," by efficient learning of measured signal/pattern characteristics via HMM/SVM methods. The SVM and HMM/SVM application areas described here include cheminformatics, biophysics, and bioinformatics. The cheminformatics application examples pertain to channel current analysis on the alpha-hemolysin nanopore detector.

Markov-based statistical profiles, in a log likelihood discriminator framework, can be used to create a fixed-length feature vector for SVM based classification [1, 3]. Part of the idea of the method is that whenever a log likelihood discriminator can be constructed for classification on stochastic sequential data, an alternative discriminator can be constructed by "lifting" the log likelihood components into a feature vector description for classification by SVM. Thus, the feature vector uses the individual log likelihood components obtained in the standard log likelihood classification effort, the individual-observation log odds ratios, and "vectorizes" them rather than sums them. The individual-observation log odds ratios are themselves constructed from positionally defined Markov Models (pMM's), so what results is a pMM/SVM sensor method. This method may have utility in a number of areas of SSA, including splice-site recognition and other types of gene-structure identification, file recovery in computer forensics ("file carving"), and speech recognition.

The multiple-convergence initialized SVM-clustering approach to unsupervised learning provides another nonparametric means to clustering. In preliminary work, it is found that the SVM-based clustering method also offers prospects for inheriting the very strong performance of standard SVMs from the supervised classification setting. This offers a remarkable prospect for knowledge discovery and enhancing the scope of human cognition – the recognition of patterns and clusters without the limitations imposed by assuming a parametric mode and "fitting" to it, where resolution of the identified clusters can be at an accuracy comparable to the supervised setting (i.e. where cluster identities are already specified).

The new approach is to first obtain *multiple* SVM convergences from separate initializations (two might suffice, for example, in many situations) and thereby obtain the confidence magnitudes on data points, and their nearest neighbors (if repeatedly have the same neighbors, have high linkage to them). This is used to inform a label-flipping process to arrive at an improved clustering solution on further iterations and analysis. For example, one approach is to establish a high-linkage high-confidence label set (labels retained or flipped accordingly) and a low-linkage, low-confidence, label set (some, according to criteria, may be flipped as well, or dropped). The magnitude comparison in the simplest "multiple" convergence result would involve two convergences, with the difference in confidence value for a particular training instance producing a line segment, and for all the training instances, their two-convergent point-differences would provide a collection of line-segments. The most stable part of the line-segment "field", that of the high-linkage high-confidence data instances, can then be used, for example, to provide indication of structure to guide tuning efforts and label-flipping criteria.

SVMs are fast, easily trained, discriminators, for which strong discrimination is possible without over-fitting complications. SVMs are firmly grounded as variational-calculus based optimization methods that are constrained to have SRM, unlike neural net classifiers, such that they provide noise tolerant solutions for pattern recognition. An SVM determines a hyperplane that optimally separates one class from another, while the SRM criterion manifests as the hyperplane having a thickness, or "margin," that is made as large as possible in the process of seeking a separating hyperplane. The SVM approach thereby encapsulates model fitting and discriminatory information in the choice of kernel in the SVM, and a number of novel kernels have been shown here. SVMs are good at both classifying data and evaluating a confidence in the classifications given, which leaves an opening for use of metaheuristics to bootstrap into a clustering capability, as explored in a number of algorithmic variations in this paper. SVM's use in clustering appears to be a very robust platform and, from initial results shown here, promises to be one of the best clustering approaches. In this chapter, we have shown a number of SVM classification algorithms and new SVM/metaheuristics bootstrap algorithms for clustering.

10.8.5 SVM-External Clustering – Algorithmic Variants

10.8.5.1 Multiple-Convergence Initialized (Steepest Ascent) SVM-Clustering

The Multiple-convergence initialized SVM-clustering approach to unsupervised learning provides another nonparametric means to clustering. In preliminary work we have found that the SVM-based clustering method also offers prospects for inheriting the very strong performance of standard SVMs from the supervised classification setting. This offers a remarkable prospect for knowledge discovery and enhancing the scope of human cognition – the recognition of patterns and clusters without the limitations imposed by assuming a parametric mode and "fitting" to it, where resolution of the identified clusters can be at an accuracy comparable to the supervised setting (i.e. where cluster identities are already specified).

The new approach is to first obtain *multiple* SVM convergences at initialization (two might suffice, for example, in many situations) and thereby obtain the confidence magnitudes on data points, and their nearest neighbors (if repeatedly have the same neighbors, have high linkage to them). This is used to inform a label-flipping process to arrive at an improved clustering solution on further iterations and analysis. For example, one approach is to establish a high-linkage high-confidence label set (labels retained or flipped accordingly) and a low-linkage, low-confidence, label set (some, according to criteria, may be flipped as well, or dropped). The magnitude comparison in the simplest "multiple" convergence result would involve two convergences, with the difference in confidence value for a particular training instance producing a line segment, and for all the training instances, their two-convergent point-differences would provide a collection of line-segments. The most stable part of the line-segment "field", that of the high-linkage high-confidence data instances, can then be used, for example, to provide indication of structure to guide tuning efforts and label-flipping criteria.

10.8.5.2 Projection Clustering – Clustering in Decision Space

SVM methods for clustering are described that are based on the SVM's ability to not only classify, but also to give a confidence parameter on its classifications. Even without modifying the label information (passive clustering), there is often strong clustering information in an SVM training solution. One such instance occurs when one set, the positives, are a known signal species (or collection of species). If you have mixture data with known and unknown signal species, and wish to identify (i.e. cluster) the unknown species, then an SVM training attempt with the mixture taken as negatives leads to a cluster identification method via an

SVM "projection-score" histogram. (i.e. cluster partitioning in Decision Space). Real channel blockade data has been examined in this way, biotinylated DNA hairpin blockades comprised the positives, and scored as a sharp peak at around 1.0. The mixture signals seen after introduction of streptavidin cluster with scores around 0.5, corresponding to (unbound) biotinylated DNA hairpin signals, and signals that score <-1.0, corresponding to the streptavidin-bound biotinylated DNA hairpins. SVM projection clustering can be a very powerful clustering tool in and of itself as can be seen in this cheminformatics application.

10.8.5.3 SVM-ABC

New subtleties of classification-separation are possible with SVMs via their direct handling and direct identification of data instances. Individual data points, in some instances, can be associated with "SV" at the boundaries between regions. By operating on labels of SV and focusing on training on certain subsets, the SVM-ABC algorithm offers the prospect to delineate highly complex geometries and graph-connectivity:

Split the clustering data into sets A, B and C

- A: Strong negatives
- C: Strong positives
- B: Weak negatives and weak positives
- Train an SVM on Data from A (labeled negative) and B (labeled positive)
- The SV: SV_AB
- Similarly train a new SVM on Data from C (labeled positive) and B (labeled negative)
- The SV: SV_CB
- Our objective is that the SV_AB and SV_CB sets have their labels flipped to be set A and C
- Regrow set A and C into the weak "B" region.
- If an element of SV_AB is also in SV_BC, then the intersection of these sets are the elements that should be flipped to class B (if not already listed as class B).
- Stop at the first occurrence of any of these events
- Set B becomes empty
- Set B does not change

The SVM_ABC Algorithm may offer recovery of subtle graph-like connectedness between cluster elements, a weakness of manifold-like separability approaches such as parametric-based clustering methods.

10.8.5.4 SVM-Relabeler

The SVM classification formulation is used as the foundation for clustering a set of feature vectors with no *a priori* knowledge of the feature vector's classification.

The non-separable SVM solution guarantees convergence at the cost of allowing misclassification. The extent of slack is controlled through the regularization constant, C, to penalize the slack variable, ξ. If the random mapping $((x_1, y_1), ..., (x_m, y_m)) \in X^m \times y$ is not linearly separable when ran through a binary SVM, the misclassified features are more likely to belong to the other cluster. Moreover, by relabeling those heavily misclassified features and by repeating this process we arrive at an optimal separation between the two clusters. The basics of this procedure is presented in Algorithm 1, where $\hat{y}$ is the new cluster assignment for x and θ contains ω, α, y'.

Algorithm 1: SVM-Relabeler

Require: m, x

1) $\hat{y} \leftarrow$ Randomly chosen from $\{-1, +1\}$
2) repeat
3) $\theta \leftarrow doSVM(x, \hat{y})$
4) $\hat{y} \leftarrow doRelabel(x, \theta)$
5) until $\hat{y}$ remains constant

The $doSVM()$ procedure can be any standard and complete implementation of an SVM classifier with support for nonlinear discriminator function. The idea is that $doSVM()$ has to converge regardless of the geometry of the data, in order to provide the $doRelabel()$ procedure with the hyperplane and other standard SVM outputs. After this procedure, $doRelabel()$ reassigns some (or all) of the misclassified features to the other cluster. If $D(x_i, \theta)$ is the distance between x_i feature and the trained SVM hyperplane, then heavily misclassified feature, $x_{j \in J}$ could be selected by comparing $D(x_j, \theta)$ to $D(x_{j'}, \theta)$ for all $j' \in J$. Algorithm 2 clarifies the basic implementation of this procedure.

Algorithm 2: doRelabel() Procedure

Require: Input vector: x

Cluster labeling: $\hat{y}$

SVM model: θ

Confidence Factor: α Identify misclassified features:

$x'^+ \leftarrow K$ misclassified features with $\hat{y} = +1$

$x'^- \leftarrow L$ misclassified features with $\hat{y} = -1$

1) for all ith component of x'^+ do
2) if $i/K \sum_{j=1}^{K} D(x'^+_j, \theta) < \alpha D(x'^+_j, \theta)$ then
3) $\hat{y}_i^+ \leftarrow -1$
4) end if
5) end for
6) for all ith component of x'^- do
7) if $1/L \sum_{j=1}^{L} D(x'^-_j, \theta) < \alpha D(x'^-_j, \theta)$ then

8) $\hat{y}_i^- \leftarrow +1$
9) end if
10) end for

10.8.5.5 SV-Dropper

In most applications of clustering, the dataset is composed of *leverage* and *influential* points. *Leverage* points are subsets of the dataset that are highly deviated from the rest of the cluster, and removing them does *not* significantly change the result of the clustering. In contrast, *influential* points are those in the highly deviated subset whose inclusion or removal significantly changes the decision of the clustering algorithm. Effective, identification of these special points is of interest to improve accuracy and correctness of the clustering algorithm. A systematic way to manage these deviants is given by the SV-Dropper algorithm.

As depicted in Algorithm 3, SVM is initially trained on the clustered data; the weakest of the cluster data – those closest to the hyperplane, i.e. the SV – are dropped thereafter. This process is repeated until the desired ratio of accuracy and number of data dropped is achieved.

Algorithm 3: SV-Dropper Algorithm
Require: Input vector: x
Cluster labeling: $\hat{y}$
let:
$x^+ \leftarrow K$ features with $y = +1$
$x^- \leftarrow L$ fatures with $\hat{y} = -1$

1) repeat
2) $\theta \leftarrow doSVM(x, y)$
3) for all features, x_j,
4) drop feature, x_j, if $|D(x_j, \theta)| < 1$ end for
5) until desired ratio of SSE and number of data dropped

10.8.5.6 Rayleigh's Criterion Clustering Algorithm

Rayleigh's criterion, also known as the Rayleigh Limit, is used for the resolution of two light sources. In the case of two laser beam sources falling upon a single slit, the resolution limit is defined by the single-slit interference pattern where one source's maximum falls on the first minimum of the diffraction pattern of the second source. This definition is used for resolving distant stars as singletons or identification of binary star systems, etc. In the case of laser optics, the resolution of two sources can be pushed *beyond* the Rayleigh limit due to tracking the statistics of the individual photons that arrive. The relevance of all of this is that resolving two sources is equivalent to saying that a binary clustering solution exists, i.e. that the data is separable to some degree. If the clustering algorithm tracks the data instances individually, as it does with our SVM-external approach, we have a scenario analogous to the resolution in laser optics beyond the "Rayleigh limit."

10.9 Exercises

10.1 Write code that generates two clusters, then write code that scores the SSE of the two clusters.

10.2 Implement K-means (using sample Perl code given) and use it to resolve the two clusters generated in (Exercise 10.1).

10.3 Implement K-NN (using sample Perl code given) and use it to perform five-fold cross-validation on a labeled set of test data. Create the labeled test data with one cluster generation getting one label (+), and a separate cluster generation (perhaps partly overlapping) getting a different label (−). From the cross-validation compute average of {TP,TN,FP,FN} and resultant {SN,SP}.

10.4 Prove that the perceptron finds a solution (a local minima) in a finite number of learning steps (even though exploring an infinite solution space). Following notation introduced earlier and using the steps outlined in [109]:

Suppose we have solution, a set of weights, w^*, that separates the data. Let us start at $w(0) = 0$ and show that $w(t)$ gets more aligned with w^* with every learning update:

(a) Let $\rho = \min_{1 \le n \le N} y_n(w^* \cdot x_n)$, show $\rho > 0$.

(b) Show that $w(t) \cdot w^* \ge w(t-1) \cdot w^* + \rho$, and conclude that $w(t) \cdot w^* \ge t\rho$ (use induction)

(c) Show that $\|w(t)\|^2 \le \|w(t-1)\|^2 + \|x(t-1)\|^2$ ($x(t-1)$ was misclass by $w(t-1)$...)

(d) Show by induction that $\|w(t)\|^2 \le tR^2$, where $R = \max_{1 \le n \le N} \|x_n\|$

(e) Using (b) and (d) show that: $w(t) \cdot w^*/\|w(t)\| \ge \sqrt{t}\frac{\rho}{R}$ and $t \le R^2 \|w^*\|^2/\rho^2$

Typically convergence happens much quicker than indicated by $t \le R^2 \|w^*\|^2/\rho^2$, often by a factor of 100, the bad news is that you neither know ρ in advance nor the number of iterations until convergence. This is a big problem because with significant iterations without convergence, you do not know if more iterations are required, or if you are dealing with data that is non-separable.

10.5 Re-derive Eqs. (10.2)–(10.5).

10.6 Re-derive analysis for Eqs. (10.7)–(10.8).

10.7 Re-derive equations in Section 10.3.7.

10.8 Re-derive equations in Section 10.4.1.

10.9 For the following questions begin by first getting the SVM code (in Perl) provided in Section 10.4.5 working. Then do the following:

10.10 Run the code as is by entering at the prompt: ./Binary_Classifier.pl (which uses SVM_Discriminator.pm), you should see the learning process of the code, and eventually a score result is appended to the "scores" file. Then directly modify and change the *kp* value from 0.1 in the script to ten other values, generate scores for each, plot the 11 results of your "brute force" tuning effort. What was your best (SN,SP) score pair?

10.11 Let us automate the brute-force tuning. Run the following:
```perl
#!/usr/bin/perl
use strict;
my @kernels =
("absdiff0.5","gaussian","sentropic","poly");
my @sigmas = (0.0001, 0.001, 0,005, 0.01, 0.05,
0.1, 0.5, 0.7, 1.0, 2.0);

my $kernel;
my $sigma;
foreach $kernel (@kernels) {
    foreach $sigma (@sigmas) {

        `./Binary_Classifier.pl $kernel $sigma`;
    }
}
```

Note the use of perl's "backtick" operator to execute shell commands (perl can be used as a shell environment called, you guessed it, a "perl shell").

Describe the tuning outcomes. Is there a kernel that always works (where all sigmas result in a convergent training process and thus get scored with a line appended in the "scores" file)? Is there a kernel that rarely works for choice of sigma?

10.12 Let us demonstrate the "bag learning" capabilities of the SVM:

 i) in the training data file, change half of the positive label data to have negative label – thereby create a smaller (pure) positive set, and a mixed "negative" set. Run the SVM with the optimal Kp value identified in (Exercise 10.11) with the first half of the data.

 ii) with the second half of the data as test, and get a histogram on confidence scores.

 iii) identify a cutoff for identifying "true negatives", thereby select out this true negative set.

 iv) use the true negatives and the pure positive set to train a new SVM (with cleaned data), then use the old (original) test data to see how well it performs.

11

Search Metaheuristics

Many search methods have been encountered in the context of "tuning" on the previous acquisition, classification, and clustering methods (especially tuning on the SVM in Chapter 14). Tuning is *searching* for an optimal configuration. Methods and metaheuristics to perform searches are now described in a more general context.

Numerous prior book, journal, and patent publications by the author are drawn upon extensively throughout the text [1–68]. Almost all of the journal publications are open access. These publications can typically be found online at either the author's personal website (www.meta-logos.com) or with one of the following online publishers: www.m-hikari.com or bmcbioinformatics.biomedcentral.com.

11.1 Trajectory-Based Search Metaheuristics

If you have a configuration that you need to optimize, and for any configuration you can evaluate its "score," or "fitness," then a variety of metaheuristics have been developed for configuration selection or model tuning, both by Man and by Nature. If the configuration fitness can be determined from a *differentiable* function of its configuration parameters, then classic gradient ascent (or descent) can be used to optimize the configuration by making learning steps that climb (for maximization type optimization, see Figure 11.1). If the fitness function has a second derivative, then an improved version has been known for over 400 years, Newton's method, which involves calculation of the Hessian to get a higher order correction that avoids overshooting on reaching optima. In practice, local minima "traps" require repeated attempts with randomly initialized and repeated gradient ascent efforts in order to seek a global optimum – enter the modern era where use of computers now makes such problems directly addressable, and often fully resolvable.

Informatics and Machine Learning: From Martingales to Metaheuristics, First Edition.
Stephen Winters-Hilt.

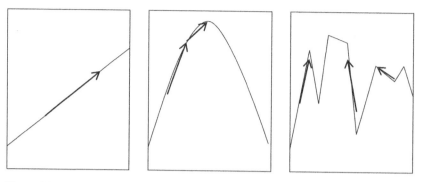

Figure 11.1 Gradient Ascent, Newton's Ascent, Newton's Ascent with restart.

11.1.1 Optimal-Fitness Configuration Trajectories – Fitness Function Known and Sufficiently Regular

Suppose the fitness function, $F(x)$, is known, we can then simply choose to move, from a current configuration "position" in the direction that fitness is increasing the most.

11.1.1.1 Metaheuristic #1: Euler's Method – First-Order Gradient Ascent

If the configuration's fitness can be determined from a differentiable function of its configuration parameters, then gradient ascent (or descent) can be used to optimize the configuration (see Figure 11.2).

(1) Calculate $x_{i+1} = x_i + \eta_i \nabla (F(x))$, where η_i is the "learning rate," assumed constant for the discussion to follow, but could vary with further iterations such as

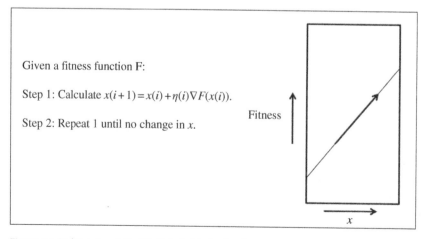

Given a fitness function F:

Step 1: Calculate $x(i+1) = x(i) + \eta(i) \nabla F(x(i))$.

Step 2: Repeat 1 until no change in x.

Fitness

x

Figure 11.2 Metaheuristic #1: Euler's Method – first-order gradient ascent.

in some simulated annealing approaches. For small enough learning rate must have $F(x_{i+1}) \geq F(x_i)$. Choosing a learning rate that is large enough to converge quickly, yet not overshoot badly, is balanced against the smaller learning rate that can more precisely optimize. This is a data dependent learning rate selection and often requires simple tuning tests on the choice of learning rate (before initialization).

(2) repeat (1) until no change in x.

Problem: can easily overshoot, and tends to oscillate around local minima without halting simply.

11.1.1.2 Metaheuristic #2: Newton's Method – Second-Order Gradient Ascent

If the fitness function has a second derivative, then an improved version has been known for over 400 years, Newton's method, which involves calculation of the Hessian to get a higher order correction that avoids overshooting on reaching optima (see Figure 11.3).

(1) Calculate $x_{i+1} = x_i + \eta_i \nabla F(x)/[H\,F(x)]$, where η_i is the "learning rate," assumed constant for the discussion to follow, and H $F(x)$ is the Hessian of $F(x)$. For small enough learning rate must have $F(x_{i+1}) \geq F(x_i)$. As before, this is a data dependent learning rate situation and often requires simple tuning tests on the choice of learning rate (before initialization).

(2) repeat (1) until no change in x.

Problem: local minima "traps" require repeated attempts with randomly initialized and repeated gradient ascent efforts in order to seek a global optimum.

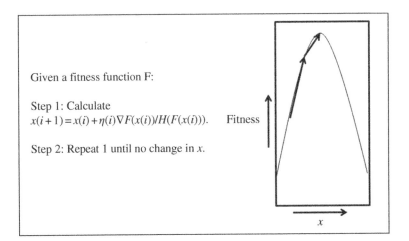

Given a fitness function F:

Step 1: Calculate
$x(i+1) = x(i) + \eta(i)\nabla F(x(i))/H(F(x(i)))$.   Fitness

Step 2: Repeat 1 until no change in x.

x

Figure 11.3 Metaheuristic #2: Newton's Method – second-order gradient ascent.

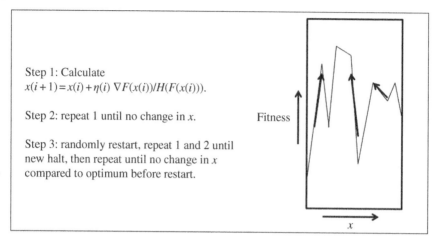

Step 1: Calculate
$x(i+1) = x(i) + \eta(i) \; \nabla F(x(i))/H(F(x(i)))$.

Step 2: repeat 1 until no change in x.

Step 3: randomly restart, repeat 1 and 2 until new halt, then repeat until no change in x compared to optimum before restart.

Fitness

x

Figure 11.4 Metaheuristic #3: Newton's Method – second-order gradient ascent.

11.1.1.3 Metaheuristic #3: Gradient Ascent with (Random) Restart

The use of computers now makes algorithmic aspects of the solution, such as "random restart" (according to some specification), accessible in a way that was impossible before having modern computational tools (see Figure 11.4):

(1) Calculate $x_{i+1} = x_i + \eta_i \nabla F(x)$, where η_i is the "learning rate," assumed constant for the discussion to follow, but could vary with further iterations such as in some simulated annealing approaches. For small enough learning rate must have $F(x_{i+1}) \geq F(x_i)$. Choosing a learning rate that is large enough to converge quickly, yet not overshoot badly, is balanced against the smaller learning rate that can more precisely optimize. This is a data dependent learning rate selection and often requires simple tuning tests on the choice of learning rate (before initialization).
(2) repeat (1) until no change in x.
(3) randomly restart, repeat (1)–(2) until overlapping global coverage is achieved. Problem: local minima often cannot be overcome with random restart methods (reduces to randomly searching too large a space).

11.1.2 Optimal-Fitness Configuration Trajectories – Fitness *Function* not Known

Suppose you have a score (or fitness), but the function is not known. To adopt the terminology of [219], a configuration "tweak" will refer to randomly changed configuration settings by some perturbatively bounded amount. In other words, we always start with some configuration that is "live," in that it has a perturbatively defined score or fitness at the indicated configuration setting, and we want to evolve the configuration towards one with an improved score or fitness via a series of "tweaks." Suppose we tweak the configuration and the new configuration has a better score, we can then take the better configuration as our working optimum and iterate with another tweak, if now the new configuration has worse score, we would not switch to the new configuration, but stay with what we have got, and then iterate again, this is what is known as the "Hill Climbing" optimization algorithm.

The standard Hill Climbing algorithm should more aptly be named "blind hill climbing" because there's no equivalent of "looking around," it is akin to gradient ascent in that there is clear choice of up or down hill, but it is not akin to Newton's refinement as there is no ability to "look ahead" (with a second derivative correction in the case of Newton's method). This can be improved by generating multiple tweak configurations on a given parent configuration and selecting the one providing the best score from that test set. This is known as steepest ascent hill climbing, where the sampled steepest ascent approach serves as an approximation to gradient ascent (when a differentiable fitness function is known), and in some cases will also approximate the correction indicated in Newton's method (the steepest ascent sampling and maximal selection precludes the overshooting complication). Of course, the expense of this approximately matching performance between steepest ascent and gradient ascent is that steepest ascent requires multiple tweaks at each step, while gradient ascent, informed as it is by the knowledge of the differentiable fitness function, only requires one tweak at each step. What levels the playing field between the differentiable function and not-known function scenarios, however, is that even with a differentiable function the function may admit numerous local minima, and thus require that its global optima be sought by further refinements, "random restart" in particular, where we randomly restart the whole algorithm in a different part of the configuration space and hope for a better result. The fact that even the differentiable case will typically require random restart, however, means that the multiple tweak steepest ascent is not so onerous a burden over that of a single-tweak differentiably extrapolatable formulation. This will be especially so, if the "looking around" burden of the multiple tweaks at each steepest ascent step can be handled by metaheuristics in part of a directed (partially) random restart.

11.1.2.1 Metaheuristic #4: (Blind) Hill Climbing

The specifications of the methods that follow are pure algorithmic... there is no calculus (Figure 11.5).

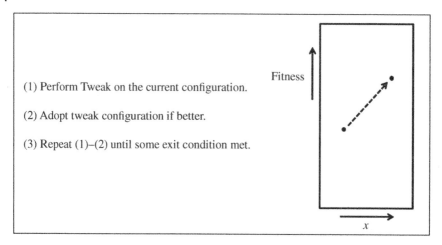

Figure 11.5 Metaheuristic #4: (blind) hill climbing.

(1) Perform Tweak on the current configuration. In other words, generate a new configuration that is a random, locally bounded, perturbation of the current configuration (involving in one or more perturbed parameters).
(2) If tweak configuration scores better then current best case, then make the tweak configuration the new best case, otherwise ignore.
(3) Repeat (1)–(2) until some exit condition met (no improvements for MAX tries, for example).
Problem: has similar weaknesses to Method #1.

11.1.2.2 Metaheuristic #5: Steepest Ascent Hill Climbing
Again, the specification of the methods is pure algorithmic... there is no calculus (Figure 11.6).

(1) Generate N Tweak configurations from the current configuration.
(2) If the best tweak configuration score is better then current best case, then make the tweak configuration the new best case, otherwise ignore.
(3) Repeat (1)–(2) until some exit condition met (no improvements for MAX tries, for example).
Problem: has similar weaknesses to Method #2.

11.1.2.3 Metaheuristic #6: Steepest Ascent Hill Climbing with Restart
In gradient ascent with random restarts the idea is to find a global optimum by eventually restarting in the region of configuration space covered by the global optimum's footprint. Now we have less information, so cannot perform a Newton's

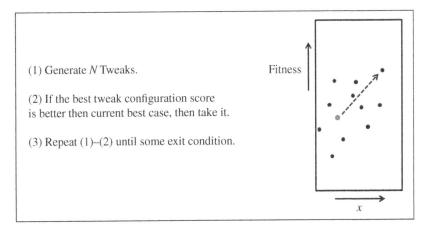

Figure 11.6 Metaheuristic #5: steepest ascent hill climbing.

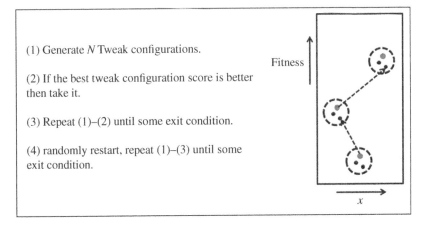

Figure 11.7 Metaheuristic #6: steepest ascent hill climbing with restart.

method refinement as easily, but we still have the random restart and instead of gradient ascent we have steepest ascent hill climbing. So, what is often the case is that the challenge is not the diffentiability of the fitness function, if it is even known, but in finding a global optimum when local optima are numerous (see methods shown in Figures 11.7–11.10).

(1) Generate *N* Tweak configurations from the current configuration.
(2) If the best tweak configuration score is better then current best case, then make the tweak configuration the new best case, otherwise ignore.

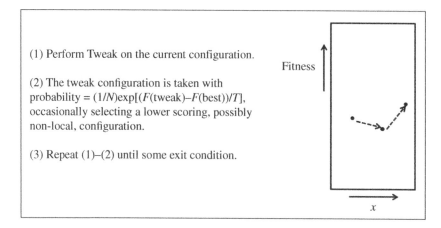

(1) Perform Tweak on the current configuration.

(2) The tweak configuration is taken with probability $= (1/N)\exp[(F(\text{tweak})-F(\text{best}))/T]$, occasionally selecting a lower scoring, possibly non-local, configuration.

(3) Repeat (1)–(2) until some exit condition.

Figure 11.8 Metaheuristic #7: simulated annealing hill climbing.

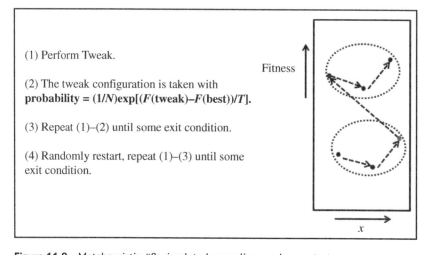

(1) Perform Tweak.

(2) The tweak configuration is taken with **probability $= (1/N)\exp[(F(\text{tweak})-F(\text{best}))/T]$.**

(3) Repeat (1)–(2) until some exit condition.

(4) Randomly restart, repeat (1)–(3) until some exit condition.

Figure 11.9 Metaheuristic #8: simulated annealing random restart.

(3) Repeat (1)–(2) until some exit condition met (no improvements for MAX tries, for example).

(4) randomly restart, repeat (1)–(3) until overlapping global coverage is achieved. Problem: has similar weaknesses to Method #3.

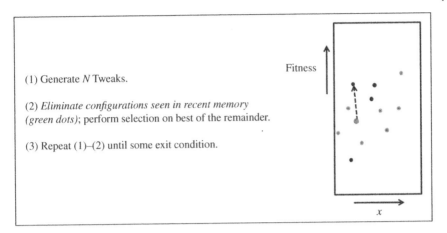

(1) Generate *N* Tweaks.

(2) *Eliminate configurations seen in recent memory (green dots)*; perform selection on best of the remainder.

(3) Repeat (1)–(2) until some exit condition.

Figure 11.10 Metaheuristic #9: taboo search.

11.1.3 Fitness Configuration Trajectories with Nonoptimal Updates

Our ability to assess a score, or assign a fitness, allows for a collection of metaheuristics that basically reduce to "look around and take the best way forward" via a series of tweaks. This is not possible for some problems, however, because the "looking around" part isn't that informative, e.g., the fitness landscape has sections that are at a fixed level (with noise variations about that level, for example). This is the larger problem of the simple globalization algorithm, via random restart: if the fitness landscape or configuration space is too large random restart would not offer a solution, even if it can, in a reasonable amount of time. This is where more clever metaheuristics are involved, to extend to a global optimization algorithm.

Global Optimization

One of the weaknesses of the brute force random restart approach mentioned so far is that the tweak involved is with a *bounded perturbative* change, which may *already* exclude the possibility of reaching the solution sought (given the computational resources and a reasonable amount of time). So one generalization is to allow for tweaks that are unbounded, but in some perturbatively stable way, such as with a Boltzmann factor for regularization, and in doing so we arrive at the Simulated Annealing approach:

11.1.3.1 Metaheuristic #7: Simulated Annealing Hill Climbing

(1) Perform Tweak on the current configuration.

(2) The tweak configuration is taken with probability $= (1/N)\exp[(F(\text{tweak}) - F(\text{best}))/T]$, where N is a normalization factor, and T is a "temperature",

e.g., a Boltzmann probability factor for occasionally selecting a lower scoring, possibly nonlocal, configuration.

(3) Repeat (1)–(2) until some exit condition met (no improvements for MAX tries, for example).

Problem: has similar weaknesses to Method #6 and #3.

11.1.3.2 Metaheuristic #8: Simulated Annealing Hill Climbing with Random Restart

Now redo with random restart:

(1) Perform Tweak on the current configuration.
(2) The tweak configuration is taken with

$$\textbf{probability} = (\mathbf{1/N})\,\textbf{exp}\,[(\textbf{F(tweak)} - \textbf{F(best)})/\textbf{T}],$$

where N is a normalization factor, and T is a "temperature", e.g., a Boltzmann probability factor for occasionally selecting a lower scoring, possibly nonlocal, configuration.

(3) Repeat (1)–(2) until some exit condition met (no improvements for MAX tries, for example).
(4) randomly restart, repeat (1)–(3) until overlapping global coverage is achieved.

Problem: has similar weaknesses to Method #3.

11.1.3.3 Metaheuristic #9: Taboo Search

Taboo search can be seen as a form of simulated annealing in that lower scores can be selected, here by an exclusion rule where configuration areas visited are "taboo" from revisiting again "too soon." Taboo search can also be extended into a variant that operates on the configuration components, known as component-based taboo search.

Simulated annealing and Taboo search enhancements to the steepest ascent hill climbing algorithm involve use of multiple configuration samples (tweaks), where the steepest ascent sample space is constrained according to choice of learning heuristic. If implemented efficiently what can result is a sampling that is focused on the allowed sample space region (without wasting time generating configurations to reject due to taboo, or reject due to below-threshold probability).

(1) Generate N Tweak configurations from the current configuration.
(2) ***Eliminate configurations seen in recent memory***; perform selection on best of the remainder.
(3) Repeat (1)–(2) until some exit condition met (no improvements for MAX tries, for example).

Problem: has similar weaknesses to Method #3.

11.1.3.4 Metaheuristic #10: Tabu Search with Restart

Taboo search can be seen as a form of simulated annealing in that lower scores can be selected, here by an exclusion rule where configuration areas visited are "taboo" from revisiting again "too soon."

(1) Generate N Tweak configurations from the current configuration.
(2) Eliminate configurations seen in recent memory; perform selection on best of the remainder.
(3) Repeat (1)–(2) until some exit condition met (no improvements for MAX tries, for example).
(4) randomly restart, repeat (1)–(3) until overlapping global coverage is achieved.
Problem: has similar weaknesses to Method #3.

11.1.3.5 Metaheuristic #11: Component-Based Tabu Search

Taboo search can also be extended into a variant that operates on the configuration **components**, known as component-based taboo search.

(1) Generate N Tweak configurations from the current configuration.
(2) Eliminate configurations involving component usage seen in recent memory; perform selection on best of the remainder.
(3) Repeat (1)–(2) until some exit condition met (no improvements for MAX tries, for example).
Problem: has similar weaknesses to Method #3.

11.1.3.6 Metaheuristic #12: Component-Based Tabu Search with Restart

Taboo search can also be extended into a variant that operates on the configuration **components**, known as component-based taboo search.

(1) Generate N Tweak configurations from the current configuration.
(2) Eliminate configurations involving component usage seen in recent memory; perform selection on best of the remainder.
(3) Repeat (1)–(2) until some exit condition met (no improvements for MAX tries, for example).
(4) randomly restart, repeat (1)–(3) until overlapping global coverage is achieved.
Problem: has similar weaknesses to Method #3.

11.2 Population-Based Search Metaheuristics

In seeking global optimization metaheuristics we are starting to see more sophisticated configuration selection at the component level, on the one hand, and at the population level, on the other hand, especially as we work with the probabilistic simulated annealing approach and the history-based taboo approach, especially with the component-based versions of the latter. This is because the population and history aspects point to a general metaheuristics that operates on populations of configurations (or populations of "agents" that interact as intermediaries to determining a configuration selection). The notion of "history" also must address the conveyance of this information or "artifact". In the case of ACO, described in what follows, this will be via stygmergy.

11.2.1 Population with Evolution

A fixed-size (or size otherwise constrained) population of configurations can have a birth/death cycle or be static. If it has a birth/death cycle, one popular method is the evolutionary computation approach (Darwinian evolution; asexual reproduction) (see Figure 11.11):

11.2.1.1 Metaheuristic #13: Evolutionary Optimization (Darwinian Evolution; Asexual Reproduction)

(1) Starting population of parent configurations ("parents") undergoes initial selection according to cutoff (truncation selection) that is chosen.

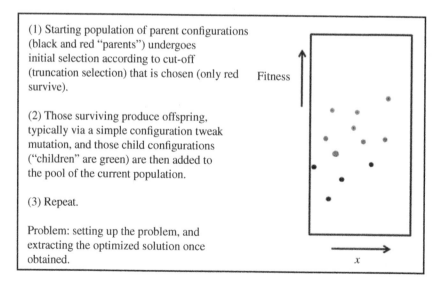

(1) Starting population of parent configurations (black and red "parents") undergoes initial selection according to cut-off (truncation selection) that is chosen (only red survive).

(2) Those surviving produce offspring, typically via a simple configuration tweak mutation, and those child configurations ("children" are green) are then added to the pool of the current population.

(3) Repeat.

Problem: setting up the problem, and extracting the optimized solution once obtained.

Figure 11.11 **Metaheuristic #13: evolutionary optimization (darwinian evolution; asexual reproduction).** Depending on algorithm, the parents may be selected against generationally also (such as for salmon). The reproduction step can be done by the population all at once (generationally), or individually.

(2) Those surviving produce offspring, typically via a simple configuration tweak mutation, and those child configurations ("children") are then added to the pool of the current population.

(3) Repeat.

Depending on algorithm, the parents may be selected against generationally also (such as for salmon). The reproduction step can be done by the population all at once (generationally), as described here, or individually, out-of-phase, as commonly done with the GA's given next.

 As we work with the probabilistic simulated annealing approach and the history-based taboo approach it becomes apparent that the population and history aspects of these methods point to a general metaheuristics that operates on populations of configurations, or populations of "agents" that interact as intermediaries to determining a configuration selection. The notion of "history" also must address the conveyance of this information via "artifact." In the case of ACO, described in what follows, this will be via stygmergy.

11.2.1.2   Metaheuristic #14: Genetic Algorithm (Darwinian Evolution; Sexual Reproduction – Binary Interaction)

(1) Starting population of parent configurations (black and red "parents") undergoes initial selection according to cutoff (truncation selection) that is chosen (only red survive).

(2) Those surviving produce offspring, typically via both simple configuration tweak mutation and nonlocal configuration component-level swapping, and those child configurations ("children" shown green) are then added to the pool of the current population.

(3) Repeat.

Problem: setting up the problem, and extracting the optimized solution once obtained.

11.2.1.3   Evolutionary Algorithm Parameters

Both of the evolutionary algorithms have tuning parameters in their manner of iteration that shift between less reliance on what has been learned (random search) and strong use of existing information (highly directed gradient ascent). These parameters can be summarized as follows:

Population size: if population is one, then we are back to simpler hill climbing optimization, with just one configuration tweak under consideration. If the population is unduly large, then we are just looking at everything, e.g., we have random search. If the population is maintained at a more optimal medium size, we arrive at a form of steepest ascent.

Selection pressure: if the selection pressure is high only the very best survive, if only one this reduces to steepest ascent hill climbing. If the selection pressure is low, most survive, and thereby arrive at population that "walks" from current population via series of configuration "tweaks", thus have a random walk search.

Generational pressure: if there are many children per parent(s), then we obtain many samples with mutations in the neighborhood of those parents, a form of steepest ascent hill climbing. If there are only a few children have a optimization procedure more akin to simple hill climbing.

Mutation rate: If the mutation rate is large, we return to a random search since no memory or learning is retained in the decision process. If mutation rate is small, we can fine-tune the optimization on solution. If the mutation is local we may need to use random restart methods to recover global optimization capabilities, even minimally. If the mutation is nonlocal, such as with simulated annealing, and with sexual reproduction in GA with swapped sections of configuration components allowed, then have improved ability to probe global configuration space.

Populations with interactions (and subpopulations, e.g., speciation)

Once we consider that the configurations in a population may interact with one another, we have a situation where different subpopulations may be given (and now not trivially de-couple), i.e., speciation is possible in the evolutionary population. From there whole ecologies of evolutionary complexity can be developed.

11.2.2 Population with Group Interaction – Swarm Intelligence

With population-based interactions have possible direct coordination between agents (sexual reproduction providing cross-over mutation in GA's, as mentioned above, and swarm activity that provides global information to all agents with action defined accordingly to desired local and global swarm behavior, as defined in what follows:

11.2.2.1 Metaheuristic #15: Particle Swarm Optimization (PSO) (Lamarckian Evolution)

Particle swarm optimization (PSO) also takes its cue from Biology, but not from evolutionary model, but from a swarm model. Here the population is static (the other case than the birth/death cycle case), and there is no selection of any kind. Now the configurations in the population are themselves directly tweaked in response to new information obtained. This is a form of directed mutation and is part of a Lamarckian evolutionary paradigm. The configurations are often viewed as describing particles in a space and the configurations undergo directed mutation, "motion", in the configuration space, with motion towards the best known configuration, where three levels of knowledge are weighed in the balance: (i) the fittest configuration ascertained by a particular during its history; (ii) the fittest configuration ascertained by the informants of a particular particle (often just a randomly chosen set of particles); and (iii) the fittest configuration discovered by any particle.

11.2.3 Population with Indirect Interaction via Artifact

With population-based interaction also have possible *indirect* interactions between agents. One indirect interaction between agents in a population is stygmergy, such as leaving a mark, as with ants and their pheromone markers. In a broader sense, have interaction of a population with an artifact in general. This broader category includes the refinement of tools as artifacts and places human evolution in the context of coevolution with self via artifact (with evolution in the artifact as well, e.g., our improved arrowhead production capability is one of the distinguishing marks of modern human over the Neanderthals). A critical new tool for humans is the computer. We have seen the rapid evolution of the computer's capabilities and sophistication in our lifetimes. Perhaps the original self-replication life-cycle used RNA as an artifact of its operation, giving rise to "RNA World"-based life. Perhaps protein began as an artifact of RNA World evolution that co-evolved into a critical part of the organism. Similarly, it not clear where coevolution with computer as artifact will lead for humans, but it will clearly tap into newfound potential as a new coevolutionary niche is now open for exploration. The role of artifact is thus, subtle, and can change in the coevolutionary process. Perhaps someday we will have a "Pinocchio effect" whereby a computer-based artifact becomes self-sufficient (and eventually self-replicating). In this evolutionary history of computer-based life, perhaps Man's role will also be forgotten, but more likely is that we would probably still be there, as a cog in the machine, as with prior coevolutionary partners displaced by their artifacts, but still extant, especially with a role that is in the information-packaging process chain).

11.2.3.1 Metaheuristic #16: Ant Colony Optimization (ACO) (Swarm Intelligence; Stygmergy; Have Coevolution with Artifact)
"no ants"!

(1) Population is based on "ant trails," made by selecting components one-by-one based on their pheromones (can be viewed as a component weight).
(2) Evaporate the pheromones a little (reduce the component weights a little).
(3) Assess fitness of each trail, update weights (pheromone) of components on the trail according to the trails fitness.

11.2.3.2 Other Population-Based Search Metaheuristics

Sometimes even more involved population metaheuristics are needed, with greater handling of constraints, and the methodology blends into the FSA-based modeling with greater complexity, and we must fall back on the standard information measures to help guide, which brings us full circle to involving the FSA-based methods described in Chapter 2, and they, in turn, are more clearly defined in their own internal tuning needs given the advance model search ("tuning") that can be encompassed with the metaheuristics described here.

Hybridization of any of the methods shown are found to be even more powerful → **co-evolutionary algorithms** are a specific type of hybridization known to allow the "No free lunch theorem" to be violated, opening the door to a proof of the "feasibility of search" to parallel the breakthrough proof (basis for machine learning) that shows the "feasibility of (statistical) learning."

11.3 Exercises

11.1 Implement Euler's Method ("GD" but typically maximization in this context for "ascent") and show its operation in seeking a maximum on a fitness function that you choose.

11.2 In (Ex. 11.1), choose a learning rate and fitness function that demonstrates the overshoot oscillation instability in the algorithm. Then, implement Newton's Method and show how it resolves the problem.

11.3 Implement gradient ascent with random restart (building off of the solution to (Ex. 11.1)).

11.4 Implement the hill climbing algorithm to obtain a maximal fitness on a dataset of your choice.

11.5 Implement the steepest ascent algorithm to obtain a maximal fitness on a dataset of your choice.

11.6 Redo (Ex. 11.5) with random restart added.

11.7 Implement the simulated annealing algorithm to obtain a maximal fitness on a dataset of your choice.

11.8 Redo (Ex. 11.7) with random restart added.

11.9   Implement the taboo search algorithm to obtain a maximal fitness on a dataset of your choice.

11.10   Redo (Ex. 11.9) with random restart added.

11.11   Redo (Ex. 11.9) but doing component-based taboo.

11.12   Redo (Ex. 11.11) with random restart added.

11.13   Implement an evolutionary population algorithm with Darwinian selection and asexual reproduction, and apply to a dataset of your choice.

11.14   Implement the genetic algorithm search (with Darwinian selection and sexual reproduction), and apply to a dataset of your choice.

11.15   Implement a PSO algorithm and apply to a dataset of your choice.

11.16   Implement the ACO algorithm and apply to a dataset of your choice.

11.17   Implement one of the following population algorithms, and apply to a dataset of your choice: Bee; Harmony; Firefly; Bat; Cuckoo; Bacterial Foraging Optimization.

11.18   Implement one of the following cell population algorithms, and apply to a dataset of your choice: Clonal selection algorithm; Negative selection algorithm; Artificial immune recognition system; Immune network algorithm; Dendritic Cell algorithm.

11.19   Implement one of the following neural learning population algorithms, and apply to a dataset of your choice: Hopfield Network; Learning Vector Quantization; Self-organizing Map (SOM).

12

Stochastic Sequential Analysis (SSA)

A protocol has been developed for the discovery, characterization, and classification of localizable, approximately stationary, statistical signal structures in stochastic sequential data, such as the channel current data to be described in Chapter 14.

The stochastic sequential analysis (SSA) methods described in what follows, provide a robust and efficient means to make a device or process "smart" (e.g., a "smartening" is possible with a software AI effecting the operational role of a Maxwell Demon on the flow of information, instead of the classic example of flow of hot/cold matter in a preferential manner), with possible enhancement to device (or process) sensitivity and productivity and efficiency, as well as possibly enabling new capabilities for the device or process (via transduction coupling, for example, as with the nanopore transduction detector [NTD] platform in Chapter 14). The SSA Protocol can work with existing device or process information flows, or can work with additional information induced via modulation or introduction via transduction couplings (comprising carrier references [90, 91], among other things, see Chapter 14). Hardware device-smartening may be possible via introduction of modulations or transduction couplings, when used in conjunction with the SSA Protocol implemented to operate on the appropriate timescales to enable real-time experimental or operational control, where real-time adaptive pattern recognition is critical.

Numerous prior book, journal, and patent publications by the author are drawn upon extensively throughout the text [1–68]. Almost all of the journal publications are open access. These publications can typically be found online at either the author's personal website (www.meta-logos.com) or with one of the following online publishers: www.m-hikari.com or bmcbioinformatics.biomedcentral.com.

Informatics and Machine Learning: From Martingales to Metaheuristics, First Edition.
Stephen Winters-Hilt.
© 2022 John Wiley & Sons, Inc. Published 2022 by John Wiley & Sons, Inc.

12.1 HMM and FSA-Based Methods for Signal Acquisition and Feature Extraction

Central to the SSA Protocol method are hidden Markov models (HMMs) [1, 3, 60]. To realize the potential of the HMM-based methods, however, there must be a means to directly acquire the signals from the "raw" form. Eventually an HMM can be trained for the acquisition task as well, but it typically must have previously acquired data to do that initial training. Thus the need for a critical, *ad hoc*, front-end to the signal processing to do signal acquisition even if that acquisition is to eventually be done with an HMM. HMMs and support vector machines (SVMs) [1, 3, 44, 59, 60] are core methods in signal processing and pattern recognition. Both HMMs and SVMs are well-founded in the sense that they are mathematically well-defined and have robust learning properties, and, not surprisingly, they are core methodologies in machine learning-based signal processing and pattern recognition.

Even if the signal sought is well understood, and a purely HMM-based approach is possible, this is often needlessly computationally intensive (and slow), especially in areas where there is no signal. To address this there are numerous hybrid FSA/ HMM approaches (such as BLAST [127]) that benefit from the $O(L)$ complexity on length L signal with FSA processing, with more targeted processing at $O(LN^2)$ complexity with HMM processing (where there are N states in the HMM model). This is not to say that HMMs cannot be used in a signal discovery role in their own right, via use of Viterbi and Baum-Welch algorithms with a "generic" HMM feature extraction using a statistical modal analysis. HMM approaches on stochastic signals, however, like their periodic signal counterparts (Fourier transform-based) from classic electrical engineering signal processing, usually involve preprocessing that assumes linear system properties or assumes observation is frequency band limited and not time limited (via stationarity assumption), etc., and indirectly inherits similar time-frequency uncertainty relations, Gabor limit, and Nyquist sampling relations. FSA methods can be used to recover (or extract) signal features missed by HMM or classical electrical engineering signal processing.

Many signal features of interest are time limited and not band limited in the observational context of interest, such as noise "clicks," "spikes," or impulses. To acquire these signal features a time-domain FSA (tFSA) is often most appropriate. Human hearing, for example, is a nonlinear system that thereby circumvents the restrictions of the Gabor limit to allow musical geniuses, for example, with "perfect pitch" whose time-frequency acuity surpasses what should be possible by linear signal processing alone [116], such as with Nyquist sampled linear response recording devices that are bound by the limits imposed by the Fourier uncertainty principle (or Benedick's theorem) [117]. Thus, even when the powerful HMM feature extraction methods are utilized to full advantage, there is often a sector of the signal analysis that is only conveniently accessible to analysis by way of FSA's (without significant oversampling), such that a parallel processing with both HMM and FSA methods is often needed (results demonstrating this in the context of channel current analysis [1, 3] will be briefly discussed). Not all of the methods employed at the FSA processing stage derive from standard signal processing approaches, either, some are purely statistical such as with oversampling [118] (used in radar range oversampling [119, 120]) and dithering [121] (used in device stabilization and to reduce quantization error [122, 123]).

Hidden Markov models, unlike tFSAs, have a straightforward mathematical and computational foundation at the nexus where Bayesian probability and Markov models meet dynamic programming. To properly define or choose the HMM model in a machine learning context, however, further generalization is usually required. This is because the "bare-bones" HMM description has critical weaknesses in most applications, which are summarized in what follows, along with their "fixes." Fortunately, each of the standard HMM weaknesses can be addressed in computationally efficient ways as shown in Chapter 7. The generalized HMMs allows for a generalized Viterbi Algorithm and generalized Baum–Welch Algorithm. The generalized algorithms retain path probabilities in terms of a sequence of likelihood ratios, which satisfy Martingale statistics under appropriate circumstances [102], thereby having Martingale convergence properties (where convergence is associated with "learning" in this context). Thus, HMM learning proceeds via convergence to a limit state that provably exists in a similar sense to that shown with the Hoeffding inequality [104], via its proven extension to Martingales [108]. The Hoeffding inequality is a key part of the VC Theorem in Machine Learning, whereby convergence for the Perceptron learning process to a solution is proven to exist in a finite number of learning steps [109].

The generalizations that encompass the hidden semi-Markov models (HMMBD), described in what follows, cannot be "rolled into" an equivalent dynamic Bayesian network (DBN), and while the large-clique generalized meta-HMM can be rolled into a theoretically equivalent DBN, that DBN still has to express the (fully connected) large-clique generalization, and adopt the same meta-state constraints as the meta-HMM. When standard HMMs are compared to DBN in character recognition [136], the HMM is found to outperform the DBN in terms of both higher recognition rate and lower complexity. In the large-clique generalization the direct HMM dynamic programming implementation of the meta-states and their transitions, for both the Viterbi and Baum-Welch algorithms, this would lead to an even greater performance gap between the generalized HMM and DBN methods. DBN can offer insights into novel HMM embeddings in the meta-HMM, however, but that falls outside the scope of this work, so DBNs would not be discussed further.

A description of the SSA Protocol and SCW communications will now be given, where stochastic phase modulation (SPM), a simple form of SCW communication, has been used in the engineered nanopore transduction detector experiments in

Chapter 14 and [1, 3, 31]. A description for a pragmatic distributed HMM generalization and implementation is given in Chapter 7. A discussion of how the HMM with binned duration (HMMBD) and meta-HMM algorithmic methodologies enable practical SCW encoding/decoding then follows, where SCW signal processing can be used in a number of settings in science and nanotechnology.

12.2 The Stochastic Sequential Analysis (SSA) Protocol

The SSA protocol is shown in Figures 12.1–12.4, where Figure 12.1 shows a general signal-processing flow topology (see Left Panel), and specialized variants for channel current cheminformatics (Center) and kinetic feature extraction based on blockade-

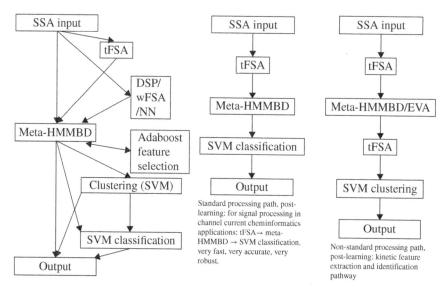

Figure 12.1 **Left.** The general stochastic sequential analysis flow topology. **Center.** The general signal processing flow in performing channel current analysis is typically Input → tFSA → Meta-HMMBD → SVM → Output. **Right.** Notable differences occur in channel current cheminformatics during state discovery when EVA-projection (emission variance amplification projection), or a similar method, is used to achieve a quantization on states, then have Input → tFSA → HMMBD/EVA (state discovery) → meta-HMMBD-side → SVM → Output. While, in gene-finding just have: Input → meta-HMMBD-sisde → Output. In gene-finding, however, the HMM internal "sensors" are sometimes replaced, locally, with profile-HMMs [1, 3] (equivalent to position-dependent Markov Models, or pMM's, see Methods), or SVM-based profiling [1, 3], so the topology can differ not only in the connections between the boxes shown, but in their ability to embed in other boxes as part of an internal refinement. *Source:* Based on Winters-Hilt [1, 3].

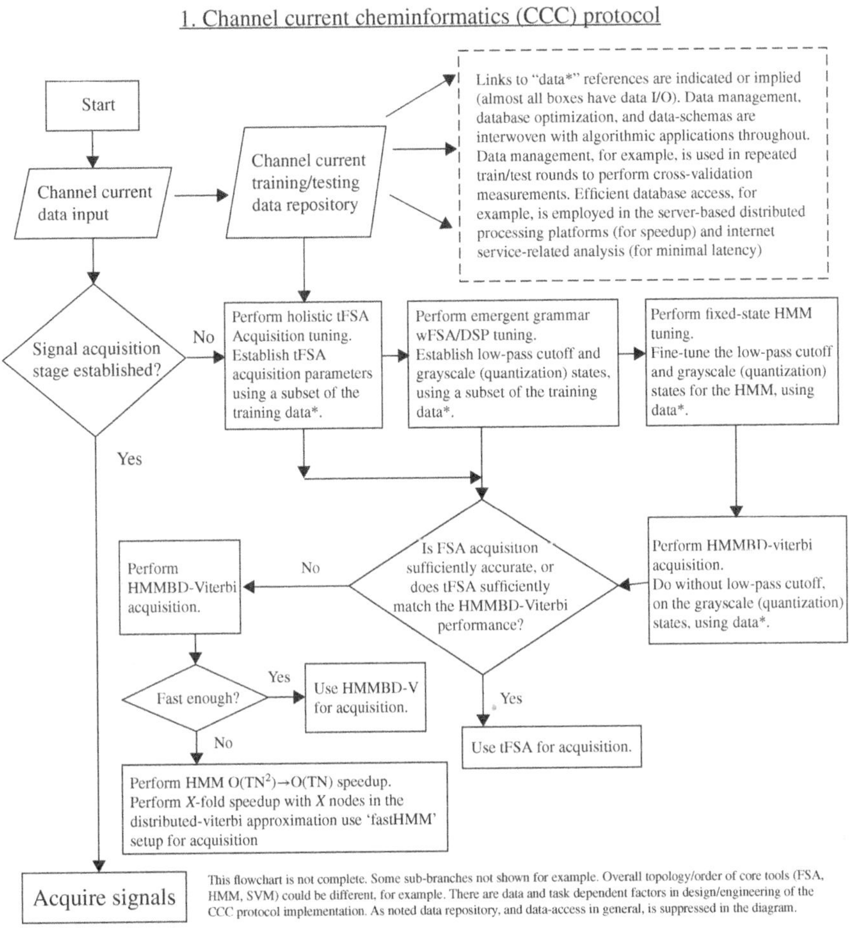

1. Channel current cheminformatics (CCC) protocol

Figure 12.2 CCC protocol flowchart (part 1).

level duration observations (Right). The SSA Protocol allows for the discovery, characterization, and classification of localizable, approximately-stationary, statistical signal structures in channel current data, or genomic data, or sequential data in general. The core signal processing stage in Figure 12.1 is usually the feature extraction stage, where central to the signal processing protocol is a generalized Hidden Markov model. The SSA Protocol also has a built-in recovery protocol for weak signal handling, outlined next, where the HMM methods are complemented by the strengths of other Machine Learning methods.

The sequence of algorithmic methods used in the SSA Protocol, for the information-processing flow topology shown in Figure 12.1, comprise a weak signal handling

II. Channel current cheminformatics (CCC) protocol

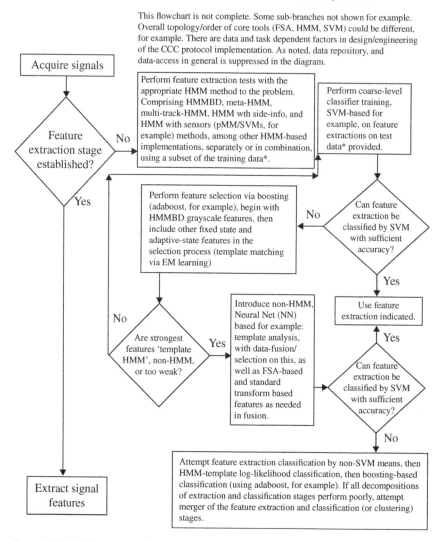

This flowchart is not complete. Some sub-branches not shown for example. Overall topology/order of core tools (FSA, HMM, SVM) could be different, for example. There are data and task dependent factors in design/engineering of the CCC protocol implementation. As noted, data repository, and data-access in general is suppressed in the diagram.

Acquire signals

Feature extraction stage established? — No → Perform feature extraction tests with the appropriate HMM method to the problem. Comprising HMMBD, meta-HMM, multi-track-HMM, HMM wth side-info, and HMM with sensors (pMM/SVMs, for example) methods, among other HMM-based implementations, separately or in combination, using a subset of the training data*.

Perform coarse-level classifier training, SVM-based for example, on feature extractions on test data* provided.

Yes

Perform feature selection via boosting (adaboost, for example), begin with HMMBD grayscale features, then include other fixed state and adaptive-state features in the selection process (template matching via EM learning)

No

Can feature extraction be classified by SVM with sufficient accuracy?

Yes

No

Are strongest features 'template HMM', non-HMM, or too weak? — Yes → Introduce non-HMM, Neural Net (NN) based for example: template analysis, with data-fusion/ as well as FSA-based and standard transform based features as needed in fusion.

Use feature extraction indicated.

Yes

Can feature extraction be classified by SVM with sufficient accuracy?

No

Extract signal features

Attempt feature extraction classification by non-SVM means, then HMM-template log-likelihood classification, then boosting-based classification (using adaboost, for example). If all decompositions of extraction and classification stages perform poorly, attempt merger of the feature extraction and classification (or clustering) stages.

Figure 12.3 CCC protocol flowchart (part 2).

protocol as follows: (i) the weakness in the (fast) Finite State Automaton (FSA) methods will be shown to be their difficulty in nonlocal structure identification, for which HMM methods (and tuning metaheuristics) are the solution. (ii) for the HMM, in turn, the main weakness is in local sensing "classification" due to conditional independence assumptions. Once in the setting of a classification problem, however, the problem can

III. Channel current cheminformatics (CCC) protocol

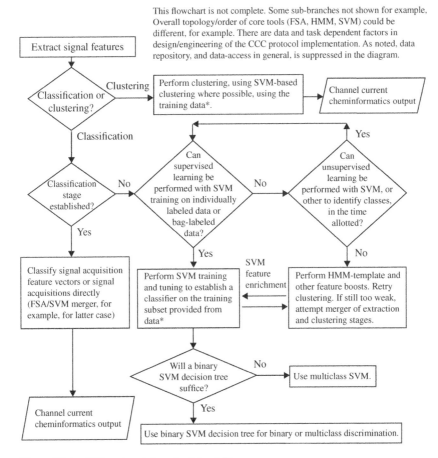

This flowchart is not complete. Some sub-branches not shown for example, Overall topology/order of core tools (FSA, HMM, SVM) could be different, for example. There are data and task dependent factors in design/engineering of the CCC protocol implementation. As noted, data repository, and data-access in general, is suppressed in the diagram.

Extract signal features

Classification or clustering?

Clustering

Perform clustering, using SVM-based clustering where possible, using the training data*.

Channel current cheminformatics output

Classification

Yes

Classification stage established?

No

Can supervised learning be performed with SVM training on individually labeled data or bag-labeled data?

No

Can unsupervised learning be performed with SVM, or other to identify classes, in the time allotted?

Yes

Yes

Yes

No

Classify signal acquisition feature vectors or signal acquisitions directly (FSA/SVM merger, for example, for latter case)

Perform SVM training and tuning to establish a classifier on the training subset provided from data*

SVM feature enrichment

Perform HMM-template and other feature boosts. Retry clustering. If still too weak, attempt merger of extraction and clustering stages.

Will a binary SVM decision tree suffice?

No

Use multiclass SVM.

Channel current cheminformatics output

Yes

Use binary SVM decision tree for binary or multiclass discrimination.

Figure 12.4 CCC protocol flowchart (part 3).

be solved via incorporation of generalized SVM methods. If facing only classification task (data already preprocessed), the SVM will also be the method of choice in what follows. (iii) The weakness of the SVM, whether used for classification or clustering, but especially for the latter, is the need to optimize over algorithmic, model (kernel), chunking, and other process parameters during learning. This is solved via use of meta-heuristics for optimization such as simulated annealing, genetic algorithm optimization, and particle swarm optimization. (iv) The main weaknesses in the metaheuristic effort is partly resolved via use of the "front-end" methods, like the FSA, and partly resolved by a knowledge discovery process using the SVM clustering methods. The

SSA Protocol weak signal acquisition and analysis method thereby establishes a robust signal processing platform.

The HMM methods are the central methodology or stage in the SSA Protocol, particularly in the channel current cheminformatics (CCC) protocol or implementation, in that the other stages can be dropped or merged with the HMM stage in many incarnations. For example, in some CCC analysis situations the time-domain Finite State Automaton (tFSA) methods could be totally eliminated in favor of the more accurate (but time consuming) HMM-based approach to the problem, with signal states defined or explored in much the same setting, but with the optimized Viterbi path solution taken as the basis for the signal acquisition.

The HMM features, and other features (from neural net, wavelet, or spike profiling, etc.) can be fused and selected via use of various data fusion methods, such as a modified Adaboost selection (from [1, 3]). The HMM-based feature extraction provides a well-focused set of "eyes" on the data, no matter what its nature, according to the underpinnings of its Bayesian statistical representation. The key is that the HMM not be too limiting in its state definition, while there is the typical engineering trade-off on the choice of number of states, N, which impacts the order of computation via a quadratic factor of N in the various dynamic programming calculations (comprising the Viterbi and Baum-Welch algorithms among others).

The HMM "sensor" capabilities can be significantly improved via switching from profile-MM (pMM) sensors to pMM/SVM-based sensors, as indicated in [1, 3, 44], where the superior performance and generalization capability of this approach was demonstrated.

In standard band-limited (and not time-limited) signal analysis with periodic waveforms, sampling is done at the Nyquist rate to have a fully reproducible signal capability. If the sample information is needed elsewhere, it is then compressed (possibly lossy) and transmitted (a "smart encoder"). The received data is then decompressed and reconstructed (by simply summing wave components, e.g., a "simple" decoder). If the signal is sparse or compressible, then compressive sensing [189] can be used, where sampling and compression are combined into one efficient step to obtain compressive measurements (the simple encoding in [189] since a set of random projections are employed), which are then transmitted. On the receiving end, the decompression and reconstruction steps are, likewise, combined using an asymmetric "smart" decoding step. This progression toward asymmetric compressive signal processing can be taken a step further if we consider signal sequences to be equivalent if they have the same stationary statistics. What is obtained is a method similar to compressive sensing, but involving stationary-statistics generative-projection sensing, where the signal processing is nonlossy at the level of stationary statistics equivalence. In the SCW signal analysis the signal source is generative in that it is describable via use of a hidden Markov model, and the HMM's Viterbi-derived generative projections are used to describe the sparse components contributing to the signal source.

In SCW encoding the modulation of stationary statistics can be man-made or natural, with the latter in many experimental situations involving a flow phenomenology that has stationary statistics. If the signal is man-made, usually the underlying stochastic process is still a natural source, where it is the changes in the stationary statistics that is under the control of the man-made encoding scheme. Transmission and reception are then followed by generative projection via Viterbi-HMM template matching or via Viterbi-HMM feature extraction followed by separate classification (using SVM). So in the SCW approach the encoding is even simpler (possibly nonexistent, directly passing quantized signal) and is applicable to any noise source with stationary statistics (the case for many experimental observations). The decoding must be even "smarter," on the other hand, in that generalized Viterbi algorithms are used, and possibly other machine learning methods as well, SVMs in particular. An example of the stationary statistics sensing with a machine learning-based decoder is described in application to channel current cheminformatics studies in what follows.

12.2.1 (Stage 1) Primitive Feature Identification

This stage is typically finite-state automaton based, with feature identification comprising identification of signal regions (critically, their beginnings and ends), and, as-needed, identification of sharply localizable "spike" behavior in any parameter of the "complete" (non-lossy, reversibly transformable) classic EE signal representation domains: raw time-domain, Fourier transform domain, wavelet domain, etc. (The methodology for spike detection is shown applied to the time-domain in Chapter 3.) Primitive feature extraction can be operated in two modes: off-line, typically for batch learning and tuning on signal features and acquisition; and on-line, typically for the overall signal acquisition (with acquisition parameters set – e.g., no tuning), and, if needed, "spike" feature acquisition(s).

The FSA method that is primarily used in the channel current cheminformatics (CCC) signal discovery and acquisition is to identify signal-regions in terms of their having a valid "start" and a valid "end," with internal information to the hypothesized signal region consisting, minimally, of the duration of that signal (e.g., the duration between the hypothesized valid "end" and hypothesized valid "start"). One approach along these lines is a signal "fishing" protocol "...constraints on valid 'starts' that are weak (with prominent use of 'OR' conjugation) and constraints on valid 'ends' that are strong (with prominent use of 'AND' conjugation)." Another approach to the signal analysis involves identifying anomalously long-duration regions. Identification of anomalously long duration regions in the more sophisticated Hidden Markov model (HMM) representation would require use of a

HMM-with-duration to not lose the information on the anomalous durations, which is one of the application areas for the HMMBD method (as discussed in Chapter 7).

Once identification rules, often threshold-based, are established for the signal start's and signal end's, then those definitions can be explored/used in signal acquisition. As those definitions are tuned over, by exploring the different signal acquisition results obtained with different parameter settings, the signal acquisition counts can undergo radical phase transitions, providing the most rudimentary of the holistic tuning methods on the primitive feature acquisition FSA (Chapter 3 for example). By examining those phase transitions, and the stable regimes in the signal counts (and other attributes in more involved holistic tuning), the recognition of good parameter regimes for accurate acquisition of signal can be obtained. As more internal signal structure is modeled by the FSA, the holistic tuning can involve more sophisticated tuning recognition of emergent grammars on the signal sub-states. The end-result of the tuning is a signal acquisition FSA that can operate in an on-line setting, and very efficiently (computation on the same order as simply reading the sequence) in performing acquisition on the class of signals it has been "trained" to recognize. Online learning is possible via periodic updates on the batch learning state/tuning process. For typical SSA (and CCC) applications, the tFSA is used to recognize and acquire "blockade" events (which have clearly defined start and stop transitions).

12.2.2 (Stage 2) Feature Identification and Feature Selection

This stage in the signal processing protocol is typically Hidden Markov model (HMM)-based, where identified signal regions are examined using a fixed state HMM feature extractor or a template-HMM (states not fixed during template learning process where they learn to "fit" to arrive at the best recognition on their train-data, the states then become fixed when the HMM-template is used on test data). The Stage 2 HMM methods are the central methodology/stage in the CCC protocol in that the other stages can be dropped or merged with the Stage 2 HMM in many incarnations. For example, in some data analysis situations the Stage 1 methods could be totally eliminated in favor of the more accurate HMM-based approach to the problem, with signal states defined/explored in much the same setting, but with the optimized Viterbi path solution taken as the basis for the signal acquisition structure identification. The reason this is not typically done is that the FSA methods sought in Stage 1 are usually only $O(T)$ computational expense, where "T" is the length of the stochastic sequential data that is to be examined, and "$O(T)$" denotes an order of computation that scales as "T" (linearly in the length of the sequence). The typical HMM Viterbi algorithm, on the other hand, is $O(TN^2)$, where "N" is the number of states in the HMM. Stage 1 provides a faster, and often more flexible, means to acquire signal, but it is more hands-on. If the core HMM/ Viterbi method can be approximated such that it can run at $O(TN)$ or even $O(T)$ in certain data regimes, for example, then the non-HMM methods in stage 1 could be phased out. Such HMM approximation methods present a data-dependent branching in the most efficient implementation of the protocol. If the data is sufficiently regular, direct tuning and regional approximation with HMM's may allow Stage 1 FSA methods to be avoided entirely. For general data, however, some tuning and signal acquisition according to Stage 1 will be needed (possibly off-line) if only to then bootstrap (accelerate) the learning task of the HMM approximation methods.

The HMM emission probabilities, transition probabilities, and Viterbi path sampled features, among other things, provide a rich set of data to draw from, for feature extraction (to create "feature vectors"). The choice of features is optimized according to the classification or clustering method that will make use of that feature information. In typical operation of the protocol, the feature vector information is classified using a Support Vector Machine (SVM). This is described in Stage 3 to follow. Once again, however, the Stage 3 classification could be totally eliminated in favor of the HMM's log likelihood ratio classification capability at Stage 2, for example, when a number of template HMMs are employed (one for each signal class). This classification approach is inherently weaker and slower than the (off-line trained) SVM methodology in many respects, but, depending on the data, there are circumstances where it may provide the best performing implementation of the protocol.

12.2.2.1 Stochastic Carrier Wave Encoding/Decoding

Using HMMBD we have an efficient means to establish a new form of carrier-based communications where the carrier is not periodic but is stochastic, with stationary statistics. The HMMBD algorithmic methodology [36], enables practical stochastic carrier wave (SCW) encoding/decoding with this method.

Stochastic carrier wave (SCW) signal processing is also encountered at the forefront of a number of efforts in nanotechnology, where it can result from establishing or injecting signal modulations so as to boost device sensitivity. The notion of modulations for effectively larger bandwidth and increased sensitivity is also described in [90, 91]. Here, we choose modulations that specifically evoke a signal type that can be modeled well with a HMMD but not with a HMM. This is a generally applicable approach where conventional, periodic, signal analysis methods will often fail. Nature at the single-molecule scale may not provide a periodic signal source, or allow for such, but may allow for a signal modulation that is stochastic with stationary statistics, as in the case of the nanopore transduction detector (NTD).

12.2.3 (Stage 3) Classification

This stage is typically Support Vector Machine (SVM)-based. SVMs are a robust classification method. If there are more classes to discern than two, the SVM can either be applied in a Decision Tree construction with binary-SVM classifiers at each node, or the SVM can internally represent the multiple classes. Depending on the noise attributes of the data, one or the other approach may be optimal (or even achievable). Both methods are typically explored in tuning, for example, where a variety of kernels and kernel parameters are also chosen, as well as tuning on internal KKT handling protocols. Simulated annealing and genetic algorithms have been found to be useful in doing the tuning in an orderly, efficient, manner. If the feature vectors produced correspond to complete data information/profiling in some manner, such is explicitly the case in a probability feature vector representation on a complete set of signal event frequencies (where all the feature "components" are positive and sum to 1), then kernels can be chosen that conform to evaluating a measure of distance between feature vectors in accordance with that notion of completeness (or internal constraint, such as with the probability vectors), e.g., the entropic kernel. Use of entropic kernels with probability feature vectors in proof-of-concept experiments have been found to work well with channel blockade analysis and is thought to convey the benefit of having a better pairing of kernel and feature vector, here the kernels have probability distribution measures (with Kulback–Leibler information divergences), for example, and the feature vectors are (discrete) probability distributions.

12.2.4 (Stage 4) Clustering

This stage is often not performed in the "real-time" operational signal processing task as it is more for knowledge discovery, structure identification, etc., although there are notable exceptions, one such comprising the jack-knife transition detection via clustering consistency with a causal boundary that is described in what follows. This stage can involve any standard clustering method, in a number of applications, but the best performing in the channel current analysis setting is often found to be an SVM-based external clustering approach (see [1, 3, 59]), which is doubly convenient when the learning phase ends because the SVM-based clustering solution can then be fixed as the supervised learning set for a SVM-based classifier (that is then used at the operational level).

A computationally "expensive" HMM signal acquisition at Stage 1 may be necessary for very weak signals, for example, if the typical Stage 1 methods fail. In this situation the HMM will probably have a very weak signal differential on the different signal classes if it were to attempt direct classification (and eliminate the need for a separate Stage 3). In this setting, the HMM would probably be run in the finest grayscale generic-state mode (see Chapter 7), with a number of passes with different window sample sizes to "step through" the sequence to be analyzed. Then, there are two ways to proceed: (i) with a supervised learning "bias", where windows on one side of a "cut" are one class, and those on the other side the other class, can a the SVM classify at high accuracy on train/test with the labeled data so indicated? If so, a transition is identified. In (ii), the idea is to use an unsupervised learning SVM-based clustering method where we look for a strong knife-edge split on clustered populations along the sequence of window samples. When this occurs, there is a strong identification of a transition. Since regions are identified (delineated) by their transition boundaries, we arrive at a minimally informed means for state and state-transition discovery in stochastic sequential data involving HMM/SVM-based channel current signal processing.

12.2.5   (All Stages) Database/Data-Warehouse System Specification

The adaptive HMM (AHMM) and modified SVM systems require implementation-specific data schema designs, for both input and output. The signal processing algorithms depend on information, represented structurally in the data, the algorithms are both process driven and data driven – these components impact the implementation of the algorithms.

The data schemas are typically implemented for optimal read time and ease of reuse and deployment, and have system dependencies that can be very significant, such as with client data-services involving distributed data access. The data schemas are typically implemented using flat files, low level operating system specific system calls to map data onto virtual memory, Relational Database Management Systems (RDBMS), and Object Database Management Systems (ODBMS). The database schemas are defined in two system contexts, (i) real time data acquisition, which includes feature recognition (AHMM) and classification (SVM), and, (ii) data warehousing for client data-service, and for further analysis that can be computationally intensive and requires substantial data processing.

The real-time data acquisition systems associated with the signal processing can be implemented using flat file systems and operating system specific virtual memory management interfaces. These interfaces are optimized to be scalable and high-bandwidth, to meet the requirements of high speed, real-time, data acquisition and storage. The data schemas allow for real-time signal processing such as feature recognition and classification, as well as local storage for subsequent export to a data warehouse, which can be implemented using industry standard RDBMS and ODBMS systems.

12.2.6 (All Stages) Server-Based Data Analysis System Specification

The data warehouse data schemas are optimized for applications-specific analysis of the signal processing tools in a distributed, scalable environment where substantial computing power can extend the analysis beyond what is possible in real-time. The local data acquisition systems produce and identify structure in real-time, storing the data locally, while another process can stream the data to an off-site data warehouse for subsequent analysis. The database uses data modeling tools to identify data schemas that work in tandem with the signal processing algorithms. The structure of the data schemas are typically integral to efficient implementation of the algorithms. Substantial off-line data preprocessing, for example, is used to create data structures based on inherent structure identified in the data. An internet-based user interface allows for access to the stored data and provides a suite of server-based, application-specific analysis, and data mining tools.

12.3 Channel Current Cheminformatics (CCC) Implementation of the Stochastic Sequential Analysis (SSA) Protocol

NTD, with the channel current cheminformatics (CCC) implementation of the SSA protocol described in Chapter 14, provides proof-of-concept examples of the SSA methods utilization, and can be used as a platform for finite state communication. From the CCC/NTD starting point it is easier to convey the unique signal boosting capabilities when working with real-time capable HMMBD signal processing [36] and other SSA methods. In the larger sense, recognition of stationary statistics transitions allows one to generalize to full-scale encoding/decoding in terms of stationary statistics "phases," i.e., stochastic phase modulation, a form of stochastic carrier-wave (SCW) communications. Many of the Proof-of-concept experiments described in what follows involve SSA applications, in a CCC implementation or a context for the NTD platform. The SSA Protocol is a general signal processing paradigm for characterizing stochastic sequential data.

In the Nanopore Transduction Detector (NTD) experiments the molecular dynamics of a (single) captured transducer molecule provides a unique stochastic reference signal with stable statistics on the observed, single-molecule blockade, channel current, somewhat analogous to a carrier signal in standard electrical engineering signal analysis. Changes in transient blockade statistics, coupled to SSA signal processing protocols, enables the means for a highly detailed characterization of the interactions of the transducer molecule with binding cognates in the surrounding (extra-channel) environment.

The transducer molecule is specifically engineered to generate distinct signals depending on its interaction with the target molecule. Statistical models are trained for each binding mode, bound and unbound, for example, by exposing the transducer molecule to zero or high (excess) concentrations of the target molecule. The transducer molecule is engineered so that these different binding states generate distinct signals with high resolution. Once the signals are characterized, the information can be used in a real-time setting to determine if trace amounts of the target are present in a sample through a serial, high-frequency sampling process.

Thus, in NTD applications of the SSA Protocol, due to the molecular dynamics of the captured transducer molecule, a unique reference signal with stationary (or approximately stationary) statistics is engineered to be generated during transducer blockade, analogous to a carrier signal in standard electrical engineering signal analysis. The adaptive machine learning algorithms for real-time analysis of the stochastic signal generated by the transducer molecule offer a "lock and key" level of signal discrimination. The heart of the signal processing algorithm is an adaptive Hidden Markov Model (AHMM)-based feature extraction method, implemented on a distributed processing platform for real-time operation. For real-time processing, the AHMM is used for feature extraction on channel blockade current data while classification and clustering analysis are implemented using a Support Vector Machine (SVM). In addition, the design of the machine learning-based algorithms allow for scaling to large datasets, real-time distributed processing, and are adaptable to analysis on any channel-based dataset, including resolving signals for different nanopore substrates (e.g. solid state configurations) or for systems based on translocation technology. The machine learning software has also been integrated into the nanopore detector for "real-time" pattern-recognition informed (PRI) feedback [1, 3, 35]. The methods used to implement the PRI feedback include *distributed* HMM and SVM implementations, which enable the processing speedup that is needed.

A mixture of two DNA hairpin species {9TA, 9GC} is examined in an experimental test of the PRI system. In separate experiments, data is gathered for the 9TA and 9GC blockades in order to have known examples to train the SVM pattern recognition software. A nanopore experiment is then run with a 1 : 70 mix of 9GC : 9TA, with the goal to eject 9TA signals as soon as they are identified, while keeping the 9GC's for a full five seconds (when possible, sometimes a channel-dissociation or melting event can occur in less than that time). The results showing the successful operation of the PRI system is shown in Figure 12.5 as a 4D plot, where the radius of the event "points" corresponds to the duration of the signal blockade (the fourth dimension). The result in Figure 12.5 demonstrates an approximately 50-fold speedup on data acquisition of the desired minority species.

To provide enhanced, autonomous reliability, the NTD is self-calibrating: the signals are normalized computationally with respect to physical parameters

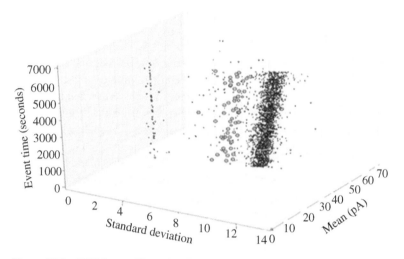

Figure 12.5 PRI Mixture Clustering Test with 4D plot [35]. The vertical axis is the event observation time, and the plotted points correspond to the standard deviation and mean values for the event observed at the indicated event time. The radius of the points correspond to the duration of the corresponding signal blockade (the fourth dimension). Three blockade clusters appear as the three vertical trajectories. The abundant 9TA events appear as the thick band of small-diameter (short duration, ~100 ms) blockade events. The 1 : 70 rarer 9GC events appear as the band of large-diameter (long duration, ~5 seconds) blockade events. The third, very small, blockade class corresponds to blockades that partially thread and almost entirely blockade the channel. *Source:* From Eren et al. [35].

(e.g. temperature, ph, salt concentration, etc.) eliminating the need for physical feedback systems to stabilize the device. In addition, specially engineered calibration probes have been designed to enable real-time self-calibration by generating a standard "carrier signal." These probes are added to samples being analyzed to provide a run-by-run self-calibration. These redundant, self-calibration capabilities result in a device which may be operated by an entry level lab technician.

Although the nanopore transduction detector can be a self-contained "device" in a lab, external information can be used, for example, to update and broaden the operational information on control molecules ("carrier references"). For the general "kit" user (see [1, 3]), carrier reference signals and other systemically engineered constructs can be used, for example, for a wide range of thin-client arrangements (where they typically have minimal local computational resource and knowledge resource). The paradigm for both device and kit implementations involve system-oriented interactions, where the kit implementation may operate on more of a data service/data repository level and thus need "real-time" (high bandwidth) system processing of data-service requests or data-analysis requests. Although not as system-dependent on database-server linkages, the more self-contained "device" implementation will still typically have, for example, local networked (parallelized) data-warehousing, and fast-access, for distributed processing speedup on real-time experimental operations.

12.4 SCW for Detector Sensitivity Boosting

Since an HMM template match can be weighted by its Viterbi path probability, the HMM's generative projection might offer improved efficiency in the compressive sensor method [189] when working with stochastic data with stationary statistics [67]. The HMM's generative capability derives from its ability to be "run in reverse," e.g., given the learned parameterization of a particular state model, an HMM can produce stochastic data with the same stationary statistics as that which originally was used in the statistical learning process. (The terminology "emission probability" relates to this generative perspective.) The SCW approach is similar to the compressive sensing approach [189] in that it involves a random (stochastic) projection, but in the SCW approach the stochastic projection is constrained to have stationary statistics and the main area of application is to stochastic sequential analysis. Compressive sensing makes use of a sequence of random projections, while in SCW a sequence of generative projections are done by using template HMMs and compressing stationary statistical parameters according to the generalized Viterbi algorithms. See [1, 3, 67, 218] for further variations (some using SVMs). Stationary statistical signals that are truly modulatory, e.g., not simply at a fixed level with Gaussian noise fluctuations about that level, will have a sparse data representation in terms of generative feature sets (which is used by SVMs for some classification tasks in [1, 3, 67, 218]).

New HMM implementations, allow a new form of carrier-based communications where the carrier is not periodic but is stochastic, with stationary statistics. The "stochastic carrier wave" (SCW) approach is not only a means to understand the messages Nature provides (in near-equilibrium flow phenomenologies with stationary statistics), but also provides a hidden carrier method, enabling security and making signal jamming much more difficult. An algorithmic methodology allows for 100-fold, or faster, implementation of a Hidden Markov model with duration (the HMMBD algorithm [34, 36, 56]), is critical to this encoding/decoding method. SCW communications are found at the forefront of a number of efforts in nanotechnology. This is because nature at the single-molecule scale has a signal modulation that is stochastic, sometimes with stationary statistics. Such is the case with the signal analysis in a nanopore transduction detector (NTD). Thus, further developments with stochastic carrier wave methods would serve both communications efforts and biosensing efforts.

If states have self-transitions with a notably nongeometric distribution on their self-transition "durations," then a fit to a geometric distribution in this capacity, as will be forced by the standard HMM, will be weak, and HMMD modeling will serve better. In engineered communications protocols, or in engineered, modulated, nanopore transduction detector (NTD) signals, highly nongeometric distributions can be sought. One encoding scheme that is strongly nongeometric in same-state duration distribution is the familiar *long* open-reading-frame (ORF) encoding found in genomic data. This suggests a similar ORF-like encoding scheme to establish a carrier duration peak in the self-transition distribution's tail region, e.g., a second peak in the duration distribution (perhaps one even more skewed from the geometric distribution than the heavy-tail distributions found for ORFs).

The NTD signal analysis demonstrates the simplest stochastic carrier wave utilization, in a biophysics experimental setting. A minor elaboration on the signal analysis, to go from a simple two-state (bound/unbound) signal recognition to a lengthy two-state telegraph signal, then yields the rudimentary implementation for stochastic carrier communications purposes.

12.4.1 NTD with Multiple Channels (or High Noise)

The nanopore transduction detection (NTD) platform involves functionalizing a standard nanopore detector platform in a new way that is cognizant of signal processing and machine learning capabilities and advantages, such that a highly sensitive biosensing capability is achieved. In the NTD functionalization of the standard nanopore detector we design a molecule that can be drawn into the channel (by an applied potential) but be too big to translocate, instead becoming stuck in a bistable "capture" such that it modulates the ion-flow in the *single* nanopore channel established in a distinctive way. An approximately two-state "telegraph signal" can be engineered. If the channel modulator is bifunctional, in that one end is meant to be captured and modulated while the other end is linked to an aptamer or antibody for specific binding, then we have the basis for a remarkably sensitive and specific biosensing capability. The biosensing task is reduced to the channel-based recognition of bound or unbound NTD modulators. Preliminary results demonstrate successful application of this method in a streptavidin (toxin) detection scenario using a biotinylated DNA hairpin. In typical NTD biosensing there is only one (nanometer-scale) channel established in the detector apparatus, however, where other channels bridging the same membrane (bilayer) would do so in parallel with the first (single) channel. In a naïve setting, additional channel noise sources degrade sensitivity and offset gains from having multiple channel "receptors." *In the stochastic carrier wave encoding/decoding with HMMD we aim to have multiple channels but avoid signal degradation such that the full benefits of a multiple receptor gain can be realized.*

In the NTD platform, sensitivity increases with observation time in contrast to translocation technologies where the observation window is fixed to the time it takes for a molecule to move through the channel. The key to the sensitivity and versatility of the NTD platform is the unique ability to couple real-time adaptive signal processing algorithms to the complex blockade current signals generated by the captured transducer molecule. The NTD approach can provide exquisite sensitivity and can be deployed in many applications where trace level detection is required.

In the NTD experiments the molecular dynamics of the captured transducer molecule provides a unique stochastic reference signal with (typically or approximately) stable statistics that is generated from the modulating blockade current, analogous to a carrier signal in standard electrical engineering signal analysis. By extension, changes in transient blockade statistics, coupled with sophisticated signal processing protocols, provide the means for a highly detailed characterization of the interactions of the transducer molecule with molecules in the surrounding (extra-channel) environment.

Consider the case where 100 parallel channels arc in operation, a scenario that has the potential to increase the sensitivity of the NTD 100-fold, but the signal analysis typically becomes more challenging, and sensitivity gains limited, since there are 100 parallel noise sources. The HMMD recognition of a transducer signal's stationary statistics, however, is analogous to "time integration" heterodyning a radio signal with a periodic carrier in classic electrical engineering in that there is improved carrier-signal recognition with longer observation time. In order to introduce a "time integration" benefit in the recognition of a transducer signal, periodic (or stochastic) modulations may be introduced to the transducer environment. In a high noise background, modulations may allow some of the transducer states to have heavy-tailed, or multimodal, self-transition duration distributions. With these modifications to the signal processing software a single transducer molecule signal is recognizable in the presence of 100's of channels. Increasing the number of channels by 100 and retaining the capability of recognizing a single transducer blockading one of those channels provides a direct gain in sensitivity according to the number of channels (e.g., 100 channels would provide a sensitivity boost of 100). It is important to note that the increase in sensitivity is mostly implemented computationally and does not add complexity or cost to the NTD device itself.

The adaptive machine learning algorithms for real-time analysis of the stochastic signal generated by the NTD transducer molecule are critical to realizing the increased sensitivity of the NTD and offer a "lock and key" level of signal discrimination. The transducer molecule is specifically engineered to generate distinct signals depending on its interaction with the target molecule. Statistical models are trained for each binding mode, bound and unbound, by exposing the transducer molecule to high concentrations of the target molecule. The transducer molecule has been engineered so that these different binding states generate distinct signals with high resolution. In operation, the NTD-biosensing process is analogous to giving a bloodhound a distinct memory of a human target by having it sniff a piece of clothing. Once the signals are characterized, the information is used in a real-time setting to determine if trace amounts of the target are present in a sample through a serial, high frequency sampling process.

The algorithms which describe the stochastic channel current modulations are a pure form of Machine Learning (a branch of Artificial Intelligence) in that there is no assumption of an underlying probability distribution; the statistical representation is directly generated from the data being produced by the molecular dynamics of the transducer molecule. A statistical model that is based on direct observation of the transducer molecule's dynamics eliminates the need for a parameterized statistical model, resulting in a higher resolution of discrimination.

12.4.2 Stochastic Carrier Wave

General methods are proposed for (i) stochastic sequential analysis; (ii) stochastic carrier-wave communications; (iii) holographic HMM extensions; and (iv) distributed HMM implementations. In method (ii), in particular, it is possible to establish a new type of communication process where the carrier wave is a stochastic observation sequence that obeys stationary statistics. In standard periodic carrier wave signal processing convolving with the carrier frequency allows the signal modulations of that carrier to be obtained. Here we have something analogous, but we have a carrier with stationary statistics, not fixed frequency, and can recognize different phases of stationary statistics via HMM methods for class-independent feature extraction, with Support Vector Machines (SVMs) for sparse data classification, or via HMM methods for class-dependent HMM generative projection (as mentioned in earlier comments).

In standard signal analysis with periodic waveforms, sampling is done at the Nyquist rate and the data compressed and transmitted (a "smart" encoder). The received data is then decompressed and reconstructed (by simply summing wave components, e.g., a "simple" decoder). If the signal is sparse or compressible, then compressive sensing [189] can be used, where sampling and compression are combined into one efficient step to obtain compressive measurements (referred to as "dumb" encoding in [189] since a set of random projections are employed), which are then transmitted. On the receiving end, the decompression and reconstruction steps are, likewise, combined using an asymmetric "smart" decoding step. This progression toward asymmetric compressive signal processing can be taken a step further if we consider signal sequences to be equivalent if they have the same stationary statistics. What is obtained is a method similar to compressive sensing, but involving stationary-statistics generative-projection sensing, where the signal processing is non-lossy at the level of stationary statistics equivalence. In the SCW signal analysis the signal source is generative in that it is describable via use of a hidden Markov model, and the HMM's Viterbi-derived generative projections are used to describe the sparse components contributing to the signal source. In SCW encoding the modulation of stationary statistics can be man-made or natural, with the latter in many experimental situations that involve flow phenomologies that have stationary statistics. If the signal is man-made, usually the underlying stochastic process is still a natural source, where it is the changes in the stationary statistics that is under the control of the man-made encoding scheme. Transmission and reception are then followed by generative projection via Viterbi-HMM template matching or via Viterbi-HMM feature extraction followed by separate classification (using SVM). So in the SCW approach the encoding is even "dumber" in that it can be any noise source with stationary statistics (the case for many experimental observations), with stationary statistics phase modulation for encoding. The decoding must be even "smarter," on the other hand, in that generalized Viterbi algorithms are used to perform a generative projection (and possibly other machine learning methods as well, SVMs in particular). An example of the stationary statistics sensing with a machine learning-based decoder is described in application to channel current cheminformatics studies in what follows.

In the standard HMM [126, 128], when a state "i" is entered, that state is occupied for a period of time, via self-transitions (with self-transition probability denoted as a_{ii}), until transiting to another state "j" (with probability a_{ij}). If the same-state interval is given as "d," the standard HMM description of the probability distribution on state intervals is implicitly given by a geometric distribution (see Chapter 7 for more details). The best-fit geometric distribution, however, is inappropriate in many cases. The standard HMMD replaces the above equation with a $p_i(d)$ that models the real duration distribution of state i. In this way, explicit knowledge about the duration of states is incorporated into the HMM.

The original description of an explicit HMMD required computation of order O $(TN^2 + TND^2)$ (where T is the sequence length to be examined, N is the number of states in the HMM/HMMD model, and D is the maximum duration length allowed in the HMMD model). The "D^2" term made the original approach prohibitively computationally expensive in practical, real-time, operations, and introduced a severe maximum-duration constraint on the duration-distribution model. Improvements via hidden semi-Markov models to computations of order O $(TN^2 + TND)$ are described in Chapter 7, where the maximum-interval constraint is still employed. In [36] we show that $O(TN^2 + TND*)$ is possible with the HMMBD algorithm, where $D*$ is the number of binned length states. The HMMBD implementation brings the HMMD modeling within the range of computational viability for many applications. In the HMMBD approach we also eliminate the maximum-duration constraint. We can often reduce to a bin representation with $D* < 10$, such that $D* \ll N$ in many situations, in which case that the HMMBD requires computations of order $O(TN^2)$, the same as for the HMM alone.

One important application of the HMM-with-duration (HMMD) method used in [1, 3, 36] includes kinetic feature extraction from EVA projected channel current data (the HMM-with-Duration is shown to offer a critical stabilizing capability in an example in [1, 3] and Chapter 7). The EVA-projected/HMMD processing offers a hands-off (minimal tuning) method for extracting the mean dwell times for various blockade states (the core kinetic information on the blockading molecule's channel interactions).

The HMM-with-Duration implementation, described in [1, 3] and Chapter 7, has been explored in terms of its performance at parsing synthetic blockade signals. In [1, 3] experiment the synthetic data was designed to have two levels, with lifetime in each level determined by a governing distribution (Poisson and Gaussian distributions with a range of mean values were considered.) The results clearly demonstrate the superior performance of the HMMD over the simpler HMM formulation on data with nongeometrically distributed same-state interval durations. With use of the EVA-projection method this affords a robust means to obtain kinetic feature extraction. The HMM with duration is critical for accurate kinetic feature extraction, and the results in [1, 3] suggest that this problem can be elegantly solved with a pairing of the HMM-with-Duration stabilization with EVA-projection.

In Figure 12.6 we show state-decoding on synthetic data that is representative of a biological-channel two-state ion-current decoding problem, or an encode/decode software radio signal. For this problem 120 data sequences were generated that have two states with channel blockade levels set at 30 and 40 pA (a typical scenario in practice). Every data sequence has 10 000 samples. Each state has emitted values in a range from 0 to 49 pA. The maximum duration of states is set at 500. The mean duration of the 40 pA state is given as 200 samples (typically have one sample every 20 ms in actual experiments), while the 30 pA level has mean duration set at 300 samples. The task is to train using 100 of the generated data sequences and attempt state-decoding on the remaining 20 data sequences. Example sequences are shown in Figure 12.6, along with their decoding when an HMM or an HMMD is employed. The performance difference is stark: the exact and adaptive HMMD decodings are 97.1% correct, while the HMM decoding is only correct 61% of the time (where random guessing would accomplish 50%, on average, in a two-state system). Three parameterized distributions were examined: geometric, Gaussian, and Poisson. Distributions that were segmented and "messy" were also examined. In all cases the HMMD performed robustly, similar to the above, and in

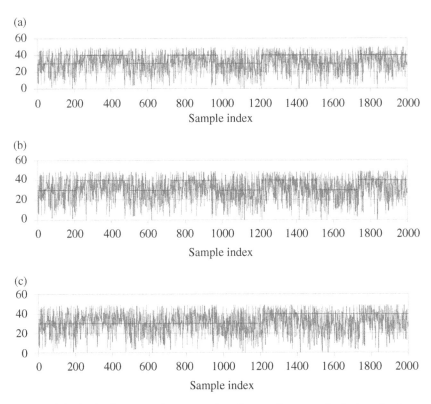

Figure 12.6 In the figure we show state-decoding results on synthetic data that is representative of a biological-channel two-state ion-current decoding problem. Signal segment (a) (at the top) shows the original two-level signal as the dark line, while the noised version of the signal is shown in gray. Signal segment (b) (in the middle) shows the noised signal in red and the two-state denoised signal according to the HMMD decoding process (whether exact or adaptive), which is stable (97.1% accurate) allowing for state-lifetime extraction (with the concomitant chemical kinetics information that is thereby obtained in this channel current analysis setting). Signal segment (c) (at the bottom) shows the standard HMM signal resolution, and its failure to properly resolve the desired level-lifetime information.

all cases the adaptive HMMD optimization performed comparably to the more computationally expensive exact HMMD.

Stochastic Carrier Wave (SCW) signal processing occurs in both natural and engineered situations. Whenever Nature is observed with a sequence of observations that have stationary statistics (associated with equilibrium and near-equilibrium flow situations, for example), then the basis for SCW signal processing arises. SCW also parallels all electrical engineering carrier-wave methodologies where periodic wave methods are used in some modulation scheme, thus the number of engineering applications is enormous. AM heterodyning, for example, can be replaced with stochastic carrier wave with pattern recognition informed (PRI) heterodyning. Also have phase modulation equivalence: the standard periodic carrier wave approach has a coherent phase reference, while SCW introduces a stochastic carrier wave with stationary statistics "phase." Have similar capabilities as with phased-locked loop (PLL), for example, where the phase tracking is done on SCW encoded information.

12.5 SSA for Deep Learning

The SSA Protocol allows for the discovery, characterization, and classification of localizable, approximately stationary, statistical signal structures in sequential data. In this situation we have leveraged Machine Learning (ML) to provide a solution to the "Big Data" problem, where a vast amount of data needs to be distilled down to its information content. The ML solution sought usually starts by performing some signal acquisition on the raw data. In doing so, ML solutions are strongly favored where a clear elucidation of the features is revealed. This then allows a more standard engineering design cycle to be accessed, where the stronger features thereby identified may play a stronger role, or guide the refinement to better features, to arrive at an improved classifier (for example). The SSA Protocol tackles the ML-based data analysis problem by tackling different stages: (i) the acquisition of signal, where only its boundaries (i.e., start and end) need to be discerned; (ii) the feature identification and extraction from the identified signal regions; (iii) classification or clustering based on that information.

The SSA Protocol "deep learning," thus, involves known constructs in each layer or stage of processing (according to the ML method optimal at that stage of the analysis). A problem can arise, however, if acquisition (stage 1) or feature identification (stage 2) fails. Even with Big Data to support the learning, the ML methods thus far, that can trace to each stage's learning to known constructs (features, etc.), can fail. The typical situation where this can occur is where the explicit feature set that must be realized is too large to handle. Language translation, is such a complicated situation. Just a five-word phrase is an explosion of possibilities – we will see that six or seven decades and a lot of human-structured paradigms for language processing (akin to SSA Protocol) will not match what can be done in about one page of neural net learning code (that calls a neural net implementation module) that will be described in Chapter 13.

Neural Net Deep Learning, to be discussed in the next Chapter 13, is optimal for those special circumstances where the feature set complexity demands it, or where an understanding of the underlying features identified by the model can be ignored. Unlike the learning in the SSA Protocol, here we have a single construct, the neural net, also arranged in layers (e.g., stages) but now the learning process is part of one unified learning optimization. It is found, however, that if that NN optimization is done on all of the network parameters (connection weights) simultaneously, the different layers have different learning rates, and the process is not efficient. So the novelty is to do training in the NN in stages, "freezing" the learning on weights in other layers while learning focuses on one layer at a time (there are a variety of strategies for this). What results is an ML solution that might not have been obtainable otherwise, but further refinements are difficult since, unlike the SSA Protocol (deep learning implicit), with NN learning in general we lack a direct understanding of what the features are.

12.6 Exercises

12.1 Write a report on how you would perform signal acquisition, where the signal properties are unknown, would you use an HMM or an FSA approach?

12.2 Write a report on the SSA Protocol and how it is related to a signal acquisition and processing methodology of your choice.

12.3 Describe the CCC implementation of the SSA protocol and the computational complexity of the various SSA stages.

12.4 Describe how SCW encoding can enhance detector sensitivity.

12.5 Describe SSA for (supervised) Deep Learning and compare with NN Deep Learning.

13

Deep Learning Tools – TensorFlow

The examples in Section 13.2 follow tutorials presented at the Google TensorFlow website www.tensorflow.org

13.1 Neural Nets Review

The background material on neurons and neural nets (NNs) is given in Chapter 9. For convenience a brief review follows.

13.1.1 Summary of Single Neuron Discriminator

See Section 9.1.1 for further details and where a key example (the Perceptron) is discussed. Here, we start with the neuron with a sigma activation function shown in Figure 13.1. The sigma activation function is differentiable, such that a stable backpropagation learning process can be implemented.

13.1.2 Summary of Neural Net Discriminator and Back-Propagation

In Chapter 9, we focused on ways to make a single neuron as enhanced as possible for classification and learning. This eventually led to the support vector machine (SVM) in Chapter 10. We now explore ways to make a collection of neurons, arranged as a layered NN (see Figure 13.2), into the best performing classifier and learning possible.

The core rule for training the NN (updating its weights) is backpropagation:

$$\omega = -\eta \nabla L \quad \text{and} \quad \nabla L = \delta_j^L \nabla z_j^L$$

Informatics and Machine Learning: From Martingales to Metaheuristics, First Edition.
Stephen Winters-Hilt.
© 2022 John Wiley & Sons, Inc. Published 2022 by John Wiley & Sons, Inc.

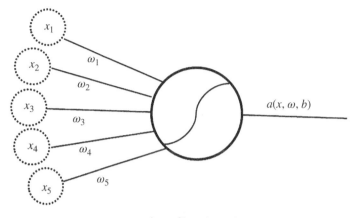

$$a(x, \omega, b) = \sigma(\omega \cdot x - b)$$

Figure 13.1 Single neuron. Sigma activation function: the inputs x_k are multiplied by the weights ω_k, and possibly have a bias applied, and the result is passed through a function that is monotonically increasing, differentiable, with asymptotes (−1 and +1 typically): $a(x,\omega,b) = \sigma(\omega \cdot x - b)$.

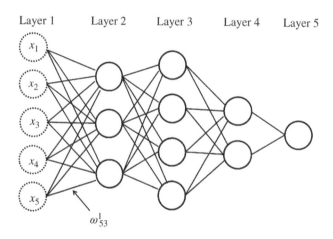

Figure 13.2 A five layer neural net.

where, back-propagation gives:

$$\delta_j^l = \frac{\partial L}{\partial z_j^l} = \sum_k \delta_k^{l+1} \omega_{kj}^{l+1} \sigma'\left(z_j^l\right)$$

See Chapter 9 for further details.

Note how the recursion with one layer only dependent on an adjacent layer reduces the weight-update computational complexity from exponential in weights to polynomial in weights (similar to what happens with the key Viterbi relation in the SVM dynamic programming table construction described in Chapter 10). Coupled with vectorized CPU/GPU in Section 13.2, this provides a directly accessible technology to solve learning problems. There are two barriers to this method simply solving everything: (i) The training for the NN can be tricky because the layers effectively learn at different rates under backpropagation [220, 221]. With a lot of time and data, however, this problem can be either be overcome by "brute-force" training, or by training with more finesse, where different layers are frozen during training in various ways [221], as will be discussed in the Tensor-Flow NN application in Section 13.2. (ii) Having sufficient data to have good training.

13.2 TensorFlow from Google

TensorFlow from Google is a library with an application programmer's interface (API) in Python and C++. In what follows we describe the developmental aspects, so only the Python API will be presented. For faster speed and separate front end development, by way of TensorBoard (Section 13.2.4) for example, a commercial production would probably switch to the C++ API. Other deep learning libraries include Caffe, Theano, and Torch, but only TensorFlow is discussed in what follows.

Implementing code from scratch for a neuron, all connected in a net for weights like in Figure 13.2 above, is not difficult and might even be a homework assignment (see Exercises in Section 13.3). So why is the interest in the deep learning libraries? It is because they have code implementations that tap into the GPU (and full CPU) capabilities of your computer to enable "vectorized" computation, where, crudely speaking, vector–vector, and vector–matrix products are done in the time of a single scalar–scalar product. This vastly accelerates the learning speed of the NN and allows different configurations to be tried in reasonable time-frames to arrive at a well-performing net. Furthermore, the mixed-mode computations (integer and floating point described in Section 4.5) are also directly accessible in the "Tensor Cores" addition to the TensorFlow library, to enable that speedup as well. This leaves little else to optimize but the hardware itself and high-level learning strategies.

TensorFlow makes NN training easy, and allows trickier, layered, training to be done in the NN more easily. The code-work is largely done and it is "simply" a matter of preparing the data in the proper format. On most projects, 90% of the work is acquiring the data in the proper format. Initial efforts at finding structure, such as two classes of data (clustering), then provide feedback on the more informative parameters and refine the data acquisition.

13.2.1 Installation/Setup

In what follows I will outline the installation trajectory I had for a linux computer. TensorFlow2 is available with a stable release (CPU and GPU support) for Ubuntu and Windows. System requirements include: Python3.5-3.8; pip 19.0+, and Ubuntu 16.04+.

Let us now proceed assuming you have a linux computer. (Very similar instructions are outlined at tensorflow.org for other hardware.) To see if your versions of Python and pip (also venv) are sufficient:

```
>python3 -version
>pip3 --version
```

Assuming you need updates:

```
>sudo apt update
>sudo apt install python3-dev python3-pip python3-venv
```

Now to isolate the TensorFlow and pip package installation from the system (to not cause damage). For this reason all that follows will now be done by switching to a python virtual environment using venv and using python's pip. First we create the home for a virtual environment via the python interpreter and making a directory./venv:

```
>python3 -m venv -system-site-packages ./venv
```

The virtual environment is then activated using a command that depends on the shell environment you have running. To echo your shell:

```
>echo $SHELL
```

My shell is bash, so I activate the virtual environment by:

```
>source ./venv/bin/activate
```

The shell prompt now switches (while in virtual) to (venv) >. Now proceed with the installation in the host environment without fear of causing damage to the system settings:

```
(venv) >pip install -upgrade pip
(venv) >pip install -upgrade tensorflow
```

All of the examples that follow are run via the Python Interpreter in the virtual environment. When done, the virtual session is ended with the command:

```
(venv) >deactivate
```

13.2.1.1 Tensors

Tensors in TensorFlow are not mathematical tensors from differential geometry (and relativistic Physics) which have transformational properties (covariance), so are not "simply" multi-dimensional arrays. In TensorFlow tensors are implementations of (immutable) multidimensional arrays...... but not "simply." This is because the TensorFlow implementation taps into the full CPU/GPU capabilities of your computer.

From a Python implementation perspective, there is a direct compatibility between TensorFlow operations and standard numpy operations. Tensors are backed by vectorized (accelerated) processing with advanced CPU/GPU handling and do this while allowing auto convert between numpy arrays and tensors and the reverse.

13.2.2 Example: Character Recognition

We are using the TensorFlow Python API, so the examples that follow are in Python. In TensorFlow the basic building block of the NN is the layer, which is usually part of a network consisting of sequentially connected layers. The Sequential model, consisting of a stack of layers, has the following schematic for a three-layer NN:

```
tensorflow as tf
from tensorflow import keras
from tensorflow.keras import layers
# Define Sequential model with 3 layers
model = keras.Sequential(
  [
    layers.Dense(2, activation="relu", name="layer1"),
    layers.Dense(3, activation="relu", name="layer2"),
    layers.Dense(4, name="layer3"),
  ]
)
```

Let us start with a classic case of image recognition, using the Modified National Institute of Standards and Technology (MNIST) character dataset [222]. The character MNIST dataset consists of 60 000 training images and 10 000 testing images.

Each scanned handwritten character image is fit into a 28 × 28 pixel image and anti-aliased, inducing grayscale values at each pixel. Each pixel is then treated as its own input, thus there are 28 × 28 = 784 inputs in the NN that follows. Extensive comments are given below the code

```
import tensorflow as tf
mnist = tf.keras.datasets.mnist
(x_train, y_train), (x_test, y_test) = mnist.load_data()
x_train, x_test = x_train / 255.0, x_test / 255.0
model = tf.keras.models.Sequential([
  tf.keras.layers.Flatten(input_shape=(28, 28)),
  tf.keras.layers.Dense(128, activation='relu'),
  tf.keras.layers.Dropout(0.2),
  tf.keras.layers.Dense(10)
])
predictions = model(x_train[:1]).numpy()
tf.nn.softmax(predictions).numpy()
loss_fn = tf.keras.losses.SparseCategoricalCrossentropy
(from_logits=True)
loss_fn(y_train[:1], predictions).numpy()
model.compile(optimizer='adam',
              loss=loss_fn,
              metrics=['accuracy'])
model.fit(x_train, y_train, epochs=5)
model.evaluate(x_test,  y_test, verbose=2)
probability_model = tf.keras.Sequential([
  model,
  tf.keras.layers.Softmax()
])
probability_model(x_test[:5])
```

OUTPUT (reduced):

```
Epoch 1/5:  1/1875 accuracy: 0.0312 → 1875/1875
accuracy: 0.9127.
Epoch 2/5:  1/1875 accuracy: 0.9375 → 1875/1875
accuracy: 0.9569.
Epoch 3/5:  1/1875 accuracy: 0.9688 → 1875/1875
accuracy: 0.9675.
Epoch 4/5:  1/1875 accuracy: 0.9688 → 1875/1875
accuracy: 0.9730.
```

```
Epoch 5/5:   1/1875 accuracy: 0.9062 → 1875/1875
accuracy: 0.9762.
```

In arriving at the NN code above, there were three stages: I Specify the Net; II Compile the Net; III Train the Net (usually with a final generalization learning Test automatic):

I) *Specify the node/network model:* The first layer simply reformats ("flattens") the 28×28 2D array data as a 1D array with 784 elements. The layers with the specification "Dense" are fully connected (as shown in the example in Figure 13.2). Here we see that the first dense layer has 128 nodes while the second (output) layer has 10 nodes (each output being a score indicating the image, a number, belongs to one of ten classes).

II) *Compile the model:* From the theory, establishing the network topology and the activation functions of the nodes is not a complete theory – a learning action must be specified (such as gradient descent) to arrive at the update (learning) rule on the weights. If wrapped in a variational formulation with a loss function (a regularizer), the theory can be completed in a way more amenable to theoretical analysis (as done in Chapter 9). For the square loss function, for example, one gets the standard gradient descent. The performance metric is here specified to be calling accuracy, and this is shown in the summarized output lines produced. With these code specification, the network is ready to be trained.

III) *Train the model:* Training is the classic supervised learning process of stepping through each instance of data input and output in the training set and perform the learning rule for each sequentially. For a training set consisting of 1875 instances, we see in the output how the calling accuracy improves from the first to last (1875th) instance in the training set. (epoch 1). This cycle through the training data is then repeated four more time (epoch 2 thru 5) with steady improvement as shown in the Output summary above.

Training is started with call the model.fit method, here specified to have five epochs. Once training is complete, a true test, with separate test data, can be done, with the model.evaluate method, for which we will generally see less than the training performance (due to overtraining).

Finally, to get a probability vector output and to show a simple example of adding another layer, here the softmax function layer, we declare probability_model as shown in the code. The ability to graft additional layers onto existing, learned, layers (that are "frozen") can be elaborated into a concept known as transfer learning.

13.2.2.1 Transfer Learning

Transfer learning freezes the bottom layers in a model and only trains the top layers. Here are two common transfer learning blueprints from tensorflow.org involving Sequential models. To freeze the weights at a given level, set `layer.trainable = False`, like this:

```
model = keras.Sequential([
  keras.Input(shape=(784))
  layers.Dense(32, activation='relu'),
  layers.Dense(32, activation='relu'),
  layers.Dense(32, activation='relu'),
  layers.Dense(10),
])

# load weights already learned
model.load_weights(...)

# freeze all but last layer
for layer in model.layers[:-1]:
  layer.trainable = False

# recompile and train:
model.compile(...)
model.fit(...)
```

A typical transfer learning workflow implementation for Keras:

1) Instantiate a base model and load pretrained weights into it.
2) Freeze all layers in the base model by setting `trainable = False`.
3) Create a new model on top of the output of one (or several) layers from the base model.
4) Train your new model on your new dataset.

13.2.2.2 Fine-tuning

Once your model has converged on the new data, you can try to unfreeze all or part of the base model and retrain the whole model end-to-end with a very low learning rate.

13.2.3 Example: Language Translation

In this section is described the amazing NN application to language translation. Assuming it is not a dead language (and even then, with a lot of written records...), there is as much data as we want. Thus, a key condition for NNs to play to their strength has been met. Other methods in ML and other, human constructed, language translators, with access to this same data, would often fail in complex translations, while the NN method does impressively well. With about one page of code using TensorFlow NNs this is solved. The key is to not only have the data, but have a good model for the data (at least for NN learning purposes), and many such models exist, including the attention model [223] that is used in the code here. Language datasets are available at http://www.manythings.org/anki/, where for English – Spanish you would see a line with the pair:

```
May I borrow this book? ¿Puedo tomar prestado este libro?
```

To use with the TensorFlow language learning with the attention model we need to prepare the data, according to tensorflow.org:

1) Add a *start* and *end* token to each sentence.
2) Clean the sentences by removing special characters.
3) Create a word index and reverse word index (dictionaries mapping from word → id and id → word).
4) Pad each sentence to a maximum length.

The code for data formatting and attention model implementation is at tensorflow.org, the NN training is then along the lines of what has been shown, so would not be repeated here.

13.2.4   TensorBoard and the TensorFlow Profiler

For NN training, the training process can lead to overtraining, or be slowed down for a variety of data-flow engineering reasons, so we must track performance as we train. Also, closely tracking this training process often offers suggestions for improvement (for faster training). In this section we show how to access a tool, TensorBoard, for getting performance measurements and visualizations. In what follows we will reuse the MNIST character recognition example, and show how to activate the TensorBoard log process and then run the TensorBoard visualizer on those logs. Here is how it is shown at tensorflow.org:

```
# Place the logs in a timestamped subdirectory
model = create_model()
model.compile(optimizer='adam',
          loss='sparse_categorical_crossentropy',
          metrics=['accuracy'])
log_dir = "logs/fit/" + datetime.datetime.now().strftime
("%Y%m%d-%H%M%S")
```

```
tensorboard_callback = tf.keras.callbacks.TensorBoard
(log_dir=log_dir, histogram_freq=1)
```

```
model.fit(x=x_train,
          y=y_train,
          epochs=5,
          validation_data=(x_test, y_test),
          callbacks=[tensorboard_callback])
```

Once the training completes, start the TensorBoard viewer at the command line. This launches a visualization GUI which graphically displays the learning process in terms of accuracy, loss, etc.

To optimize training performance a profiler is also needed on computational resource use and timing. The TensorFlow Profiler does this and requires the latest versions of TensorFlow and TensorBoard ($>=2.2$). The code to activate this (still in the virtual environment):

```
(venv) >pip install -U tensorboard_plugin_profile
```

The python script would run in the above *venv* environment and we'd create a tensorboard callback as before. The example of the MNIST character recognition is continued in this code example (from tensorflo.org):

```
import tensorflow as tf

# Confirm that TensorFlow can access the GPU.
device_name = tf.test.gpu_device_name()
if not device_name:
 raise SystemError('GPU device not found')
print('Found GPU at: {}'.format(device_name))

# Train an image classification model with TensorBoard
callbacks
import tensorflow_datasets as tfds
tfds.disable_progress_bar()

(ds_train, ds_test), ds_info = tfds.load(
    'mnist',
    split=['train', 'test'],
    shuffle_files=True,
    as_supervised=True,
    with_info=True,
)
```

```
    # Preprocess by normalizing pixel values to be between 0
and 1.
    def normalize_img(image, label):  """Normalizes images:
`uint8` -> `float32`."""  return tf.cast(image, tf.
float32) / 255., label

ds_train = ds_train.map(normalize_img)
ds_train = ds_train.batch(128)

    ds_test = ds_test.map(normalize_img)
ds_test = ds_test.batch(128)

    # Create the image classification model using Keras.
    model = tf.keras.models.Sequential([
    tf.keras.layers.Flatten(input_shape=(28, 28, 1)),
    tf.keras.layers.Dense(128,activation='relu'),
    tf.keras.layers.Dense(10, activation='softmax')
])
model.compile(
    loss='sparse_categorical_crossentropy',
    optimizer=tf.keras.optimizers.Adam(0.001),
    metrics=['accuracy']
)
    # Create a TensorBoard callback to capture performance
profiles and call it while training the model.
    # Create a TensorBoard callback
logs = "logs/" + datetime.now().strftime("%Y%m%d-%H%M%S")

tboard_callback = tf.keras.callbacks.TensorBoard(log_dir
= logs,
                        histogram_freq = 1,
                        profile_batch = '500,520')

model.fit(ds_train,
          epochs=2,
          validation_data=ds_test,
          callbacks = [tboard_callback])
```

The TensorFlow Profiler is embedded within TensorBoard. Launch Tensor-Board and navigate to the Profile tab to view performance profile:

```
(venv) >tensorboard --logdir=logs
```

In the example at tensorflow.org the MNIST analysis reveals the data processing is strongly input bound and adjustments to the data pipelining are suggested. This is then done and the profile is done again, showing a training speed improvement by a magnitude.

13.2.5 Tensor Cores

Mixed-precision computing (without sacrificing accuracy) is implemented from scratch in code in Section 4.5, in this section we note it is implemented in certain GPU architectures available since 2017. For the Tesla V100 accelerator, for example, their peak throughput was 12 times their 32-bit floating point throughput from their previous generation Tesla P100. Referred to as Tensor Cores on the Tesla V100, they are already supported in Deep Learning frameworks, including Tensor-Flow, as well as PyTorch, MXNet, and Caffe2.

According to the Tesla V100's specifications, Tensor Cores are "programmable matrix-multiply-and-accumulate units that can deliver up to 125 Tensor TFLOPS for training and inference applications. The Tesla V100 GPU contains 640 Tensor Cores: 8 per SM. Tensor Cores and their associated data paths are custom-crafted to dramatically increase floating-point compute throughput at only modest area and power costs. Clock gating is used extensively to maximize power savings. Each Tensor Core provides a $4 \times 4 \times 4$ matrix processing array which performs the operation $D = A \times B + C$, where A, B, C and D are 4×4 matrices.... The matrix multiply inputs A and B are FP16 matrices, while the accumulation matrices C and D may be FP16 or FP32 matrices."

13.3 Exercises

13.1 Implement a NN (from scratch) in Python, with back-propagation learning with gradient descent learning, and apply it to the MNIST character recognition problem.

13.2 Implement the NN as described in (Exercise 13.1) but now using Tensor-Flow (partly described in Section 13.2.2).

13.3 Implement an English – Spanish Translator according to Section 13.2.3 and the translator implementation instructions at tensorflow.org.

13.4 Implement an English → Spanish → French → English "Rephraser" by repeated application of the solution for (Exercise 13.3). How well does it work?

13.5 Implement a paragraph auto-story generation, from any of a variety of story-generation websites online, then "Rephrase" by applying the solution to (Exerxise 13.4). Is the "regularized" auto-generated story or dialogue more meaningful?

14

Nanopore Detection – A Case Study

Numerous prior book, journal, and patent publications by the author are drawn upon extensively throughout this chapter [1–68]. Almost all of the journal publications are open access. These publications can typically be found online at either the author's personal website (www.meta-logos.com) or with one of the following online publishers: www.m-hikari.com, or bmcbioinformatics.biomedcentral.com.

A nanopore detector (ND) is based on a membrane that separates two chambers of electrolytic solution with a single nanopore providing a channel across that membrane. The detector is based on ionic current observations when a potential difference is applied across the membrane. Objects drawn into the nanopore cause ionic current blockades that form the basis of the molecular observations (i.e. observations derived from the ionic current imprints, or blockades, due to captured or translocating molecules). With channel current detection, particle analysis can be done on solutions to obtain particle concentrations, solution mixture composition, and even molecular dynamics. Early channel current detectors had millimeter diameters (0.1 mm) and were used to count cell concentrations and mixture compositions [224]. Information obtained about the excluded cell volume was used in classifying blood cells as red or white, for example, the ratio of which provided important data for medical diagnostics. The 100 μm-scale pores of Coulter were devised in the early 1950s.

It was not until the early 1970s that nanometer-scale pores were examined [225–227]. Bean made a nanometer-scale channel from crystalline structures (mica) that had defective tracks (from fission events). When etched with HF the normally impervious mica is removed along the defect-track in its crystalline structure. Depending on how this process is controlled, pores have been obtained with diameters ranging down to 6 nm (50 nm diam. pores commercially available). Although this technology has been used for observations on uncharged particles (polystyrene spheres with 90 nm diameter [226]), it does not work as well with charged molecules (like DNA). Another complication is that the etching method for pore construction inevitably leads to long tunnel-like channels, which does not

Informatics and Machine Learning: From Martingales to Metaheuristics, First Edition.
Stephen Winters-Hilt.
© 2022 John Wiley & Sons, Inc. Published 2022 by John Wiley & Sons, Inc.

provide the best configuration for detector uses. Detection of biomolecules with biologically based nanometer-scale pores also showed promise at about this time with the work by Hladky and Haydon [228]. They showed that a biological channel, the bacterial antibiotic gramicidin, could self-assemble in a lipid bilayer to form a functional channel (with currents of order 1 pA). This potentially solved two of the mica-channel problems: the lipid bilayers are very thin, about 5 nm, and the protein-based, biologically functional, nanometer-scale pore seemed better suited to passing charged biomolecules. Gramicidin was too small to detect most biomolecules, however, since it could barely pass molecules the size of the water molecule. It was not until 1994 [229] that a sufficiently large pore was studied, α-hemolysin. In the 1994 paper, Bezrukov et al. studied the blockades resulting from a charge-neutral polymer: polyethylene glycol (PEG). Later modifications to the gramicidin pore permitted its use as an antibody-modulated (on–off) biosensor, while modifications to the α-hemolysin pore enabled its use as a metal biosensor [230], among other things [231, 232]. In 1995 and 1999, α-hemolysin was successfully used for DNA homopolymer translocation studies and classification [233, 234]. Nanometer-scale pores then began to be developed in solid-state media [235, 236]. Nanopores provide rich opportunities for the future because at nanometer scale a wealth of new prospects arise, from characterizing just about anything that can form a colloidal suspension in electrolytic solutions, to polymers like DNA, to the molecular motions that indicate molecular identity [67, 93, 95].

One of the key strengths of NDs is that they analyze populations of single molecules. With signal processing and pattern recognition, this information enables a new type of cheminformatics based on channel current measurements. Single molecule observations are also of interest in biophysics; binding/conformational changes on captured dsDNA end regions, for example, might be tracked and understood using the nanopore blockade signal. DNA regions away from the ends may eventually be studied in a similar manner, using pore-translocation confinement to reveal distinctive conductance/binding properties on those bases threading the pore's limiting aperture constriction. Single molecule classifications permit a number of technical innovations. For sequencing, the single molecule basis of measurement may permit Sanger-type sequencing [237, 238] on DNA molecules separated by capillary electrophoresis. If DNA can be translocated slowly enough, through a limiting aperture with dominant contributions to resistance spanning only two or three nucleotides length (about 20 Å for ssDNA, 10 Å for dsDNA), then DNA sequencing of a single molecule may eventually be possible. For single nucleotide polymorphism (SNP) identification, small sample volumes can be used, such that PCR amplification may not be needed. Expression analysis and disease identification (for individualized therapeutics) are just a few of the possibilities. Non-PCR expression analysis may even offer a new level of experimentation on live cells using patch-clamp methods.

14.1   Standard Apparatus

Nanopore detection is based on a nanometer-scale ion channel that can report on the channel-interactions of individual, nanometer-scale, biomolecules. The reporting is via measurements of ion flow through the channel when there is only a single channel, i.e. there is only one conductance path.

Each nanopore experiment described in what follows was conducted using one α-hemolysin channel inserted into a diphytanoyl-phosphatidylcholine/hexadecane bilayer, where the bilayer was formed across a 20-μm diameter horizontal Teflon aperture (see Figure 14.1). The bilayer separates two 70-ml chambers containing 1.0 M KCl buffered at pH 8.0 (10 mM HEPES/KOH). A completed bilayer between the chambers was indicated by the lack of ionic current flow when a voltage was applied across the bilayer (using Ag–AgCl electrodes). Once the bilayer was in place, a dilute solution of α-hemolysin (monomer) was added to the *cis* chamber. Self-assembly of the α-hemolysin heptamer and insertion into the bilayer results in a stable, highly reproducible, nanometer-scale channel with a steady current of 120 pA under an applied potential of 120 mV at 23 °C (±0.1 °C using a Peltier device). Once one channel is formed, further pores were prevented from forming by thoroughly perfusing the *cis* chamber with buffer. Molecular blockade signals were then observed by mixing analytes into the *cis* chamber.

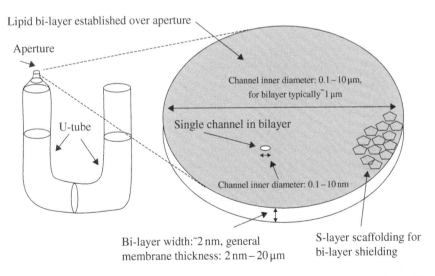

Lipid bi-layer established over aperture

Aperture

U-tube

Channel inner diameter: 0.1 – 10 μm, for bilayer typically ~1 μm

Single channel in bilayer

Channel inner diameter: 0.1 – 10 nm

Bi-layer width: ~2 nm, general membrane thickness: 2 nm – 20 μm

S-layer scaffolding for bi-layer shielding

Figure 14.1   A schematic for the U-tube, aperture, bilayer, and single channel, with possible S-layer modifications to the bilayer.

14.1.1 Standard Operational and Physiological Buffer Conditions

The standard buffer condition for the ND is 1.0 M KCl with a pH of 8.0. This buffer was found to be most conducive to channel formation and to channels that do not gate. At significantly lower pH the channel is known to gate, if it even forms in the first place, which complicates use of the ND at physiological conditions (pH 7.0, 100 mM NaCl). Since the pH of blood is usually in the range 7.35–7.45, and channel formation has been observed at 250 mM KCl, nanopore operation at the high pH and high salt end of the physiological range, relevant for antibody function in the bloodstream, may be possible with minimal alteration to the experimental parameters. Evaluation of antibody/antigen binding efficacy in a physiological buffer environment is particularly important if the nanopore/antibody detector is to be used for clinically relevant screening on the efficacy of antibodies to a given antigen. Biochemically relevant screening on enzyme activity, for example, requires working with physiological buffer testing.

14.1.2 α-Hemolysin Channel Stability – Introduction of Chaotropes

The α-hemolysin channel is stable up to high salt concentrations ($MgCl_2$ above 2 M and KCl up to 4 M) and presence of some other additives (urea up to 7 M in some experiments, glycerol 5%) at pH around 8.0. Typical pattern of current rise with increase in background electrolyte, KCl is observed. Specifically, the current versus KCl concentration is obtained in running buffer with composition 1 M KCl, 20 mM HEPES (pH 8.0), with HEPES concentration maintained constant as content of KCl is increased.

A limitation in the utility of the nanopore/antibody antigen-binding tester (similarly for a nanopore/aptamer tester) is that once antigen is bound by a channel-captured antibody it is very difficult to effect the release of that antigen. This is a complicating issue in acquiring a large sample of antibody–antigen binding observations. A buffer-based solution to this problem is already known from purifying antibodies through a column containing antigen, where the release of antibodies bound in the column is effected by perfusion with 1.0 M $MgCl_2$. This presents the possibility of weakening the antibody–antigen binding by some choice of buffer in order to obtain large sample sets of binding events. The limitation of this is that the parameters will have likely deviated substantially from the physiological norm. Alternatively, a balanced stoichiometric ratio of antibody to antigen could be rapidly sampled, with lengthy sampling acquisitions only on antibody captures that occur without bound antibody and that then wait to observe antigen binding. Further details on buffer modulation, particularly with chaotropes, is given in [2, 31].

14.2 Controlling Nanopore Noise Sources and Choice of Aperture

The accessible detector bandwidth is limited by noise resulting from $1/f$ (flicker) noise, Johnson noise, Shot noise, and membrane capacitance noise. In Figure 14.2, upper left, the current spectral density is shown for the typical bilayer, an open α-hemolysin channel, and a channel with DNA hairpin blockade. For 1.0 M KCl at 23 °C, the α-hemolysin channel conducts 120 pA under an applied potential of 120 mV. The thermal noise contribution at the 1 GΩ channel resistance has an RMS noise current of 0.4 pA, consistent with Figure 14.2. Shot noise is the result of current flow based on discrete charge transport. During nanopore operation with 120 pA current (with 10 kHz bandwidth) there is, similarly, about 0.6 pA noise due to the discreteness of the charge flow. As with Johnson noise, the Shot noise spectrum is white, consistent with Figure 14.2. The specific capacitance of lipid bilayers is approximately 0.8 $\mu F/cm^2$ (very large due to molecular dimensions), and the specific conductance is approximately $10^{-6}\ \Omega^{-1}/cm^2$. In order for bilayer conductance to produce less RMS noise current than fundamental noise sources (under the conditions above), the leakage current must be a fraction of a pA. This problem is solved by reducing to less than a 500 μm^2 bilayer area, for which less than 0.6 pA leakage current results and for which total bilayer capacitance is at most 4 pF. This indicates that a decrease in bilayer area by another magnitude is about as far as this type of noise reduction can go. Preliminary attempts to do this, however, led to a very unpredictable toxin intercalation rate (possibly due to surface tension factors), among other difficulties.

The five DNA hairpins shown in Figure 14.2 have been studied in [1, 3, 67, 93, 95, 217], where they have been carefully characterized, and are used in other experiments as highly sensitive controls. Use of the controls entails testing a channel, especially an oddly behaving channel, with a known nine base-pair DNA hairpin control. If the familiar, visibly discernible, control blockade signals do not occur, the channel's viability is then looked into further. The nine base-pair hairpin molecules examined in the prototype experiment share an eight base-pair hairpin core sequence, with addition of one of the four permutations of Watson–Crick base-pairs that may exist at the blunt end terminus, i.e. 5′-GC-3′, 5′-CG-3′, 5′-TA-3′, and 5′-AT-3′. Denoted 9GC, 9CG, 9TA, and 9AT, respectively. The full sequence for the 9CG hairpin is 5′ <u>CTTCGAACG</u>TTTT <u>CGTTCGAAG</u> 3′, where the base-pairing region is underlined. The eight base-pair DNA hairpin is identical to the core nine base-pair subsequence, except the terminal base-pair is 5′-GC-3′. The prediction that each hairpin would adopt one base-paired structure was tested and confirmed using the DNA mfold server.

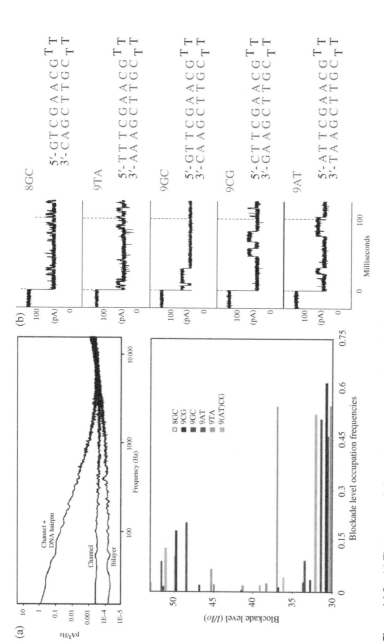

Figure 14.2 (a) The top panel shows the power spectral density for signals obtained. The bottom panel shows the dominant blockades, and their frequencies, for the different hairpin molecules. (b) **DNA hairpin controls and their diagnostic signals.** The secondary structure of the DNA hairpins is shown on the right, with their highest scoring diagnostic signals shown on the left. Each signal trace starts at approximately120 pA open channel current and all blockades are in a range 40–60 pA upon "capture" of the associated DNA hairpin. Even so, the signal traces have discernibly different blockade structure, which is extracted using an HMM. The signals are aligned at their blockade starts and the demarked time-trace is for 100 ms. *Source:* Based on Winters-Hilt et al. [67]; Winters-Hilt [217].

14.3   Length Resolution of Individual DNA Hairpins

The α-hemolysin geometry was probed using a series of hairpins, with stem lengths ranging from three to nine base-pairs. The six-base-pair hairpin is described in what follows, while those hairpins longer than six base pairs shared the six-base-pair stem/head at their core. Those hairpins with stems less than six base pairs were constructed by removing base pairs from the six base-pair core sequence. Starting from the three-base-pair hairpin, each base pair addition resulted in a measurable increase in median blockade shoulder lifetime that correlated with the calculated $\Delta\Delta G°$ of hairpin formation (Figure 14.3).

Standard free energy of hairpin formation was calculated using the mfold DNA server (see Figure 14.3b), and correlated with median duration of hairpin shoulder blockades (solid circles). Each point represents the median blockade duration for a given hairpin length acquired using a separate α-hemolysin pore on a separate day. Median blockade durations and $\Delta G°$ for the equivalent of the 6 bp hairpin with a single mismatch (6 bpA$_{14}$, Figure 14.3b) are represented by open squares. All experiments were conducted in 1.0 M KCl at $22 \pm 1\,°C$ with a 120 mV applied potential. (Increasing stem length also resulted in a 10 μs increase in median duration of the terminal spike, consistent with longer, but still microsecond time-scale, ssDNA translocation on the dissociated hairpin.) A downward trend in shoulder current amplitude was also observed from I/I_o equal to 68% for a 3 bp stem to I/I_o equal to 32% for a 9 bp stem (see Figure 14.3b).

The time-domain finite state automaton (FSA) passed 529 of the six-base-pair hairpin events to the support vector machine (SVM) and 3185 of all other events (see Figure 14.4a). Blockade events caused by six-base-pair hairpins were classified against blockades caused by 3, 4, 5, 7, and 8 base-pair hairpins (see Figure 14.4b). Because selectivity was relaxed at the FSA, there were many ambiguous signals with SVM scores near zero. Using an additional set of independent data, the SVM can be trained to exclude these by introducing a rejection region (the region between dashed lines in Figure 14.4b). The events that were rejected were primarily fast blockades similar to those caused by dumbbell hairpin (which cannot translocate, see [95] for further details) or acquisition errors caused by the low selectivity threshold of the FSA. When 20% of the events were rejected in this manner, the SVM scores for the six-base-pair hairpin discrimination achieved a sensitivity of 98.8% and a specificity of 98.8%. Sensitivity is defined as true positives/(true positives + false negatives), and specificity is defined as true positives/(true positives + false positives). (A true positive is an event in the test data that comes from the positive class and is assigned a positive value; a false positive occurs when the SVM assigns a positive score to an event in the test data when that event actually comes from the negative class. A false negative is an event that is assigned a negative value, but actually

(a)

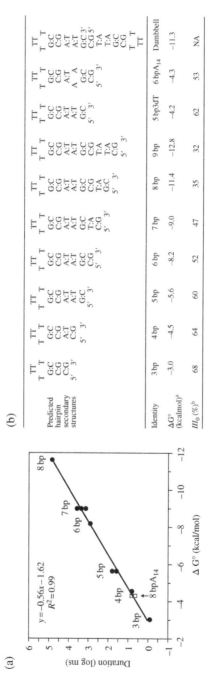

(b)

Predicted hairpin secondary structures	3bp	4bp	5bp	6bp	7bp	8bp	9bp	5bp3dT	6bpA14	Dumbbell
Identity	3 bp	4 bp	5 bp	6 bp	7 bp	8 bp	9 bp	5 bp3dT	6 bpA₁₄	Dumbbell
$\Delta G°$ (kcal/mol)[a]	−3.0	−4.5	−5.6	−8.2	−9.0	−11.4	−12.8	−4.2	−4.3	−11.3
III_0 (%)[b]	68	64	60	52	47	35	32	62	53	NA

Figure 14.3 (a) Standard free energy of hairpin formation vs. shoulder blockade duration. (b) DNA hairpins studied on the prototype nanopore detector. *Source:* (b) Winters-Hilt [217].

(a)

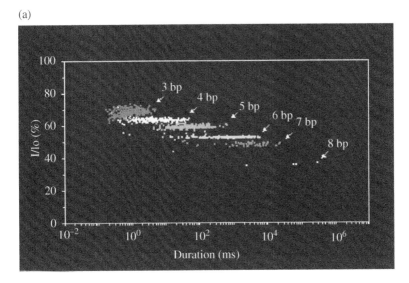

(b)

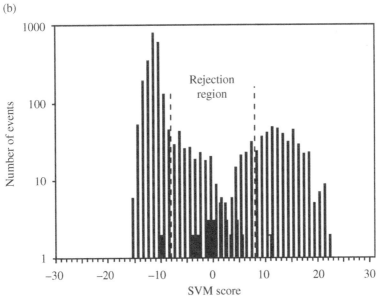

Figure 14.4 (a) Event diagram for DNA hairpins with three to eight base-pair stems. Events were selected for adherence to the shoulder-spike signature. Each point represents the duration and amplitude of a shoulder blockade caused by one DNA hairpin captured in the pore vestibule. The data for each hairpin are from at least two different experiments run on different days. Median I/I_o values for each type of hairpin varied by at most 2%. The duration of the 9 bp hairpin blockade shoulders were too long for us to record a statistically significant number of events. Control oligonucleotides with the same base compositions as

comes from the positive class.) Similar results were obtained for each class of hairpins depicted in Figure 14.4a. Overall, the SVM achieved an average sensitivity of 98% and average specificity of 99%. Thus, the stem length of an individual DNA hairpin was determined at single base-pair resolution.

14.4 Detection of Single Nucleotide Differences (Large Changes in Structure)

DNA hairpins with single nucleotide differences were examined in a context where those differences were expected to lead to significant differences in the hairpin molecule's secondary structure. The first hairpin considered involved alterations to the loop of the standard five-base-pair hairpin with a 4-deoxythymidine loop (**5bp4dT** in Figure 14.3b) to one with a 3-deoxythymidine loop (**5bp3dT** in Figure 14.3b). The **5bp3dT** hairpin caused pore blockades in which the shoulder amplitude was increased ~2 pA and the median shoulder duration (21 ms) was reduced threefold relative to the same hairpin stem with a 4-deoxythymidine loop (**5bp** in Figure 14.3b). Typical events are illustrated in Figure 14.5. The FSA acquired 3500 possible five-base-pair hairpin signals from 10 minutes of recorded data. The SVM classification for this data set (Figure 14.5) gave sensitivity and specificity values of 99.9% when 788 events were rejected as the unknown class. The second example involved the hairpin stem. Introduction of a single base-pair mismatch into the stem of a six-base-pair hairpin ($T_{14} \rightarrow A_{14}$, **6bpA$_{14}$** in Figure 14.3b) caused an approximately 100-fold decrease in the median blockade shoulder duration relative to a hairpin with a perfectly matched stem (**6bp** in Figure 14.3b). Typical events are shown in Figure 14.5. This difference in duration is consistent with the effect of a mismatch on $\Delta G°$ of hairpin formation (Figure 14.3a), and it permitted a 90% separation of the two populations using the manually applied shoulder-spike diagnostic. When analysis was automated, the FSA acquired 1031 possible events from 10 minutes of recorded data (Figure 14.5). With the aid of wavelet features that characterized the low frequency

---

Figure 14.4 (Continued) the DNA hairpins, but scrambled, caused blockade events that were on average much shorter than the hairpin events and that did not conform to the shoulder-spike pattern. *Source*: Winters-Hilt [217]. (b) Classification of the 6 bp hairpin (solid bars) versus all other hairpins (open bars) by SVM. Note the log scale on the y-axis. The dashed lines mark the limits of the rejection region. The boundaries of the rejection region were determined by independent data, not *post hoc*, on the data shown. The events that were rejected were primarily fast blockades similar to those caused by loops on the dumbbell hairpin (shown in the table in the right panel in Figure 14.3) or acquisition errors caused by the low selectivity threshold of the FSA.

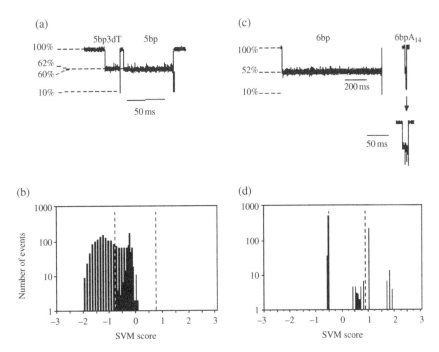

Figure 14.5 Detection of single nucleotide differences between DNA hairpins. (a) Comparison of typical current blockade signatures for a 5 bp hairpin and a 5 bp hairpin with a three-dT loop. The standard 5 bp hairpin event has a 2% deeper blockade than the 5 bp3dT hairpin. (b) Histogram of SVM scores for 5 bp hairpins (filled bars) versus 5 bp hairpins with three-dT loops (clear bars). (c) Comparison of typical current blockade signatures for a standard 6 bp hairpin and a 6 bp hairpin with a single dA_3-dA_{14} mismatch in the stem. The 6bpA$_{14}$ event is expanded to show the fast downward spikes. These rapid, near-full blockades and the much shorter shoulder durations are the main characteristics identified and used by SVM to distinguish 6bpA$_{14}$ hairpin events from 6 bp hairpin events. (d) Histogram of SVM scores for 6 bp hairpins (filled bars) versus 6bpA$_{14}$ hairpins (clear bars). *Source*: Winters-Hilt [217].

noise within the shoulder current, the SVM was able to discriminate the standard six-base-pair hairpin from the mismatched six-base-pair hairpin with sensitivity 97.6% and specificity 99.9% while rejecting only 42 events.

14.5 Blockade Mechanism for 9bphp

More involved than the classification or sequencing of a molecule is the actual understanding of that molecule's kinetic behavior as revealed by the ionic current blockade information measured by the ND. In the discussion of blockade mechanism that follows, for the nine-base-pair hairpins, the remarkable sensitivity of the nanopore device becomes apparent. This indicates that the α-hemolysin ND is likely to be an important tool for single-molecule observation and manipulation.

Ionic flow through the α-hemolysin channel was strongly modulated by the terminal base-pair on DNA hairpins with stem-length nine or more base-pairs. This modulation was most apparent on the nine base-pair hairpins, where the blockade states were discerned with the shortest time-constants (lifetimes). The lower level (LL) blockade states for the DNA hairpins are found to have lifetimes that correlate with the energy of dissociation on the terminal bond [93]. An anticorrelation with terminal bond energy is found for the density of LL blockade spikes.

A working model has been developed to explain the mechanisms underlying the current transitions for the observed 9 bp hairpin blockades (Figure 14.6). The model requires that the 9 bp duplex stem is long enough so that the terminal base pair can interact with amino acids in the vestibule wall and that a frayed end can reach the limiting aperture (at lysine-147). This is a reasonable assumption because circular dichroism assays indicate that the 9 bp hairpin stem is a B form duplex in bulk phase. The length per base-pair of B form DNA is 3.38 Å, therefore the total stem length is approximately 30.4 Å. The distance between the narrowest part of the vestibule mouth at threonine-9 and the pore limiting aperture at lysine-147 is 33 Å. Therefore, if the hairpin loop is perched at the ring formed by threonine-9, the 9 bp stem would reach within 3 Å of the limiting aperture. Given the uncertainty about the exact position of the hairpin loop and the 1.9 Å precision of the α-hemolysin X-ray crystal structure [239], not to mention effects from waters of hydration and ion fixed-layers, this distance is probably accurate within ± 1 bp. Upon capture of a 9 bp blunt hairpin, the initial conductance state ($I_{IL}/I_o = 35\%$, where IL stands for intermediate level) is caused by orientation and immobilization (on the millisecond time scale) of the hairpin due to an electrostatic bond formed between the terminal base pair of the hairpin stem and residues in the vestibule wall. The predominant interaction is binding between the nucleotide in the 3′ position and the protein. This state initiates virtually all events because it is entropically favored. The dwell time, τ_{IL}, for the intermediate conductance state is largely independent of base pair identity or orientation because the bases are hydrogen bonded to one another and the interaction with the surface is due to the terminal 3′ phosphodiester anion. If, however, the 3′ nucleotide is unpaired (i.e. a dangling nucleotide) its identity does matter in terms of the duration of τ_{IL}. Preliminary results show that single nucleotide 3′ overhangs have dwell times in the IL state with order dA > dC. This suggests that the unpaired bases hydrogen bond or stack against residues in the pore vestibule. The IL state invariably transitions to the upper conductance state, upper level (UL). This state ($I_{UL}/I_o = 48\%$) corresponds to desorption of the terminal base pair from the protein wall and orientation of the hairpin stem along the axis of

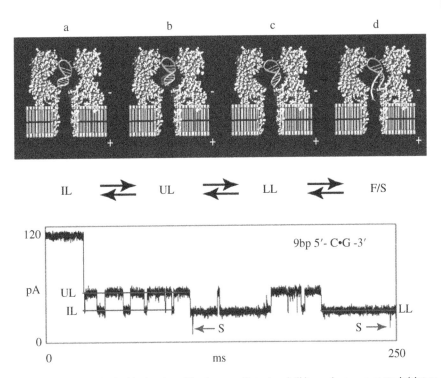

Figure 14.6 Blockade Mechanism. The intermediate level (IL) conductance state initiates most blockades and always transitions to the upper level conductance state (UL). This is explained by binding of the hairpin terminus to the vestibule interior (IL) followed by desorption of the DNA from the protein wall and orientation of the stem along the axis of the electric field (UL). Transitions from the UL state were either back to the IL state or to the lower level conductance state (LL). From the LL state there were brief transitions to nearly full blockade, denoted by F/S for fray/spike conductance state. The LL and F/S states are both thought to involve binding between the hairpin's terminal 5′ base and the pore's limiting aperture. The brief F/S state behavior is explained by a terminus-fraying event that is accompanied by extension by the terminal 3′ base into the limiting aperture. Part of the evidence for this is a strong spike (fraying) frequency correlation with the different terminus binding energies. Asymmetric base addition or phosphorylation (at the terminal 3′ and 5′ positions) is part of the evidence for the asymmetric roles for 5′ binding (LL and F/S) and 3′ fraying/extension (F/S). *Source*: Based on Winters-Hilt [1]; Winters-Hilt [3]; Winters-Hilt [217]

the electric field and the axis of ionic flow. Current is higher in this state because the low resistance path along the major groove leads relatively unimpeded from the pore mouth to the limiting aperture. From the UL conductance state, the hairpin may return to the IL state or it may transition into a third conductance state, LL, where the residual current is equal to 32% of the open channel current. Residence time in this state is dependent upon terminal base pair identity and orientation. In this state, it is hypothesized that the nucleotide at the 5′ end of the duplex stem is adsorbed to the pore wall so that the 3′ nucleotide is positioned directly over the pore-limiting aperture. Thus, when the duplex end frays, the 3′ strand may extend and penetrate the limiting aperture resulting in the transient spikes.

Beyond hydrogen bonding and steric considerations, some of the terminus dynamics was influenced by the nearest-neighbor base-pair (i.e. stacking energies), which indicated that the penultimate base-pair might be readable as well. Next-to-nearest-neighbor influence on the terminal base-pair dynamics was thought to be much less, and this was consistent with the minor changes in blockade signatures on hairpins whose ends were the same but that have different base-pairings further up their stems. This drop-off in sensitivity bodes well for terminus classification on *generic* duplex DNA. At the same time, gross changes in stem base-pairs, such as a change from a Watson–Crick base pair to a Hoogsteen (or wobble) base-pair, resulted in distinctly different blockade signatures [93], which bodes well for SNP assaying schemes.

Residual channel current decreases as blockading DNA hairpins increase their stem length from 3 to 8 base-pairs. For DNA hairpins with stems shorter than eight base-pairs, multiple states were not clearly discernible, presumably because the hairpins were too short to bind to the channel favorably or interact with the current/force constriction near the limiting aperture. For nine base-pair hairpins, and longer, a clear $1/f$ noise (flicker noise) is discernible (Figure 14.2, top) – a preliminary indication of the single-molecule binding kinetics described in Figure 14.6 (and in detail in [93]). Hidden Markov model (HMM)/expectation-maximization (EM) characterization on the five classes of hairpin signatures revealed the existence of two major conductance blockade levels, one minor level intermediate between them, and one to three other statistically relevant levels depending on the hairpin (a pre-processed form, found by HMM/EM level identification, and use of emission variance amplification (EVA)-projection, is shown in Figure 14.2, bottom). By examining the transition probabilities between the various levels it was found that blockades typically began in the less common intermediate level and from there almost always transitioned to the UL blockade level. The mechanism described in Figure 14.6 hypothesizes that the UL blockade state is unbound. A result that strengthens this hypothesis, is that the UL blockade levels are approximately the same for 8, 9, 10, 11, and 12 base-pair DNA hairpins. This plateau occurs well before that of the other blockade levels – the LL blockades, for example, continue to become greater as the hairpin stem length is increased from 8 base-pairs to 10, beyond which it plateaus as well. Beyond 10 base-pairs the hairpin is simply longer than the depth of the channel's *cis*-vestibule, so further base addition causes it to "stick-out further," but cause negligibly greater occlusion of flow than that caused by the fully blockaded *cis*-vestibule. With base-pair addition, however, there arrives greater residual charge, thus greater force drawing the molecule into the vestibule (and slightly deeper channel blockades are seen consistent with this) and dominance of the LL state in the blockade ("toggling") signal. The explanation for the early UL plateau centers on the tight flow geometry between channel and captured hairpin. In such a geometry, much of the ionic flow is confined to be in or near the grooves of the captured DNA molecule. For the unbound molecule, this groove flow can be directed towards the limiting aperture by appropriate orientation of the hairpin molecule. The unbound molecule, thus, appears to cause a gap junction "short circuit" effect, where the contribution to the ionic current is not significantly altered as the hairpin is extended across a three base-pair (approx. 1 nm) gap separating the hairpin terminus from the vestibule's limiting aperture.

A critical understanding derived from the nine-base-pair DNA hairpin analysis is that if the UL blockade state is unbound at its terminus there is the possibility that conformational kinetics might be observable at the pore-captured polymer end. This motivated examination of a set of dsDNA termini that had already been examined using NMR. Results (in [93]) show agreement with NMR via number of low energy conformational states observed [93].

14.6 Conformational Kinetics on Model Biomolecules

Two conformational kinetic studies have been done, one on DNA hairpins with HIV-like termini, the other on antibodies. The objective of the DNA HIV-hairpin conformational study was to systematically test how DNA dinucleotide flexibility (and reactivity) could be discerned using channel current blockade information (see [57], for a complete description of the results pertaining to this study). The structural and physical properties of DNA depend upon nucleotide sequence, as is manifest in differences in three dimensional structure and anisotropic flexibility. Despite the multitude of crystallographic studies conducted on DNA, however, it is still difficult to translate the sequence-directed curvature information obtained through these tools to actual systems found in solution. Information on the DNA molecules' variation in structure and flexibility is important to understanding the dynamically enhanced DNA complex formations that are found with strong affinities to other, specific, DNA and protein molecules. An important example of this is the HIV attack on cells:one of the most critical stages in HIV's attack is the enzyme mediated insertion of viral into human DNA, which is influenced by the dynamic-coupling induced high flexibility of a CA dinucleotide step positioned precisely two base-pairs from the blunt terminus of the duplex viral DNA. This flexibility appears to be critical to allowing the HIV integrase to perform its DNA modifications. The CA dinucleotide presence is also a universal characteristic of retroviral genomes [240, 241]. The behavior of the DNA hairpins containing the CA dinucleotide at different positions relative to their blunt-end termini, is studied in [57] using a ND. The ND feature extraction makes use of HMM-based feature extraction and SVM-based classification/clustering of "like" molecular kinetics. We hypothesized that the DNA hairpin with CA dinucleotide, positioned two base-pairs from the blunt terminus, would have "outlier" channel current statistics qualitatively differentiable from the other DNA hairpin variants. This is found to be the case, where the UL state, corresponding to the unbound terminus state, has shortest life for hairpin labeled CA_3 (with the CA dinucleotide step two basepairs from the blunt terminus). Since the UL state is hypothesized to be unbound, the fact that it has the shortest lifetime on average is an indication of the associated molecule's propensity to be bound to the channel (binding site is unknown at this time, although the work in [93] suggests some likely binding sites on the channel). In other words, CA_3 has strongest interaction with channel (and surroundings), as hypothesized, and neighboring variants (CA_2, CA_4), that have GC pairs shifted one base-pair shifted closer and further from the terminus, share this property to a lesser extent. Note: the "CA" notation refers to a dinucleotide step along the backbone of the self-annealed ssDNA strand in the hairpin molecule, while the GC base-pair described is part of that strand annealing, with the "G" base-paired to the "C" referred to in the CA-step. The molecules with GC pairs that are more than 1 base-pair distant behave similarly to eachother; the DNA hairpin with no GC pair also separates with its own characteristic curve.

14.7 Channel Current Cheminformatics

14.7.1 Power Spectra and Standard EE Signal Analysis

Typical power spectra for captured nine-base-pair DNA hairpins are shown in Figure 14.7, along with a spectrum for the open channel. Below 10 kHz, the current fluctuation caused by the captured DNA molecule (i.e. the blockade noise) is greater than all other noise sources. Such blockade noise typically arises from changes in DNA conformation (molecular structure), changes in DNA configuration (molecular orientation, including waters of hydration), and changes in chemical bonds (internally or with surrounding channel). The power spectra for all the signals examined in [67] had approximately Lorentzian profiles, indicative of a predominately two-state switching process (seen as random telegraph noise). Discriminating between the DNA hairpins on the basis of their power spectral (or other Fourier transform properties, or wavelet properties) is possible for small sets of hairpins. For larger sets of hairpins, or for very similar hairpins like here, the

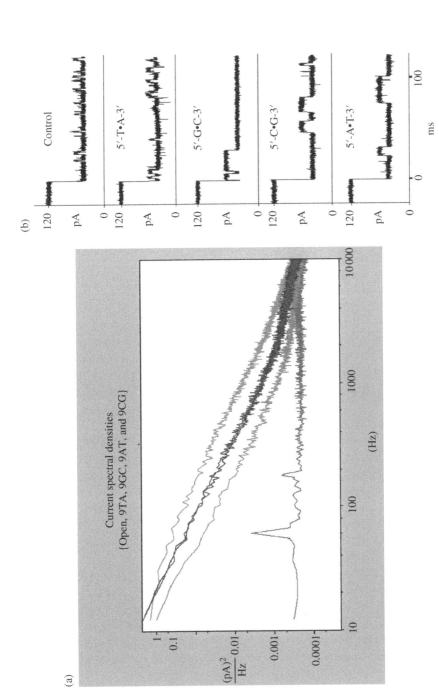

Figure 14.7 (a) Typical power spectra for captured nine-base-pair DNA hairpins and the open channel. *Source:* Winters-Hilt [217]. (b) Typical blockade signatures for each of the five classes of DNA hairpins. The nine base-pair hairpins differ in only their terminal base pairs.

HMM-based feature extraction proved critical, due to their strengths at extracting features from aperiodic (stochastic) sequential data. HMMs can be used for classification as well as feature extraction. Here, HMMs are used for feature extraction in conjunction with a fast, highly accurate, pattern recognition method, known as a SVM. The resulting signal processing and pattern recognition architecture enabled real-time single molecule classification on blockade samplings of only 100 ms [67]. With modern computational methods and hardware it is possible to extract the resolving power of the nanopore instrument for real-time classification and handling.

14.7.2 Channel Current Cheminformatics for Single-Biomolecule/ Mixture Identifications

Using the testing protocol, we were able to determine which of five species of DNA hairpin had been added to the *cis* chamber of the nanopore device. This was achieved in less than 6 seconds with 99.6% accuracy. The five species of DNA hairpins consisted of a control hairpin and four hairpins that differed only in their terminal base-pairs (Figure 14.7). The variants were chosen to include the two possible Watson–Crick base pairs and the two possible orientations of those base pairs at the duplex ends. The core 8 bp stem and 4 dT loop were identical with the primary sequence 5'-<u>TTCGAACG</u>TTTT<u>CGTTCGAA</u>-3', where the base-paired compliments are underlined. The eight base-pair hairpin that was used as a control had the primary sequence 5'-<u>GTCGAACG</u>TTTT<u>CGTTCGAC</u>-3'. *These results were for test data drawn from nanopores established on days other than those used to generate the training data (shown on next page).*

Figure 14.8 shows the scoring for multiple observation days, with the number of single molecule sampling/classifications ranging from 1 to 30. At 75% weak signal rejection, approximately 15 classification attempts were needed to classify the type of single-species solution being sampled; final solution classification was obtained in 6 seconds on average. If training and testing were done on data drawn from the same set of days of nanopore operation, albeit different samples, 99.9% calling was obtained with 15% rejection, and throughput was about one call every half second.

Identification of two hairpins in mixtures was also attempted. Figure 14.9 shows the percentage of 9TA classification in a 3 : 1 mixture of 9TA to 9GC. (Although the mixture preparations are estimated to be ±10% of their stated mixture ratios, calibration and testing of aliquots from the same mixture compensates for such common error.) The assay on 9TA concentration asymptotes to 75% ± 1%, consistent with the 3 : 1 ratio, and the assay error drops to 1% after approximately 100 individual molecule classification attempts (completed in 40 seconds).

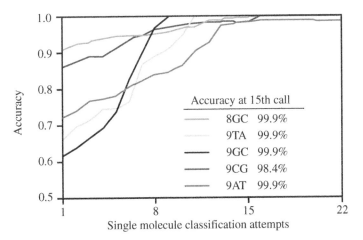

Figure 14.8 Accuracy for classification of single-species solutions of 9TA, 9GC, 9CG, 9AT, and 8GC. By the 15th classification attempt single-species solutions can be identified with high accuracy (inset). *Source*: Winters-Hilt [217].

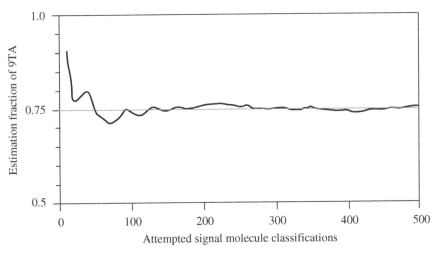

Figure 14.9 Classification on a 3 : 1 mixture of 9TA and 9GC hairpin molecules as a function of single molecule acquisitions. The 3 : 1 mole ratio is accurately identified within 1% error after 100 observations (about 40 seconds). *Source*: Winters-Hilt [217].

HMM/EM characterization on the five classes of hairpin signatures revealed the existence of two major conductance blockade levels, one minor level intermediate between them, and one to three other statistically relevant levels depending on the hairpin. By examining the transition probabilities between the various levels it was found that blockades typically began in the less common intermediate level and from there almost always transitioned to the greater conductance blockade level. (See Chapter 7 for further details.)

With completion of FSA preprocessing, an HMM is used to remove noise from the acquired signals, and to extract features from them. The HMM configuration used for control probe validation is implemented with 50 states that correspond to current blockades in 1% increments ranging from 20% residual current to 69% residual current [1, 3]. In this HMM application the HMM states, numbered 0–49, corresponded to the 50 different current blockade levels in the sequences that are processed. The standard "grayscale" HMM, or "generic HMM", feature extraction setup is then done: the state emission parameters of the HMM are initially set so that the state j, $0 <= j <= 49$ corresponding to level $L = j + 20$, can emit all possible levels, with the probability distribution over emitted levels set to a discretized Gaussian with mean L and unit variance. All transitions between states are possible, and initially are equally likely. Each blockade signature is de-noised by five rounds of EM training on the parameters of the HMM. After the EM iterations, 150 parameters are extracted from the HMM. The 150 feature vectors obtained from the 50-state HMM–EM/Viterbi implementation are: the 50 dwell percentage in the different blockade levels (from the Viterbi trace-back states), the 50 variances of the emission probability distributions associated with the different states, and the 50 merged transition probabilities from the primary and secondary blockade occupation levels (fits to two-state dominant modulatory blockade signals). Variations on the HMM 50 state implementation are made as necessary to encompass the signal classes under study.

The 150-component feature vector extracted for each blockade signal is then classified using a trained SVM. The SVM training is done off-line using data acquired with only one type of molecule present for the training data (bag learning). Further details on the SVM and overall channel current cheminformatics signal processing are detailed in [1, 3].

14.7.3 Channel Current Cheminformatics: Feature Extraction by HMM

The HMM-based profiling used for feature extraction provided better discrimination than the wavelet-based profiling used in previous efforts [95]. The improved signal resolution on channel blockades with HMMs is not new [242]. (The wavelet-domain FSA that generates the blockade-level profiling does have the advantage, however, of being hundreds of times faster than the HMM processing in this instance.) The better performance with HMM processing indicated that signal analysis benefited from parsing structural information in the stochastic sequence of blockade-states. Parsing structures in stochastic data is a familiar problem in gene prediction, where HMMs have been used to great advantage [243]. Typically with gene prediction, however, HMMs are operated at a high level that parses coding starts and stops, etc., with feature scoring on starts and stops performed at a LL by neural net or related statistical methods. For channel current analysis, the HMM extracts structural features without identifying them, effectively operating at the LL, and used with EM [126], accomplishing de-noising on the blockade-state structure [242] prior to extracting those features.

A single HMM/EM process was used to perform the feature extraction in the experiments that follow. If separate HMMs were used to model each species, the HMM/EM processing could also be operated in a discriminative mode. This requires multiple HMM/EM evaluations (one for each species) on each unknown signal as it is observed. Increased computational burden would thus be added at the worst place: the expensive feature extraction stage. Semi-scalable, species-specific processing could be considered for the HMM/EM in an indirect manner, by using prior HMM/EM characterization of the species to identify a reduced set of features relevant to each species. The reduced feature set relates to physical characterizations of the captured molecule, such as level states, their time-constants, and allowed level transitions.

Samples using blockade signatures of longer duration (prior to truncation) require fewer rejections to achieve the same signal classification accuracy. A situation that would probably favor longer signal samples than the 100 ms used here was seen in attempts to read more of the DNA hairpin end-sequence than the terminal base pair. Preliminary indications are that the penultimate base pairs can probably also be identified using longer signal samples (17 species with control). Scaling the classification task from 5 to 17 species may also require refinements to the feature extraction, such as the species specific HMM feature extractions mentioned above.

Tests with mixtures of hairpins required an added calibration due to the nanopore's different acceptance rates for different hairpins (i.e. there are different free energy barriers to capture). This finding was consistent with a model for hairpin capture (see below) in which hairpins are captured by an entropically accessible binding site. It is also in agreement with the brief intermediate level state typically observed at the start of the signal blockades.

14.7.4 Bandwidth Limitations

Nanopore-based detection is limited by the kinetic time-scale of the molecular blockade states, where the molecular blockade states typically correspond to binding and dissociation (analyte-channel binding, or antibody–antigen binding, for example), or due to internal conformational flexing. It is hypothesized that it is possible to probe higher frequency realms than those directly accessible at the operational bandwidth of the channel current-based device, or due to the time-scale of the particular analyte interaction kinetics, by modulated excitations. This can be accomplished by chemically linking the analyte or channel to an "excitable object," such as a magnetic bead, excited by laser pulsations, for example. In one configuration, the excitable object can be chemically linked to the analyte molecule to modulate its blockade current by modulating the molecule during its blockade. In another configuration, the excitable object is chemically linked to the channel, to provide a means to modulate the passage of ions through that channel. Studies involving the first, analyte modulated, configuration, indicate that this approach can be successfully employed to keep the end of a long strand of duplex DNA from permanently residing in a single blockade state (see next section). Similar study of magnetic beads linked to antigen may be used in the nanopore/antibody experiments if similar single blockade level, "stuck," states occur with the captured antibody (at physiological conditions, for example).

Examples of excitable objects include microscopic beads (magnetic and non-magnetic), fluorescent dyes, charged molecules, etc. Bead attachments can couple in excitations passively from background thermal (Brownian) motions, or actively by laser pulsing and laser-tweezer manipulation. Dye attachments can couple excitations via laser or light (UV) excitations to the targeted dye molecule. Large, classical, objects, such as microscopic beads, provide a method to couple periodic modulations into the single-molecule system. The direct coupling of such modulations, at the channel itself, avoids the low Reynolds number limitations of the nanometer-scale flow environment. For rigid coupling on short biopolymers, the overall rigidity of the system also circumvents limitations due to the low Reynolds number flow environment. Similar consideration also come into play for the dye attachments, except now the excitable object is typically small, in the sense that it is usually the size of a single (dye) molecule attachment. Excitable objects such as dyes must contend with quantum statistical effects (at the single-molecule level), so their application may require time averaging or ensemble averaging (where the ensemble case might involve multiple channels observed simultaneously). In both of the experimental configurations, a multi-channel platform may be used to obtain rapid ensemble information. In all cases the modulatory injection of excitations may be in the form of a stochastic source (such as thermal background noise), a directed periodic source (laser pulsing, piezoelectric vibrational modulation, etc.), or a chirp (single laser pulse or sound impulse, etc.). If the modulatory injection coincides with a high frequency resonant state of the system, informative low frequency excitations may result, i.e. excitations that can be monitored in the usable bandwidth of the channel detector. Increasing the effective bandwidth of the nanopore device greatly enhances its utility in almost every application, particularly those, such as

DNA sequencing, where the speed with which blockade classifications can be made is directly limited by bandwidth restrictions.

14.8 Channel-Based Detection Mechanisms

14.8.1 Partitioning and Translocation-Based ND Biosensing Methods

The standard ND detection paradigm, that is predominantly translocation (or dwell-time) based, is shown in Figure 14.10 side-by-side with the Nanopore *transduction* detector paradigm. Figure 14.11 elaborates on the possible ND detection platform topologies possible with translocation-based approaches, where the difference in translocation times is often the critical information that is used. The difference in dwell times can depend on the off-binding time of the target binding entity (possibly in a high strain environment), where binding failure allows polymer (ssDNA) translocation to complete (and the channel blockade to end). By this mechanism, and its variants, bound probes can be distinguished from unbound. There are specificity limits on the melting-based detection, however, that are not a problem in the nanopore transduction detection (NTD) approach.

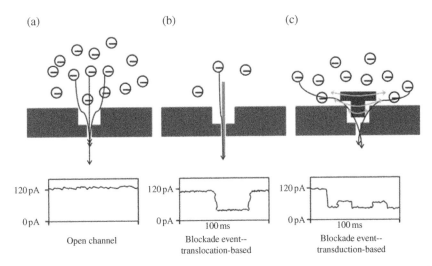

Figure 14.10 Translocation information and transduction information. (a) Open channel. (b) A channel blockade event with feature extraction that is typically dwell-time based. A single-molecule coulter counter. (c) Single-molecule transduction detection is shown with a transduction molecule modulating current flow (typically switching between a few dominant levels of blockade, dwell time of the overall blockade is not typically a feature – many blockade durations will not translocate in the time-scale of the experiment, for example, active ejection control is often involved). *Source*: Winters-Hilt et al. [31].

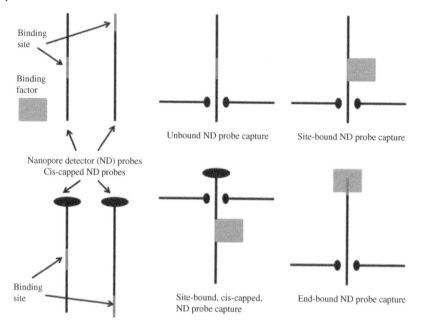

Binding site

Binding factor

Nanopore detector (ND) probes
Cis-capped ND probes

Binding site

Unbound ND probe capture

Site-bound ND probe capture

Site-bound, cis-capped, ND probe capture

End-bound ND probe capture

Figure 14.11 Nanopore detector detection topologies involving polymer translocation or threading [244–276]. The detection event is given by polymer (ssDNA in [244-262, 276]) translocations that are delayed if bound (side and end configurations shown). If bound entity is on the trans side (with cis-side capped, or vice versa), and bound entity is a processive DNA enzyme, then sequencing may be possible as described in [252, 263–266]. *Source*: Winters-Hilt et al. [31].

14.8.2 Transduction Versus Translation

There are two ways to functionalize measurements of the flow (of something) through a "hole": (i) translocation sensing; and (ii) transduction sensing. The translocation methods in the literature are typically a form of a "Coulter Counter," with a wide range of channel dimensions allowable, that typically measures molecules non-specifically via pulses in the current flow through a channel as each molecule translocates. The transduction biosensing method, on the other hand, requires nanopore sizes that are much more restricted, to the 1–10 nm inner diameters that might capture, and not translocate, most biomolecules. Transduction functionalization uses a channel flow modulator that also has a specific binding moiety, the transducer molecule. In transduction, the transducer molecule is used to measure molecular characteristics *indirectly*, by using a transducer/reporter molecule that binds to certain molecules, with subsequent distinctive blockade by the bound, or unbound, molecule complex. One such transducer, among many studied, was a channel-captured dsDNA "gauge" that was covalently bound to an antibody. The transducer was designed to provide a blockade shift upon antigen binding to its exposed antibody binding sites. In turn, the dsDNA-antibody transducer platform then provides a means for directly observing the *single molecule* antigen-binding affinities of any antibody in single-molecule focused assays, in addition to detecting the presence of binding target in biosensing applications.

There are two approaches to utilizing a nanopore for detection purposes: translocation/dwell-time (T/DT)-based approaches, which strongly relies on blockade DTs, and NTD-based approaches, which functionalizes the nanopore by utilizing an engineered blockade molecule with blockade features typically not including DT.

Translocation/DT methods introduce different states to the channel via use of the frequency of channel blockades, from a series of individual molecular blockades (often during their translocation). The strongest feature employed in translocation/DT discrimination, and often the only feature, is the blockade DT where the DT is typically engineered to be associated with the lifetime until a specific bond failure occurs. Other feature variations include time *until* a bond-formation occurs, or simply measuring the approximate length of a polymer according to its translocation "dwell"-time.

14.8.3 Single-Molecule Versus Ensemble

When the extra-channel states correspond to bound or unbound, there are two protocols for how to set up the NTD platform: (i) observe a sampling of bound/unbound states, each sample only held for the length of time necessary for a high accuracy classification. Or, (ii), hold and observe *a single* bound/unbound system and track its history of bound/unbound states. The single molecule binding history in (ii) has significant utility in its own right, especially for observation of critical conformational change information not observable by any other methods (critical information for understanding antibodies, allosteric proteins, and many enzymes). The ensemble measurement approach in (i), however, is able to benefit from numerous further augmentations, and can be used with general transducer states, not just those that correspond to a bound/unbound extra-channel states.

Fundamentally, the weaknesses of the standard ensemble-based binding analysis methods are directly addressed with the single-molecule approach, even if only to do a more informed type of ensemble analysis. The role of conformational change during binding, in particular, could potentially be directly explored in this setting. This approach also offers advantages over other single-molecule translation-based nanopore detection approaches in that the transduction-based apparatus introduces two strong mechanisms for boosting sensitivity on single-molecule observation: (i) engineered enhancement to the device sensitivity via the transduction molecule itself; and (ii) machine learning-based signal stabilization with highly sensitive state resolution. NTD used in conjunction with recently developed pattern recognition informed (PRI) sampling capabilities [35] greatly extends the usage of the single-channel apparatus. For medicine and biology, NTD and machine learning methods may aid in understanding multicomponent interactions (with cofactors), and aid in designing cofactors according to their ability to result in desired binding or modified state.

In ensemble *single-molecule* measurements (via serial detection process), the PRI sampling on molecular populations provides a means to accelerate the accumulation of kinetic information. PRI sampling over a population of molecules is also the basis for introducing a number of gain factors. In the ensemble detection with PRI approach [35], in particular, one can make use of antibody capture matrix and ELISA-like methods [1, 3], to introduce two-state NTD modulators that have concentration-gain (in an antibody capture matrix) or concentration-with-enzyme-boost-gain (ELISA-like system, with production of NTD modulators by enzyme cleavage instead of activated fluorophore). In the latter systems the NTD modulator can have as "two-states," cleaved and uncleaved binding moieties. UV- and enzyme-based cleavage methods on immobilized probe-target can be designed to produce a high-electrophoretic-contrast, non-immobilized, NTD modulator, that is strongly drawn to the channel to provide a "burst" NTD detection signal [1, 3].

14.8.4 Biosensing with High Sensitivity in Presence of Interference

Clinical studies have shown an abundance of protein-based disease markers that accumulate in the blood of patients suffering from chronic kidney disease. In the case of the Bioscience PXRF01marker the stage of kidney disease is linearly correlated ($r = 0.83$) indicating that the more severe the disease, the greater the accumulation of the marker in the bloodstream of patients. The NTD biosensing platform provides a tool for quantifying the relationship between PXRF01 and its biosystem interactants with an unparalleled fidelity. With higher quantification of PXRF01 a more accurate characterization of the disease biomarker and kidney disease progression can be established. Greater sensitivity translates directly to earlier diagnosis and improved outcomes. The electrophoretic nature of the biosensing platform also allows for significant advantage in dealing with interference agents, whether in the blood sample itself, say, or due to contaminants, since the reporter molecule can be designed to have a charge that easily separates it from the interference agents. (This is why blood can be scraped off the dirty floor at a crime scene and still accurately report on the identity or identities of those present.)

14.8.5 Nanopore Transduction Detection Methods

Transduction methods introduce different states to the channel via observations of changes in blockade statistics on a single molecular blockade event that is modulatory. This is a specially engineered arrangement involving a partially captured, single-molecule, channel modulator, typically with a binding moiety for a specific target of interest linked to the modulator's extra-channel portion. The modulator's "state" changes according to whether its binding moiety is bound or unbound. For further comparative analysis, see Table 14.1.

The NTD platform involves functionalizing a ND platform in a new way that is cognizant of signal processing and machine learning capabilities and advantages, such that a highly sensitive biosensing capability is achieved. The core idea in the NTD functionalization of the ND is to design a molecule that can be drawn into the

Table 14.1 Comparative analysis of the translocation/dwell-time (T/TD) and nanopore transduction detection (NTD) approaches.

(1) *Feature space.* The T/DT approach typically has a single feature, the dwell time. Sometimes a second feature, the fixed blockade level observed, is also considered, but usually not more features sought (or engineered) than that. The NTD approach has multiple features, e.g. blockade HMM parameters, etc., with number and type according to modulator design objectives.

(2) *Versatility.* T/DT: highly engineered/pre-processed for detection application to a particular target. NTD: requires minimal preparation/augmentation to the transduction platform via use of separately provided binding moieties (antibody or aptamer, for example) for particular target or biomarker (which are then simply linked to modulator).

(3) *Speed.* T/DT: Slow: entire detection "process" is at the channel, and typically restricted on processing speed to the average time-scale feature (dwell-time) for the longest-lived blockade signal class. NTD: Fast: feature extraction not dependent on dwell-time. Very low probability to get a false positive.

(4) *Multichannel.* T/DT: method not amenable to multichannel gain with single-potential platform (cannot resolve single-channel blockade signal with multichannel noise). NTD: have multichannel gain due to rich signal resolution capabilities of an engineered modulator molecule.

(5) *Feature refinement/engineering.* T/DT: No buffer modifications or off-channel detection extensions via introduction of substrates; the weak feature set limited to dwell-time does not allow such methods to be utilized. NTD: have "lock-and-key" level signal resolution. The introduction of off-channel substrates in the buffer solution can increase sensitivity.

(6) *Multiplex capabilities.* T/DT: Each modified channel is limited to detect a single analyte or single bond-change-event detection, so no multiplexing without brute force production of arrays of T/DT detectors in a semiconductor production setting. NTD: supports multi-transducer, multi-analyte detection from a single sample. Supports multichannel with a single aperture.

channel (by an applied potential) but be too big to translocate, instead becoming stuck in a bistable "capture" such that it modulates the ion-flow in a distinctive way (Figure 14.2 shows some controls). An approximately two-state "telegraph signal" has been engineered for a number of NTD modulators. If the channel modulator is bifunctional in that one end is meant to be captured and modulated while the other end is linked to an aptamer or antibody for specific binding, then we have the basis for a remarkably sensitive and specific biosensing capability. The biosensing task is reduced to the channel-based recognition of bound or unbound NTD modulators (or formed/unformed NTD modulators if target is ssDNA).

In order to have a *capture* state in the channel with a *single* molecule, a true nanopore is needed, not a micropore, and to establish a coherent capture-signal exhibiting non-trivial stationary signal statistics, which is the modulating-blockade desired, the nanopore's limiting inner diameter typically needs to be sized at approximately 1.5 nm for duplex DNA channel modulators (precisely what is found for the α-hemolysin channel). The modulating-blockader is captured at the channel for the time-interval of interest by electrophoretic means, which is established by the applied potential that also establishes the observed current flow through the nanopore.

The NTD molecule providing the channel blockade has a second functionality, typically to specifically bind to some target of interest, with blockade modulation discernibly different according to binding state (DNA annealing examples are shown in what follows). NTD modulators are engineered to be bifunctional: one end is meant to be captured and modulate the channel current, while the other, extra-channel-exposed end, is engineered to have different states according to the event detection. Examples include extra-channel ends linked to binding moieties such as antibodies, antibody fragments, or aptamers. Examples also include "reporter transducer" molecules with cleaved/uncleaved extra-channel-exposed ends, with cleavage by, for example, UV or enzymatic means. By using signal processing with pattern recognition to manage the streaming channel current blockade modulations, and thereby track the molecular states engineered into the transducer molecules, a biosensor or assayer is enabled.

NTD works at a scale where physics, chemistry, and biomedicine methodologies intersect. In some applications the NTD platform functions like a biosensor, or an artificial nose, at the single-molecule scale, e.g. a transducer molecule rattles around in a single protein channel, making transient bonds to its surroundings, and the binding kinetics of those transient bonds is directly imprinted on a surrounding, electrophoretically driven, flow of ions. The observed channel current blockade patterns are engineered or selected to have distinctive stationary statistics, and changes in the channel blockade stationary statistics are found to occur for a transducer molecule's interaction moiety upon introduction of its interaction target. In other applications the NTD functions like a "nanoscope", e.g. a device that can observe the states of a single molecule or molecular complex. With the NTD apparatus the observation is not in the optical realm, like with the microscope, but in the molecular-state classification realm. NTD, thus, provides an unprecedented new technology for characterization of transient complexes. The nanopore detection method uses the stochastic carrier wave signal processing methods developed and described in prior work [1, 3], and comprises machine learning methods for pattern recognition that can be implemented on a distributed network of computers for real-time experimental feedback and sampling control [35]. Details on engineering NTD transducers are given in [15] and what follows.

14.8.5.1 Things to "Contact" with the Channel: Aptamers

Aptamers are synthetically-derived, single-stranded, RNA or DNA molecules up to ~80 oligonucleotides in length with a high affinity towards bonding to specific targets. In 1990, a new method dubbed Systematic Evolution of Ligands by EXponential Enrichment (SELEX) provided a process of producing aptamers from random DNA or RNA libraries [267, 268]. Application of real-time PCR in the production of aptamers has contributed to the growing effectiveness of aptamers in a variety of research areas today [269, 270]. The main advantages of aptamers over antibodies are that aptamers are more durable (i.e. longer shelf life, do not require *in vivo* conditions, can sustain high immune response and toxins), are more obtainable (i.e. cost effective, quicker to make, easily modified, uniformity due to synthetic origin), and have greater specificity and sensitivity (i.e. the degree of binding target recognition, lack of cross-species overlap) [269, 270]. Aptamers may bind with anything from dyes, drugs, peptides, proteins, metal ions, antibodies, and enzymes. The values of Kd range between ~pML^{-1} to ~nML^{-1}, better than that of antibodies [269, 271, 272]. Aptamers are now replacing antibodies as detection reagents, in particular, due to having several advantages over antibodies: versatility, the creation of a lab-on-a-chip to process, low detection limits, simpler reactions to perform, diversity and specificity of aptamer-target binding properties [270]. The use of aptamer beacons has been used in flow cytometry [272], in place of antibody-based assays [270, 273] and most abundantly in studies of specific proteins [271, 274, 275].

14.8.5.2 Things to "Contact" with the Channel: Immunoglobulins

The immunoglobulin molecule IgG is often described as a bifunctional molecule: one region for binding to target antigen, the other region for mediating effector function. Effector functions include binding of the antibody to host tissues, to various cells of the immune system, to some phagocytic cells, and to the first component (C1q) of the classical complement system. Activation of the immune system in response to a specific antigen is an amazing example of how a series of protein phosphorylation and dephosphorylation reactions convert a cell surface event to changes in DNA transcription and cell replication.

The structure of the IgG antibody forms three globular regions that are attached to each other in the middle of its grouping. The overall shape of the structure forms a Y configuration. At the base of this structure is the Constant (Fc) region where the effector functions take place and at the tips of the two arms, both referred to as the variable region (Fab), are the antigen binding sites. These variable regions are tethered to the trunk of the Y shaped molecule by a flexible hinge which allows for a high degree of arm movement. The relative size of the antibody is about three times the size of the α-hemolysin channel. Its length from base (Fc) to arm tip (Fab) is 25 nm and the width of each globular arm ranges from 6 to 10 nm.

The forces binding antigen to antibody are an important and difficult area of study. Hydrophobic bonds, in particular, are very difficult to characterize by existing crystallographic and other means, and often contribute half of the overall binding strength of the antigen–antibody bond. Hydrophobic groups of the biomolecules exclude water while forming lock and key complementary shapes. The importance of the hydrophobic bonds in protein–protein interactions, and of critically placed waters of hydration, and the complex conformational negotiation whereby they are established, may be accessible to direct study using nanopore detection methods in future developments of this technology.

14.9 The NTD Nanoscope

Nanopore event transduction is done using single-molecule biophysics, engineered information flows, and nanopore cheminformatics. NTD is a unique platform, or "nanoscope," for detection and analysis of single molecules. Proof-of-Concept experiments shown in what follows indicate a promising approach for SNP detection, and other biosensing, for clinical diagnostics. This is accomplished via use of the channel-blockade signals produced by engineered event-transducers or by signal-profiled channel modulators in general. The transducer molecule is a bi-functional molecule: one end is captured in the nanopore channel while the other end is outside the channel. This extra-channel end is typically engineered to bond to a specific target: the analyte being measured. When the outside portion is bound to the target, the molecular changes (conformational and charge) and environmental changes (current flow obstruction geometry and electro-osmotic flow) result in a change in the channel-binding kinetics of the portion that is captured in the channel. The change in channel interaction kinetics generates a change in the channel blockade current (which is engineered to have a signal unique to the target molecule). The transducer molecule is, thus, a bi-functional molecule which is engineered to produce a change in its stationary-statistics channel-blockade profile upon binding to cognate. For detection of DNA molecules, the binding can itself lead to NTD modulator *formation*, including formation of the modulator function itself (a duplex DNA molecule annealed to form a Y-branching, for example).

NTD methods for SNP detection alone offers the tantalizing prospect of medical diagnostics and cancer screening by highly accurate assaying of targeted genomic regions. Common methods for SNP detection are typically PCR-based, thus inherit the PCR error rate (0.1% in some situations). The percentages of minority SNP population might be 0.1%, or less, in instances of clinical interest, thus the PCR error rate is critically limiting in the standard approach. Although standard methods for SNP detection have high sensitivity, they typically lack high specificity and versatility. As will be shown, the nanopore transduction detector is a unique platform with both high sensitivity and high specificity.

An interdisciplinary perspective is important to understanding the experimental approach, so initial background describes nanopore electrochemistry and single-molecule biophysics and how the biophysics information flows can result in stationary statistics observations. Then details are provided on the use of engineered stationary statistics signal processing in device enhancement, and communication, as inferred from the selected ND blockade sensing experimental results that are shown.

14.9.1 Nanopore Transduction Detection (NTD)

The NTD platform [1, 3] comprises a single nanometer scale channel that allows a single ionic current flow across a membrane and an engineered, or selected, channel blockading molecule. The channel blockading molecule is engineered or selected such that it provides a current modulating blockade in the detector channel when drawn into the channel. The channel is chosen such that it has inner diameter at the scale of that molecule or one of its molecular-complexes. For most biomolecular analysis implementations this leads to a choice of channel that has inner diameter in the range 0.1–10 nm (see Figure 14.1). Given the channel's size it is referred to as a nanopore in what follows.

The NTD platform [1, 3] includes a single nanometer scale channel and an engineered, or selected, channel blockading molecule. The channel blockading molecule is engineered to provide a current modulating blockade in the detector channel when drawn into the channel, and held, by electrophoretic means. The channel has inner diameter at the scale of that molecule. For most biomolecular analysis implementations this leads to a choice of channel that has inner diameter in the range 0.1–10 nm to encompass small and large biomolecules, where the inner diameter is 1.5 nm in the α-hemolysin protein-based channel used in the

(a)　　　　　　　　　　　　　(b)　　　　　　(c)

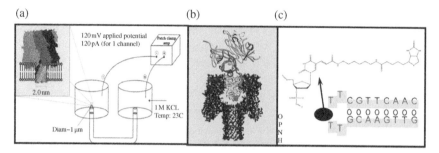

Figure 14.12 Schematic diagram of the nanopore transduction detector [1, 3]. (a) The nanopore detector consists of a single pore in a lipid bilayer which is created by the oligomerization of the staphylococcal α-hemolysin toxin in the left chamber, and a patch clamp amplifier capable of measuring pico Ampere channel currents located in the upper right-hand corner. (b) A biotinylated DNA hairpin molecule captured in the channel's *cis*-vestibule, with streptavidin bound to the biotin linkage that is attached to the loop of the DNA hairpin. (c) shows the biotinylated DNA hairpin molecule (Bt-8gc). *Source*: Winters-Hilt et al. [31].

results that follow (see Figure 14.12). Given the channel's size, it is referred to as a nanopore in what follows. In efforts by others "nanopore" is sometimes used to describe 100–1000 nm range channels, which are here referred to here as micropores.

In order to have a *capture* state in the channel with a *single* molecule, a nanopore is needed. In order to establish a coherent capture-signal exhibiting non-trivial stationary signal statistics the nanopore's limiting inner diameter typically needs to be sized at approximately 1.5 nm for duplex DNA channel modulators (precisely what is found for the α-hemolysin channel). The modulating-blockader is captured at the channel for the time-interval of interest by electrophoretic means.

The NTD molecule providing the modulating blockade in what follows has a second functionality, to specifically bind to some target of interest such that its blockade modulation is discernibly different according to binding state (see the DNA annealing examples in [50–52]). Thus, the NTD modulators are engineered to be bifunctional in that one end is meant to modulate the channel current, while the other end is engineered to have different states according to the event detection, or event-reporting, of interest. Examples include extra-channel ends linked to binding moieties such as antibodies or aptamers. Examples also include "reporter transducer" molecules with cleaved/uncleaved extra-channel-exposed ends, with cleavage by UV or enzymatic means [1, 3]. By using pattern recognition to process the channel current blockade modulations, and thereby track the molecular states, a biosensor is thereby enabled.

With the NTD apparatus the observation is not in the optical realm, like with the microscope, but in the molecular-state classification realm. NTD, thus, provides a technology for characterization of transient complexes. The nanopore detection method uses the stochastic carrier wave signal processing methods developed and described in prior work [1, 3], and comprises machine learning methods for pattern recognition that can be implemented on a distributed network of computers for real-time experimental feedback and sampling control [35].

In assaying applications the ND offers two types of analysis: (i) direct glycoform assaying according to blockade modulation produced directly by the analyte interacting with the ND, which works on negatively charged glycosylation and glycation profiling best; and (ii) indirect isomer assaying by means of surface feature measurements using a specifically binding intermediary, such as with the antibody used in HbA1c testing. A mixture of the direct and indirect assaying methods may be necessary for complex problems of interest.

One of the most challenging nanopore assaying applications is for discriminating between isomers, approximately mass equivalent molecular variants, or aptamers [1, 3, 17]. Other nanopore-based efforts include DNA sequencing applications, and nanopore device physics studies in general, including with channels other than α-hemolysin.

The components comprising the NTD platform [12, 13] include a single nanometer scale channel that allows a single ionic current flow across a membrane and an engineered, or selected, channel blockading molecule. The channel blockading molecule is engineered or selected such that it provides a current modulating blockade in the detector channel when drawn into the channel, and held, by electrophoretic means. The channel is chosen such that it has inner diameter at the scale of that molecule or one of its molecular-complexes. For most biomolecular analysis implementations this leads to a choice of channel that has inner diameter in the range 0.1–10 nm to encompass small and large biomolecules and molecular complexes, where the inner diameter of 1.5 nm is utilized in the α-hemolysin protein-based channel used in the results that follow (see Figure 14.12).

The NTD molecule providing the modulating blockade has a second functionality, typically to specifically bind to some target of interest such that its blockade modulation is discernibly different according to binding state. Thus, the NTD modulators are engineered to be bifunctional in that one end is meant to be captured, and modulate the channel current, while the other, extra-channel-exposed end, is engineered to have different states according to the event detection, or event-reporting, of interest. By using signal processing to process the channel current blockade modulations, and thereby track the molecular states engineered into the transducer molecules, a biosensor or assayer is thereby enabled. By tracking transduced states of a coupled molecule undergoing conformational changes, such as an antibody, or a protein with a folding-pathway associated with disease, direct examination of cofactor, and other, influences on conformation can also be assayed at the single-molecule level.

The nanopore detection method uses the stochastic carrier wave signal processing methods developed and described in prior work [1, 3, 50–52], and comprises machine learning methods for pattern recognition that can be implemented on a distributed network of computers for real-time experimental feedback and sampling control [35]. PRI sampling capabilities greatly extends the usage of the single-channel apparatus, including learning the avoidance of blockades associated with channel failure when contaminants necessitate, and nanomanipulation of a single-molecule under active control in a nanofluidics-controlled environment.

The NTD system, deployed as a biosensor platform, possesses highly beneficial characteristics from multiple technologies: (i) the specificity of antibody binding, aptamer binding, or nucleic acid annealing; (ii) the sensitivity of an engineered channel modulator to specific environmental change; and (iii) the robustness of the electrophoresis platform in handling biological samples.

A critical component in the NTD system is the transducer molecule. A NTD transducer is typically a compound molecule that serves to transduce the conformational or binding state of a molecule of interest into different channel current modulations. A NTD transducer can often be constructed by covalently tethering a molecule of interest to a nanopore channel modulator. In previous work, using inexpensive (commoditized) biomolecular components, such as DNA hairpins, as channel-modulators, and antibodies as specific binding moieties (with inexpensive immuno-PCR linkages to DNA), experiments were done to analyze individual antibodies and DNA molecules, their conformations, glycosylations, and their binding properties. It was found that in many applications the DNA-based transducers worked well, but in efforts to extend the methodology to biosensing and glycosylation profiling the DNA modulators often had too short a lifetime until melting. To make matters worse, the DNA-based modulators often had internal conformational freedom of their own that complicated analysis of any linked molecule's conformational changes. Worst of all, sometimes the DNA modulators only modulated when unbound (and the NTD method works best with clearly different modulatory states). Efforts to fix the non-modulatory aspect were partly solved by using a laser-tweezer apparatus to drive distinctive stochastic modulatory blockades in the DNA modulator. This was accomplished by introducing a periodic laser-tweezer "tugging" on channel-modulator variants that had a biotinylated portion that was bound to a streptavidin-coated magnetic bead (another commoditized component). With modulations "reawakened," however, the number of types of blockade signal appeared to proliferate significantly, and it was not clear if an automated signal analysis could be implemented as had been done previously [1, 3].

14.9.1.1 Ponderable Media Flow Phenomenology and Related Information

Flow Phenomenology

The first step in the NTD methodology is to have a stable ion-channel *sized* such that it can be modulated by a single, non-translocating, molecule, where the channel is significantly blockaded, and not at a fixed level. The next step is to *establish a modulated ion-flow* through the ion-channel with a NTD transducer molecule, usually with the molecule electrophoretically drawn into the channel.

A single molecule's blockading interaction upon capture in an ion channel can be "self-modulating" upon capture (i.e. without a dominant interaction state), and this has been found in a number of experiments [1, 3, 217]. Self-modulatory blockaders each have unique blockade signatures that can be resolved to very high confidence over time. Given the engineering freedom to design the self-modulatory molecules, and the generalizations in the standard periodic carrier-based signal processing to stationary statistics carrier-based signal processing [221, 222], we arrive at a means to leverage ponderable media flow phenomena, and interaction kinetics, into a stochastic carrier wave signal processing problem that can be solved by efficient dynamic programming table computational methods [40, 223].

At the nanometer-scale of the nanopore experiment the Reynold's number of the flow is incredibly small (10^{-10}). Thus the flow environment is not fluid-like in a familiar sense. The fluid strongly damps transverse vibrations, for example, so no string-like-motion on polymers. The motions are strongly driven by electrostatic forces and steric constraints and have significant thermal energy contributions, such that a stochastic process is effectively obtained in typical measurements.

14.9.2 NTD: A Versatile Platform for Biosensing

The use of a channel modulator introduces significant, engineered, signal analysis complexity, that we resolve using artificial intelligence (machine learning) methods. The benefit of this complication is a significant gain in sensitivity over T/TD, that uses a "sensing" moiety covalently attached to the channel itself, where they have a T/TD-type blockade "lifetime" event, with minimal or no internal blockade structure engineered. The NTD approach, on the other hand, has significant improvement in versatility, e.g. we can "swap out" modulators on a given channel, in a variety of ways, since they are not covalently attached to the channel. The improvements in sensitivity derived from the measurable stationary statistics of the channel blockades (and how this can be used to classify state with very high accuracy). The overall improvement in versatility is because all that needs to be redesigned for a different NTD experiment (or binding assay) is the linkage-interaction moiety portion of the bifunctional molecules involved. There is also the versatility that *mixtures* of different types of transducers can be used, a method that cannot be employed in single-channel devices that use covalently bound binding moieties (or that discriminate by DT in the channel).

At the nanopore channel one can observe a sampling of bound/unbound states, each sample only held for the length of time necessary for a high accuracy classification. Or, one could hold and observe a single bound/unbound system and track its history of bound/unbound states or conformational states. The *single* molecule detection, thus, allows measurement of molecular characteristics that are obscured in ensemble-based measurements. Ensemble averages, for example, lose information about the true diversity of behavior of individual molecules. For complex *bio*molecules there is likely to be a tremendous diversity in behavior, and in many cases this diversity may be the basis for their function. There can also be a great deal of diversity via post-translational modifications (PTMs), as well, such as with heterogeneous mixtures of protein glycoforms that typically occur in living organisms (e.g. for TSH and hemoglobin proteins in blood serum and red blood cells, respectively). The hemoglobin "A1c" glycoprotein, for example, is a disease diagnostic (diabetes), and for TSH, glycation is critical component in the TSH-based regulation of the endocrine axis. Multicomponent regulatory systems and their variations (often sources of disease) could also be studied much more directly using the NTD approach, as could multicomponent (or multi-cofactor) enzyme systems. Glycoform assays, characterization of single-molecule conformational variants, and multicomponent assays are significant capabilities to be developed further with the NTD approach, further details on NTD assaying will follow in a later section.

In NTD applications we seek DNA modulators with specific, nonlinear, topologies, such as Y-shaped DNA duplexes, to obtain molecules whose non-translocating blockades modulate the channel. We include shorter nucleic acids, with channel modulating and simple, DNA-complement, annealing properties, in the collection of DNA-based "NTD aptamers" described in the NTD biosensor applications that follows. This is because the detection of ssDNA can enable the NTD-transducer's channel-modulatory formation, for direct signal validation, as will be described in what follows.

Nanopore transduction detection provides an inexpensive, quick, accurate, and versatile method for performing medical diagnostics. It is hypothesized that NTD biomarkers can be developed for early stage disease detection with femtomolar to attomolar sensitivity (see Table 14.2) for doing the standard clinical tests of the future. The potentially incredible sensitivity of the NTD targeting on biomarkers also provides a significant new tool for public health and biodefense in general.

In the preliminary results shown in what follows, we first demonstrate a 0.17 μM streptavidin sensitivity in the presence of a 0.5 μM concentration of detection probes with a 100 seconds detection window. The detection probe is a biotinylated

Table 14.2 Sensitivity limits for detection in the streptavidin-biosensor model system.

Method	SN
Low-probe concentration, 100 s obs.	100 nM
High probe conc, 100 s observation	100 pM
High probe conc, long observation (~1 day)	100 fM^a
TARISA (conc. gain), 100 s observation	100 fM
TERISA (enzyme gain), 100 s obs.	100 aM^b
Electrophoretic contrast gain, 100 s	1.0 aM

^{a} We have done 1–1.5 day long experiments in other contexts, but not longer. Thus, current capabilities, with no modifications to the NTD platform for specialization for biosensing, can achieve close to 100 fM sensitivity by pushing the device limits and the observation window.
^{b} Only a slow enzyme turnover of 10 per second is assumed.

DNA-hairpin transducer molecule (Bt-8gc) [1, 3]. In repeated experiments we see the sensitivity limit ranging inversely to the concentration of detection probes. If taken to its limits, with established PRI sampling capabilities [35], and with stock Bt-8gc at 1 mM concentration conveniently available, we believe it is possible to boost probe concentration almost three magnitudes. In doing so, we would boost sensitivity by similar measure, until the minimal observation time needed to reject limits this gain mechanism (see Table 14.2).

Detection in the attomolar regime (see Table 14.2) is critical for early discovery of type I diabetes destructive processes and for early detection of Hepatitis B. Early PSA detection currently has a 500 µM sensitivity. For some toxins, their potency, even at trace amounts, precludes their usage in the typical antibody-generation procedures (for mAb's that target that toxin). In this instance, however, aptamer-based NTD probes can still be obtained.

14.9.3 NTD Platform

The components comprising the NTD platform include an engineered molecule that can be drawn, by electrophoretic means (using an applied potential), into a channel that has inner diameter at the scale of that molecule, or one of its molecular-complexes, a means to establish a current flow through that nanopore (such as an ion flow under an applied potential), a means to establish the molecular capture for the timescale of interest (electrophoresis, for example), and the computational means to perform signal processing and pattern recognition (see Figures 14.13 and 14.14).

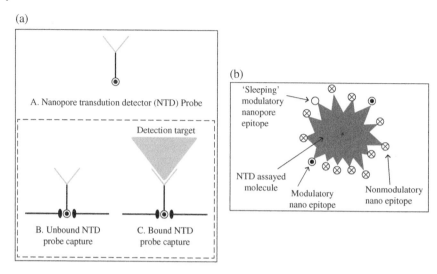

Figure 14.13 (a) Nanopore transduction detector (NTD) probe – a bifunctional molecule (A), one end channel-modulatory upon channel-capture (and typically long-lived), the other end multi-state according to the event detection of interest, such as the binding moieties (antibody and aptamer, schematically indicated in bound and unbound configurations in (B) and (C)), to enable a biosensing and assaying capability. (b) NTD assayed molecule (a protein, or other biomolecule, for example) Antibodies (proteins) are NTD assayed in the Proof-of-Concept experiments, for example. Nanopore epitopes may arise from glyocprotein modifications and provide a means to measure surface features on heterogeneities mixture of protein glycoforms (such mixtures occur in blood chemistry, commercially available test on HbA1c glycosylation common, for example). A molecule may be examined via NTD sampling assay upon exposure to nanopore detector, (or molecular complex including molecule of interest).

The channel is sized such that a transducer molecule, or transducer-complex, is too big to translocate, instead the transducer molecule is designed to get stuck in a "capture" configuration that modulates the ion-flow in a distinctive way.

The NTD modulators are engineered to be bifunctional in that one end is meant to be captured, and modulate the channel current, while the other, extra-channel-exposed end, is engineered to have different states according to the event detection, or event-reporting, of interest. Examples include extra-channel ends linked to binding moieties such as antibodies, antibody fragments, or aptamers. Examples also include "reporter transducer" molecules with cleaved/uncleaved extra-channel-exposed ends, with cleavage by, for example, UV or enzymatic means. By using signal processing to track the molecular states engineered into the transducer molecules, a biosensor or assayer is thereby enabled. By tracking transduced states of a coupled molecule undergoing conformational changes, such as an antibody, or

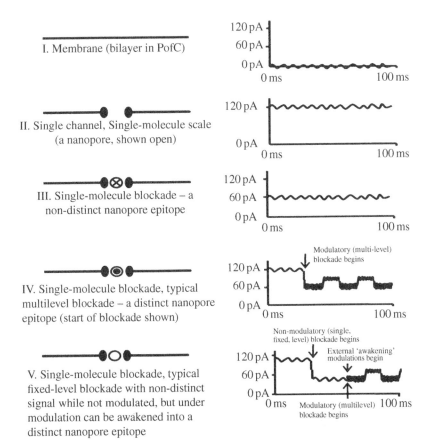

Figure 14.14 The various modes of channel blockade are shown: I. No channel – e.g. a Membrane (bilayer). II. Single channel, single-molecule scale (a nanopore, shown open). III. Single-molecule blockade, a brief interaction or blockade with fixed-level with non-distinct signal – a non-modulatory nanopore epitope. IV. Single-molecule blockade, typical multilevel blockade with distinct signal modulations (typically obeying stationary statistics or shifts between phases of such). V. Single-molecule blockade, typical fixed-level blockade with non-distinct signal while not modulated, but under modulation can be awakened into distinct signal, with distinct modulations.

a protein with a folding-pathway associated with disease, direct examination of cofactor, and other, influences on conformation can also be assayed at the single-molecule level. The channel blockade modes in a NTD experiment thus make special use of channel current modulation scenarios (with stationary statistics), see Figures 14.13–14.17 for further details.

14.9.4 NTD Operation

When the extra-channel states correspond to bound/unbound, there are two protocols for how to set up the NTD platform: (i) observe a sampling of bound/unbound states, each sample only held for the length of time necessary for a high accuracy classification. Or, (ii), hold and observe a single bound/unbound system and track its history of bound/unbound states. The single molecule binding history in (ii) has significant utility in its own right, especially for observation of critical conformational change information not observable by any other methods. The ensemble measurement approach in (i), however, is able to benefit from numerous further augmentations, and can be used with general transducer states (see Figure 14.15), not just those that correspond to a bound/unbound extra-channel states.

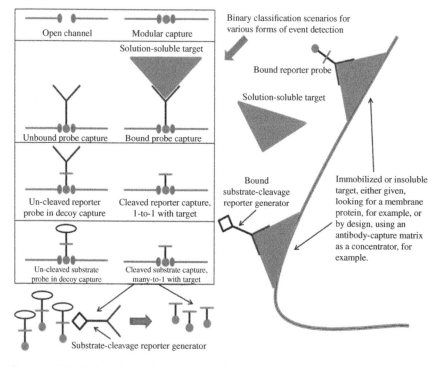

Figure 14.15 Probes shown: bound/unbound type and uncleaved/cleaved type.

(a)

Nanopore epitope based protein glycoform assayer, etc.

(b)

Nanopore population assayer with/without buffer shift

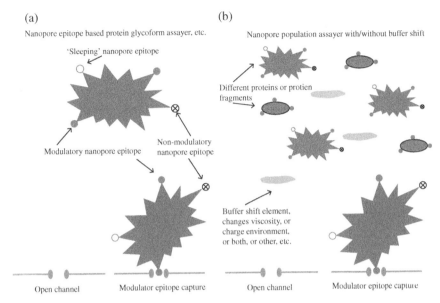

'Sleeping' nanopore epitope

Modulatory nanopore epitope

Non-modulatory nanopore epitope

Different proteins or protien fragments

Buffer shift element, changes viscosity, or charge environment, or both, or other, etc.

Open channel   Modulator epitope capture   Open channel   Modulator epitope capture

Figure 14.16 (a) Nanopore epitope assay (of a protein, or a heterogenous mixture of related glycoprotein, for example, via glycosilation that need not be enzynatically driven, as occurs in blood, for example). (b) **Gel-shift mechanism.** Electrophoretically draw molecules across a diffusionally resistive buffer, gel, or matrix (PEG-shift experiments). If medium in buffer, gel, or matrix is endowed with a charge gradient, or a fixed charge, or pH gradient, etc., isoelectric focusing effects, for example, might be discernable.

The PRI sampling "acceleration," in ensemble-based measurements, for example, provides a means to accelerate the accumulation of kinetic information in most situations [35]. Furthermore, the sampling over a population of molecules is the key to a number of other gain factors that may be realized. In the ensemble detection with PRI approach [35], in particular, one can make use of antibody capture matrix and ELISA-like methods [45], to introduce two-state NTD modulators that have concentration-gain (in an antibody capture matrix) or concentration-with-enzyme-boost-gain (ELISA-like system, with production of NTD modulators by enzyme cleavage instead of activated fluorophore production). (Note that in the latter systems the NTD modulator is simply specified as "two-state," where here we typically do not have bound/unbound, but cleaved/uncleaved instead.) In the ensemble evaluations, with the aforementioned off-channel-engineered event gain factors, we can introduce a NTD probe substrate that thoroughly probes the

(a) (b)

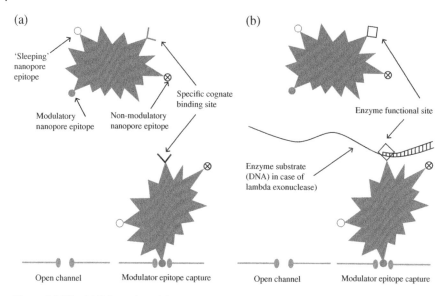

Figure 14.17 (a) Oriented modulator capture on protein (or other) with specific binding (an antibody for example). (b) Oriented modulator capture on protein (or other) with enzymatic activity (lambda exonuclease for example).

sample presented if some element of the probe-target system is immobilized, or significantly reduced in mobilization (see Figure 14.15). In this circumstance, UV- and enzyme-based cleavage methods on immobilized probe-target can be designed to produce a high concentration, or concentration burst, of NTD modulators, that will be strongly drawn to the channel and provide a UV-event correlated "burst" concentration detection signal.

A multi-channel implementation of the NTD can be utilized if a distinctive-signature NTD-modulator on one of those channels can be discerned (the scenario for trace, or low-concentration, biosensing, see Figure 14.18). In this situation, other channels bridging the same membrane (bilayer in case of α-hemolysin-based experiment) are in parallel with the first (single) channel, with overall background noise growing accordingly. In the stochastic carrier wave encoding/decoding with HMM with Duration (HMMD) (Chapters 7 and 8), we retain strong signal-to-noise, such that the benefits of a multiple-receptor gain in the multi-channel NTD platforms can be realized.

Multichannel, one
transducer source

Figure 14.18 Multichannel scenario, with only one blockade present (at low concentration, for example).

14.9.5 Driven Modulations

It is possible to probe higher frequency realms than those directly accessible at the operational bandwidth of the channel current-based device (~200 kHz), or due to the time-scale of the particular analyte interaction kinetics, by introducing modulated excitations. This can be accomplished by chemically linking the analyte or channel to an excitable object, such as a magnetic bead, under the influence of laser pulsations. In one configuration, the excitable object can be chemically linked to the analyte molecule to modulate its blockade current by modulating the molecule during its blockade. In another configuration, the excitable object is chemically linked to the channel, to provide a means to modulate the passage of ions through that channel. In a third experimental variant, the membrane is itself modulated (using sound, for example) in order to effect modulation of the channel environment and the ionic current flowing though that channel. Studies involving the first, analyte modulated, configuration (Figures 14.19 and 14.20), indicate that this approach can be successfully employed to keep the end of a long strand of duplex DNA from permanently residing in a single blockade state. Similar study of magnetic beads linked to antigen may be used in the nanopore/antibody experiments if similar single blockade level, "stuck," states occur with the captured antibody (at physiological conditions, for example). Likewise, this approach can be considered for increasing the antibody–antigen dissociation rate if it does not occur within the time-scale of the experiment. It may be possible, with appropriate laser pulsing, or some other modulation, to drive a captured DNA molecule in an informative way even when not linked to a bead, or other macroscopic entity (Figure 14.21).

(a)

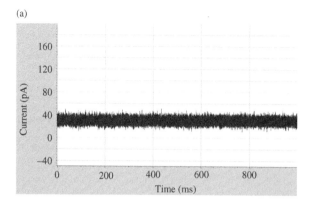

(b)

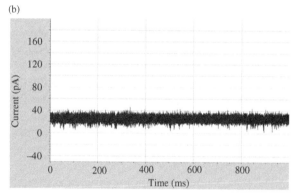

(c)

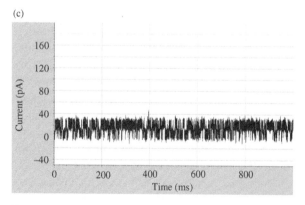

Figure 14.19 (a) Channel current blockade signal where the blockade is produced by 9GC DNA hairpin with 20 bp stem. (b) Channel current blockade signal where the blockade is produced by 9GC 20 bp stem with magnetic bead attached. (c) Channel current blockade signal where the blockade is produced by c9GC 20 bp stem with magnetic bead attached and driven by a laser beam chopped at 4 Hz. Each graph shows the level of current in picoamps over time in milliseconds.

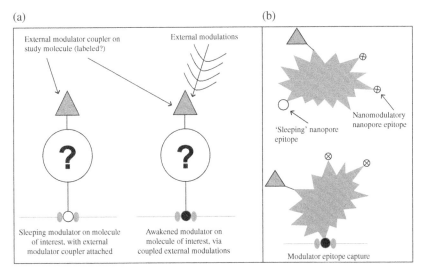

Figure 14.20 (a) Study molecule with externally driven modulator linkage to awaken modulator signal. (b) Study molecule with externally driven modulator linkage to awaken modulator signal, with epitope-selection to obtain sleeping epitope, then determine its identity, and based on known modulator-activation driving signals, proceed with driving the system to obtain a modulator capture linkage.

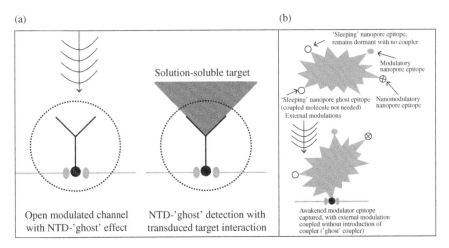

Figure 14.21 (a) Same situation as in cases with linked-modulator, but more extensive range of external modulations explored, such that, in some situations, a sleeping nanopore epitope is "awakened" (modulatory channel blockades produced), and the target molecule does not require a coupler attachment., e.g. using external modulations with no coupler, may be able to obtain "ghost" transducers in some situations. (b) "Sleeping" Nanopore Ghost Epitope (coupled molecule not needed).

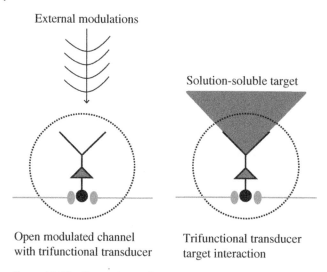

External modulations

Solution-soluble target

Open modulated channel
with trifunctional transducer

Trifunctional transducer
target interaction

Figure 14.22 External modulations with transducer with coupler, a trifunctional molecule.

14.9.6 Driven Modulations with Multichannel Augmentation

The *Staphylococcus aureus* α-hemolysin pore-forming toxin that is used to produce our single-channel nanopore-detector self-oligomerizes to derive the energetics necessary to create a channel through the bi-layer membrane. In the nanopore construction protocol, the process is limited to the creation of a single channel. It is possible to allow the process to continue unabated to create 100 channels or more. The 100 channel scenario has the potential to increase the sensitivity of the NTD, but the signal analysis becomes more challenging since there are 100 parallel noise sources. The recognition of a transducer signal is possible by the introduction of "time integration" to the signal analysis akin to heterodyning a radio signal with a periodic carrier in classic electrical engineering. In order to introduce a "time integration" benefit in the transducer signal, periodic (or stationary stochastic) modulations can be introduced to the transducer environment. In a high noise background, modulations can be introduced such that some of the transducer level lifetimes have heavy-tailed distributions. With these modifications to the signal processing software a single transducer molecule signal could be recognizable in the presence of 100 channels or more. Increasing the number of channels by 100 and retaining the capability of recognizing a single transducer blockading one of those channels provides a direct gain in sensitivity according to the number of channels (e.g. 100 channels would provide a sensitivity boost of two orders of magnitude). It is important to note that this type of increase in sensitivity

(a)

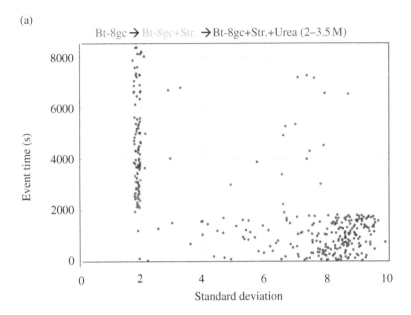

(b)

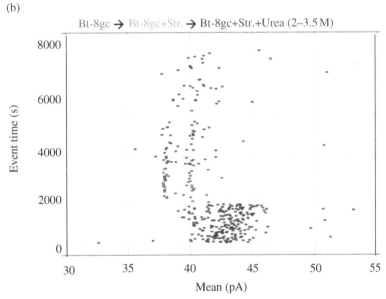

Figure 14.23 (a) Observations of individual blockade events are shown in terms of their blockade standard deviation (x-axis) and labeled by their observation time (y-axis). The standard deviation provides a good discriminatory parameter in this instance since the transducer molecules are engineered to have a notably higher standard deviation than typical noise or contaminant signals. At $T = 0$ seconds, 1.0 μM Bt-8gc is introduced and event

(a)

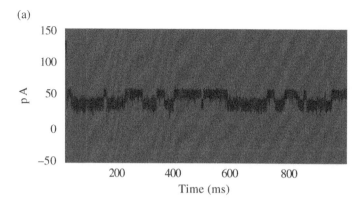

(b)

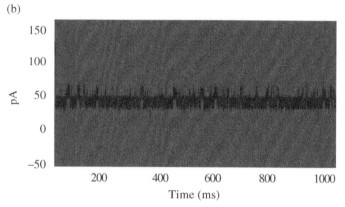

Figure 14.24 Pseudo-aptamer: DNA overhang binding complement – signal blockades. (a) Before introduction of 5-base ssDNA complement. (b) After introduction of complement.

Figure 14.23 (Continued) tracking is shown on the horizontal axis via the individual blockade standard deviation values about their means. At T = 2000 seconds, 1.0 µM Streptavidin is introduced. Immediately thereafter, there is a shift in blockade signal classes observed to a quiescent blockade signal, as can be visually discerned. The new signal class is hypothesized to be due to (Streptavidin)-(Bt-8gc) bound-complex captures. (b) As with the left panel on the same data, a marked change in the Bt-8gc blockade observations is shown immediately upon introducing streptavidin at T = 2000 seconds, but with the mean feature we clearly see two distinctive and equally frequented (racemic) event categories.

Introduction of chaotropic agents degrades first one, then both, of the event categories, as 2.0 M urea is introduced at T = 4000 seconds and steadily increased to 3.5 M urea at T = 8100 seconds. *Source*: Based on Winters-Hilt et al. [31]; Winters-Hilt [50].

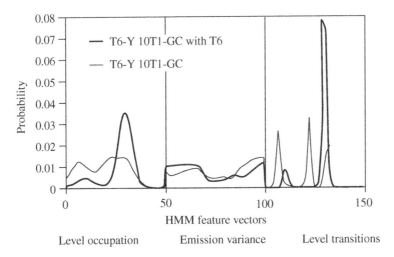

Figure 14.25 Y-aptamer with DNA overhang that binds complement. (a) Signal profiles before and after binding. (b) The dwell-time distributions on the three dominant levels indicated in the *unbound* blockade signal. The profiles aresurprisingly different, the bound case, with annealed complement, appears to be more "stable", with only two dominant blockade levels. This is consistent with it being a molecule with fewer degrees of freedom (with 6T overhang now annealed to 6A complement).

is mostly implemented computationally and does not add complexity or cost to the NTD device.

The single-channel biosensing methods used here can be generalized to where many channels are present, where each channel offers parallel conductance paths for the ionic current, and where each channel is augmented with antibody (or aptamer) to establish a background collection of channel/antibody signals that is modifiable in the presence of antigen. Such "passive" multichannel methods offer similar capabilities to surface plasmon resonance approaches for characterizing binding affinity. Multiple antibody (aptamer) species can be present in this multichannel operation. Anything that can evoke an antibody response (or SELEX selection, for aptamers) can be taken as the antigen or collection of antigens for which the bio-sensing is designed.

Multichannel with modulation is shown in Figure 14.22, where modulation forces a population inversion, such that state durations are strongly nongeometrically distributed. Even without such modulations, however, there may be a strong enough signal recognition with the HMM methods without duration modeling enhancements.

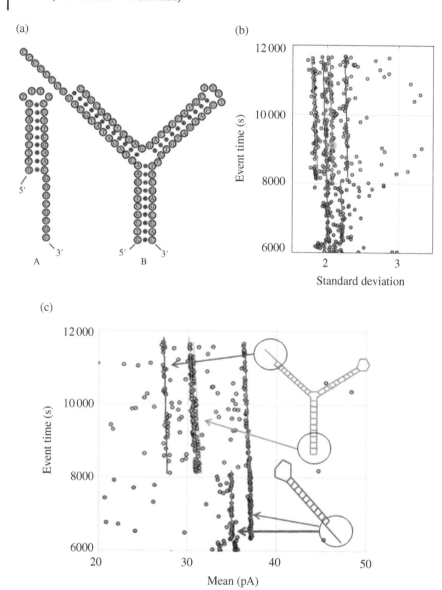

Figure 14.26 Eight-base annealing using a NTD Y-transducer. (a) The DNA hairpin and DNA Y-nexus transducer secondary structures with sequence information shown. (b) and (c) Y-shaped DNA transducer with overhang binding to DNA hairpin with complementary overhang. *Source*: Based on Winters-Hilt et al. [31].

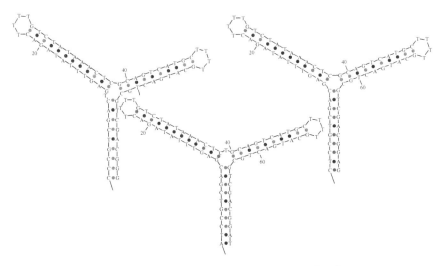

Figure 14.27 Shown are Y-shaped aptamers that have shown they have capture states with the desired blockaded toggling. *Source*: Winters-Hilt et al. [31].

14.10 NTD Biosensing Methods

NTD biosensing methods can involve a DNA modulator with linkages to an apta-mer, antibody, or some other binding moiety, including simply a ssDNA overhang. The linkages needed to connect a DNA-based channel-modulator to a DNA-based aptamer involves a trivial join of the underlying ssDNA sequences involved. The linkage needed to connect a DNA-based channel-modulator to an antibody *could* involve use of linker technology, and this has been used in the past with dsDNA hairpins [53], but another, more commoditized route to be discussed, easily acces-sible with use of the nanopore-detector directed (NADIR) refined Y-shaped DNA channel modulators [50, 51], is that the antibody need merely be "tagged" with the appropriate ssDNA strand, e.g. where the DNA sequence is complement to part of the "Y" shaped DNA channel modulator, and antibody tagging with DNA is a standard service for use in immuno-PCR. Proof-of-concept biosensing experiments are described for the streptavidin–biotin and DNA annealing model systems, a pathogen/SNP detection prototype, and for aptamer and antibody-based detection.

14.10.1 Model Biosensor Based on Streptavidin and Biotin

A biotinylated DNA-hairpin that is engineered to generate two signals depending on whether or not a streptavidin molecule is bound to the biotin (see Figure 14.23). Results in Figure 14.23b suggest that the new signal class on binding is actually a

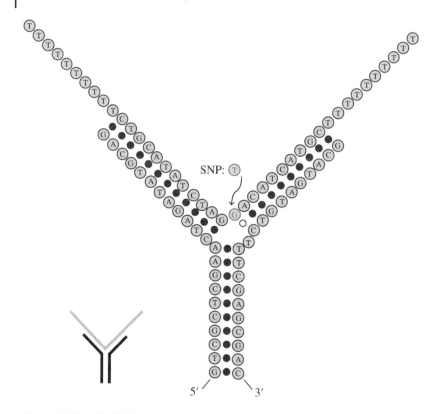

Figure 14.28 The Y-SNP with test complex is shown at the base-level specification and at the diagrammatic level, where a SNP base is as indicated. If the SNP is its variant form (typically only one other base possibility is common), then a base-pairing will not occur at the nexus of the Y-SNP shown (with the red base becoming a "T" in the variant as indicated). This allows discrimination between the annealed forms with high accuracy, while also discerning from the signals produced by the non-annealed Y-SNP, where there is no target-bound, or only nonspecific molecular interactions imparting much less conformational structure as occurs with the matching (or mostly matching) annealing interaction. *Source:* Winters-Hilt et al. [31].

racemic mixture of two hairpin-loop twist states. At $T = 4000$ urea is introduced at 2.0 M and gradually increased to 3.5 M at $T = 8100$.

The transducer molecule in the NTD "Streptavidin Toxin Biosensor" configuration consists of a bi-functional molecule: one end is captured in the nanopore channel while the other end is outside the channel. This exterior-channel end is engineered to bond to a specific target: the analyte being measured. When the outside portion is bound to the target, the molecular changes (conformational and charge) and environmental changes (current flow obstruction geometry and electro-osmotic flow)

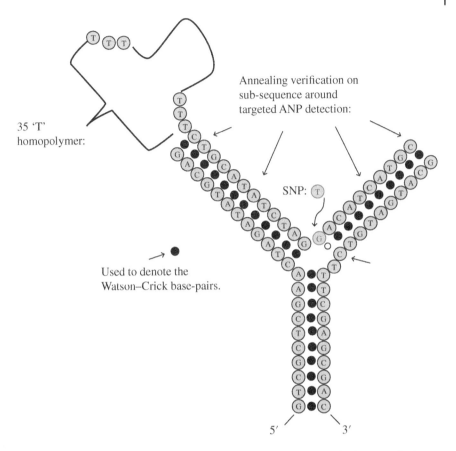

35 'T' homopolymer:

Annealing verification on sub-sequence around targeted ANP detection:

SNP: T

Used to denote the Watson–Crick base-pairs.

5'   3'

Figure 14.29 The Y-SNP test complex with 35 dT length overhang is shown at the base-level specification, where a SNP base is as shown. If the SNP is its variant form (typically only one other base possibility is common), then a base-pairing will not occur at the nexus of the Y-SNP shown. This allows discrimination between the annealed forms with high accuracy, while also discerning from the signals produced by the non-annealed Y-SNP, where there is no target-bound, or only nonspecific molecular interactions imparting much less conformational structure as occurs with the matching (or mostly matching) annealing interaction. *Source*: Winters-Hilt et al. [31].

result in a change in the channel-binding kinetics of the portion that is captured in the channel. This change of kinetics generates a change in the channel blockade current which represents a signal unique to the target molecule.

Some of the transducer molecule results from [31, 50] are shown in Figure 14.23, for a biotinylated DNA–hairpin that is engineered to generate two unique signals depending on whether or not a streptavidin molecule is bound.

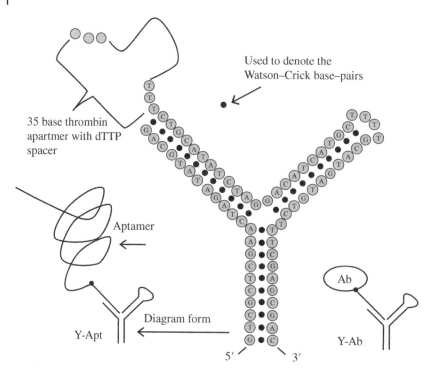

Figure 14.30 The thrombin aptamer from [277] is 5′-CACTGGTAGGTTGGTGTGGTTGGGGCCAGTG-3′. *Source*: Winters-Hilt et al. [31].

In the NTD platform, sensitivity increases with observation time [42] in contrast to translocation technologies where the observation window is fixed to the time it takes for a molecule to move through the channel. Part of the sensitivity and versatility of the NTD platform derives from the ability to couple real-time adaptive signal processing algorithms to the complex blockade current signals generated by the captured transducer molecule. If used with the appropriately designed NTD transducers, NTD can provide excellent sensitivity and specificity and can be deployed in many applications where trace level detection is desired. The monoclonal antibody-based NTD system, deployed as a biosensor platform, possesses highly beneficial characteristics from multiple technologies: the specificity of monoclonal antibody binding, the sensitivity of an engineered channel modulator to specific environmental change, and the robustness of the electrophoresis platform in handling biological samples. In combination, the NTD platform can provide trace level detection for early

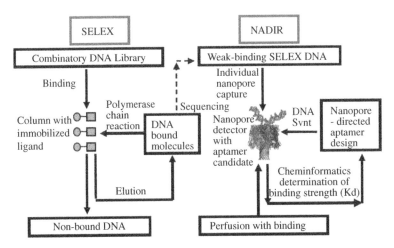

Figure 14.31 The determination of aptamers can be done (or initiated) via Systematic Evolution of Ligands by Exponential Enrichment (SELEX), as shown schematically on the left. What is proposed here is a linkage to a *na*nopore-detector *dir*ected (NADIR) search for aptamers that is based on bound-state lifetime measurements. NADIR complements and augments SELEX in usage: SELEX can be used to obtain a functional aptamer, and NADIR used for directed modifications (for stronger binding affinity, for example).

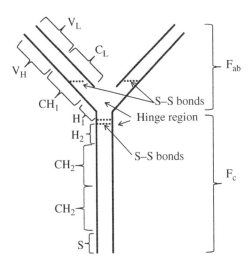

Figure 14.32 The standard antibody schematic. Standard notation is shown for the constant heavy chain sequence ("CH", "H", and "S" parts), variable heavy chain region ("VH" part), the variable light chain region ("VL" part), and constant light chain region ("CL" part). The full heavy chain sequence is derived from recombination of the VH part and {CH,H,S} parts (where the secretory region S is also called CH4). The long and short chains are symmetric from left to right,

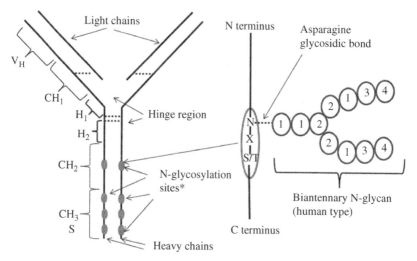

Figure 14.33 Typical antibody N-glycosylation. A schematic for typical antibody N-glycosylation is shown (drawn from results on the equine IGHD gene [278, 279]), where one possible N-glycosylation site is indicated in region CH2, and three possible N-glycosylation sites are indicated in region CH3. N-glycosylation consists of a covalent bond (glycosidic) between a biantennary N-glycan (in humans) and asparagine (amino acid "N", thus N-glycan). The covalent glycosidic bond is enzymatically established in one of the most complex post translational modifications on protein in the cell's ER and Golgi organelles, and usually only occurs in regions with sequence "NX(S/T) – C-terminus" where X is anything but proline and the sequence is oriented with the C-terminus as shown. Licensed therapeutic antibodies typically display 32 types of biantennary N-glycans, consisting of N-acetyl-glucosamine residues (GlcNAc, regions "1"); mannose residues (Man, regions "2"); galactose residues (Gal, regions "3"), and sialic acid residues (NeuAc, regions "4"). The N-glycans are classified according to their degree of sialylation and number of galactose residues: if disialylated (shown) have A2 class. If asymmetric and monosialylated have A1 class. If not sialylated then neutral (N class). If two galactose residues (shown) then G2 class, if one, then G1 class, if zero, then G0 class. If there is an extra GlcNAc residue bisecting between the two antennae +Bi class (–Bi shown). If a core fucose is present (location near GlcNAc at base), then +F (–F shown). So the class shown is G2-A2. The breakdown on the 32 types is as follows: 4 G2-A2; 8 G2-A1; 4 G1-A1; 4 G2-A0; 7 G1-A0; 4 G0-A0. The N-glycans with significant acidity (A2 and A1) are 16 of the 32, so roughly half of the N-glycans enhance acidity. The other main glycosylation, involving O-glycans, occurs at serine or threonine (S/T). The main non-enzymatic glycations occur spontaneously at lysines ("K") in proteins in the blood stream upon exposure to glucose via the reversible Maillard reaction to form a Schiff Base (cross-linking and further reactions can be irreversible). *Source:* Based on Wagner et al. [278].

Figure 14.32 (Continued) their glycosylations, however, are generally not symmetric. Critical di-sulfide bonds are shown connecting between chains, each of the VH and CH regions typically have an internal disulfide bond as well. The lower portion of the antibody is water soluble and can be crystallized (denoted Fc). The upper portion of the antibody is the antigen binding part (denoted Fab). *Source:* Based on Wagner et al. [278].

diagnosis of disease as well as quantify the concentration of a target analyte or the presence and relative concentrations of multiple distinct analytes in a single sample.

In [31, 50] a 0.17 μM streptavidin sensitivity is demonstrated in the presence of a 0.5 μM concentration of detection probes, with only a 100 seconds detection window. The detection probe is the biotinylated DNA-hairpin transducer molecule (Bt-8gc) described in Figure 14.12. In repeated experiments, the sensitivity limit ranges inversely to the concentration of detection probes (with PRI sampling) or the duration of detection window. The stock Bt-8gc has 1 mM concentration, so a 1.0 mM probe concentration is easily introduced. (Note: The higher concentrations of transducer probes need not be expensive on the nanopore platform because the working volume can be very small: *cis* chamber volume is 70 μl, and could be reduced to 1.0 μl with use of microfluidics.) In [31, 50] the selectivity of the detector in the presence of interference agents, such as albumin and sucrose and a variety of antibodies (without specific binding to biotin or the channel) was also examined, and a control transducer molecule with the same six-carbon linker arm from the DNA hairpin, but without the biotin "fishing lure" binding site, was introduced, where it was shown that no interaction (via change of blockade signal) was observed upon introduction of streptavidin, as expected.

14.10.2 Model System Based on DNA Annealing

14.10.2.1 Linear DNA Annealing Test

Proof-of-Concept experiments for DNA annealing were initially tested for detection of a specific 5-base ssDNA molecule, where we have a linear molecule with a bulge in the center. To one side of the bulge is the blunt-ended stem sequence like that used in one of our DNA hairpin controls, where the bulge is now in the position of the hairpin's loop. To the other side of the bulge is a cap-section of base-pairs followed by an overhang section of length five bases. A similar set of experiments is performed with the "Y-aptamer", a Y-shaped DNA complex with one arm of the Y with an overhang (6Ts), while the other arm is capped with a 4dT loop. The base of the Y is a stem of 10 base-pairs length, prior to the Y-nexus of the molecule. Here the Y-nexus is in the place of the bulge, or the hairpin loop. Nine or ten base-pairs are approximately the length in dsDNA from the mouth of the channel to its limiting aperture. The significance of this length in the modeling is due to its delicate placement of the end of the captured molecule over the high electrophoretic field strength zone near the limiting aperture of the channel, permitting operation in transduction model. The overhang's binding strength can be adjusted by tailoring its length in both of these experiments, and in future work this will also permit a highly precise study of DNA annealing.

(a)

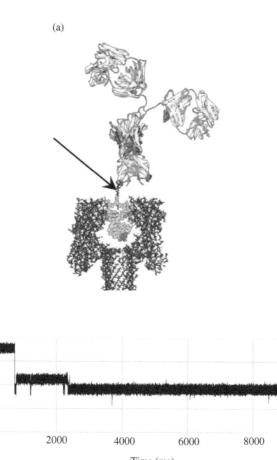

(b)

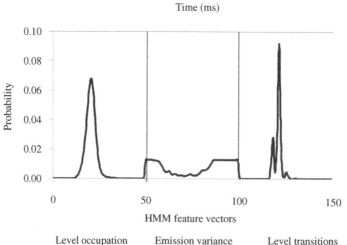

Time (ms)

HMM feature vectors

Level occupation   Emission variance   Level transitions

Figure 14.34 (a) **DNA hairpin bound to antibody via an EDC-linker.** Approximately shown to scale. Arrow points to the internal amino thymine modification with primary amine on a six carbon spacer arm. Primary amine can be crosslinked using 1-ethyl-3-(3-

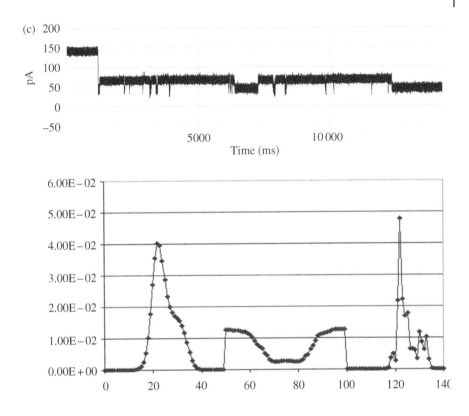

Figure 14.34 (Continued) dimethylaminopropyl) carbodiimide hydrochloride (EDC) to the peptide carboxyl terminus of the antibody heavy chain. This crosslinkage results in a covalent bond between the primary amine and the carboxyl. (b) **Antibody linked to DNA-hairpin blockade signal and HMM profile.** (See [30] for description of HMM profile.) (c) **Antibody linked to DNA-hairpin, now bound to its target antigen (biotin) – new blockade signal, and associated HMM profile.** Antigen binding to an EDC-linked Antibody/DNA-hairpin, where stem of the hairpin is captured in the nanopore detector. (See [53] for description of HMM profile.)

The linear duplex DNA molecule, with bulge, and ssDNA overhang, is given below. Examples of the signals that occur when a properly annealed duplex is captured are shown in Figure 14.24. Figure 14.24 compares signal traces before/after in terms of their standard 150-component feature set. The linear aptamer with bulge consists of annealing the following two ssDNA strands:

1) 5′-GAGGCTTGG TTT CAATAGGTA-3′
2) 5′-ATTG TTT CCAAGCCTC-3′

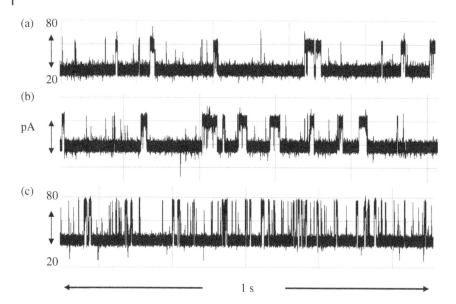

Figure 14.35 DNA-hairpin signals. (a) No biotin concentration. (b) Low-to-high biotin concentration (1000-fold excess). (c) Low urea concentration.

The complementary 5 nucleotide ssDNA sequence (3):

3) 5′-TACCT-3′

14.10.2.2 "Y" DNA Annealing Test

The Y-aptamer DNA molecule consists of a three-way DNA junction created by annealing two DNA molecules:

1) 5′-CTCCGTCGAC GAGTTTATAGAC TTTTTT-3′
2) 5′-GTCTATAAACTCGCAGTCATGCTTTTGCATGACTGCGTCGACGGAG-3′

For the resulting Y-aptamer, one of the junctions' arms terminate in a 4dT-loop and the other arm has a 6T overhang in place of a 4dT-loop. Preliminary results are shown in Figure 14.25. The blunt ended arm has to be carefully designed such that when it is captured by the nanopore it produces a toggling blockade. One of the arms of the Y-shaped aptamer (Y-aptamer) has a TATA sequence, and is meant to be a binding target for TBP. In general, any transcription factor binding site

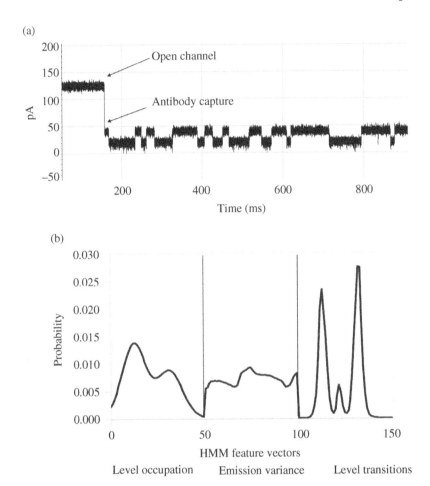

(a)

(b)

Figure 14.36 **Example that provides a very clear, stable, blockade direct by an Ab.** (a) A toggle signal is generated as a channel-captured region of the molecule (IgG) wiggles above the limiting aperture of the α-hemolysin channel varying the ionic current between two transient states. (b) Antibody Toggle HMM signal profile. The 150 feature vectors obtained from the 50-state HMM–EM/Viterbi implementation in [1, 3, 53] are: the 50 dwell percentage in the different blockade levels (from the Viterbi trace-back states), the 50 variances of the emission probability distributions associated with the different states, and the 50 merged transition probabilities from the primary and secondary blockade occupation levels (fits to two-state dominant modulatory blockade signals). *Source*: Based on Winters-Hilt [1]; Winters-Hilt [3]; Winters-Hilt et al. [53].

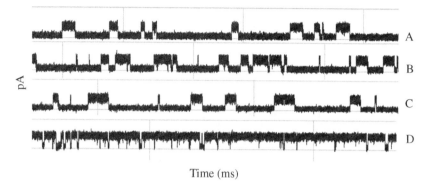

pA

Time (ms)

Figure 14.37 Antibody–Antigen binding – clear example from specific capture orientation. Each trace shows the first 750 ms of a three minute recording, beginning with the blockade signal by an antibody molecule that has inserted (some portion) into the α-hemolysin channel to produce a toggle signal (A). Antigen is introduced at the beginning of frame (A). Changes to the toggle signal are discernible in frame D, indicating the binding event between the antibody and antigen has taken place.

could be studied (or verified) in this manner. Similarly, transcription factor could be verified, or the efficacy of a synthetic transcription factor could be examined.

14.10.3 Y-Aptamer with Use of Chaotropes to Improve Signal Resolution

In the nucleic acid annealing studies on the NTD platform described in [31, 50] (see Figure 14.26), the introduction of chaotropes allows for improved nucleic acid annealing identification.

The ability of the of the NTD apparatus to tolerate high chaotrope concentration, up to 5 M urea, was demonstrated in [23]. DNA hairpin control molecules have demonstrated a manageable amount of isoform variation even at 5 M urea.

In Figure 14.26, only a portion of a repetitive validation experiment is shown, thus time indexing starts at the 6000th second. From time 6000 to 6300 seconds (the first 5 minutes of data shown) only the DNA hairpin (sequence details in [31]) is introduced into the analyte chamber, where each point in the plots corresponds to an individual molecular blockade measurement. At time 6300 seconds, urea is introduced into the analyte chamber at a concentration of 2.0 M. The DNA hairpin with overhang is found to have two capture states (clearly identified at 2 M urea). The two hairpin channel-capture states are marked with the green and red lines, in both the plot of signal means and signal standard deviations. After 30

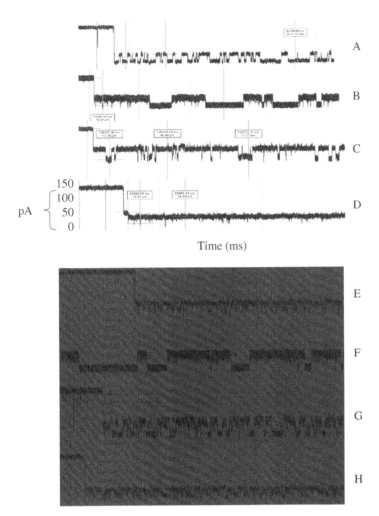

Figure 14.38 Antibody signal classes and Ab-antigen signal classes. A–D: various IgG region captures and their associated toggle signals (1 second traces). E–F: various IgG +Antigen region captures and their associated toggle signals (1 second traces). Each blockade signal was identified visually and represents a commonly observed signal class. Note the changes in dwell times for the upper and lower current levels in each signal class. We find a higher current level bias in the level occupancy as a result of binding with the antigen molecule.

minutes of sampling on the hairpin+urea mixture (from 6300 to 8100 seconds), the Y-shaped DNA molecule is introduced at time 8100. Observations are shown for an hour (8100–11 700 seconds). A number of changes and new signals now are observed: (i) the DNA hairpin signal class identified with the green line is no

(a)

(b)

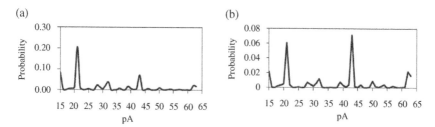

Figure 14.39 Multivalent antigen binding. (a) First antibody–antigen binding – 1st 50 feature components extracted from the HMM. (b) Shifts in the values of these 1st 50 HMM feature components indicate a possible second antibody–antigen binding (same molecular capture). The first 50 components of the 150 feature vectors obtained from the 50-state HMM–EM/Viterbi implementation are the dwell percentages in the different blockade levels from the Viterbi trace-back states (approximately the Histogram in that range).

longer observed – this class is hypothesized to be no longer free, but annealed to its Y-shaped DNA partner; (ii) the Y-shaped DNA molecule is found to have a bifurcation in its class identified with the yellow lines, a bifurcation clearly discernible in the plots of the signal standard deviations; (iii) the hairpin class with the red line appears to be unable to bind to its Y-shaped DNA partner, an inhibition currently thought to be due to G-quadruplex formation in its G-rich overhang; (iv) the Y-shaped DNA molecule also exhibits a signal class (blue line) associated with capture of the arm of the "Y" that is meant for annealing, rather than the base of the "Y" that is designed for channel capture.

14.10.4 Pathogen Detection, miRNA Detection, and miRNA Haplotyping

In clinical diagnostics, as well as in biodefense testing, patient blood samples can be drawn for the purpose of assaying the DNA content. Obviously there will be a preponderance of human DNA in such a sample, but if there is infection then trace amounts of the associated viral or bacterial DNA will be present as well. The question then arises as to how to detect unique elements of bacterial DNA sequence that are singled-out for detection, with very high sensitivity and specificity. This may be possible in the NTD approach, with annealing-based detection along the lines described earlier and in [31, 50, 51], where ssDNA sequences are targeted for detection of approximate length 22 base sub-sequences. A 22-mer is shown in Figure 14.26, "B"-labeled secondary structure, in the leftmost, linear, ssDNA segment. The Y-shaped secondary structure in Figure 14.26 ("B") shows the blueprint for a NTD ssDNA probe for any targeted ssDNA segment, upon "recognition" (annealing-based), a Y-shaped channel modulator is engineered to occur. If not the correct modulator, due to a few mis-matches or inserts (particularly at the Y-nexus), then the difference can be discerned with high discrimination. All that is needed is a specific set of enzyme digestion steps on the DNA sample to "chop" it into shorter segments, and leave targeted regions at the ends of (some) of the resulting ssDNA digests➔so as to obtain-dsDNA annealed targets with probe match as in Figure 14.26 ("B"), where the excess ssDNA length (beyond the 22-mer match template) is left to dangle off of one end, as shown for the eight-base segment shown in Figure 14.26 ("B"). In ongoing work target-segment annealing with high specificity is being explored in the presence of large polymer extensions to the annealing target.

In clinical diagnostics, as well as in biodefense testing, patient blood samples can be drawn for the purpose of assaying the DNA and glycoprotein contents. In the case of DNA there will be a preponderance of the individual's own genomic DNA in such a sample, but if there is infection then trace amounts of the associated viral or bacterial DNA will be present as well. One of the questions that then arises is how to detect unique elements of bacterial DNA sequence with very high sensitivity and specificity. In [1, 3] annealing-based detection is explored, where Y-shaped NTD transducer results are shown for tests involving an eight base ssDNA target [31, 50]. The method can be extended to other lengths of targeted ssDNA, using annealing-based recognition. For longer lengths we can arrive at interesting detection scenarios for pathogens or for miRNA's (some possibly pathogenic). The known pathogen ssDNA targets could be longer, 15–25 bases say, to enable unique identifiers respective to a particular pathogen. For miRNA detection probes could be designed for ssDNA target annealing that is in the 7–15 base range.

MicroRNA detection follows a similar approach to the pathogen detection problem, but now typically working with a much shorter length nucleic acid detection target, a miRNA sequence-based annealing target. In this setting often have similar "informed" analysis to pathogen detection analysis.

The detection of SNPs via annealing is demonstrated with the Y-shaped DNA transduction molecule that is minimally altered, and such that the SNP variant occurs in the Y-nexus region. In preliminary work with Y-transducers [31, 50] we demonstrate how *single-base insertions or modifications at the nexus of the Y-shaped molecule can provide clearly discernible changes in channel-blockade signals.* The design of the Y-transducer for SNP detection was similar to the process mentioned in [31, 50] for NADIR searches for aptamers based on bound-state lifetime measurements. The NTD method provides a viable prospect for SNP variant detection to very high accuracy – possibly equaling the accuracy with which the NTD can discern DNA control hairpins that only differ in terminal base-pair (greater than 99.999% for sufficiently long observation time).

Y-DNA modulator platforms for biosensing can also provide a simple linker platform for use with antibody binding moieties, where a "linker" aptamer can be used that is covalently linked to the common base of the antibody (IgG) molecule (using a DNA tagged antibody approach). Aptamer tuning can also be enhanced in the nanopore setting using nanopore directed SELEX (referred to as NADIR in [31, 50]), where binding strength can be selected to be not too strong or weak according to the desired tuning on the observed binding lifetimes, as seen in the state durations of the observed state noise.

Linkage of ssDNA to antibody is commonly done in immuno-PCR preparations, so another path with rapid deployment is to make use of a linkage technology that is already commoditized, e.g. a good NTD signal can then be produced with immuno-PCR tagged antibodies that are designed to anneal to another DNA molecule to form a NTD "Y-transducer."

The explosive geographic expansion of the Zika virus provides another reminder that rapid diagnostic tools for new viral infections is an ever increasing need. The rapid deployment of a fast diagnostic tool in the example of the Zika virus is all the more pertinent given that the virus has been shown to be the cause of microcephaly in the fetuses of exposed pregnant women, along with results indicating possible brain damage (Guillain–Barre reaction) to a significant fraction of those exposed. A rapid development, deployment, and evaluation of a Zika virus diagnostic would afford the patient the critical time needed to undergo aggressive prophylactic measures. Similarly, certain fungal infections need to be diagnosed as early as possible (cryptococcus neoformans, for example, can disrupt and cross the blood–brain barrier). The treatments for many fungal infections are highly toxic, however, such that they will only be undertaken if infection is highly likely.

Pathogens that are suspected can potentially be probed in a matter of hours using a NTD platform with the methods described here using probes designed according to the pathogen's genomic profile. Unknown pathogens would first need to either have their genomes sequenced (less than a day) if sufficient DNA already available, or a sample directly measured via a test assay template (same procedure as for biomarker discovery) for assay-level fingerprint determination, then testing for that pathogen fingerprint in the patient.

The NTD platform can be enhanced to be a rapid annealing-based detection platform due a recently established ability to operate under high chaotropic conditions (up to 5 M urea), which allows measurement of collective binding interactions such as nucleic acid annealing with other simpler binding and related complexes thereby eliminated and effectively filtered from the analysis task.

14.10.5 SNP Detection

The proposed test of DNA SNP annealing is with the Y-shaped DNA transduction molecule shown in Figure 14.26 ("B") that is minimally altered, and such that the SNP variant occurs in the Y-nexus region. For the case where digestion can not conveniently provide extension only to one-side, a Y-shaped annealed dsDNA molecule can still be obtained, but such that the ssDNA extensions outside the annealed region are now free to extend on both arms of the Y-molecule.

SNP variant detection is reduced to resolving the signals of two Y-shaped duplex DNA molecules, one with mismatch at SNP, one with Watson–Crick base-pairing match at SNP. In preliminary studies of Y-shaped DNA molecules, numerous Y-shaped DNA molecules were considered. Three variants that successfully demonstrated the easily discernible, modulatory, channel blockade signals are shown in Figure 14.27 [31]. In those variants we considered the Y-nexus with and without an extra base (that is not base-paired). And if an extra base is inserted we explore the three positions at the Y (left and middle inserts shown in the left and center Y-molecules shown in Figure 14.27.

The DNA molecule design we are currently using consists of a three-way DNA junction created: 5′-CTCCGTCGAC GAGTTTATAGAC TTTT GTCTATAAACTC GCAGTCATGC TTTT GCATGACTGC GTCGACGGAG-3′. Two of the junctions' arms terminate in a 4T-loop and the remaining arm, of length 10 base-pairs, is usually designed to be blunt ended (sometimes shorter with an overhang). The blunt ended arm has to be carefully designed such that when it is captured by the nanopore it produces a toggling blockade. One of the arms of the Y-shaped aptamer (Y-aptamer) has a TATA sequence, and is meant to be a binding target for TBP. In general, any transcription factor binding site could be studied (or verified) in this manner. Similarly, transcription factor could be verified by such constructions, or the efficacy of a synthetic transcription factor could be examined. The other Y-aptamer, used in the integrase binding analysis, is shown in Figure 14.27 (both sequence and secondary structure).

A preliminary test of DNA SNP annealing can be done with the Y-shaped DNA transduction molecule shown in Figure 14.28, which is minimally altered (e.g. mostly common sequence identity) from the Y-annealing transducer introduced in Figure 14.27.

Once the Y-SNP transducer has been tested on a single-species of short overhang length test molecules the next experimental challenge will be to detect SNP variants using the Y-SNP transducer probe in the presence of a heterogeneous length mixture (some with target SNP region of interest), with overhang as shown in Figure 14.29.

The value of 35 "T"s on the extension is to also match the approximate extension, with same "Y"-sequence (except for a 4 dT cap) as the previously "blunt-ended" annealed conformation. SNP variant detection is reduced to resolving the signals of two Y-shaped duplex DNA molecules, one with mismatch at SNP, one with Watson–Crick base-pairing match at SNP. From the above it is clear that the NTD method provides a viable prospect for SNP variant detection to very high accuracy (possibly the accuracy with which the NTD can discern DNA control hairpins that only differ in terminal base-pair, greater than 99.999%). SNP detection via *translocation-based* methods, on the other hand, must discern between two SNP variants according to the different dwell times of the complement-template annealed SNPs, until dissociation from the template allows translocation of the blockading dsDNA annealed conformation.

14.10.6 Aptamer-Based Detection

Aptamers are especially appropriate for study by nanopore detection due to the fact they can be designed with an end to be captured and modulate a nanopore (i.e. the captured end is dsDNA) while other parts of the aptamer are intended to bind a specific target. This directly provides a NTD transducer if one or both of the bound/unbound states (captured in the channel at the dsDNA end) provides distinctive channel modulations. The binding statistics derived from the study of aptamers in a ND can also be used in the design of the aptamer itself, e.g. NADIR selection instead of further SELEX-based selection [31]. In Figure 14.30 we see the first aptamer test case to be considered, where we seek to detect thrombin [277] in one case, and IgG [280] in another. We use the thrombin aptamer found by Ikebukuro et al. [277], it is selected via SELEX and EMA and is a 31-mer, linked by a 4 dT spacer to link to the Y-transducer (see Figure 14.30).

14.10.6.1 NaDir SELEX

In using the NADIR refinement process to arrive at the Y-transducer used in the DNA annealing test [31], we have demonstrated how *single-base insertions or modifications at the nexus of the Y-shaped molecule can have clearly discernible changes in channel-blockade signal.* Y-molecules as DNA probes with single point mutations discernible at the Y-nexus are explored in [31] (see Figures 14.27 and 14.31). What is described in [31] is a linkage to a NADIR search for aptamers that is based on bound-state lifetime measurements (or some other selection criterion of interest). NADIR complements and augments SELEX in usage.

14.10.7 Antibody-Based Detection

Linkage of ssDNA to antibody is commonly done in immuno-PCR preparations, so another path with rapid deployment is to make use of a linkage technology that is already commoditized, e.g. the molecules required for the antibody-based biosensing with this approach are simple (non-specialty) molecular components. The core issue to be tested here is whether a good NTD signal can be produced with immuno-PCR tagged antibodies that are designed to anneal to another DNA molecule to form a NTD "Y-transducer" (see Figure 14.30, lower right). From previous efforts [53], with more complicated 1-ethyl-3-(3-dimethylaminopropyl) carbodiimide hydrochloride (EDC) linkages between a modified thymine and an antibody (see Figure 14.34), it is clear that there are strong prospects for success with this method. What is sought is not just further validation of the method, however, but a less expensive, accessible, platform from which to refine and develop NTD-based systems.

Some mAb blockades produce a very clean toggling between two levels (see Figure 14.38 for antibody description and some typical blockade signals). The mAb interference modulatory signals are easily discerned from a modulatory signal of interest, however, especially with increased observation time as needed. Aside from being an interference agent, antibodies offer a direct means for having a NTD transducer since their modulatory blockade signals are observed to change upon introduction of antigen. The problem with using an antibody directly as a transducer in a biosensor arrangement is that the antibody produces multiple blockade signal types (a dozen or more) just by itself (without binding). This weakness for use directly as a biosensor (they can still be linked indirectly as in [30]) is because the antibody is a glycoprotein that has numerous heterogeneous glycosylations and glycations, with many molecular side-groups that might be captured by the ND to produce modulatory blockades. If the purpose is to study the PTMs themselves, a glyco-profile of the antibody in other words, then the numerous signal types seen are precisely the information desired. A more complete analysis of antibody blockades on the ND is beyond the scope of this paper, and will be in a separate paper. Some further details on the Antibody structure and its direct glyco-profiling is still given next, however, since similar PTMs can be analyzed on other proteins of critical biomedical interest.

14.10.7.1 Managing Antibodies as Easily Identifiable Interference or Transducer

Antibodies are the secreted form of a B-cell receptor, where the difference between forms is in the C-terminus of the heavy chain region. Figure 14.32 shows the standard antibody schematic. Standard notation is shown for the constant heavy chain sequence ("CH," "H," and "S" parts), variable heavy chain region ("VH" part), the variable light chain region ("VL" part), and constant light chain region ("CL" part). The equine IGHD gene for the constant portion of the heavy chain has exons corresponding with each of the sections CH1,H1,H2,CH2,CH3,CH4(S), and for the membrane-bound form of IGHD, there are two additional exons, M1 and M2 for the transmembrane part, thus, CH1, H1, H2, CH2, CH3, CH4(S), M1, M2 [278]. In Figure 14.32, the long and short chains are symmetric from left to right, their glycosylations, however, are generally not symmetric. Critical di-sulfide bonds are shown connecting between chains, each of the VH and CH regions typically have an internal disulfide bond as well. The lower portion of the antibody is water soluble and can be crystallized (denoted Fc). The upper portion of the antibody is the antigen binding part (denoted Fab).

Figure 14.33 shows a typical antibody N-glycosylation (exact example for equine IGHD [43]). One possible N-glycosylation site is indicated in region CH_2, and three possible N-glycosylation sites are indicated in region CH_3. N-Glycosylation consists of a covalent bond (glycosidic) between a biantennary N-glycan (in humans) and asparagine (amino acid "N", thus N-glycan). The covalent glycosidic bond is enzymatically established in one of the most complex post translational modifications on protein in the cell's ER and Golgi organelles, and usually only occurs in regions with sequence "NX(S/T) – C-terminus" where X is "anything but proline" and the sequence is oriented with the C-terminus as shown. Licensed therapeutic antibodies typically display 32 types of biantennary N-glycans [278], consisting of N-acetyl-glucosamine residues (GlcNAc, regions "1"); mannose residues (Man, regions "2"); galactose residues (Gal, regions "3"), and sialic acid residues (NeuAc, regions "4"), as shown in Figure 14.33. The N-glycans are classified according to their degree of sialylation and number of galactose residues: if disialylated (shown) have A2 class. If asymmetric and monosialylated have A1 class. If not sialylated then neutral (N class). If two galactose residues (shown) then G2 class, if one, then G1 class, if zero, then G0 class. If there is an extra GlcNAc residue bisecting between the two antennae +Bi class (−Bi shown). If a core fucose is present (location near GlcNAc at base), then +F (−F shown). So the class shown is G2-A2. The breakdown on the 32 types is as follows: 4 G2-A2; 8 G2-A1; 4 G1-A1; 4 G2-A0; 7 G1-A0; 4 G0-A0 [278]. The N-glycans with significant acidity (A2 and A1) are 16 of the 32, so roughly half of the N-glycans enhance acidity. The other main glycosylation, involving O-glycans, occurs at serine or threonine (S/T). The main non-enzymatic glycations occur spontaneously at lysines ("K") in proteins in the blood stream upon exposure to glucose via the reversible Maillard reaction to form a Schiff Base (cross-linking and further reactions, however, are irreversible and associated with the aging process).

The base of the antibody plays the key role in modulating immune cell activity. The base is called the Fc region for "fragment, crystallizable", which is the case, and to differentiate it from the Fab region for "fragment, antigen-binding" that is found in each of the arms of the Y-shaped antibody molecule (see Figure 14.32). The Fc region triggers an appropriate immune response for a given antigen (bound by the Fab region). The Fab region gives the antibody its antigen specificity; the Fc region gives the antibody its class effect. IgG and IgA Fc regions can bind to receptors on neutrophils and macrophages to connect antigen with phagocyte, known as opsonization (opsonins attach antigens to phagocytes). This key detail may explain the modulatory antibody interaction with the nanopore channel. IgG, IgA, and IgM can also activate complement pathways whereby C3b and C4b can act as the desired opsonins. The C-termini and Fc glycosylations of an antibody's heavy chain, especially for IgG, is thus a highly selected construct that appears to be what is recognized by immune receptors, and is evidently what is recognized as distinct channel modulator signals in the case of the NTD. Using NTD we can co-opt the opsonization receptor-binding role of the Fc glycosylations (and mAB glycations and glycosylations in general), and C-terminus region, to be a channel modulating role. This may also permit a new manner of study of the critical opsonization role of certain classes of antibodies (and possibly differentiate the classes in more refined ways) by use of the ND platform. The channel may provide a means to directly measure and characterize antibody Fc glycosylations, a critical quality control needed in antibody therapeutics to have correct human-type glycosylation profiles in order to not (prematurely) evoke an immunogenic response.

14.10.7.2 Small Target Antibody-Based Detection (Linked Modulator)

IgG antibodies may vary in net charge but are nowhere near as negatively charged as the DNA hairpin molecules examined in [50, 53]. Differences in channel interaction are often attributed to its net charge and its electrophoretic mobility. To improve the antibody's affinity for the channel and to aid in signal classification, a complex of antibody and DNA hairpin is sometimes used. The result is the increase in channel affinity and a significant reduction in capture class configurations (see Figure 14.34 for further details), while still retaining binding detection sensitivity. The *small*-antigen biosensing results (described here) complements those for *large*-antigen biosensing (presented in the next section). The large antigen study done here is also notable in that it involves direct use of the antibody as a bifunctional reporter molecule. This leads to complications with capture, and uniqueness in the orientation of that capture, but may offer a more sensitive detection approach since there is no linkage separating the bound/unbound complex from the channel flow environment.

A DNA hairpin with EDC linkage to an antibody is shown in Figure 14.34, and examined in [53]. When the DNA portion of this linked complex inserts itself into the α-hemolysin channel it creates a definable toggle signal that serves as reliable "carrier signal" for monitoring any changes of molecular state (such as binding). In our first study of DNA-hairpin linked antibody complexes [53], we used an anti-biotin-antibody (Stressgen) as our binding element linked to our DNA hairpin. (Note, as one of many control tests, we see that the blockade toggle signal is relatively unchanged after addition of excess biotin.)

The clarity of the current blockade signal for Ab-antigen binding and Ab-pore interaction was examined by varying the composition of working buffer in presence of urea and $MgCl_2$. In one series of experiments, mentioned above, we used free antibody molecule interacting with the ND, where the antibody (anti-biotin) molecule is introduced to our nanopore device to produce the characteristic two-state telegraph signal (Figure 14.34). The blockade signal for the antigen is practically unaltered by excess antigen: even 100-fold excess of biotin does not change the blockade signal considerably (Figure 14.35). The signal changes greatly in the presence of urea, however, in a relatively small concentration. Here the duration of any event to occupy upper state level becomes shorter and the total probability value of UL decreases with urea concentration rise.

14.10.7.3 Large Target Antibody-Based Detection

For large-antigen antigen-binding studies, different versions of copolymer (Y, E)-A—K ("large" targets) were originally prepared to allow study of the effect of antigenic mass and valency of binding upon the observations in [53]. In this and other studies involving direct antibody interactions with the channel, however, we found that antibodies themselves typically produce a variety of "long-lived" blockades at the channel themselves, sometimes modulatory, even possibly producing clear "toggle" signals as shown in the study cases in Figure 14.36.

It is found that the antibody blockade signal alters shortly after introduction of antigen, as Figure 14.37 shows upon addition of a moderately high concentration (100 µg/ml) of 200 kD multivalent synthetic polypeptide (Y, E)-A—K. Presumably, these changes are the result of antibody binding to antigen. The time before the blockade signal is altered is also interesting; it ranges from seconds to minutes (not shown). This presumably is a reflection of antibody affinity.

Direct antibody nanopore blockades are examined further in Figure 14.38a, where the different capture signals provided by a single antibody species provide a "nanopore epitope" mapping or assay of the antibody's surface features, including glycations and nitrosilations, as described in the following section. Typical captures seen after introduction of antigen are shown for the same system in Figure 14.38b. Figure 14.39 shows a possible indication of a multivalent binding signal (the Ab being bivalent).

14.11 Exercises

14.1 Describe the device physics of the α-hemolysin nanopore detector apparatus.

14.2 Detail the nanopore noise sources.

14.3 Describe the length resolution of individual DNA hairpins.

14.4 Describe the results on detection of single nucleotide differences.

14.5 Explain the blockade mechanism that produces the 9 bphp blockade signal.

14.6 Explain how the nanopore detector can be used to obntain conformational kinetics on single molecules.

14.7 Describe the channel current cheminformatics architecture.

14.8 Describe the main channel-based detection mechanisms.

14.9 Describe the NTD nanoscope.

14.10 Describe how the NTD nanoscope can be used as a biosensor.

Appendix A

Python and Perl System Programming in Linux

A.1 Getting Linux and Python in a Flash (Drive)

The operating system (OS) environment is part of the programming environment in what follows since *systems* level programming techniques for using the Python code will be used, including using Python as a "shell" scripting language (all of this terminology will be reviewed in what follows).

If you have a Mac (OS X) then you are already running a type of Linux behind the scenes. There are many pitfalls for a beginner programmer with using Macs, however, including use of many of the text editors. So using the Mac environment requires care with the choice of editors, such as using Sublime Text.

If you are running an older Windows machine (not Windows 10) then I suggest a bootable USB drive that has some type of Linux (mine has Mint on 128G and cost $30). The procedure for preparing a bootable USB drive will be outlined below. If you have Win 10 then, as with the Mac, you can run a version of Linux on the machine while Win 10 is running – you just need to learn how to access it. If you have a Chromebook, you have a version of Linux running behind the scenes (like the Mac's) – here, however, saving work on the computer can be tricky when accessing Linux, so an inexpensive USB drive is still needed as a simple/safe memory.

If you have most Win8, any Win7, or anything earlier than Win7, then you can get Linux via a bootable USB drive. This approach requires that you change the boot order in the BIOS of your computer so that it tries to boot from USB drive first. Plugged into the USB drive will be a bootable USB drive prepared according to the steps below.

If you do not have a computer, and want both Windows and Linux capabilities cheaply, then you might consider getting a bootable USB drive (with Linux) and a Windows laptop. You will want to get, at a minimum, a used Win7 laptop, with at

Informatics and Machine Learning: From Martingales to Metaheuristics, First Edition.
Stephen Winters-Hilt.
© 2022 John Wiley & Sons, Inc. Published 2022 by John Wiley & Sons, Inc.

least 4 GB RAM (where BIOS access is not difficult, a Dell, for example), and also get a 32 GB or greater flash drive.

Set up a Linux OS USB drive, aka flash-drive (also called a pendrive or thumb-drive), by going to www.pendrivelinux.com. This approach changes the boot order in the BIOS to try to boot from USB first.

Any system where you can access the BIOS, you can take over the system via flash drive. A minicomputer option for Linux is also possible, known as the Raspberry Pi ($35) that plugs into your TV or monitor via HDMI. Free software for simulating a Linux environment, called "VirtualBox," can also be set up to run on any OS, and this provides an alternative means to have a Linux OS. To get VirtualBox go to www.virtualbox.org, and follow instructions from there.

A.2 Linux and the Command Shell

The Linux (Unix) OS:
- provides an extensive set of computational utilities and maintenance utilities (several hundred) – an ideal software development environment
- takes full advantage of available hardware (can run efficiently on a micro-, mini-, or super-computer)
- 95% of the OS is written in C, a portable, machine-independent, language (only the drivers are machine-dependent, as they should be)
- multitasking was part of the original design, not an afterthought
- effective data sharing/privacy also part of the original design

Device Driver Level
- a separate program exists for interacting with each device
- the drivers execute on behalf of the kernel only (no user access)
- all devices appear as files to Unix programs, this permits a standardized device interface and enhances inter-process communication (piping, etc.)

The Kernel
- performs process management and inter-process communication (IPC)
- performs file management
- performs main memory management (RAM optimization with AI)
- performs disk management (cache optimization with AI)
- performs peripherals management

The System Call Interface
- Unix access at its lowest level, consists of function calls known as system calls
- system calls allow the user to manipulate processes, files, and other system resources.

- system calls are the most reliable function calls on the system (having stood the test of time and heavy usage) and are more secure than other function calls for the same reason.

The Language Libraries: prewritten/pretested functions exist for most programming languages. The relevant libraries and system calls define the application programmer's interface (API).

The UNIX Shell: starts at logon, interprets commands entered.

Applications: Programs that run via a combination of library calls and system calls. Since library calls are themselves composed of system calls, use of library calls can often be slower.

Directory and File Manipulations:

"ls" tells you the contents of your "working directory"

"ls –la" lists all content including hidden files

Your working directory when you first log in is your "home directory" (with name tied to your username). To change directories to a different working directory use the "cd" command to change to one of the directories shown when you enter "ls –la.."

"cd .." changes to parent directory for your "working directory"

"cd child_directory" changes to the subdirectory indicated

"mv file_a file_b" renames file_a as file_b

"mv file_a dir_x" moves file_a to dir_x

"cp file_a file_b" copies file_a to file_b

"cp file_a dir_x" copies file_a to dir_x

"less file_a" opens file_a in readonly mode (q to quit)

"vi file_a" opens file_a in vi edit mode (some cmds are like less)

Ubuntu and other Linux typically have a simple GUI editor "gedit"

Some UNIX commands: cut, bg and fg, nice, nohup, set and setenv, umask, less.

Some UNIX utilities: cat, chmod, chown, df, finger, ftp, grep, head, more, ping, rsh and ssh, sty, telnet, top, which.

A.3 Perl Review: I/O, Primitives, String Handling, Regex

Let us get started, try the following:

1) Login to Linux a Linux system.
2) Open a file named program1.pl in your favorite editor (use gedit as default) and type three lines:

```perl
#!/usr/bin/perl -w
# your basic "Hello World" program.
print "Hello World!\n";
```

3) Save the edited file.
4) Enter on the command line: "chmod u+x program1.pl."
5) Execute the program by typing ./program1.pl at the command prompt and hitting enter.

Something like that should work....... and print "Hello World" to standard out (STDOUT), which is the terminal screen, and we have thereby demonstrated we have Perl and print capability.

Let us learn more Perl syntax and then write some more programs:

Most computer languages have a sequence of lines of commands, each line ending in ";" (Python is the main exception, with no ";" end-of-line syntax).

```
$x = "Hello ";
$y = "World\n";
$sentence = $x . $y;
```

In the above example you see the introduction of scalar variables $x and $y, that are set to the string values indicated in quotes, and each command is terminated with a semicolon. The scalar variable $sentence is then set to the value of $x concatenated with the value of $y. The concatenation is indicated by the period operator ".".
The command

```
print "$sentence";
```

will print "Hello world" to the terminal screen as before. Notice how this print statement involves a (scalar) variable, $sentence, inside the quotes. Perl performs what is known as variable interpolation in this instance. Simply put, it substitutes "Hello World\n" for what should be printed as in the previous example.

Suppose your scalar variable holds a number (integer) instead:

```
$x = 8;
```

We can increase the value held in the variable $x by 7 by the command

```
$x = $x + 7;
```

The new value held in $x will now be 15. Notice that the line of code is not an algebraic statement! If it were it would be a false statement as we would be saying $x = x + 7$, and if we cancel the x's we get $0 = 7$, which is false. So in programming instructions in general (not just Perl syntax examples), it must be remembered that the code is for an operational instruction.

Often there is a close correspondence between an algebraic expression and its code implementation, but there is a fundamental difference just the same. The variable on the left side of the equals sign is known as an "lvalue" for this reason in programming languages jargon, where "lvalue" stands for "left value." What's occurring operationally is the computer performs the algebraic manipulations indicated on the right hand side of the equals sign, and the result is placed in the memory location of the variable indicated on the left-hand side of the equals sign.

It turns out that the above example is very common in programming, where a variable is often modified, and then the new value placed back into that same variable. Many languages have a shorthand syntax to emphasize precisely this type of operation:

```
$x += 7; is equivalent to the above line of code: $x=$x+7;
```

In Perl, if you have a string that you want to concatenate strings to, you could have something like:

```
$x = "c";
$sentence = "ab";
$sentence .= $x;
```

where the last instruction can be "unpacked" to mean:

```
$sentence = $sentence . $x;
```

Which would yield an updated value for $sentence of "abc."

Consider the list, or array, of values: (first, second, and third). In Perl this is simply coded with the following syntax:

```
@vars = ("first," "second," "third");
```

Let us now try to access the information and print it to the screen:

```
print "$vars[1] is after $vars[0] and before $vars[2]\n";
```

The individual variables in the array are simply accessed by using the array name followed by square brackets with a number inside the brackets that indicates which individual variable is being accessed.

Notice two important details. First, the indexing into the array starts at an index value "0" not "1"! So $vars[0] is accessing the string "first" that is listed first in the array. Second, since the indexed variable is single valued, it is a scalar, thus must be referenced as "$vars[0]" with a "$" symbol at the start, not "@"! The result of the print statement would thus read:

"second is after first and before third" (where the quotes are not shown in the actual printout).

Array indexing that starts at 0 is an annoying artifact that results from how the array variable information is represented in memory, and it is a shared convention for all computer languages that the indexing in arrays always starts at 0.

The reason for this is actually quite straightforward. Think of variables as actually being containers that hold the values they are "set equal to" in their container (memory) space. In the case of an array, the different values in the array are held in a contiguous sequence of memory locations, where the memory location of the first element of the array is simply that indicated when referencing the array. The second element of the array is one position over from the start of that contiguous memory block for the array, so is indexed with value "1," similarly for the indexing into the other array positions.

Perl is very friendly with array data. Perl allows two arrays to be concatenated:

```
@combined = (@first_array, @second_array);
```

Perl allows elements to be added to the front or back of an array with similar syntax:

```
@new_array_front = ($new_element, @old_array);
@new_array_back = (@old_array, $new_element);
```

Perl allows arrays to be created that range over values in an obvious numerical or alphabetical sequence, where:

```
@alpha = ('a' .. 'z');
```

results in the array holding the letters of the alphabet (all lower case). Perl allows sections, or a "slice," of an existing array to be taken by indexing the individual positions, or a range of positions, such as with

```
@slice = @alpha[4, 10 .. 15];
```

Which results in an array ('e','k','l','m','n','o','p'). You have seen what print does when presented with a scalar variable, what about when presented with an array variable? Very similar, but with an array you have multiple elements to print, and the question arises as to whether there is any delimiter between the printing of the individual values. Perl has a special variable, "$,", that allows you to specify that delimiter. The default for $, is a space, but you may want to specify it directly to be sure by simply including the command:

```
$, = " ";
```

If we now do the print:

```
print @slice;
```

we get as output to the terminal screen:

```
"e k l m n o p"
```

Perl also allows the elements of an array to be "joined" to form a string, where a delimiter can be specified in the join command:

```
join '+', ('apple','orange','banana');
```

evaluates to: "apple+orange+banana"

The reverse of this can be done as well, where a string can be split into its constituent parts according to a specified delimiter. If no delimiter is given, the splitting occurs at the individual character level:

```
$seq = "acgtag";
@arrayseq = split //, $seq;
```

The delimiter for the splitting is given between the "/" symbols, and as nothing is given in this example, the splitting defaults to individual character level, with the result that the array @arrayseq is now ('a','c','g','t','a','g').

When dealing with an associative array, or hash variable, the process is much the same. We might declare a hash variable with the name "hash" with the code:

```
%hash;
```

We might then set up an associative memory between the strings "key" and "value" by:

```
$hash{"key"}="value";
```

Notice again that when setting an individual value or accessing an individual hash entry, the special character in the hash variable changes from "%" to "$" similar to what occurred with the array variable entries. Consider the following print statement:

```
print "Each key is associated with a $hash{"key"}.\n";
```

which results in the output:

```
"Each key is associated with a value."
```

Most languages do not have a hash data primitive. Convenient hash variable primitives make Perl very effective in informatics applications where associative memory is often a fundamental aspect of the problem.

Now that we have seen the different variable types, the next thing to consider is operations between them. There are operators that involve one variable (unary), two variables (binary), and three variables (trinary). The most common unary operators are negation (use of the negative sign) and increment. There are two forms of the increment operator:

```
++$a;
```

and

```
$a++;
```

When given simply as shown these are both the same as "$a+=1;" which is the same as "$a=$a+1;", e.g., incrementing by one. The increment operators can be used in compound commands, however, and that's when they act differently:

```
$b=++$a;  → increments $a first, then $b is assigned that value
$b=$a++;  → $b is assigned $a's value, then $a is incremented.
```

If either of the above operations is done a significant amount in your programming, then either is good programming style. If there is not a lot of increment/assign operations then the combined compound expressions are bad style, and a good, readable, style convention is to simply use $a++ always for increment, and avoid compound increment forms entirely.

Binary operations are the most prevalent, including standard mathematical operations such as addition, multiplication, etc., as well comparison evaluation, equality evaluation, bitwise operations, and assignment operations:

Comparison Operators: <, >, <=, >=, <=>, lt, gt, le, ge, cmp

Equality Operators: ==, !=, eq, ne

Logical Operators: (||,OR), (&&,AND), (!,NOT), XOR

Bitwise Operators: <<, >>, &, |, ^, ~

Assignment Operators: =, +=, -=, *=, /=, %=, **=, .=, x=, and more, where op1 operator= op2;→op1=op1 operator op2; such as $a += 1;→$a=$a+1;Variables, and operations involving them, are accessed in programs according to control structures (if-then conditionals) and (repeated) loop structures. The group of commands, or code "blocks," that is run conditionally or in some loop, then forms the core of the imperative programming design. Sometimes variables only "exist," or are only made use of, while in a particular code block. Where a variable "lives" and can be accessed is referred to as the variable's "scope." A variable introduced in a particular block of code, usually only can be accessed inside that block. Before the advent of object oriented coding, the careful management of code blocks and variable scopes within those blocks, was a key aspect of software engineering to have "safe" code with minimal errors and maximal, safe, reusability in other programs.

Examples of the syntax for conditionals and loops:

Basic Conditional construct:

```
if (EXPR1) BLOCK_1
[elsif (EXPR2) BLOCK_2] …
[else BLOCK_N]
```

Basic Loop constructs:

```
"while (cond-expr) BLOCK", which is open ended on a
conditional, like a file read operation.
"for ([init-expr]; [cond-expr]; [loop-expr]) BLOCK", which
exits after a max number of iterations, if not sooner.
"foreach [ [my] $loop_var] (list) BLOCK", which loops in
association with the items of a list (when possible,
"foreach" is preferred over "for").
```

Appendix B

Physics

B.1 The Calculus of Variations

As Scientists we seek explanations (hypotheses) and subject them to tests and refinements. Elaborate explanations are more easily refuted, such that refinement typically leads to pared-down, simple explanations. The most efficient way to proceed given this well-known interplay between hypothesis complexity and robust modeling is to seek *simple* physical explanations before more complicated ones. The latter, pragmatic, notion is often referred to as "Occam's Razor," for the William of Occam phrase "causes shall not be multiplied beyond necessity." (From "nunquam ponenda est pluralitas sine necesitate" [311].) From the success of the scientific method with Occam's Razor it appears that nature effects an "economy of means" [312]. The hypotheses that underpin most of physical theory, in fact, can be described in concise, elegant, terms via extremal hypotheses. The study of extremal properties advanced greatly with the theological, philosophical, and mathematical explorations of Gottfried Leibniz, who postulated that our world is organized such that it is "the best of all possible worlds." To pursue the study of extremal properties, Leibniz and Newton, separately, invented calculus to have the mathematical framework they needed. Leibniz, in particular, described how variation in an "action" would be extremal in describing a motion. An exact definition of action was not given by him, however, or by Maupertuis who followed (who popularized only the minimum action version of the principle). The exact definition of the action only came with the rigorous explorations by Euler and Lagrange roughly 30 years later. Our legacy from these early explorers is a remarkably succinct description of physical properties using the calculus of variations.

The classic variational calculus application to physics describes moving bodies using the Lagrangian or Hamiltonian formalisms (which are related by a Legendre transformation). The Lagrangian variational formulation is based on "configuration space" where elements are described in terms of positions x_i and their

Informatics and Machine Learning: From Martingales to Metaheuristics, First Edition.
Stephen Winters-Hilt.
© 2022 John Wiley & Sons, Inc. Published 2022 by John Wiley & Sons, Inc.

corresponding velocities $\mathbf{v_i}$. Variation of a function, the "Lagrangian," then yields the equations of motion when the variation is extremized. The Lagrangian, $\mathbf{L}$, is often written in terms of kinetic energy contributions, $\mathbf{T}$, and potential energy contributions, $\mathbf{U}$: $\mathbf{L(x_i, v_i, t) = T(x_i, v_i, t) - U(x_i)}$. The classic case is where the kinetic energy is $\mathbf{mv^2/2}$, for which variation leads to $\mathbf{F = ma}$, Newton's Second Law. In general, the kinetic aspect of the motion is defined to be solely a description of the motion itself, devoid of explanation or causation. There is still substantial information in the kinetic representation, however, as this is where one's geometric assumptions enter the problem [318] (here a Galilean frame of reference on time). The $\mathbf{m}$ term then encodes the inertial concept and serves double duty as the dimensionful parameter linking acceleration to force. The dynamics, then, is whatever makes the description work (balances the equations). Different objects may share similar geometric "backgrounds," i.e., the kinetic energy part, but their dynamical description may differ in any way conceivable – encoded as force (or potential energy when working from the Lagrangian). Not all descriptions of motion need be fundamentally empirical in the dynamical parts of their descriptions, however, as Einstein's theory of gravitation attests. When expressed in a Lagrangian variational context, as first done by Hilbert, Einstein's theory can be shown to give a full description of both the kinematical and dynamical components of the motion entirely within a geometric construction (where geometry itself is dynamical). What is particularly satisfying about this presentation of Einstein's theory is that it depends on a minimum amount of structure (Occam's Razor): if a covariant tensorial theory is presumed to exist (the unstated kinematic structure), and if one disallows spontaneous creation of energy, disallows rotational instabilities (change in rotation with no torque applied), assumes three space and one time dimension (locally Lorentzian manifold exhibited by Maxwell's electromagnetic equations), and assumes the algebraic topology notion that the boundary of a boundary is zero (for example the boundary of a flat disk is taken to be the circle at its perimeter, and the boundary of that circle, a line with no ends, is zero), then one obtains, *uniquely*, Einstein's theory of gravitation [319].

Appendix C

Math

C.1 Martingales [102]

Martingale Definition

A stochastic process $\{X_n; n = 0, 1, ...\}$ is martingale if, for $n = 0, 1, ...$,

1) $E[|X_n|] < \infty$
2) $E[X_{n+1}|X_0, ..., X_n] = X_n$

Def.: Let $\{X_n; n = 0, 1, ...\}$ and $\{Y_n; n = 0, 1, ...\}$ be stochastic processes. We say $\{X_n\}$ is martingale with respect to (w.r.t) $\{Y_n\}$ if, for $n = 0, 1, ...$:

1) $E[|X_n|] < \infty$
2) $E[X_{n+1}|Y_0, ..., Y_n] = X_n$

Examples of Martingales:

a) Suns of independent random variables: $X_n = Y_1 + \cdots + Y_n$.
b) Variance of a Sum $X_n = \left(\sum_{k=1}^{n} Yk\right)^2 - n\sigma^2$
c) Have induced Martingales with Markov Chains! ....
d) For HMM learning, sequences of likelihood ratios are martingale....

The asymptotic equipartition theorem (AEP) and Hoeffding Inequalities (critical in Chapter 11) have both been generalized to Martingales.

Informatics and Machine Learning: From Martingales to Metaheuristics, First Edition.
Stephen Winters-Hilt.
© 2022 John Wiley & Sons, Inc. Published 2022 by John Wiley & Sons, Inc.

Induced Martingales with Markov Chains [102]

Let $\{Y_n; n = 0, 1, ...\}$ be a Markov Chain (MC) process with transition probability matrix $P = ||P_{ij}||$. Let f be a bounded right regular sequence for P:

$f(i)$ is non-negative and $f(i) = \sum_{k=1}^{n} P_{ij} f(j)$. Let $X_n = f(Y_n) \rightarrow E[|X_n|] < \infty$ (since f is bounded). Now have:

$$
\begin{aligned}
& E[X_{n+1} \mid Y_0, ..., Y_n] \\
& = E[f(Y_{n+1}) \mid Y_0, ..., Y_n] \\
& = E[f(Y_{n+1}) \mid Y_n] \quad \text{(due to MC)} \\
& = \sum_{k=1}^{n} P Y_n, _{j} f(j) \quad \text{(def.of } P_{ij} \text{ and } f) \\
& = f(Y_n) \\
& = X_n
\end{aligned}
$$

In HMM Learning Have Sequences of Likelihood Ratios, Which Is a Martingale, Proof

Let Y_0, Y_1, ... be iid rv.s and let f_0 and f_1 be probability density functions. A stochastic process of fundamental importance in the theory of testing statistical hypotheses is the sequence of likelihood ratios:

$$
X_n = \frac{f_1(Y_0)f_1(Y_1)\cdots f_1(Y_n)}{f_0(Y_0)f_0(Y_1)\cdots f_0(Y_n)}, \quad n = 0, 1, ...
$$

Assume $f_0(y) > 0$ for all y:

$$
E[X_{n+1} \mid Y_0, ..., Y_n] = E\left[X_n \left(\frac{f_1(Y_{n+1})}{f_0(Y_{n+1})}\right) \middle| Y_0, ..., Y_n\right] = X_n E\left[\frac{f_1(Y_{n+1})}{f_0(Y_{n+1})}\right]
$$

When the common distribution of the Y_k's (used in the "E" function) has f_0 as its probability density, have:

$$
E\left[\frac{f_1(Y_{n+1})}{f_0(Y_{n+1})}\right] = 1
$$

So, $E[X_{n+1} \mid Y_0, ..., Y_n] = X_n$

So likelihood ratios are martingale when the common distribution is f_0.

Supermartingales and Submartingales [102]

Let $\{X_n; n = 0, 1, ...\}$ and $\{Y_n; n = 0, 1, ...\}$ be stochastic processes. Then $\{X_n\}$ is called a *supermartingale* with respect to $\{Y_n\}$ if, for all n:

i) $E[X_n^-] > -\infty$, where $x^- = \min\{x,0\}$
ii) $E[X_{n+1} \mid Y_0, ..., Y_n] \leq X_n$
iii) X_n is a function of $(Y_0, ..., Y_n)$ [explicit due to inequality in (ii)]

The stochastic process $\{X_n; n = 0, 1, ...\}$ is called a *submartingale* w.r.t $\{Y_n\}$ if, for all n:

i) $E[X_n^+] > -\infty$, where $x^+ = \max\{x,0\}$
ii) $E[X_{n+1}|Y_0, ..., Y_n] \geq X_n$
iii) X_n is a function of $(Y_0, ..., Y_n)$

With Jensen's inequality for convex function φ and conditional expectations have:

$$E[\varphi(X) \mid Y_0, ..., Y_n] \geq \varphi(E[X \mid Y_0, ..., Y_n])$$

So, have means to construct submartingales from martingales (with supermartingales the same aside from a sign flip).

Martingale Convergence Theorems [102]

Under very general conditions, a martingale X_n will converge to a limit random variable X as n increases.

Theorem
a) Let $\{X_n\}$ be a submartingale satisfying

$$\sup_{n \geq 0} E[|X_n|] < \infty$$

Then there exists a r.v. X_∞ to which $\{X_n\}$ converges with probability one:

$$\text{Prob}\left(\lim_{n \to \infty} X_n = X_\infty \right) = 1$$

b) If $\{X_n\}$ is a martingale and is uniformly integrable, then, in addition to the above, $\{X_n\}$, converges in the mean:

$$\lim_{n \to \infty} E[|X_n - X_\infty|] = 0$$

And $E[X_\infty] = E[X_n]$, for all n.
A sequence is uniformly integral if:

$$\lim_{c \to \infty} \sup_{n \geq 0} E[|X_n|I\{|X_n| > c\}] = 0$$

where I is the indicator function: 1 if $|X_n| > c$, and 0 otherwise.

"Maximal" Inequalities for Martingales [102]

Chebyshev's inequality applied to a sequence can be "tightened" to a finer inequality known as the Kolmogorov inequality in terms of the maximum of the sequence. This carries over to Martingales:

Let $\{X_n; n = 0, 1, ...\}$ be iid rvs with $E[X_i] = 0 \,\forall\, i$ and $E[(X_i)^2] = \sigma^2 < \infty$. Define $S_0 = 0$, $S_n = X_1 + \cdots + X_n$, for $n \geq 1$. From Chebyshev's Inequality:

$$\varepsilon^2 \mathrm{Prob}(|S_n| > \varepsilon) \leq n\sigma^2, \quad \varepsilon > 0$$

A finer inequality is possible:

$$\varepsilon^2 \mathrm{Prob}\left(\max_{0 \leq k \leq n} |S_n| > \varepsilon\right) \leq n\sigma^2, \quad \varepsilon > 0$$

Known as the Kolmogorov inequality, it can be generalized to provide a maximal inequality on submartingales:

Lemma 1   Let $\{X_n\}$ be a submartingale for which $X_n \geq 0$ for all n. Then for any positive λ:

$$\lambda \,\mathrm{Prob}\left(\max_{0 \leq k \leq n} |X_k| > l\right) \leq E[X_n]$$

Lemma 2   Let $\{X_n\}$ be a non-negative supermartingale then for any positive λ:

$$\lambda \,\mathrm{Prob}\left(\max_{0 \leq k \leq n} |X_k| > l\right) \leq E[X_0]$$

Mean-Square Convergence Theorem for Martingales [102]

Let $\{X_n\}$ be a submartingale w.r.t $\{Y_n\}$ satisfying, for some constant k, $E[(X_n)^2] \leq k < \infty$, for all n. Then $\{X_n\}$ converges as $n \to \infty$ to a limit r.v. X_∞ both with probability one and in mean square:

Prob($\lim_{n \to \infty} X_n = X_\infty$) = 1, and $\lim_{n \to \infty} E[|X_n - X_\infty|^2] = 0$,
where $E[X_\infty] = E[X_n] = E[X_0]$, for all n.

Martingales w.r.t σ-Field Formalism

Review of axiomatic probability theory, have three basic elements:

1) The sample space, a set Ω whose elements ω correspond to the possible outcomes of an experiment;
2) The family of elements, a collection F of subsets A of Ω (the sigma fields). We say that the event A occurs if the outcome ω of the experiment is an element of A;
3) The probability measure, a function P defined on F and satisfying:

   i) $0 = [\emptyset] \leq P[A] \leq P[\Omega] = 1$ for $A \in F$

   ii) $P[A_1 \cup A_2] = P[A_1] + P[A_2] - P[A_1 \cap A_2]$ for $A_i \in F$

   iii) $P[\bigcup_{n=1}^{\infty} A_n] = \sum_{n=1}^{\infty} P[A_n]$ if $A_i \in F$ are mutually disjoint.

Then, the triple (Ω, F, P) is called a probability space.

Backwards Martingale Definition (w.r.t Sigma Sub-fields)

Let $\{Z_n\}$ be rv's on a probability space (Ω, F, P) and let $\{G_n; n = 0, 1, ...\}$ be a decreasing sequence of sub sigma-fields of F, viz.,

$F \supset F_n \supset F_{n+1}$,   for all n.

Then $\{Z_n\}$ is called a backward martingale w.r.t. $\{G_n\}$ if for $n = 0, 1, ...$:

i) Z_n is G_n-measurable
ii) $E[|Z_n|] < \infty$, and
iii) $E[Z_n|G_{n+1}] < Z_{n+1}$

$\{Z_n\}$ is a backwards martingale, iff $X_n = Z_{-n}$, $n = 0, -1, -2, ...$ forms a martingale w.r.t $F_n = G_{-n}$, $n = 0, -1, -2, ...$

Backwards Martingale Convergence Theorem

Let $\{Z_n\}$ be a backwards martingale w.r.t a decreasing sequence of sub sigma-fields $\{G_n\}$. Then:

$\text{Prob}(\lim_{n \to \infty} Z_n = Z) = 1$,   and $\lim_{n \to \infty} E[|Z - Z_n|] = 0$,
and $E[Z_n] = E[Z]$, for all n.

Strong Law of Large Numbers Proof

Let $\{X_n; n = 1, 2, ...\}$ be iid rvs with $E[|X_1|] < \infty$. Let $\mu = E[X_1]$, $S_0 = 0$, and $S_n = X_1 + \cdots + X_n$, for $n \geq 1$. Let G_n be the sigma field generated by $\{S_n, S_{n+1}, ...\}$. We can derive the strong law of large numbers from the observation that $Z_n = S_n/n$ $(Z_0 = \mu)$, forms a backward martingale w.r.t G_n. Have $E[|Z_n|] < \infty$ and Z_n is G_n-measurable by construction, so just need relation (iii):

$S_n \equiv E[S_n \mid S_n] = E[S_n \mid S_n, S_{n+1}, ...] = E[S_n \mid G_n] = \sum_{k=1}^{n} E[X_k \mid G_n] = nE[X_k \mid G_n]$,

with the last equality for $1 \leq k \leq n$, thus:

$Z_n = S_n/n = E[X_k \mid G_n]$

So,

$E[Z_{n-1} \mid G_n] = (n-1)^{-1} E[S_{n-1} \mid G_n] = (n-1)^{-1} \sum_{k=} 1^{n-1} E[X_k \mid G_n] = Z_n!!!$

Now use backward martingale convergence theorem to show the strong law:

$\text{Prob}\left(\lim_{n \to \infty} \frac{S_n}{n} = \mu \right) = 1$

Stationary Processes

A *stationary* process is a stochastic process $\{X(t), t \in T\}$ with the property that for any positive integer "k," and any points $t_1, ..., t_k$, and h in T, the joint distribution of $\{X(t_1), ..., X(t_k)\}$ is the same as the joint distribution of $\{X(t_1 + h), ..., X(t_k + h)\}$.

An ergodic theorem gives conditions under which an average over time

$$\overline{x_n} = \frac{1}{n}(x_1 + \cdots + x_n)$$

of a stochastic process will converge as the number n of observed periods becomes large. The strong law of large numbers is one such ergodic theorem.

Stationary processes provide a natural setting for generalization of the law of large numbers since for such processes the mean value is a constant $m = E[X_n]$, independent of time. Just as there are strong and weak laws of large numbers, there are a variety of ergodic theorems....

Strong Ergodic Theorem [102]

Let $\{X_n;\ n = 0, 1, ...\}$ be a strictly stationary process having finite mean $m = E[X_n]$. Let

$$\overline{X_n} = \frac{1}{n}(X_0 + \cdots + X_{n-1})$$

be the sample time average. Then, with probability one, the sequence $\{\overline{X_n}\}$ converges to some limit rv denoted $\overline{X}$:

$$\text{Prob}\left(\lim_{n \to \infty} \overline{X_n} = \overline{X}\right) = 1, \quad \text{and} \quad \lim_{n \to \infty} E\left[|\overline{X} - \overline{X_n}|\right] = 0, \quad \text{and}$$
$$E\left[\overline{X_n}\right] = E\left[\overline{X}\right] = m.$$

Asymptotic Equipartition Property (AEP)

$$\lim_{n \to \infty}\left[-\frac{1}{n}\log p(X_0, ..., X_{n-1})\right] = H(\{X_n\})$$

With probability one, provided $\{X_n\}$ is ergodic.

Proof: For $\{X_n\}$ a stationary ergodic finite Markov chain use relation that:

$H(\{X_n\}) = \lim_{k \to \infty} H(X_k \mid X_1, ..., X_{k-1})$ Or $H(\{X_n\}) = \lim_{l \to \infty} \frac{1}{l} H(X_1, ..., X_l)$

$H(X_n \mid X_0, ..., X_{n-1}) = -\sum_{i,j} \pi(i) P_{ij} \log P_{ij}$, where $\pi(i)$ is the prior on X_i and P_{ij} is the transition probability to go from X_i to X_j. Thus

$H(\{X_n\}) = -\sum_{i,j} \pi(i) P_{ij} \log P_{ij}$,

while,

$$-\frac{1}{n}\log p(X_0,...,X_{n-1}) = \frac{1}{n}\sum_{i=0}^{n-2}W_i - \frac{1}{n}\log \pi(X_0), \text{ where } W_i = -\log P_{i,i+1}$$

The ergodic theorem applies:

$$\lim_{n\to\infty}\left[-\frac{1}{n}\log p(X_0,...,X_{n-1})\right] = E[W_0] = -\sum_{i,j}\pi(i)P_{ij}\log P_{ij} = H(\{X_n\})$$

The general AEP proof uses the backwards martingale convergence theorem instead of the ergodic theorem.

De Finetti's Theorem

Let $\{X_n; n = 0, 1, ...\}$ be an infinite sequence of rv's. They are said to be exchangeable if for any finite cardinal number n and any two finite sequences $i_1, ..., i_n$ and $j_1, ...,j_n$ (with each of the i's distinct and each of the j's distinct), the two sequences

$$X_{i_1},...,X_{i_n} \quad \text{and} \quad X_{j_1},...,X_{j_n}$$

both have the same joint probability distribution.

iid→exchangeable, but not the reverse.

C.2   Hoeffding Inequality

· Hoeffdisng's inequality provides an supper bound on the probability that the sum of random variables deviates from its expected value (Wassily Hoeffding [104]). It is generalized to martingale differences by Azuma [108] and to functions of random variables $\{X_n\}$ with bounded differences [where function is empirical mean of the sequence of variables: $\overline{X} = \frac{1}{n}(X_1 + \cdots + X_n)$ recovers the special case of Hoeffding].

Recall:

Let $X_1, ..., X_n$ be independent random variables. Assume that the X_i are almost surely bounded: $P(X_i \in [a_i,b_i]) = 1$. Define the empirical mean of the sequence of variables as:

$$\overline{X} = \frac{1}{n}(X_1 + \cdots + X_n)$$

Hoeffding [104] proves the following:

$$P(\overline{X} - E[\overline{X}] \ge k) \le \exp\left(-\frac{2n^2k^2}{\sum_{i=1}^{n}(b_i - a_i)^2}\right)$$

$$P(|\overline{X} - E[\overline{X}]| \ge k) \le 2\exp\left(-\frac{2n^2k^2}{\sum_{i=1}^{n}(b_i - a_i)^2}\right)$$

For each X almost surely bounded have another relation if $E(X) = 0$ known as the Hoeffding Lemma:

$$E\left[e^{\lambda X}\right] \leq \exp\left(\frac{\lambda^2(b-a)^2}{8}\right)$$

The proof begins with showing the Lemma as the hard part.......

Hoeffding Lemma Proof

Since $e^{\lambda X}$ is a convex function, we have

$$e^{\lambda X} \leq \frac{b-X}{b-a}e^{\lambda a} + \frac{X-a}{b-a}e^{\lambda b}, \quad \forall \quad a \leq x \leq b$$

So,

$$E\left[e^{\lambda X}\right] \leq E\left[\frac{b-X}{b-a}e^{\lambda a} + \frac{X-a}{b-a}e^{\lambda b}\right] = \frac{b}{b-a}e^{\lambda a} + \frac{-a}{b-a}e^{\lambda b} \quad \text{(last is since } E[X] = 0)$$

The convexity method involves a line interpolation, let us shift to those parameters with $p = -a/(b-a)$, and introduce $hp = -a\lambda$ [so have $h = \lambda(b-a)$]:

$$\frac{b}{b-a}e^{\lambda a} + \frac{-a}{b-a}e^{\lambda b} = e^{\lambda a}\left[1 - p + pe^{\lambda(b-a)}\right] = e^{-hp}\left[1 - p + pe^h\right]$$

$$E\left[e^{\lambda X}\right] \leq e^{L(h)},$$

where,

$$L(h) = -hp + \ln\left(1 - p + pe^h\right) \rightarrow L(0) = 0.$$
$$L'(h) = -p + pe^h/\left(1 - p + pe^h\right) \rightarrow L'(0) = 0.$$
$$L''(h) = p(1-p)e^h \rightarrow L''(0) = p(1-p).$$
$$L^{(n)}(h) = p(1-p)e^h > 0$$

Using Taylor series for $L(h)$:

$$L(h) = L(0) + hL'(0) + \frac{1}{2}h^2 L''(0) + (\text{more positive terms at higher order in } h)$$

$$L(h) \leq \frac{1}{2}h^2 p(1-p)$$

Since we have $E[X] = 0$, have $p = -a/(b-a)$ is $\in [0,1]$, so classic logistic function, where the maximum value of $p(1-p)$ on range $[0,1]$ is ¼ (when $p = 1/2$), so:

$$L(h) \leq \frac{1}{8}h^2 \quad \text{and} \quad E\left[e^{\lambda X}\right] \leq e^{\frac{1}{8}\lambda^2(b-a)^2}$$

Hoeffding Inequality Proof (for Further Details, See [104])

Consider Sum on iid X_i, where $S_m = m\overline{X}$ where $\overline{X}$ has m terms in its empirical average:

$$P(S_m - E[S_m] \geq k) \leq e^{-tk}E\left[e^{t(S_m - E[S_m])}\right] \quad \text{(Chernoff Bounding Technique)}$$

$$= \prod_{i=1}^{m} e^{-tk}E\left[e^{t(X_i - E[X_i])}\right] \quad (\{X_n\} \text{ are iid})$$

$$\leq \prod_{i=1}^{m} e^{-tk}e^{\frac{1}{8}t^2(b_i - a_i)^2} \quad \text{(Hoeffding Lemma)}$$

$$= e^{-tk}e^{\frac{1}{8}t^2\sum_{i=1}^{m}(b_i - a_i)^2}$$

Have $f(t) = -tk + \frac{1}{8}t_2\sum_{i=1}^{m}(b_i - a_i)^2$; Choose $t = 4k/\sum_{i=1}^{m}(b_i - a_i)^2$ to minimize the upper bound to get:

$$P(S_m - E[S_m] \geq k)e^{-2k^2/\sum_{i=1}^{m}(b_i - a_i)^2}$$

$$P(\overline{X} - E[\overline{X}] \geq k)e^{-2m^2k^2/\sum_{i=1}^{m}(b_i - a_i)^2}$$

Chernoff Bounding Technique:

$P[X \geq k] = P[e^{tX} \geq e^{tk}] \leq e^{-tk}E[e^{tX}]$ (Chernoff uses Markov Inequality on last).

References

1 Winters-Hilt, S. (2019). *Data Analytics, Bioinformatics, and Machine Learning.* USA: Golden Tao Publishing. ISBN: 978-0-578-22302-5.

2 Winters-Hilt, S. (2019). *The Nanoscope.* USA: Golden Tao Publishing. ISBN: 978-0-578-22307-0.

3 Winters-Hilt, S. (2011). *Machine-Learning Based Sequence Analysis, Bioinformatics & Nanopore Transduction Detection.* USA: Golden Tao Publishing. ISBN: 978-1-257-64525-1.

4 Winters-Hilt, S. (2019). Immune repertoire profiling using nanopore transduction. Patent Pending.

5 Winters-Hilt, S. (2019). vbN?. Patent Pending.

6 Winters-Hilt, S. (2019). Biomolecule conformation, cofactor, and binding profiling using nanopore transduction. Patent Pending.

7 Winters-Hilt, S. (2019). Molecular system profiling using multiple nanopore transduction detectors and active probing of a cell or bioreactor. Patent Pending.

8 Winters-Hilt, S. (2018). Unified propagator theory and a non-experimental derivation for the fine-structure constant. *Advanced Studies in Theoretical Physics* **12** (5): 243–255. https://doi.org/10.12988/astp.2018.8626.

9 Winters-Hilt, S. (2018). The 22 letters of reality: chiral bisedenion properties for maximal information propagation. *Advanced Studies in Theoretical Physics* **12** (7): 301–318. https://doi.org/10.12988/astp.2018.8832.

10 Winters-Hilt, S. (2017). RNA-dependent RNA polymerase encoding artifacts in eukaryotic transcriptomes. *International Journal of Molecular Genetics and Gene Therapy* **2** (1) https://doi.org/10.16966/2471-4968.108.

11 Winters-Hilt, S. and Evanilla, J. (2017). Characterization of fish stock diversity via EST based miRNA trans-regulation profiling. *International Journal of Molecular Genetics and Gene Therapy* **3** (1) https://doi.org/10.16966/2471-4968.110.

12 Winters-Hilt, S. (2017). Distributed SVM learning and support vector reduction. *International Journal of Computing and Optimization* **4** (1): 91–114.

Informatics and Machine Learning: From Martingales to Metaheuristics, First Edition. Stephen Winters-Hilt.
© 2022 John Wiley & Sons, Inc. Published 2022 by John Wiley & Sons, Inc.

13 Winters-Hilt, S. (2017). Clustering via Support Vector Machine boosting with simulated annealing. *International Journal of Computing and Optimization* **4** (1): 53–89.

14 Winters-Hilt, S. (2017). Finite state automaton based signal acquisition with bootstrap learning. *International Journal of Computing and Optimization* **4** (1): 159–186.

15 Winters-Hilt, S. (2017). Nanopore transducer engineering and design. *International Journal of Molecular Biology* **2** (1) https://doi.org/10.16966/ijmbm.108.

16 Winters-Hilt, S. (2017). Biological system analysis using a nanopore transduction detector: from miRNA validation, to viral monitoring, to gene circuit feedback studies. *Advanced Studies in Medical Sciences* **5** (1): 13–53. https://doi.org/10.12988/asms.2017.722.

17 Winters-Hilt, S. (2017). Isomer-specific trace-level biosensing using a nanopore transduction detector. *Clinical and Experimental Medical Sciences* **5** (1): 35–66. https://doi.org/10.12988/cems.2017.722.

18 Winters-Hilt, S. and Lewis, A. (2017). Alt-splice gene predictor using multitrack-clique analysis: verification of statistical support for modelling in genomes of multicellular eukaryotes. *Informatics* **4**: 3. https://doi.org/10.3390/informatics4010003.

19 Winters-Hilt, S. (2017). Channel current cheminformatics and stochastic carrier-wave signal processing. *International Journal of Computing and Optimization* **4** (1): 115–157.

20 Winters-Hilt, S. (2017). Method and system for miRNA binding site profiling using a nanopore transduction detector. Patent Pending.

21 Winters-Hilt, S. (2017). Method and system for profiling protein conformation-binding relationships using a nanopore transduction detector. Patent Pending.

22 Winters-Hilt, S. (2016). Exploring protein conformation-binding relationships and antibody glyco-profiles using a nanopore transduction detector. *Molecules & Medicinal Chemistry* **2**: e1378. https://doi.org/10.14800/mmc.1378.

23 Winters-Hilt, S. and Stoyanov, A. (2016). Nanopore event-transduction signal stabilization for wide pH range under extreme chaotrope conditions. *Molecules* **21** (3): 346.

24 Winters-Hilt, S. (2016). Method and system for single molecule analysis using nanopore transduction and stochastic carrier wave signal processing. Non-provisional patent, June.

25 Winters-Hilt, S. (2015). Feynman-Cayley path integrals select chiral bi-sedenions with 10-dimensional space-time propagation. *Advanced Studies in Theoretical Physics* **9** (14): 667–683.

26 Winters-Hilt, S. (2015). Channel current cheminformatics and bioengineering methods for immunological screening, single-molecule analysis, and single molecular-interaction analysis. Patent Awarded and Published by the European Patent Office on 24 March 2015.

27 Winters-Hilt, S. (2015). Method and system for stochastic carrier wave communications, radio-noise embedded steganography, and robust self-tuning signal discovery and data-mining. Patented June 2015.

28 Winters-Hilt, S. (2015). Method and system for miRNA profiling and haplotyping, protein post-translational modification assaying, aptamer design and optimization, protein-protein interaction analysis, and analysis of gene circuits and complex biosystems. Patented June 2015.

29 Winters-Hilt, S. (2015). Method and system for sequencing nucleic acids using nanopore transduction, laser excitations, and tracking on learned nanopore noise states. Patented June 2015.

30 Winters-Hilt, S. (2015). Method and system for noise-state transduction detection, ion channel state tracking without translocation current, and nanopore-based non-destructive live-cell cytosol assaying. Patented June 2015.

31 Winters-Hilt, S., Horton-Chao, E., and Morales, E. (2011). The NTD nanoscope: potential applications and implementations. *BMC Bioinformatics* **12** (Suppl 10): S21.

32 Winters-Hilt, S. and Adelman, R. (2011). Methods and systems for classification, clustering, pattern recognition, and nanopore detector cheminformatics, using support vector machines (SVMs). Patented.

33 Winters-Hilt, S. and Baribault, C. (2010). A Meta-state HMM with application to gene structure identification in eukaryotes. *EURASIP Journal of Advances in Signal Processing*, Special Issue on Genomic Signal Processing **2010**: 18.

34 Winters-Hilt, S., Jiang, Z., and Baribault, C. (2010). Hidden Markov model with duration side-information for novel HMMD derivation, with application to eukaryotic gene finding. *EURASIP Journal of Advances in Signal Processing*, Special Issue on Genomic Signal Processing **2010**: 11.

35 Eren, A.M., Amin, I., Alba, A. et al. Pattern recognition informed feedback for nanopore detector cheminformatics. In: *Advances in Computational Biology* (ed. H.R. Arabnia). Springer in Advances in Experimental Medicine and Biology, AEMB 2010 book series.

36 Winters-Hilt, S. and Jiang, Z. (2010). A hidden Markov model with binned duration algorithm. *IEEE Transactions on Signal Processing* **58** (2): 5.

37 Winters-Hilt, S. and Adelman, R. (2010). Method and system for characterizing or identifying molecules and mmolecular mixtures. Patented August 2010.

38 Winters-Hilt, S. (2010). Nanopore transduction of DNA sequence information using enzymes covalently bound to channel modulators. Patented February 2010.

39 Winters-Hilt, S. (2010). Nanopore-based single-molecule DNA sequencing via simultaneous, single-molecule, discrimination of dsDNA terminus identification and dsDNA strand length, via resonance modulation with appropriate choice of buffer and/or dsDNA modifications. Patented February 2010.

40 Winters-Hilt, S. (2010). Hidden Markov model based structure identification using (i) HMM-with-Duration with positionally dependent emissions and incorporation of side-information into an HMMD via the ratio of cumulants method; and/or (ii) meta-HMMs and higher-order HMMs with gap and sequence-specific (hash) interpolated Markov models and Support Vector Machine signal boosting; and/or (iii) topological structure identification; and/or (iv) multi-track, parallel, or holographic HMMs; and/or (v) distributed HMM methods via Viterbi-path based reconstruction and verification; and/or (vi) adaptive null-state binning for O(TN) computation. Patented February 2010.

41 Winters-Hilt, S. (2009). Nanopore cheminformatics based studies of individual molecular interactions. In: *Machine Learning in Bioinformatics*, chapter 19 (eds. Y. Zhang and J.C. Rajapakse). Wiley.

42 Churbanov, A. and Winters-Hilt, S. (2008). Implementing EM and Viterbi algorithms for hidden Markov model in linear memory. *BMC Bioinformatics* **9**: 228.

43 Churbanov, A. and Winters-Hilt, S. (2008). Clustering ionic flow blockade toggles with a mixture of HMMs. *BMC Bioinformatics* **9**: S9–S13.

44 Roux, B. and Winters-Hilt, S. (2008). Hybrid SVM/MM structural sensors for stochastic sequential data. *BMC Bioinformatics* **9**: S9–S12.

45 Winters-Hilt, S. (2009). Biosensing processes with substrates, both immobilized (immuno-absorbant matrices) and free (enzyme substrate): Transducer Enzyme-Release with Immuno-absorbent Assay (TERISA); Transducer Accumulation and Release with Immuno-absorbent Assay (TARISA); Electrophoretic contrast substrate. Patent filing 2008, refiling August 2009.

46 Winters-Hilt, S. (2009). Post-translational protein modification assaying and transient complex characterization via nanopore detection and nanopore transduction detection. PATENT filing 2008, refiling August 2009.

47 Winters-Hilt, S. and Zhang, J. (2009). An efficient implementation for HMM with duration. PATENT filing 2008, refiling August 2009.

48 Winters-Hilt, S. (2009). NTD-based methods for: (I) electrophoresis-separation based on nanopore acquisition rate and nanopore-based classification; (II) multi-channel sensitivity gain and affinity gain, and related architectural refinements; and (III) multicomponent and nanomanipulation refinements. PATENT filing 2008, refiling August 2009.

49 Winters-Hilt, S. (2009). Pattern recognition informed (PRI) nanopore detection for sample boosting, nanomanipulation, and device stabilization; and PRI device stabilization methods in general. PATENT filing 2008, refiling August 2009.

50 Winters-Hilt, S. (2007). The alpha-Hemolysin Nanopore Transduction Detector – single-molecule binding studies and immunological screening of antibodies and aptamers. *BMC Bioinformatics* **8**: S7–S12.

51 Thomson, K., Amin, I., Morales, E., and Winters-Hilt, S. (2007). Preliminary nanopore cheminformatics analysis of aptamer-target binding strength. *BMC Bioinformatics* **8**: S7–S14.

52 Winters-Hilt, S., Davis, A., Amin, I., and Morales, E. (2007). The nanopore cheminformatics of individual transcription factor binding site interactions. *BMC Bioinformatics* **8**: S7–S10.

53 Winters-Hilt, S., Morales, E., Amin, I., and Stoyanov, A. (2007). Nanopore cheminformatics analysis of single antibody-channel interactions and antibody-antigen binding. *BMC Bioinformatics* **8**: S7–S20.

54 Winters-Hilt, S. and Merat, S. (2007). SVM clustering. *BMC Bioinformatics* **8**: S7–S18.

55 Landry, M. and Winters-Hilt, S. (2007). Analysis of nanopore detector measurements using machine learning methods, with application to single-molecule kinetic analysis. *BMC Bioinformatics* **8**: S7–S12.

56 Winters-Hilt, S. and Baribault, C. (2007). A novel, fast, HMM-with-Duration implementation – for application with a new, pattern recognition informed, nanopore detector. *BMC Bioinformatics* **8**: S7–S19.

57 Winters-Hilt, S., Landry, M., Akeson, M. et al. (2006). Cheminformatics methods for novel nanopore analysis of HIV DNA termini. *BMC Bioinformatics* **7**: S2–S22.

58 Winters-Hilt, S. (2006). Nanopore detector based analysis of single-molecule conformational kinetics and binding interactions. *BMC Bioinformatics* **7**: S2–S21.

59 Winters-Hilt, S., Yelundur, A., McChesney, C., and Landry, M. (2006). Support vector machine implementations for classification & clustering. *BMC Bioinformatics* **7**: S2–S4.

60 Winters-Hilt, S. (2006). Hidden Markov model variants and their application. *BMC Bioinformatics* **7**: S2–S14.

61 Winters-Hilt, S. (2005). Single-molecule biochemical analysis using channel current cheminformatics. Fourth International Conference on Unsolved Problems of Noise and Fluctuations, 6–10 June 2005.

62 Winters-Hilt, S. (2005). Channel current cheminformatics and bioengineering methods for immunological screening, single-molecule analysis, and single molecular-interaction analysis. PATENT, UNO filing, 2005.

63 Winters-Hilt, S. (2004). Nanopore detection using channel current cheminformatics. SPIE Second International Symposium on Fluctuations and Noise, 25–28 May 2004.

64 Winters-Hilt, S. (2004). Channel current cheminformatics based immunological screening of pore inhibiting agents. PATENT, UNO filing, 2004.

65 Winters-Hilt, S. (2004). Channel current cheminformatics based assayer of cytosolic antigen delivery. PATENT, UNO, 2004.

66 Winters-Hilt, S. (2003). Highly accurate real-time classification of channel-captured DNA termini. Third International Conference on Unsolved Problems of Noise and Fluctuations, 2003.

67 Winters-Hilt, S., Vercoutere, W., DeGuzman, V.S. et al. (2003). Highly accurate classification of Watson-Crick base-pairs on termini of single DNA molecules. *Biophysical Journal* **84**: 967.

68 Akeson, M., Winters-Hilt, S., Vercoutere, W. et al. (2000). Methods and devices for characterizing duplex DNA molecules. PATENT, UCSC filing, 2000.

69 Winters-Hilt, S. and Morales, E. (2019). Aperture construction method for nanopore transduction detectors. Patent Pending.

70 Winters-Hilt, S. (2019). Cannamimetic profiling using nanopore transduction. Patent Pending.

71 Winters-Hilt, S. (2019). The nanodrop DNA microarray: molecular system profiling using a nanopore transduction detector and an annealing-based aptamer probe-set. Patent Pending.

72 Winters-Hilt, S. (2019). The nanodrop protein microarray: molecular system profiling using a nanopore transduction detector and a non-annealing-based aptamer probe-set. Patent Pending.

73 Winters-Hilt, S. (2019). The mAb nanodrop microarray: molecular system profiling using a nanopore transduction detector and a mAb based probe-set. Patent Pending.

74 Adelman, R. and Winters-Hilt, S. (2018). Cannabinoid profiling using nanopore transduction. Non-provisional Patent December 2018.

75 Winters-Hilt, S. and Adelman, R. (2017). Method and system for high-specificity trace-level molecular testing, biosensing, diagnostics, and therapeutic testing using a nanopore transduction detector. Patent Pending.

76 Winters-Hilt, S. and Adelman, R. (2017). Method and system for isomer assaying by use of a nanopore transduction detector. Patent Pending.

77 Winters-Hilt, S. (2017). Method for microRNA binding site transcriptome analysis for phenotype-diversity evaluation. Patent Pending.

78 Winters-Hilt, S. (2017). Method for microRNA binding site transcriptome analysis for individual transcript fingerprinting and system-level analysis. Patent Pending.

79 Winters-Hilt, S. and Adelman, R. (2017). Method and system for assaying mixtures of cannabinoids, terpenes, terpenoids, pyrethrins, and similar molecular weight biomolecules by use of a nanopore transduction detector. Patent Pending.

80 Winters-Hilt, S. (2017). Method and system for cooling positively pressurized spaces, rooms, and tents, using driven evaporation, and capillary-effect, heat-wicking and using a device controller with machine-learning based system monitoring by tracking the noise state of the system. Patent Pending.

81 Winters-Hilt, S. and Adelman, R. (2017). Method and System for pyrethrin and insecticide assaying, chelator targeting & design, and chelation-column filtration by use of a nanopore transduction detector. Patent Pending.

82 Winters-Hilt, S. and Adelman, R. (2011). Methods and systems for sequential analysis and nanopore detector signal analysis using stochastic sequential analysis (SSA) methods such as hidden Markov models (HMMs). USPTO Patent 2011.

83 Winters-Hilt, S. and Adelman, R. (2011). Methods and systems for nanopore biosensing. USPTO Patent 2011.

84 Churbanov, A., Winters-Hilt, S., Koonin, E.V., and Rogozin, I.B. (2008). Accumulation of GC donor splice signals in mammals. *Biology Direct* **3**: 30.

85 Stoyanov, A. and Winters-Hilt, S. (2008). Method of electrophoresis for biopolymer separation in gel media with immobilized charges according to molecular size or asymptotic electrophoretic mobility and its multi-dimensional applications. PATENT filing August 2008.

86 Churbanov, A., Baribault, C., and Winters-Hilt, S. (2007). Duration learning for nanopore ionic flow blockade analysis. *BMC Bioinformatics* **8**: S7–S14.

87 Iqbal, R., Landry, M., and Winters-Hilt, S. (2006). DNA molecule classification using feature primitives. *BMC Bioinformatics* **7**: S2–S15.

88 Deamer, D.W. and Winters-Hilt, S. (2005). Nanopore analysis of DNA. In: *Encyclopedia of Nanoscience and Nanotechnology*, vol. **7** (ed. H.S. Nalwa), 229–235. American Scientific.

89 Winters-Hilt, S. and Akeson, M. (2004). Nanopore cheminformatics. *DNA and Cell Biology* **23** (10): 9.

90 Winters-Hilt, S. and Pincus, S. (2004). Nanopore-based biosensing. PATENT UNO filing.

91 Winters-Hilt, S. and Pincus, S. (2004). Nanopore-based antibody characterization and antibody-antigen efficacy screening. PATENT, UNO filing.

92 DeGuzman, V., Winters-Hilt, S., Solbrig, A. et al. (2003). Sequence-dependent fraying of single DNA molecules measured in real time at 5 angstrom resolution using an ion channel. *Biophysical Journal* **84** (2): 490A. Part 2 Suppl.

93 Vercoutere, W., Winters-Hilt, S., DeGuzman, V.S. et al. (2003). Discrimination among individual Watson-Crick base-pairs at the termini of single DNA hairpin molecules. *Nucleic Acids Research* **31**: 1311–1318.

94 Akeson, M. and Winters-Hilt, S. (2003). Methods and devices for manipulating single biomolecules. PATENT, UCSC, 2003.

95 Vercoutere, W., Winters-Hilt, S., Olsen, H. et al. (2001). Rapid discrimination among individual DNA molecules at single nucleotide resolution using an ion channel. *Nature Biotechnology* **19**: 248.

96 Winters-Hilt, S., Redmount, I.H., and Parker, L. (1999). Physical distinction among alternative vacuum states in flat spacetime geometries. *Physical Reviews D* **60**: 124017.

97 Friedman, J.L., Louko, J., and Winters-Hilt, S. (1997). Reduced Phase space formalism for spherically symmetric geometry with a massive dust shell. *Physical Reviews D* **56**: 7674–7691.

98 Louko, J., Simon, J.Z., and Winters-Hilt, S. (1997). Hamiltonian thermodynamics of a Lovelock black hole. *Physical Reviews D* **55**: 3525–3535.

99 Louko, J. and Winters-Hilt, S. (1996). Hamiltonian thermodynamics of the Reissner-Nordstrom-anti de Sitter black hole. *Physical Reviews D* **54**: 2647–2663.

100 Winters-Hilt, S. (2021). *Lagrangian Physics and Unified Propagator Theory*. USA: Golden Tao Publishing.

101 Cox, R.T. (1946). Probability, frequency and reasonable expectation. *American Journal of Physics* **14**: 1.

102 Karlin, S. and Taylor, H.M. (1975). *A First Course in Stochastic Processes*, 2e. Academic Press.

103 Markov, A.A. (1954). *Theory of Algorithms*. Academy of Sciences of the USSR.

104 Hoeffding, W. Probability inequalities for sums of bounded variables. *Journal of the American Statistical Association* **58** (301): 13–30.

105 Kullback, S. (1968). *Information Theory and Statistics*. Dover.

106 Shannon, C.E. (1948). A mathematical theory of communication. *The Bell System Technical Journal* **27**: 379–423, 623–656.

107 Khinchine, A.I. (1957). *Mathematical Foundations of Information Theory*. Dover.

108 Azuma, K. (1967). Weighted sums of certain dependent random variables. *Tohoku Mathematical Journal* **19** (3): 11.

109 Abu-Mostafa, Y.S., Magdon-Ismail, M., and Lin, H.-T. (2012). *Learning from Data*. AMLBook.

110 Freund, Y. and Schapire, R.E. (1997). A decision-theoretic generalization of on-line learning and an application to boosting. *Journal of Computer and System Sciences* **55**: 21.

111 Kapur, J.N. and Kesavan, H.K. (1992). *Entropy Optimization Principles with Applications*. Academic Press.

112 Jaynes E. 1997. *Paradoxes of Probability Theory*. Internet accessible book preprint: http://omega.albany.edu:8008/JaynesBook.html.

113 Amari, S. (1991). Dualistic geometry of the manifold of higher-order neurons. *Neural Networks* **4** (4): 443–451.

114 Amari, S. (1995). Information geometry of the EM and EM algorithms for neural networks. *Neural Networks* **8** (9): 1379–1408.

115 Amari, S. and Nagaoka, H. (2000). *Methods of Information Geometry*, Translations of Mathematical Monographs, vol. **191**. Oxford University Press.

116 Oppenheim, J.N. and Magnasco, M.O. (2012). Human time-frequency acuity beats the Fourier uncertainty principle. *Physical Review Letters* **110** (4): 5.

117 Benedicks, M. (1985). On Fourier transforms of functions supported on sets of finite Lebesgue measure. *Journal of Mathematical Analysis and Applications* **106** (1): 180–183.

118 Silicon Laboratories Inc. (2013). Improving ADC resolution by oversampling and averaging. https://www.silabs.com/documents/public/application-notes/an118.pdf (accessed 17 January 2015).

119 Torres, S.M. and Zrnic, D.S. (2003). Whitening in range to improve weather Radar spectral moment estimates. Part I: formulation and simulation. *Journal of Atmospheric and Oceanic Technology* **20**: 16.

120 Torres, S.M. and Zrnic, D.S. (2003). Whitening of signals in range to improve estimates of polarimetric variables. *Journal of Atmospheric and Oceanic Technology* **20** (12): 1776–1789.

121 Pohlmann, K.C. (2005). *Principles of Digital Audio*. McGraw-Hill Professional. ISBN: 0-07-144156-5.

122 Schuchman, L. (1964). Dither signals and their effect on quantization noise. *IEEE Transactions on Communications* **12** (4): 162–165.

123 Analog Devices (1999). A technical tutorial on digital signal synthesis. http://www.analog.com/static/imported-files/tutorials/450968421DDS_Tutorial_rev12-2-99.pdf (accessed 17 January 2015).

124 Cormen, T.H., Leiserson, C.E., and Rivest, R.L. (1989). *Introduction to Algorithms*. Cambridge, USA: MIT-Press.

125 Majoros, W.H., Pertea, M., and Salzberg, S.L. (2004). TigrScan and GlimmerHMM: two open source *ab initio* eukaryotic gene-finders. *Bioinformatics* **1** (16): 2878–2879.

126 Durbin, R., Eddy, S., Krogh, A., and Mitchison, G. (1998). *Biological Sequence Analysis: Probalistic Models of Proteins and Nucleic Acids*. Cambridge, UK/New York: Cambridge University Press.

127 Altschul, S., Gish, W., Miller, W. et al. (1990). Basic local alignment search tool. *Journal of Molecular Biology* **215** (3): 403–410.

128 Rabiner, L.R. (1989). A tutorial on hidden Markov models and selected application in speech recognition. *Proceedings of the IEEE* **77**: 257–286.

129 Burset, M. and Guigo, R. (1996). Evaluation of gene structure prediction programs. *Genomics* **34**: 353–367.

130 Mathé, C., Sagot, M.-F., Schiex, T., and Rouzé, P. (2002). Current methods of gene prediction, their strengths and weaknesses. *Nucleic Acids Research* **30** (19): 4103–4117.

131 Stanke, M., Steinkamp, R., Waack, S., and Morgenstern, B. (2004). AUGUSTUS: a web server for gene finding in eukaryotes. *Nucleic Acids Research* **32**: W309–W312.

132 Stanke, M. and Waack, S. (2003). Gene prediction with a hidden Markov model and new intron submodel. *Bioinformatics* **19** (Suppl. 2): ii215–ii225.

133 Guigo, R., Agarwal, P., Abril, J. et al. (2000). An assessment of gene prediction accuracy in large DNA sequences. *Genome Research* **10**: 1631–1642.

134 Mian, S., Krogh, A., and Haussler, D. (1994). A hidden Markov model that finds genes in *E. coli* DNA. *Nucleic Acids Research* **22**: 68–78.

135 Lu, D. (2009). Motif finding. UNO MS Thesis 2009, Advisor – S. Winters-Hilt.

136 Alkhateeb, J.H., Pauplin, O., Ren, J., and Jiang, J. (2011). Performance of hidden Markov model and dynamic Bayesian classifiers on handwritten Arabic word recognition. *Knowledge-Based Systems* **24** (5): 680–688.

137 Venkataramaanan, L. and Sigworth, F.J. (2002). Applying hidden Markov models to the analysis of single ion channel activity. *Biophysical Journal* **82** (4): 1930–1942.

138 Ferguson, J.D. (1980). Variable duration models for speech. Proceedings of Symposium on the Application of Hidden Markov models to Text and Speech, 143–179.

139 Ramesh, P. and Wilpon, J.G. (1992). Modeling state durations in hidden Markov models for automatic speech recognition. Proceedings of IEEE International Conference on Acoustics, Speech and Signal Processing, 1, 381–384.

140 Yu, S.Z. and Kobayashi, H. (2003). An efficient forward-backward algorithm for an explicit-duration hidden Markov model. *IEEE Signal Processing Letters* **10**: 11–14.

141 Johnson, M.T. (2005). Capacity and complexity of hmm duration modeling techniques. *IEEE Signal Processing Letters* **12**: 407–410.

142 Ghahramani, Z. and Jordan, M. Factorial hidden Markov models. *Machine Learning* **29**: 245–273.

143 Singer, Y., Fine, S., and Tishby, N. (1998). The hierarchical hidden Markov model: analysis and applications. *Machine Learning* **32**: 41.

144 Murphy, K. and Paskin, M. (2001). Linear time inference in hierarchical HMMS. Proceedings of Neural Information Processing Systems.

145 Miklos, I. and Meyer, I. (2005). A linear memory algorithm for Baum-Welch training. *BMC Bioinformatics* **6** (231): 8.

146 Stoll, P.A. and Ohya, J. (1995). Applications of HMM modeling to recognizing human gestures in image sequences for a man-machine interface. IEEE International Workshop on Robot and Human Communication, 129–134.

147 Elmezain, J.A.M., Al-Hamadi, A., and Michaelis, B. (2008). A hidden Markov model-based continuous gesture recognition system for hand motion trajectory. IEEE Conference Proceeding.

148 Appenrodt, J., Elmezain, M., Al-Hamadi, A., and Michaelis, B. (2009). A hidden Markov model-based isolated and meaningful hand gesture recognition. *International Journal of Electrical, Computer, and Systems Engineering* **3**: 156–163.

149 Augustin, E., Knerr, S., and Price, D. (1998). Hidden Markov model based word recognition and its application to legal amount reading on french checks. *Computer Vision and Image Understanding* **70**: 404.

150 Schenkel, M. and Jabri, M. (1998). Low resolution, degraded document recognition using neural networks and hidden Markov models. *Pattern Recognition Letters* **3**: 365–371.

151 Vlontzos, J. and Kung, S. (1992). Hidden Markov models for character recognition. *IEEE Transactions on Image Processing* **1**: 15.

152 Najmi, A., Li, J., and Gray, R.M. (2000). Image classification by a two-dimensional hidden Markov model. *IEEE Transactions on Signal Processing* **48**: 17.

153 Olshen, R.A., Li, J., and Gray, R.M. (2000). Multiresolution image classification by hierarchical modeling with two-dimensional hidden Markov models. *IEEE Transactions on Information Theory* **46**: 16.

154 Wu, M.S., Huang, C.L., and Jeng, S.H. (2000). Gesture recognition using the multi-pdm method and hidden Markov model. *Image and Vision Computing* **18**: 865.

155 Garcia-Frias, J. and Crespo, P.M. (1997). Hidden Markov models for burst error characterization in indoor radio channels. *IEEE Transactions on Vehicular Technology* **46**: 15.

156 Hughes, J.P., Bellone, E., and Guttorp, P. (2000). A hidden Markov model for downscaling synoptic atmospheric patterns to precipitation amounts. *Climate Research* **15**: 12.

157 Raphael, C. (1998). Automatic segmentation of acoustic musical signals using hidden Markov models. *IEEE Transactions on Pattern Analysis and Machine Intelligence* **21**: 360.

158 Kogan, J.A. and Margoliash, D. (1998). Automated recognition of bird song elements from continuous recordings using dynamic time warping and hidden Markov models: a comparative study. *Journal of the Acoustical Society of America* **103**: 12.

159 Cortes, C. and Vapnik, V.N. (1995). Support vector networks. *Machine Learning* **20**: 273–297.

160 Vapnik, V.N. (1995). *The Nature of Statistical Learning Theory*. New York: Springer-Verlag.

161 Vapnik, V.N. (1998). *Statistical Learning Theory*. New York: Wiley.

162 Burgess, C.J.C. (1998). A tutorial on support vector machines for pattern recognition. *Knowledge Discovery and Data Mining* **2** (2): 121–167.

163 Platt, J. (1998). Sequential Minimal Optimization: A Fast Algorith for Training Support Vector Machines, *Microsoft Research Tech. Rep. MSR-TR-98-14*.

164 Platt, J.C. (1998). Fast training of support vector machines using sequential minimal optimization. In: *Advances in Kernel Methods – Support Vector Learning*, chapter 12 (eds. B. Scholkopf, C.J.C. Burges and A.J. Smola). Cambridge, USA: MIT Press.

165 Vapnik, V.N. (1999). *The Nature of Statistical Learning Theory*, 2e. New York: Springer-Verlag.

166 Graf, H.P., Cosatto, E., Bottou, L. et al. (2004). Parallel support vector machines: the cascade SVM. Proceedings NIPS, 2004.

167 Armond, K.C. Jr. (2008). Distributed support vector machine learning. University of New Orleans Master's Thesis in CS, Advisor Stephen Winters-Hilt.

168 McChesney, C. (2006). SVM-based clustering. University of New Orleans Master's Thesis in CS, Advisor Stephen Winters-Hilt.

169 Merat, S. (2008). Clustering via supervised support vector machines. University of New Orleans Master's Thesis in CS, Advisor Stephen Winters-Hilt.

170 Zhang, H. (2009). Distributed support vector machines with graphical processing units. University of New Orleans Master's Thesis in CS, Advisor Stephen Winters-Hilt.

171 Girosi, F. and Poggio, T. (1990). Regularization algorithms for learning that there are equivalent to multilayer networks. *Science* **247**: 978–982.

172 Scholkopf, B., Williamson, R.C., and Smola, A.J. (1990). Regularization algorithms for learning that there are equivalent to multilayer networks. *Science* **247**: 978–982.

173 Doursat, R., Geman, S., and Bienstock, E. (1992). Neural network and bias/variance dilemma. *Neural Computation* **4** (2): 1–58.

174 Kleinberg, J. (2002). An impossibility theorem for clustering. Proceedings of the 16th Conference on Neural Information Processing Systems, 12.

175 Bishop, C.M. (1995). *Neural Networks for Pattern Recognition*. Oxford University Press.

176 Schapire, R.E. and Freund, Y. (1997). A decision-theoretic generalization of on-line learning and an application to boosting. *Journal of Computer and System Sciences* **55** (1): 119–139.

177 Crammer, K. and Singer, Y. (2001). On the algorithmic implementation of multiclass kernel-based vector machines. *Journal of Machine Learning Research* **2**: 265–292.

178 Ratsch, G., Tsuda, K., Muller, K. et al. (2001). An introduction to kernel-based learning algorithms. *IEEE Transactions on Neural Networks* **12** (2): 181–201.

179 Osuna, E., Freund, R., and Girosi, F. (1997). An improved training algorithm for support vector machines. In: *Neural Networks for Signal Processing VII* (eds. J. Principe, L. Gile, N. Morgan and E. Wilson), 276–285. New York: IEEE.

180 Joachims, T. (1998). Making large-scale SVM learning practical. In: *Advances in Kernel Methods – Support Vector Learning*, chapter 11 (eds. B. Scholkopf, B. CJC and A.J. Smola). Cambridge, USA: MIT Press.

181 Hsu, C.W. and Lin, C.J. (2002). A comparison of methods for multi-class support vector machines. *IEEE Transactions on Neural Networks* **13**: 415–425.

182 Lee, Y., Lin, Y., and Wahba, G. (2001). Multicategory Support Vector Machines. *Technical Report 1043*, Department of Statistics, University of Wisconsin, Madison, WI, 2001. http://citeseer.ist.psu.edu/lee01multicategory.html

183 Duda, R.O., Hart, P.E., and Stork, D.G. (2001). *Pattern Classification*, 2e. New York: Wiley.

184 Keerthi, S.S., Shevade, S.K., Bhattacharyya, C., and Murthy, K.R.K. (2001). Improvements to Platt's SMO algorithm for SVM classifier design. *Neural Computation* **13**: 637–649.

185 Ben-Hur, A., Horn, D., Siegelmann, H.T., and Vapnik, V. (2001). Support vector clustering. *Journal of Machine Learning Research* **2**: 125–137.

186 Scholkopf, B., Platt, J.C., Shawe-Taylor, J. et al. (1999). Estimating the support of a high-dimensional distribution. *Neural Computation* **13**: 1443–1472.

187 Yang, J., Estivill-Castro, V., and Chalup, S.K. (2002). Support vector clustering through proximity graph modeling. Proceedings, 9th International Conference on Neural Information Processing (ICONIP'02), 898–903.

188 Fisher, R.A. (1936). The use of multiple measurements in taxonomic problems. *Annals of Eugenics* **7**: 179–188.

189 Donoho, D. (2006). Compressed sensing. *IEEE Transactions on Information Theory* **52** (4): 1289–1306.

190 DeFelice, L.J. (1981). *Introduction to Membrane Noise*. New York: Plenum Press.

191 Ziemer, R.E. and Tranter, W.H. (1985). *Principles of Communications; Systems, Modulation, and Noise*, 2e. Boston: Houghton Mifflin Company.

192 Wolpert, D.H. and Macready, W.G. (1997). No free lunch theorems for optimization. *IEEE Transactions on Evolutionary Computation* **1**: 67.

193 Wolpert, D.H. and Macready, W.G. (2005). Coevolutionary free lunches. *IEEE Transactions on Evolutionary Computation* **9** (6): 721–735.

194 Raspe, R.E. (1785). *Baron Munchhausen's Narrative of his Marvellous Travels and Campaigns in Russia*. London: Cresset Press.

195 IEEE Standard 100 (2000). *Authoritative Dictionary of IEEE Standards Terms*, 7e, 123. IEEE Press. ISBN: 0-7381-2601-2.

196 Oppenheim, A.V., Willsky, A.S., and Young, I.T. (1983). *Signals and Systems*. New Jersey: Prentice-Hall.

197 Abad, J. and Digital, T. (2008). Numero 15 – Junio 2008. La virtu segun maquiavelo: Significados Y Traducciones.

198 Perkins, M.A. (1994). *Coleridge's Philosophy: The Logos as Unifying Principle*. Clarendon Press. ISBN: 978-0198240754.

199 Coleridge, S.T. (1814). On the principles of genial criticism. *Felix Farley's Bristol Journal* **85**: 6.

200 McFarland, T. (ed.) (2002). *The Collected Works of Samuel Taylor Coleridge*, vol. **15**. Opus Maximum. ISBN: 9780691098821.

201 Kiparsky, P. Stress, syntax, and meter. *Language* **51** (3): 576–616.

202 Sullivan, A., Eladhari, M.P., and Cook, M. (2018). Tarot-based Narrative Generation. Foundations of Digital Games 2018 (FDG '18), 7–10 August 2018, Malmo, Sweden.

203 Stevens, A. (1999). *On Jung: Updated Edition*, 2e, 215. Princeton, NJ: Princeton University Press. ISBN: 069101048X.

204 Bousso, R. (2002). The holographic principle. *Reviews of Modern Physics* **74**: 825–874. http://arxiv.org/pdf/hep-th/0203101.

205 Jiang, Z. (2010). Binned HMM with duration: variations and applications. University of New Orleans PhD Dissertation in CS, Advisor Stephen Winters-Hilt.

206 Schutz, B. (1980). *Geometrical Methods of Mathematical Physics*. Cambridge University Press. ISBN: 978-0521298872.

207 Gockeler, M. and Schucker, T. (1987). *Differential Geometry, Gauge Theories, and Gravity*. Cambridge University Press. ISBN: 978-0521378215.

208 Misner, C., Thorne, K., and Wheeler, J. (1973). *Gravitation*. W.H. Freeman. ISBN: 978-0691177793.

209 Bishop, R. and Crittenden, R. (1964). *Geometry of Manifolds*. Academic Press. ISBN: 978-0821829233.

210 Cencov, N.N. (1982). *Statistical Decisions Rules and Optimal Inference*, Translations of Mathematical Monographs, vol. **53**. American Mathematical Society (reprint of obscure 1972 paper).

211 Eguchi, S. (1983). Second order efficiency of minimum contrast estimators in a curved exponential family. *Annals of Statistics* **11**: 793–803.

212 Csiszar, I. and Tusnady, G. (1984). Information geometry and alternating minimization procedures. *Statistics and Decisions* **Suppl 1**: 205–237.

213 Kivinen, J. and Warmuth, M.K. (1997). Exponential gradient versus gradient descent for linear predictors. *Information and Comutation* **132** (1): 63.

214 Warmuth, M.K. and Jagota, A.K. (1997). Continuous and discrete-time nonlinear gradient descent: relative loss bounds and convergence. Electronic Proceedings of the 5th International Symposium on Artificial Intelligence and Mathematics.

215 Bylander, T. (1998). Worst-case analysis of the perceptron and exponentiated update algorithms. *Artificial Intelligence* **106**: 335–352.

216 Zanghirati, G. and Zanni, L. (2003). A parallel solver for large quadratic programs in training support vector machines. *Parallel Computing* **29**: 535–551.

217 Winters-Hilt, S. (2003). Machine learning methods for channel current cheminformatics, biophysical analysis, and bioinformatics. PhD dissertation, UCSC.

218 Eppstein, D. and Bern, M. (1996). Approximation algorithms for geometric problems. In: *Approximation Algorithms for NP-hard Problems* (ed. D. Hochbaum), 296–345. PWS Publishing Co.

219 Luke, S. (2009). *Essentials of Metaheuristics*. Lulu.

220 Rumelhart, D.E., Hinton, G.E., and Williams, R.J. Learning representations by back-propagating errors. *Nature* **323** (6088): 533–536.

221 LeCun, Y., Bottou, L., Bengio, Y., and Haffner, P. (1998). Gradient-based learning applied to document recognition. *Proceedings of the IEEE* **86** (11): 2278–2324.

222 LeCun, Y. (2015). The MNIST DATABASE of handwritten digits. https://en.wikipedia.org/wiki/MNIST_database (accessed 17 January 2015).

223 Luong, T., Pham, H., and Manning, C.D. *Proceedings of the 2015 Conference on Empirical Methods in Natural Language Processing*, 1412–1421. Association for Computational Linguistics. https://doi.org/10.18653/v1/D15-1166.

224 Coulter, W. H. (1953). Means for counting particles suspended in a fluid. U.S. Patent No. 2.656.508, issued 20 Oct. 1953.

225 DeBlois, R.W. and Bean, C.P. (1970). Counting and sizing of submicron particles by the resistive pulse technique. *The Review of Scientific Instruments* **41**: 909–916.

226 Bean, C.P. (1972). The physics of porous membranes – neutral pores. In: *Membranes* (ed. G. Eisenman), 1–54. New York: Marcel Dekker.

227 DeBlois, R.W., Bean, C.P., and Wesley, R.K.A. (1977). Electrokinetic measurements with submicron particles and pores by the resistive pulse technique. *Journal of Colloid and Interface Science* **61**: 323–335.

228 Hladky, S.B. and Haydon, D.A. (1972). Ion transfer across lipid membranes in the presence of gramicidin A. *Biochimica et Biophysica Acta* **274**: 294–312.

229 Bezrukov, S.M., Vodyanoy, I., and Parsegian, V.A. (1994). Counting polymers moving through a single ion channel. *Nature* **370** (6457): 279–281.

230 Braha, O., Walker, B., Cheley, S. et al. (1997). Designed protein pores as components for biosensors. *Chemistry & Biology (London)* **4**: 497–505.

231 Bayley, H. (2000). Pore planning: functional membrane proteins by design. *The Journal of General Physiology* **116**: 1a.

232 Bayley, H., Braha, O., and Gu, L.Q. (2000). Stochastic sensing with protein pores. *Advanced Materials* **12**: 139–142.

233 Kasianowicz, J.J., Brandin, E., Branton, D., and Deamer, D.W. (1996). Characterization of individual polynucleotide molecules using a membrane channel. *Proceedings of the National Academy of Sciences of the United States of America* **93** (24): 13770–13773.

234 Akeson, M., Branton, D., Kasianowicz, J.J. et al. (1999). Microsecond time-scale discrimination among polycytidylic acid, polyadenylic acid, and polyuridylic acid as homopolymers or as segments within single RNA molecules. *Biophysical Journal* **77** (6): 3227–3233.

235 Li, J., McMullan, C., Stein, D. et al. (2001). Solid state nanopores for single molecule detection. *Biophysical Journal* **80**: 339a.

236 Li, J., Stein, D., McMullan, C. et al. (2001). Ion beam sculpting on the nanoscale. *Nature* **412**: 4.

237 Sanger, F., Nicklen, S., and Coulson, A.R. (1977). DNA sequencing with chain-terminating inhibitors. *Proceedings of the National Academy of Sciences of the United States of America* **74**: 5463–5467.

238 Maxam, A.M. and Gilbert, W. (1977). A new method for sequencing DNA. *PNAS* **74**: 560–564.

239 Song, L., Hobaugh, M.R., Shustak, C. et al. (1996). Structure of staphylococcal alpha-hemolysin, a heptameric transmembrane pore. *Science* **274** (5294): 1859–1866.

240 Chow, S.A., Vincent, K.A., Ellison, V., and Brown, P.O. (1992). Reversal of integration and DNA splicing mediated by integrase of human immunodeficiency virus. *Science* **255**: 723–726.

241 Scottoline, B.P., Chow, S., Ellison, V., and Brown, P.O. (1997). Disruption of the terminal base pairs of retroviral DNA during integration. *Genes & Development* **11**: 371–382.

242 Chung, S.-H. and Gage, P.W. (1998). Signal processing techniques for channel current analysis based on hidden Markov models. *Methods in Enzymology* **293**: 420–437.

243 Krogh, A., Mian, I.S., and Haussler, D. (1994). A hidden Markov model that finds genes in *E. coli* DNA. *Nucleic Acids Research* **22**: 4768–4778.

244 Ma, L. and Cockroft, S.L. (2010). Biological nanopores for single-molecule biophysics. *ChemBioChem* **11**: 25–34.

245 Nakane, J., Wiggin, M., and Marziali, A. (2004). A nanosensor for transmembrane capture and identification of single nucleic acid molecules. *Biophysical Journal* **87**: 3618.

246 Stoddart, D., Heron, A.J., Mikhailova, E. et al. (2009). Single-nucleotide discrimination in immobilized DNA oligonucleotides with a biological nanopore. *Proceedings of the National Academy of Sciences of the United States of America* **106**: 7702–7707.

247 Sánchez-Quesada, J., Saghatelian, A., Cheley, S. et al. (2004). Single DNA rotaxanes of a transmembrane pore protein. *Angewandte Chemie* **116**: 3125–3129.

248 Sánchez-Quesada, J., Saghatelian, A., Cheley, S. et al. (2004). Single DNA rotaxanes of a transmembrane pore protein. *Angewandte Chemie, International Edition* **43**: 3063–3067.

249 Amabilino, D.B. and Stoddart, J.F. (1995). Interlocked and intertwined structures and superstructures. *Chemical Reviews* **95**: 2725–2829.

250 Kay, E.R., Leigh, D.A., and Zerbetto, F. (2007). Synthetic molecular motars and mechanical machines. *Angewandte Chemie* **119**: 72–196.

251 Kay, E.R., Leigh, D.A., and Zerbetto, F. (2007). Synthetic molecular motars and mechanical machines. *Angewandte Chemie, International Edition* **46**: 72–191.

252 Cockroft, S.L., Chu, J., Amorin, M., and Ghadiri, M.R. (2008). A single-molecules nanopore device detects DNA polymerase activity with single-nucleotide resolution. *Journal of the American Chemical Society* **130**: 818–820.

253 Mitchell, N. and Howorka, S. (2008). Chemical tags facilitate the sensing of individual DNA strands with nanopores. *Angewandte Chemie, International Edition* **47**: 5476–5479.

254 Howorka, S., Movileanu, L., Braha, O., and Bayley, H. (2001). Kinetics of duplex formation for individual DNA strands within a single protein nanopore. *Proceedings of the National Academy of Sciences of the United States of America* **98**: 12996–13001.

255 Gu, L.-Q., Braha, O., Conlan, S. et al. (1999). Stochastic sensing of organic analytes by a pore-forming protein containing a molecular adapter. *Nature* **398**: 686–690.

256 Astier, Y., Braha, O., and Bayley, H. (2006). Toward single molecule DNA sequencing: direct identification of ribonucleoside and deoxyribonucleoside 5-monophosphates by using an engineered protein nanopore equipped with a molecular adapter. *Journal of the American Chemical Society* **128**: 1705–1710.

257 Clarke, J., Wu, H.-C., Jayasinghe, L. et al. (2009). Continuous base identification for single-molecule nanopore DNA sequencing. *Nature Nanotechnology* **4**: 265–270.

258 Mohammad, M.M., Prakash, S., Matouschek, A., and Movileanu, L. (2008). Controlling a single protein in a nanopore through electrostatic traps. *Journal of the American Chemical Society* **130**: 4081–4088.

259 Movileanu, L., Howorka, S., Braha, O., and Bayley, H. (2000). Detecting protein analytes that modulate transmembrane movement of a polymer chain within a single protein pore. *Nature Biotechnology* **18**: 1091–1095.

260 Xie, H., Braha, O., Gu, L.-Q. et al. (2005). Single-molecule observation of the catalytic subunit of cAMP-dependent protein kinase binding to an inhibitor peptide. *Chemistry & Biology* **12**: 109–120.

261 Cheley, S., Xie, H., and Bayley, H. (2006). A genetically-encoded pore for the stochastic detection of a protein kinase. *ChemBioChem* **7**: 1923–1927.

262 Hornblower, B., Coombs, A., Whitaker, R.D. et al. (2007). Single-molecule analysis of DNA-protein complexes using nanopores. *Nature Methods* **4**: 315–317.

263 Astier, Y., Kainov, D.E., Bayley, H. et al. (2007). Stochastic detection of motor protein-RNA complexes by single-channel current recording. *ChemPhysChem* **8**: 2189–2194.

264 Benner, S., Chen, R.J.A., Wilson, N.A. et al. (2007). Sequence-specific detection of individual DNA polymerase complexes in real time using a nanopore. *Nature Nanotechnology* **2**: 718–724.

265 Hurt, N., Wang, H., Akeson, M., and Lieberman, K.R. (2009). *Journal of the American Chemical Society* **131**: 3772–3778.

266 Wilson, N.A., Abu-Shumays, R., Gyarfas, B. et al. (2009). Electronic control of DNA polymerase binding and unbdining to single DNA molecules. *ACS Nano* **3**: 995–1003.

267 Ellington, A.D. and Szostak, J. (1990). In vitro selection of RNA molecules that bind specific ligands. *Nature* **346**: 818–822.

268 Tuerk, C. and Gold, L. (1990). Systematic evolution of ligands by exponential enrichment: RNA ligands to bacteriophage T4 DNA polymerase. *Science* **249**: 505–510.

269 Jayasena, S.D. (1999). Aptamers: an emerging class of molecules that rival antibodies in diagnostics. *Clinical Chemistry* **45**: 1628–1650.

270 Proske, D., Blank, M., Buhmann, R., and Resch, A. (2005). Aptamers – basic research, drug development, and clinical applications. *Applied Microbiology and Biotechnology* **69**: 367–374.

271 Hamaguchi, N., Ellington, A., and Stanton, M. (2001). Aptamer beacons for the direct detection of proteins. *Analytical Biochemistry* **294**: 126–131.

272 Ulrich, H., Martins, A.H., and Pesquero, J.B. (2004). RNA and DNA aptamers in cytomics analysis. *Cytometry Part A* **59A**: 220–231.

273 Brody, E.N. and Gold, L. (2000). Aptamers as therapeutic and diagnostic reagents. *Journal of Biotechnology* **74**: 5–13.

274 Yamamoto, R., Baba, T., and Kumar, P.K. (2000). Molecular beacon aptamer fluoresces in the presence of Tat protein of HIV-1. *Genes to Cells* **5**: 389–396.

275 Yamamoto, R., Katahira, M., Nishikawa, S. et al. (2000). A novel RNA motif that binds efficiently and specifically to the Tat protein of HIV and inhibits the transactivation by Tat of transcription in vitro and in vivo. *Genes to Cells* **5** (5): 371–388.

276 Howorka, S., Cheley, S., and Bayley, H. (July 2001). Sequence-specific detection of individual DNA strands using engineered nanopores. *Nature Biotechnology* **19** (7): 636–639.

277 Ikebukuro, K., Okumura, Y., Sumikura, K., and Karube, I. (2005). A novel method of screening thrombin-inhibiting DNA aptamers using an evolution-mimicking algorithm. *Nucleic Acids Research* **33** (12): 7.

278 Wagner, B., Miller, D.C., Lear, T.L., and Antczak, D.F. (2004). The complete map of the Ig heavy chain constant gene region reveals evidence for seven IgG isotypes and for IgD in the horse. *Journal of Immunology* **173**: 3230–3242.

279 Fernandes, D. (2005). Demonstrating comparability of antibody glycosylation during biomanufacturing. *European Biopharmaceutical Review*. Summer: 106–110.

280 Ding, S., Gao, C., and Gu, L.-Q. (2009). Capturing single molecules of immunoglobulin and Ricin with an aptamer-encoded glass nanopore. *Analytical Chemistry* **81**: ac9006705.

Index

Informatics and Machine Learning: From Martingales to Metaheuristics, First Edition.
Stephen Winters-Hilt.
© 2022 John Wiley & Sons, Inc. Published 2022 by John Wiley & Sons, Inc.

Printed and bound by CPI Group (UK) Ltd, Croydon, CR0 4YY